ELECTROMAGNETIC THEORY AND APPLICATIONS IN BEAM-WAVE ELECTRONICS

ELECTROMAGNETIC THEORY AND APPLICATIONS IN BEAM-WAVE ELECTRONICS

B. N. Basu

Professor of Electronics Engineering
Centre of Research in Microwave Tubes
Department of Electronics Engineering
Institute of Technology, Banaras Hindu University
Varanasi-221 005, India

World Scientific
Singapore • New Jersey • London • Hong Kong

Published by

World Scientific Publishing Co. Pte. Ltd.

P O Box 128, Farrer Road, Singapore 912805

USA office: Suite 1B, 1060 Main Street, River Edge, NJ 07661

UK office: 57 Shelton Street, Covent Garden, London WC2H 9HE

Library of Congress Cataloging-in-Publication Data

Basu, B. N., Ph.D.
Electromagnetic theory and applications in beam-wave electronics / B.N. Basu.
p. cm.
Includes bibliographical references and index.
ISBN 981022320X
1. Electron tubes. 2. Microwave tubes. 3. Electromagnetism. 4. Traveling-wave tubes. I. Title.
TK7871.7.B38 1995
621.3815'1--dc20 95-46596
CIP

British Library Cataloguing-in-Publication Data

A catalogue record for this book is available from the British Library.

Printed in Singapore by Uto-Print

Dedication

This book is dedicated to the memory of my father *Manmatha Nath Basu* whose untimely demise snatched him away from me, when I was two years old; and to the memory of my uncle *Sachchidananda Guha*, a medical practitioner, whose life was dedicated to the cause of the poor.

I also dedicate this book to my mother *Sisirkana Basu* who has taught me to enjoy life during difficult hours and to *Sukumar Bhattacharyya*, my father-in-law, who had been a constant source of inspiration to me during the course of the preparation of this book.

Special Acknowledgment

This book could not have been completed without the meticulous planning, organisation and active support of my colleague *Pradip K. Jain* who right from the word go helped me at every stage of its evolution. It is my pride to see in him my own student becoming a task master and perfectionist who put me to hard work, for the improvement of the standard of the book. I have no words to thank him.

PREFACE

The interest in the study of electromagnetic theory is increased when it is linked up with some practical applications. Here, as an area of such applications I have chosen beam-wave electronics, which, apparently, could have been a fashionable choice a number of decades ago when the transit-time electron tubes ushered in a new era in the generation and amplification of microwaves. But, interestingly, even today with the advancing technology, the performance characteristics of conventional microwave tubes are being continuously updated for both civilian and defence applications and, in recent times, there has been a remarkable resurgence of activities in this area after the advent of high-power, millimeter-wave, spiralling-beam gyro-devices. This is evidenced by the emphasis given on these devices in international conferences and special issues of journals. The two recent books by Gilmour on this subject published in quick succession is also a typical indicator of this trend. It may be mentioned that in the topics of beam-wave electronics one gets enough scope to apply both time-independent field concepts (for instance, in the formation of an electron beam) and time-dependent field concepts (for instance, in the propagation through a slow-wave structure or in the beam-wave interaction in an electron beam device), whereas in the usual topics of microwave engineering (such as components, antennas, propagation, etc.) only the time-dependent field concepts are applied.

The material of the present volume is mainly based on my experience of teaching electromagnetic theory, in the undergraduate electronics engineering course in one semester, and beam-wave electronics, more precisely, microwave tubes, in the post-graduate microwave engineering course in another semester. Clearly, the former serves as the pre-requisite background for the latter, while the latter exemplifies the former. Accordingly, I have divided this volume into two parts to deal with these subjects, respectively. I hope that, in the present self-contained treatise, junior-level students will be fascinated by the demonstration of practical applications of electromagnetic theory and that, at the same time, senior-level students, researchers, scientists and practising engineers will welcome the provision of the fundamentals of electromagnetic theory for a recapitulation.

In selecting the topics of beam-wave electronics, the various aspects of the formation and confinement of an electron beam, the propagation through an interaction structure, and the beam-wave interaction mechanism with reference to growing-wave electron beam devices have been kept in mind. Care has been taken to make the analysis rigorous and to choose the topics of current relevance. I was aware that the book would not be attractive if it had dealt with only a helix in free space in the sheath-helix model, and not with a practical helix loaded by a dielectric in a metal envelope in the tape-helix model revealing the space-harmonic effects; or if it had predicted, while synthesising a convergent-flow Pierce gun, merely the spacing between the electrodes but not their shapes; or it had dealt with only the conventional travelling-wave tubes (TWTs) and not the gyro-TWTs.

The reader of the present volume is assumed to know the material given in basic courses of general college physics, electronics and mathematics including vector algebra. The concepts, relations and expressions have been developed and examples worked out here from first principles giving elaborate intermediate steps. However, the mathematical details have been provided in the form of appendices at the ends of relevant sections so that, on the one hand, the main text is not buried in too many mathematical details and, on the other, they are available within the close reach of a reader for a ready reference. I hope that the students and researchers will be benefited by the theories developed here from first principles.

I am indebted to my mentors who have influenced my own understanding of electromagnetic theory and its applications in the study of electron beam devices in the various stages of my academic and research career. I take this opportunity to express my gratitude to the late N. C. Vaidya who motivated me to join the Centre of Research in Microwave Tubes at the Banaras Hindu University; to Nirmal B. Chakrabarti who supervised my doctoral thesis on double-stream and beam-plasma amplifiers, and, more importantly, who taught me the value of developing concepts from first principles; to Amarjit singh who invited me to work at the Central Electronics Engineering Research Institute, Pilani; to S.S.S. Agarwala who has guided my research in travelling-wave tubes and taught me to be always modest while claiming achievements; to R. K. Jha who has always inspired me by viewing my achievements under a microscope; to D.T. Swift-Hook whose pioneering work on helical slow-wave structures greatly influenced my own research; to R. G. Carter who has inspired me by considering my research to be of practical relevance; to S.K. Srivastava who has patronised my research efforts on top priority; and to Prasad Khastgir who has always helped me with ready-made answers whenever, as a practitioner of electromagnetic theory, I have approached him with some doubts.

This book could not have been written without positive suggestions and encouragement from Akhlesh Lakhtakia to whom goes my sincere acknowledgment. I also express my gratitude to Om. P. Gandhi who has perused the contents of the book and found them 'extremely interesting', and thus greatly motivated me to complete the write-up. I am also indebted to M. D. Raj Narayan, G. S. Sanyal, R. P. Shenoy, C. S. Jha, W. S. Khokle, R. N. Biswas, H. N. Bandopadhyaya, S. K. Balasubramanian, S. C. Dutta Roy, B. N. Das, M. K. Das Gupta, G. S. Sidhu, Samir K. Lahiri, A. N. Tripathi, R. S. Raju, R. K. Gupta, V. Srivastava, S. K. Kamath, S. U. M. Reddy, Manoj T. N., Lalit Kumar, Ping Wang and S. N. Joshi for their support and encouragement. Also I take this opportunity to acknowledge the previlege I have enjoyed of being taught by the brilliant teachers of electromagnetics like B. R. Nag and Lalit Kishore.

I extend my special thanks to A. J. Sangster and S. Jagadishwar Rao for their help in writing the section on the gyro-TWT dispersion relation. Thanks are also due A. K. Sinha , Sanjay Ghosh and Subrata K. Datta for plotting some of the results presented in the book.I express my appreciation to Pankaj K. Dalela for his painstaking efforts in proofreading the entire book. Thanks are due B. Ghosh and R. C. Pendharkar for their assistance in typing and in tracing figures, respectively. Thanks are also due Ms. Doreen Liu, Ms. Kim Tan and Ms. Barbara Aman for their editorial co-operation. Finally, I express my deep appreciation to my brothers (Dilip Roy Chowdhury and Jayanta Bose) and sisters (Bharati Dey and Manju Paul) and to my children (Indraneel and Priyoneel), for their affectionate encouragement, as well as to my wife Monikuntala for her loving support , forbearance, understanding and domestic adjustments.

August 1995
Varanasi, India

B. N. Basu

CONTENTS

ELECTROMAGNETIC THEORY AND APPLICATIONS IN BEAM-WAVE ELECTRONICS

CHAPTER 1

INTRODUCTION

For whatever may be said about the importance of aiming at depth rather than width in our studies, and however strong the demand of the present age may be for specialists, there will be work, not only for those who build up particular sciences and write monographs on them, but for those who open up such communications between the different groups of builders as will facilitate a healthy interaction between them.

— *James Clerk Maxwell*

Electromagnetic theory is a widely used conceptual structure in physics as well as in electrical, electronics, radio and communication engineering [1]-[7]. Take, for instance, a typical radio-communication link. It may be seen that electromagnetic theory is applied to the understanding of almost every stage of this link whether it is the generation or amplification of electromagnetic waves in a transmitting system, or the transportation of information through a transmission line from a transmitting to a radiating system (antenna), or the radiation from a transmitting antenna, or the propagation of electromagnetic waves from a transmitting to a receiving antenna, or the detection, amplification and display of information transported through a transmission line to a receiving system.

It is needless to mention that it becomes almost impossible to take up in a single treatise electromagnetic theory and all its possible applications. It would even be an unmanageable task to treat the subject as applied to all the individual stages of the example of a communication link cited above, and certainly the present work does not aim at doing that. Instead, here, only one practical application is chosen, namely, the application of electromagnetics to the subject of beam-wave electronics. The study of such a subject would enable one to gain an insight into the working of a microwave or millimeter-wave tube in which the electron beam is used as one of the constituents [8], [9].

Interestingly, in the topics of beam-wave electronics taken up in this treatise one finds both time-independent and time-dependent field concepts exemplified, unlike in many topics of microwave engineering such as waveguides, antennas, propagation etc. in which there are the examples of only time-dependent fields. Thus it is believed that electromagnetic theory when studied with reference to beam-wave electronics dealing with practical electron beam devices will become quite an interesting and enjoyable subject. At the same time it is also hoped that anyone contemplating to study beam-wave electronics with reference to microwave tubes as a separate subject would welcome the built-in background of electromagnetic theory provided in this book. Accordingly, both electromagnetic theory and its applications in beam-wave electronics have been presented in two parts, in this self-contained treatise.

The topics of both electromagnetics and beam-wave electronics are so chosen here that they become relevant to the understanding of the growing-wave mechanism of a practical electron beam device working in the microwave or millimeter-wave frequency range. In such a device, it is important on the one hand to understand the mechanism of the formation, confinement and collection of an electron beam and on the other hand to study the propagation characteristics of an electromagnetic structure in which the interaction between an electron beam and an electromagnetic wave can take place, and of course, finally, to gain an insight into the mechanism of beam-wave interaction.

It will be found that electromagnetic theory as related to static fields is useful in the study of the formation and confinement of an electron beam. Similarly, it will be seen that the theory concerning time-varying fields is an important tool in studying the behaviour of an electromagnetic structure as well as the mechanism of interaction between an electron beam and an electromagnetic wave supported by such a structure in practical growing-wave electron beam devices.

Chapters *2* through *4* constituting part *one* of the book deal with the basic principles of electromagnetic theory. Beam-wave electronics and related areas as relevant to the understanding of the principles of growing-wave devices are covered in part two comprised of chapters *5* through *8*. Thus part one may be considered as the prerequisite for the concepts developed in part two, while the latter considered as the application of the concepts developed in the former. These two parts, in fact, have a chapter-to-chapter correlation. For instance, the understanding of electron guns developed in chapter *6* (of part *two*) is based on the concepts of electrostatics developed in chapter *2* (of part *one*); the behaviour of a helical slow-wave structure is predicted in chapter *5* (of part *two*) using the time-varying field concepts developed in chapter *4* (of part *one*), and so on.

Introduction to electromagnetic theory customarily begins in chapter *2* with electrostatics or static electric fields which arise out of stationary charges. The most fundamental law of electrostatics is Coulomb's law which relates the force between two stationary point charges located in a homogeneous unbounded medium to the distance separating the charges. The law can be extended to a uniform charge distribution to find the static electric field intensity due to such a distribution. It is found convenient to apply another law called Gauss's law, which may be derived from Coulomb's law, to those problems which enjoy some sort of symmetry in physical geometry.

The work done in moving a charge against the force of a static electric field leads to the concept of electrostatic potential which is a scalar quantity. There exist numerous problems in which it is somewhat easy to determine first the scalar electrostatic potential and then derive the vector electric field intensity therefrom. Gauss's law expressed in terms of the electrostatic potential applied to a volume element yields Poisson's or Laplace's equation according as the volume element considered encloses charges or no charges, and as the volume element tends to zero. With the volume element shrinking to zero these equations, in fact, refer to the quantities at a point, and Gauss's law takes what is known as the point (or differential) form. Poisson's and Laplace's equations expressed in the Laplacian of potential find applications in electron beam problems, for instance, in the problem of an electron gun (chapter *6*), which is used to form a beam of electrons in a practical electron beam device.

In chapter *3*, the fundamental relations of static magnetic field begin with Biot-Savart's law. The role of Biot-Savart's law in finding the static magnetic field intensity due to a steady current is similar to that played by Coulomb's law in finding the static electric field intensity due to a stationary charge. Similarly, in static magnetic field problems enjoying some sort of symmetry, Ampere's law is as useful a tool as Gauss's law is in electrostatic problems. Also, it is worth expressing the former in its point form like the latter. There is yet another useful concept developed, namely, the concept of the magnetic vector potential which is useful in the study of magnetic field problems.

Just as does a charge placed in a static electric field experience a force, a conductor carrying a current placed in a static magnetic field will do so. The relevant expression for force may be derived form the basic concept of Lorentz force on a charge moving in a static magnetic field. In steady magnetic field problems, the force between conductors carrying steady currents is analogous to the Coulomb force between stationary charges. However, instead of drawing such an analogy starting from the direct electric current as a cause of magnetic field, it is sometimes worth doing so starting from the concept of magnetic charges or poles. (There is as yet no evidence of the existence of isolated magnetic poles(monopoles), however). Thus as an approach to finding the magnetic field intensity due to a permanent magnet one may treat the magnet as a magnetic dipole analogous to an electric dipole (two charges separated by a distance). It may be seen that once the concept of magnetic charges are introduced, all the relevant electrostatic laws like Coulomb's law, Gauss's law, etc. would also follow in static magnetic field problems in an analogous way.

Chapter *4* is devoted to the study of time-varying electric and magnetic fields. Two fundamental concepts in this study are the electric field due to a time-varying magnetic field, as predicted by Faraday's law, as well as the magnetic field caused by a time-varying electric field, the latter introducing the concept of the displacement current. These concepts are put together by a set of four equations, known as Maxwell's equations, which may be expressed in both the integral and the point forms. These equations which find extensive applications in electromagnetic engineering are 'simple enough to imprint on a T-shirt, and yet rich enough to provide new insights throughout a lifetime of study', to put them in the words of Whinnery [10]. (It is more than one hundred years since these equations ushered in a new era of electromagnetics, but we are still discovering exciting applications in several disciplines the impetus for which is provided by rapidly advancing technology. These applications have now become so diverse that one may venture to guess that Maxwell himself would be pleased to know that his equations are still generating new applications extending to the second century after their discovery. It is amazing to feel how 'several disciplines hang as gems on one priceless necklace which it was Maxwell's privilege and honour to recognise as capricious Nature's enduring ornament'[11]).

Maxwell's equations, in which the electric and magnetic fields are coupled, may be combined and thus decoupled to obtain a single equation in either of the electric and magnetic field intensities to obtain what is known as the wave equation. The solution of the wave equation would then give one a picture of how the fields are distributed in a given problem. However, the wave equation can be solved only subject to a set of electromagnetic boundary conditions appropriate to a given problem. A vectorial approach has been made here to obtain the general boundary conditions at the interface between two different media. Subsequently, one may specialise them for specific cases of practical interest (say, for a dielectric-dielectric or a conductor-dielectric or a conductor-conductor interface).

The usefulness of the wave equation is demonstrated here (chapter *4*) with reference to propagation of electromagnetic waves through an unbounded medium. One may find the example of its application to a bounded medium, for instance, in the study of a physical structure supporting an electromagnetic wave. (One such structure has been taken up in Chapter *5)*. The study of propagation through a conducting medium yields the important concepts of skin effects and surface resistance. These concepts may be utilised for evaluating the loss characteristics of a metallic electromagnetic structure.

The concepts of the scalar-electrostatic (chapter *2*) and vector-magnetic (chapter *3*) potentials may be extended to the time-varying field cases also. These potentials are obtained as the solutions of wave equations in these quantities. The concepts of these potentials developed here (chapter *4*) have actually been used in practice (for instance, in chapter *5* while finding the equivalent transmission line parameters of a structure supporting an electromagnetic wave).

Chapter *5* marks the beginning of the second group of chapters (constituting part *two* of the treatise) in which the study has been made of a structure which supports a slow electromagnetic wave in which an electron beam could be placed for a gainful purpose, for instance, for the transfer of energy from the beam to electromagnetic waves. In the study of such a structure, referred to as a slow-wave structure (SWS), the time-varying field concepts developed in chapter *4* may be extensively used. Though there exist various types of SWS, one particular type, namely, the helix has received special attention in this chapter. Such a structure has an unparallel bandwidth provided by its controlled dispersion. Moreover, with the advent of recent technologies, the power handling capacity of such a structure has been considerably enhanced. It will be of interest to find enough scope for electromagnetics while modelling and analysing such a structure. The helix has been treated here in a model called the sheath-helix model. The conditions for the validity of the model are also discussed. The method is also given as to how to take into account in the analysis the effects of the discrete supports for the helix and those of an overall envelope which may be made of a metal. Two analytical approaches have been made to the problem, namely, the field and the equivalent-circuit analyses. Both these approaches yield one and the same dispersion relation. The relative merits of these approaches are also worth looking into. Also, the relevant concepts of the characteristic and the interaction impedances are developed here. The characteristic impedance which can be found using the equivalent circuit analysis is important from the standpoint of matching the structure with the system of coupling power in and out of the structure. The interaction impedance can be found by integrating the complex Poynting vector over the structure cross section. The concept of interaction impedance developed in this chapter has been used later in chapter *8* to study the beam-wave interaction mechanism in a travelling-wave tube (TWT). The problem of dispersion-shaping a helix by means of anisotropic or inhomogeneous loading with a view to broadbanding a TWT may easily be dealt with using the simple sheath-helix model. Finally, the tape helix model is developed to refine the theory by taking into account the effects of the space harmonics arising from the space-periodicity of the structure, the effects which are not revealed by the sheath-helix model. Interestingly, the result of a helix in free space, studied in the tape-helix model, may be combined with those of a loaded helix, say, a dielectric-supported helix in a metal envelope, studied in the sheath-helix model, to obtain the results of a loaded helix in the tape-helix model.

In chapter *6* is developed the theory of the electron gun the function of which is to 'throw' an electron beam of defined parameters upto a certain distance beyond which, however, it may be confined by a separate focusing structure (chapter *7*).

The laws of electrostatics are applicable in the study of electron gun problems in which a simplified approach would be to treat the beam as a charge fluid of uniform charge distribution. It is assumed that the cross section of the beam at a given instant of time, at two different axial locations, or those at a given axial location, at two different instants of time, are indistinguishable. This would enable one to use Poisson's and Laplace's equations of electrostatics in the regions inside and outside the beam, respectively, in the gun region.

The solution of the one-dimensional Poisson's equation in the rectangular system of coordinates, known as the Child-Langmuir's law, is used here to predict the interelectrode spacing of a 'parallel-flow' gun derived from a flat cathode. A suitable analytic function may be sought as a solution to Laplace's equation which can be used to plot the equipotentials outside the beam. This would then enable one to predict the shape of the electrodes of the gun.

The convergent-flow gun is required when the demand is to have a relatively high beam current density for a high power or a low operating cathode current density, that is, light cathode-loading for a long cathode life. A conical beam may be formed using a spherical-cup cathode in such a gun. The beam after it passes through an aperture in the anode is launched into a field-free region. Subject to Coulomb space-charge spreading and lens effects at the anode, the electron beam would pass through a minimum cross-sectional area at a position called the beam 'waist' (or 'throat'). The distance of the waist from a reference electrode is the 'throw' of the gun. The jurisdiction of the convergent gun is upto the throw (beyond which it is the focussing structure (chapter *7*) which would take care to confine the beam inside the interaction structure).

It is of interest to synthesise a convergent gun which employs a spherical-cup cathode, to find the cup radius, the radii of curvature of the anode and the cathode, the anode-aperture radius, the distance between the anode and the cathode as well as the throw of the gun. For this purpose, four input parameters need to be specified. The three of them are related to the beam, called the beam parameters, namely, the beam voltage, current and radius (at the throat); the fourth parameter is either the operating cathode current density or the beam convergence, that is, the ratio of the cathode-cup radius to the beam waist radius. The synthesis of the gun would, however, be complete only if one could predict the shapes of the gun-electrodes. This is done here (chapter *6*) by conformal mapping the electrode shapes using a suitable transformation function.

Chapter *7* is devoted to developing the concepts of a magnetic focusing structure the function of which is to 'catch' the beam 'thrown' by the electron gun and realise conditions to confine the beam in the interaction region maintaining the axis and the desired cross section of the beam. Thus, if a conical beam is derived from a curved cathode such that beyond the anode it passes through a minimum radius at the waist, then the focusing structure would take care to see that a cylindrical beam of more or less constant radius, equal to the beam-waist radius, is maintained through a desired axial length of the interaction region.

The design of the magnetic focusing structure would depend upon whether or not the cathode is shielded from the magnetic field giving, accordingly, the two schemes, namely, the Brillouin and the confined-flow focusing. The required magnetic field may be obtained using either a solenoid or a permanent magnet (PM). The solenoid structure is rather heavy and calls for a power supply arrangement. Therefore, such a structure is used for ground-station applications. The weight of a PM may be considerably reduced if one uses a periodic permanent magnet (PPM) arrangements in which a number of small PM's with alternate polarities are arranged in an array over the desired axial length of an interaction structure. The PPM is particularly suitable for air-borne applications due to its reduced weight.

The problem of a focusing structure resolves down to solving the electronic motion in the presence of the space-charge electric field of the beam and the externally applied dc magnetic field. The object is to find the conditions for the minimum variation of the beam cross section over the desired beam length, that is, for the minimum beam-'scalloping'. This would include deriving a relation, called the Brillouin relation, between the beam parameters and the applied magnetic flux density (its rms value, in the case of a PPM structure).

Chapter *8* concerns with the interaction of an electromagnetic wave with an electron beam resulting in spatially-growing space-charge waves. A single electron beam supports the so-called 'Hahn and Ramo' space-charge waves which are as such not of growing-wave type. However, the electron beam may be made to support spatially-growing waves in a number of ways, for instance, if the beam is operated in the parametric mode to make a device called the electron-beam parametric amplifier in which an RF source is required as a 'pump' to provide energy to the signal wave; made to interact with 'slow' waves supported by an interaction structure as in a TWT; made to intimately mix and interact with another electron beam of a slightly different dc velocity in a device called the double-stream amplifier; made to penetrate through a plasma in a device known as the beam-plasma amplifier; made to pass through a space enclosed in a resistive metal boundary to make the so-called resistance-wall amplifier; made to have a periodically varying diameter while passing through a space enclosed in a conducting metal boundary, and so on.

Out of the various possible growing-wave devices mentioned, the TWTs have come out as increasingly important from the standpoint of their extensive civilian and military microwave applications. Naturally, therefore, TWT's have received more attention in this treatise than their other growing-wave counterparts (section *8.2*). For such a device one may derive a circuit equation combining the telegraphist's equations of the slow-wave structure, treated as an equivalent transmission line, in the presence of an electron beam. It is easy, for a thin beam close to the structure, to obtain the circuit equation which involves the characteristic impedance of the structure. The theory may be further refined to obtain the same equation, for a more generalised situation, in terms of a more realistic parameter, namely, the interaction impedance of the structure which may be defined in terms of power propagating down the structure. One may also obtain the electronic equation of the device starting from the force equation of an electron subject to the circuit field. One may then compare the circuit equation with the electronic equation to obtain the dispersion relation characterising the beam-wave interaction and finally deduce a gain formula for the TWT. Practical relevance of the problem may be further increased by incorporating into the theory the effects of the departure of the beam velocity from the phase velocity of the structure as well as the effects of circuit losses and space charges. Also, it is of interest to study the effects of an attenuator on the gain of the device, such an attenuator being provided with the slow-wave structure to suppress unwanted oscillations. Various considerations related to the performance improvement of a TWT have also been discussed.

Study has also been extended to a couple of growing-wave devices in which the role which the slow-wave structure plays in a TWT is taken by a medium, rather than an electromagnetic structure, namely, an electron beam in a double-stream amplifier and a plasma in a beam plasma amplifier (section *8.3*). Finally, a fast-wave (growing-wave) device, namely, the gyro-TWT is taken up for study (section *8.4*). In this device a fast-wave electromagnetic structure such as a hollow-pipe cylindrical waveguide is used unlike in a conventional TWT which uses a slow-wave structure. In a gyro-TWT the mechanism of bunching of electrons is based on the relativistic change of mass of an electron with its velocity. In this device the interaction takes place between the gyrating electrons in helical trajectories and a transverse-electric mode of a cylindrical waveguide. The dispersion relation of a gyro-TWT is derived and it is interpreted for the device gain in terms of the familiar terminology of a conventional TWT.

Thus, from the study of the selected topics covered in the chapters to follow, one should get a scope to either learn or recapitulate the fundamental concepts of electromagnetics concerning both time-independent and time dependent fields as well as apply them to a dedicated area, namely, beam-wave electronics related to the understanding of practical growing-wave electron beam devices.

Finally, it is necessary to mention here a few points about the organisation of a large number of appendices provided to substantiate various ideas, concepts, relations and examples in this volume. These appendices prevent the main text from being buried in too many mathematical details. However, for the benefit of those readers who do not like to skip them and enjoy reading them as the main text, these appendices have been dispersed throughout the text, and placed appropriately at the ends of various sections, instead of being relegated to the end of the volume or of each chapter. However, in order to distinguish them from the main body of the text, the appendices are printed in letters of a different type and size. Also, the equations of the appendices are numbered at their left-hand sides unlike those of the main text which are numbered at their right, the equation and figure numbers in the appendices being preceded by the letter 'a'.

References

[1] J.C. Maxwell, *A Treatise on Electricity and Magnetism* (Clarendon, Oxford 1873).

[2] J. A. Stratton, *Electromagnetic Theory* (Mc Graw-Hill, New York 1941).

[3] A. Lakhtakia, ed., *Essays on the Formal Aspects of Electromagnetic Theory* (World Scientific, Singapore, 1993).

[4] G.W. Carter, *The Electromagnetic Fields in its Engineering Aspects* (Longmans, London, 1967).

[5] L. Solymar, *Lectures on Electromagnetic Theory: A Short Course for Engineers* (Oxford Univ. Press, London, 1976).

[6] O.P. Gandhi, *Microwave Engineering and Applications* (Pergamon, New York, 1981).

[7] R.G.Carter, *Electromagnetic Waves / Microwave Components and Devices* (Champman and Hall, London, 1990).

[8] R.G.E. Hutter, *Beam and Wave Electronics in Microwave Tubes* (Van Nostrand, Princeton, 1960).

[9] M. Chodorow and C. Susskind, *Fundamentals of Microwave Electronics* (Mc-Graw Hill, New York, 1964).

[10] J.R. Whinnery, "The teaching of electromagnetics," *IEEE Trans. Education* **ED-33** (1990) 3-7.

[11] P. Khastgir, *Apologia,* Seminar on Electromagnetics and their Applications, 22-23 Dec. 1988, Varanasi, India.

PART ONE

PRELIMINARY CONCEPTS OF ELECTROMAGNETIC THEORY

CHAPTER 2

STATIC ELECTRIC FIELDS

2.0 Introduction

Electrostatics is that branch of electromagnetic theory that deals with static electric fields; it deals with the effects of static or stationary charges. In numerous problems one may be interested to know the static electric field intensity due to these charges distributed in a known manner in a medium or on the surface of a conductor. Also, in some static electric field problems one may have to consider *effectively* stationary charges which are actually in a state of motion. One such example is an *electron beam* which may be considered to be smeared out as a continuous fluid of electronic charges. The concepts of electrostatics may, therefore, be applied to electron beam problems, for instance, to the study of the formation of an electron beam by an electron gun — a problem that has been dealt with in details in Chapter *6*.

In one school of thought the laws of static field problems are derived as the special cases of those of more general, time-varying electromagnetic problems by putting equal to zero all the quantities representing a time-rate of variation. This results into the electric field decoupled from the magnetic field as well as the electric charge decoupled from the electric current. In another school of thought, and that is what we have chosen to adopt in this chapter for electrostatics, we traditionally start from the most fundamental Coulomb's law for the force between two point charges separated by a distance in a homogeneous medium. This law is taken as granted on the basis of experimental observations. It is from this law that all other concepts related to static electric field problems such as Gauss's law, Poisson's and Laplace's equations, etc., would follow in sequence. There is yet another important concept in static electric field problems — electrostatic potential, a concept which is to be developed from the knowledge of work done in moving a charge against an electric field, the work being stored in the form of potential energy. In some problems it is easier to obtain electrostatic potential first and then determine the electric field intensity therefrom.

2.1 Coulomb's Law and Electric Field Intensity

The force **F** on a point charge q placed in a region where an electric field exists is given by

$$\mathbf{F} = q\,\mathbf{E}\,, \tag{2.1.1}$$

where **E** is the electric field intensity at the point. It may be mentioned that (2.1.1) is valid for a charge whether it is at rest or in a state of motion. In this respect, the electric field differs from the magnetic field which exerts a force on a charge only when the latter is in a state of motion (see chapter *3*). One may use (2.1.1) to measure the electric field intensity at a point as

$$\mathbf{E} = \frac{\mathbf{F}}{q}\,, \tag{2.1.2}$$

that is, as an *effect* on a *test* point charge q. In (2.1.2), it is implied, the value of the test point charge q itself is sufficiently small, in the limit, tending to zero so that the electric field intensity to be measured is not perturbed by the presence of the test point charge. In other words, a more appropriate form of (2.1.2) should be [1]:

$$\mathbf{E} = \operatorname*{Lt}_{q\to 0} \frac{\mathbf{F}}{q}\;. \tag{2.1.3}$$

Thus, provided we have some means to find **F** we can determine **E** also with the help of (2.1.3).

In a variety of problems we may be interested to know the electric field intensity due to a point charge or a distribution of point charges. Thus, in view of (2.1.3), we would like to know the force on a test point charge due to a given point charge or a distribution of point charges. At this stage, let us then introduce the most fundamental law of electrostatics, namely, Coulomb's law for the force between two stationary point charges in an unbounded homogeneous insulating medium, which may be stated as [1], [2]:

$$\mathbf{F} = \frac{q_1 q_2}{4\pi\varepsilon r^2}\mathbf{a}_r\,, \tag{2.1.4}$$

where **F** represents the force (in Newton) on any one of the two point charges q_1 and q_2 (in Coulombs) due to the other, the charges being separated by a distance r (in meter); $\mathbf{a}_r$ is the unit vector along the line joining the two point charges and directed towards the point charge on which the force is sought from the other; ε is a quantity which refers to the medium in which the charges are placed and is called the permittivity of the medium. The *Coulomb force* has the greatest value if the charges are placed in free space which has the lowest permittivity $(\varepsilon =)\ \varepsilon_0 = 1/(36\pi)\times10^{-9} \cong 8.854\times10^{-12}$ Coulomb^2/Newton-meter^2. It may be mentioned that the force law given by (2.1.4) is expressed in the *rationalised* MKS system [3], the rationalisation factor here being (1/4π). The introduction of this rationalisation factor removes 4π that would otherwise appear in many useful expressions which are derived starting from the basic Coulomb's law. The ratio of the medium-to-free-space permittivity is called the relative permittivity ε_r (= $\varepsilon/\varepsilon_0$) or the dielectric constant of the medium. (2.1.4), as it is in the vectorial form, also implies that like charges repel and unlike charges attract.

It is sometimes useful to know the force on a point charge due to another point charge when the space coordinates of the point charges are known. Let the rectangular cartesian coordinates of the two point charges q_1 and q_2 be (x_1, y_1, z_1) and (x_2, y_2, z_2), respectively, and $\mathbf{r}$ be a vector whose magnitude r is equal to the distance between the point charges, and which is directed from q_1 to q_2. We may write

$$\mathbf{r} = (x_2 - x_1)\mathbf{a}_x + (y_2 - y_1)\mathbf{a}_y + (z_2 - z_1)\mathbf{a}_z\,, \tag{2.1.5}$$

and

$$r = [(x_2 - x_1)^2 + (y_2 - y_1)^2 + (z_2 - z_1)^2]^{1/2}\,. \tag{2.1.6}$$

Noting that $\mathbf{a}_r = \mathbf{r}\,/\,r$ and using (2.1.4) - (2.1.6) one may express $\mathbf{F}_{21}$, the force on the point charge q_2 due to q_1 as

$$\mathbf{F}_{21} = \frac{q_1 q_2\left[(x_2 - x_1)\mathbf{a}_x + (y_2 - y_1)\mathbf{a}_y + (z_2 - z_1)\mathbf{a}_z\right]}{4\pi\varepsilon\left[(x_2 - x_1)^2 + (y_2 - y_1)^2 + (z_2 - z_1)^2\right]^{3/2}}\,. \tag{2.1.7}$$

Also, one may choose to express $\mathbf{F}_{21}$ in terms of its magnitude = F_{21} and a unit vector $\mathbf{a}_{F_{21}}$ as

$$\mathbf{F}_{21} = F_{21}\mathbf{a}_{F_{21}}\,, \tag{2.1.8}$$

where

$$F_{21} = (F_x^2 + F_y^2 + F_z^2)^{1/2} \tag{2.1.9}$$

with

$$F_\upsilon = \frac{q_1 q_2 (\upsilon_2 - \upsilon_1)}{4\pi\varepsilon\left[(x_2 - x_1)^2 + (y_2 - y_1)^2 + (z_2 - z_1)^2\right]^{3/2}}\,. \tag{2.1.10}$$

$(\upsilon = x, y, z)$

$\mathbf{a}_{F_{21}}$ (= $\mathbf{F}_{21}/F_{21}$) is obtained simply by dividing (2.1.7) by (2.1.9).

<u>*Example 2.1.1*</u> Find the force on a point charge 'q' located at any corner of a cube of side 'a' due to point charges, each 'q', located at the remaining corners of the cube. Take the system of charges to be located in free space.

The coordinates of the corners where the point charges each q (coulomb) are placed are shown in Fig. 2.1.1. Let us find the individual forces acting separately on the point charge q at (0, 0, 0) due to the remaining point charges located at (a, 0, 0), (0, 0, a), (0, a, 0), (0, a, a), (a, a, a), (a, 0, a) and (a, a, 0), respectively. These forces can be found, respectively, with the help of (2.1.7) - (2.1.10) as:

$$-\,(q^2/4\pi\varepsilon_0 a^2)\,(\mathbf{a}_x,\ \mathbf{a}_z,\ \mathbf{a}_y,\ (\mathbf{a}_y + \mathbf{a}_z)/(2\sqrt{2}),\ (\mathbf{a}_x + \mathbf{a}_y + \mathbf{a}_z)/(3\sqrt{3}),\ (\mathbf{a}_x + \mathbf{a}_z)/(2\sqrt{2}) \text{ and } (\mathbf{a}_x + \mathbf{a}_y)/(2\sqrt{2}))\,.$$

Adding these forces, the total force $\mathbf{F}$ on the point charge q at (0, 0, 0) may be found and expressed in terms of the magnitude of the force and a unit vector as follows:

$$\mathbf{F} = \left[\left(\frac{q^2}{4\pi\varepsilon_0 a^2}\right)\left(1+\frac{1}{\sqrt{2}}+\frac{1}{3\sqrt{3}}\sqrt{3}\right)\right]\left[\frac{-\mathbf{a}_x-\mathbf{a}_y-\mathbf{a}_z}{\sqrt{3}}\right]. \tag{2.1.11}$$

(magnitude) (unit vector)

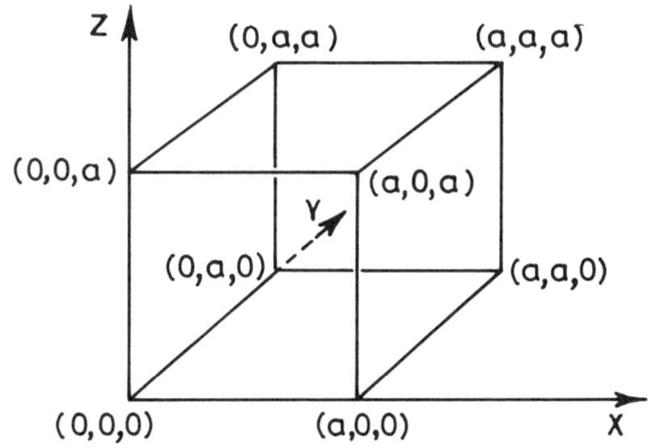

Fig. 2.1.1 Geometry of the problem in Example 2.1.1.

Now a unit vector directed from (a, a, a) to (0, 0, 0) is the length vector $(0-a)\,\mathbf{a}_x + (0-a)\,\mathbf{a}_y + (0-a)\,\mathbf{a}_z$ divided by its magnitude $\sqrt{3}a$ ($= (0-a)^2 + (0-a)^2 + (0-a)^2)^{1/2}$). This unit vector is seen to be identical with the unit vector appearing in (2.1.11). Hence, the force on the point charge q at (0, 0, 0) as given by (2.1.11) is directed along the body diagonal from (a, a, a) to (0, 0, 0).

2.1.1 Electric Field Intensity due to a Charge Distribution

Let us first find the force **F** on a test point charge q′ placed at the point where the electric field intensity (**E**) is sought. According to (2.1.3)

$$\mathbf{E} = \operatorname*{Lt}_{q'\to 0} \frac{\mathbf{F}}{q'}. \tag{2.1.12}$$

In this case **F** is given with the help of (2.1.4) by

$$\mathbf{F} = \frac{q\,q'}{4\pi\varepsilon\, r^2}\mathbf{a}_r, \tag{2.1.13}$$

which when substituted into (2.1.12) yields

$$\mathbf{E} = \frac{q}{4\pi\varepsilon\, r^2}\mathbf{a}_r, \tag{2.1.14}$$

the required expression for the electric field intensity due to a point charge. Here, we may recall, $\mathbf{a}_r$ is the unit vector directed from the point charge q to the point located at a distance r where to find the electric field intensity. It can be seen on examining (2.1.13) and (2.1.14) that this electric field intensity is numerically equal to the Coulomb force on a unit test point charge ($q' = 1$) placed at the point where the electric field intensity is sought.

Then let us take up a few illustrative problems in which we are interested to know the electric field intensity due to a uniform charge distribution. In each of these examples one may divide the given charge distribution into charge elements each regarded as a point charge. The element of electric field intensity due to each such charge element is then obtainable from (2.1.14) in which q has to be interpreted as the charge element considered. By integrating the elements of electric field intensity due to these charge elements, the electric field intensity due to the entire charge distribution can be obtained.

Example 2.1.2 Find the electric field intensity at a point due to charges sprayed uniformly over (a) a line, (b) a plane of circular boundary, (c) a sphere, and (d) a long right circular cylinder. Consider the medium to be a free space.

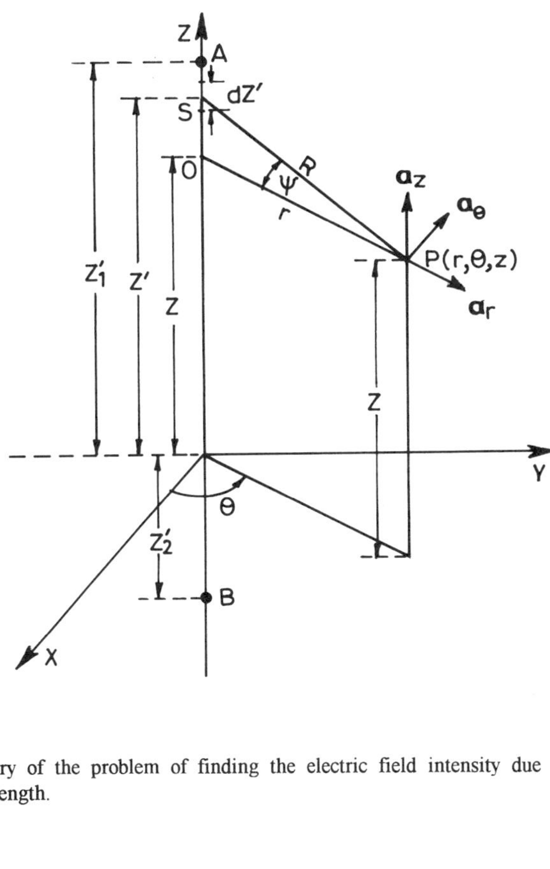

Fig. 2.1.2 Geometry of the problem of finding the electric field intensity due to a line-charge distribution of a finite length.

(a) Let the line charge be coincident with the z-axis and the point where the electric field intensity is sought be denoted by the cylindrical system of coordinates r, θ, z (see Fig. 2.1.2). If ρ_l is the charge per unit length on the line, then the charge element extending over an element of length dz′ at a point distant z′ from the origin of coordinates is $\rho_l\, dz'$. First, the element of electric field intensity **dE** at (r, θ, z) due to charge element $\rho_l\, dz'$ is found using (2.1.14). Then integrating **dE** over the entire length of the line charge is obtained the following expression for the electric field intensity due to a line charge of finite length:

$$\mathbf{E} = \frac{\rho_l}{4\pi\varepsilon_0 r}\,[\{\cos \tan^{-1}(z_1'-z)/r - \cos \tan^{-1}(z_2'-z)/r\}\mathbf{a}_z$$

$$+ \{\sin \tan^{-1}(z_1'-z)/r - \sin \tan^{-1}(z_2'-z)/r\}\mathbf{a}_r]\,, \qquad (2.1.15)$$

where z_1' and z_2' are the z-coordinates of the extreme points of the line charge. (For the details of derivation of (2.1.15) see appendix *a2.1.1*). $\mathbf{a}_z$ and $\mathbf{a}_r$ are the unit vectors in the directions of increasing z and r, respectively.

For an infinitely long line charge ($z_1' = \infty$, $z_2' = -\infty$), we have from (2.1.15) (see appendix *a2.1.1*):

$$\mathbf{E} = \frac{\rho_l}{4\pi\varepsilon_0 r}\,\mathbf{a}_r\,. \qquad (2.1.16)$$

We notice on examining (2.1.15) and (2.1.16) that, in general, due to a line charge of finite length the electric field intensity has both an axial (that is, a component parallel to its length) and a radial component. However, as the length of the line is increased, the radial component of the field dominates over the axial component, and, for an infinitely long line, only the radial component remains. Also, for both the finite and infinite lengths of line charge, the electric field intensity is azimuthally symmetric, that is, the field has no variation with θ.

(b) Let the plane of charges be bounded by a circle of radius a, say. Let also the plane be located on the XY-plane and the centre of the circle be the origin (0, 0, 0) of the rectangular cartesian coordinates, and the point where to find the electric field intensity have the coordinates (0, 0, z) (Fig. 2.1.3).

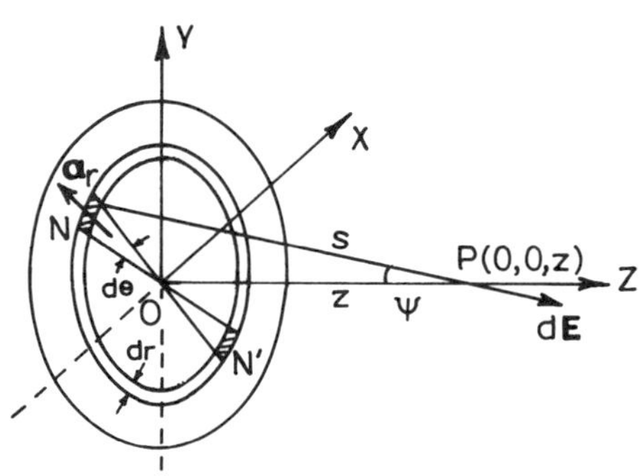

Fig. 2.1.3 Geometry of the problem of evaluating the electric field intensity due to a charge sprayed over a plane bounded by a circle.

The problem may be treated in two steps. First, one can find the element of electric field intensity **dE** at the point (0, 0, z) due to a circular ring of an arbitrary radius r ($0 < r < a$) and of infinitesimal thickness dr. The electric field intensity **E** due to the charge sprayed over the entire area of the circular plane is then found by integrating **dE** (the details of derivation are given in appendix *a2.1.2*):

$$\mathbf{E} = \left(1 - \frac{z}{(z^2 + a^2)^{1/2}}\right)\frac{\rho_s}{2\varepsilon_0}\mathbf{a}_z\,, \tag{2.1.17}$$

where ρ_s is the surface density of the charge uniformly distributed over the circular plane. The electric field intensity due to a charge distributed over an infinite plane, is obtained by putting $a \to \infty$ in (2.1.17):

$$\mathbf{E} = \frac{\rho_s}{2\varepsilon_0}\mathbf{a}_z\,, \tag{2.1.18}$$

a quantity found as independent of the distance of the point from the planar charge distribution.

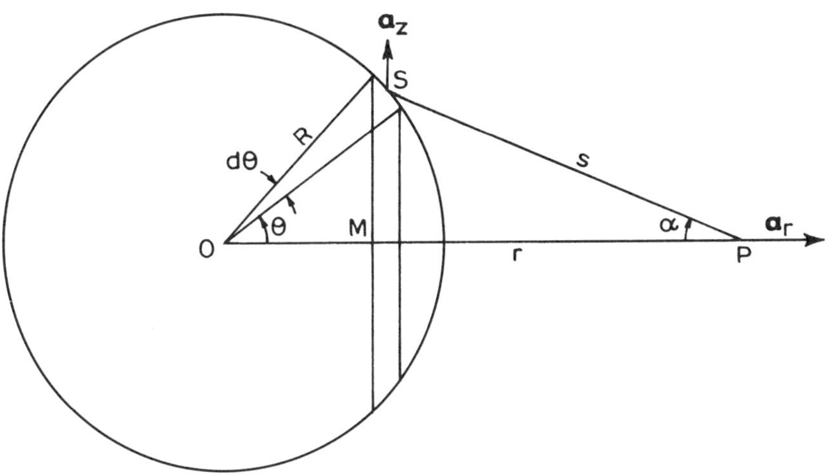

Fig. 2.1.4 Geometry of the problem of finding the electric field intensity due to a charge uniformly sprayed over a spherical surface (see appendix *a2.1.4*).

(c) Let the point where to find the electric field intensity be situated at a distance $r > R$ from the centre of the sphere of radius R over which a charge Q of uniform surface charge density ρ_s is sprayed. As in (b), let us work out the problem in two steps. First, an expression for the element of electric field intensity **dE** due to an annular belt arbitrarily considered on the spherical area is found (see Fig. 2.1.4). The contribution to the electric field intensity due to all such belts into which the entire area may be divided is then obtained by integrating **dE** as (for the details of derivation, see appendix *a2.1.3*):

$$\mathbf{E} = \frac{Q}{4\pi\varepsilon_0 r^2}\mathbf{a}_r \qquad (r > R)\,. \tag{2.1.19}$$

It follows from (2.1.14) that the expression for the electric field intensity **E** due to a point charge Q at a distance r from it in free space is:

$$\mathbf{E} = \frac{Q}{4\pi\varepsilon_0 r^2}\mathbf{a}_r \,. \tag{2.1.20}$$

The identicalness of (2.1.19) and (2.1.20) suggests that, as far the effect of a charge uniformly distributed over a spherical surface on a point outside the sphere is concerned, the charge on the sphere may be regarded as concentrated at the centre of the sphere.

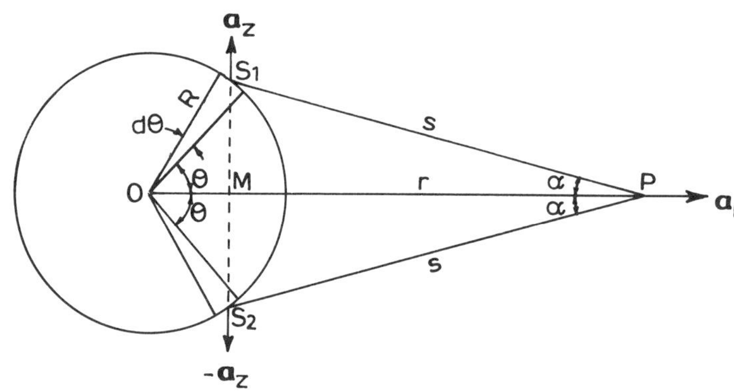

Fig. 2.1.5 Geometry of the problem of finding the electric field intensity due to a charge uniformly sprayed over a long cylinder.

(d) The charge sprayed over the surface of a long circular cylinder may similarly be divided into long elementary line charges parallel to the axis, each of infinitesimal thickness. First, evaluating the elementary contribution to the electric field intensity from an arbitrarily chosen pair of such long line charges, considered symmetrically with respect to the radial direction passing through the point where the electric field intensity is sought (see Fig. 2.1.5), and then integrating the element of electric field intensity over such pairs into which the entire charge cylinder may be divided (Fig. 2.1.5), the expression for the electric field intensity **E** due to a long charge cylinder at a point outside the cylinder may be found as (for the details of derivation, see appendix *a2.1.4*):

$$\mathbf{E} = \frac{\rho_l}{2\pi\varepsilon_0 r}\mathbf{a}_r \,, \tag{2.1.21}$$

where ρ_l is the line charge density (charge per unit length) of the cylinder and r is the radial distance of the point measured from the axis of the cylinder.

Comparing (2.1.21) and (2.1.16) we may conclude that, as far as the effects outside the cylinder over which the charge is uniformly distributed are concerned, the charge may be considered as concentrated along the axis of the cylinder with the value of the line charge density retained.

Appendix a2.1.1 Line charge distribution

Treating the charge element $\rho_l\, dz'$ *as a point charge, and using (2.1.14), we may write the element of electric field intensity* $\mathbf{dE}$ *as (see Fig. 2.1.2):*

(a2.1.1) $$\mathbf{dE} = \frac{\rho_l\, dz'}{4\pi\varepsilon_0 (SP)^2}\, \boldsymbol{a}_{SP} ,$$

where $\boldsymbol{a}_{SP}$ is a unit vector directed from S to P given by

(a2.1.2) $$\boldsymbol{a}_{SP} = \frac{\mathbf{SP}}{SP} = \frac{\mathbf{SO}+\mathbf{OP}}{SP} = \frac{-(z'-z)\,\boldsymbol{a}_z + r\,\boldsymbol{a}_r}{SP} .$$

Substituting (a2.1.2) into (a2.1.1) and writing $SP = R$, we obtain

(a2.1.3) $$\mathbf{dE} = \frac{\rho_l\left[-(z'-z)\,\mathbf{a}_z + r\,\mathbf{a}_r\right] dz'}{4\pi\varepsilon_0 R^3} .$$

The electric field intensity due to the entire line charge AB is obtained by integrating (a2.1.3):

(a2.1.4) $$\mathbf{E} = \int \mathbf{dE} = \frac{\rho_l}{4\pi\varepsilon_0}\left[\left(\int_{z_2'}^{z_1'} \frac{-(z'-z)\,dz'}{R^3}\right)\mathbf{a}_z + \left(r\int_{z_2'}^{z_1'} \frac{dz'}{R^3}\right)\mathbf{a}_r\right] .$$

In the above integration we have taken the unit vectors $\mathbf{a}_z$ and $\mathbf{a}_r$ outside the first and second integrations, respectively. This is because, for each charge element $\rho_l\, dz'$, whatever the value of z', $\mathbf{a}_z$ and $\mathbf{a}_r$ remain the same. For a similar reason, here, r is taken outside the second integration.

The evaluation of the integrals in (a2.1.4) becomes easier if we introduce a variable ψ, the angle between PS and PO, indicating the direction of the charge element from the point $P\ (r, \theta, z)$ (where to find the electric field) with respect to the radially inward direction through P. Thus one can replace z'-z, R and dz' in (a2.1.4) by:

(a2.1.5) $$z' - z = r\tan\psi, \quad R = r\sec\psi \quad \text{and} \quad dz' = r\sec^2\psi\, d\psi .$$

Making use of (a2.1.5) in (a2.1.4), and carrying out the integration, we obtain

(a2.1.6) $$\mathbf{E} = \frac{\rho_l}{4\pi\varepsilon_0 r}\left[\left(\cos\psi_1 - \cos\psi_2\right)\boldsymbol{a}_z + \left(\sin\psi_1 - \sin\psi_2\right)\boldsymbol{a}_r\right] ,$$

where ψ_1 and ψ_2 are the values of ψ $(= \tan^{-1}(z'-z)/r$; see (a2.1.4)) for the charge elements considered at the extreme points of the line, namely, at A $(z = z_1')$ and B $(z = z_2')$, respectively. In other words

(a2.1.7) $$\psi_1 = \tan^{-1}\frac{z_1' - z}{r} \quad \text{and} \quad \psi_2 = \tan^{-1}\frac{z_2' - z}{r} .$$

We may see that (a2.1.6) through (a2.1.7) is identical with (2.1.5). Now, z_1' and z_2' may have both positive and negative values. A particular case of interest is an infinitely long line for which one may put $z_1' = \infty$, $z_2' = -\infty$ in (a2.1.7) to have

(a2.1.8) $$\psi_1 = \tan^{-1}(\infty) = \pi/2 \quad \text{and} \quad \psi_2 = \tan^{-1}(-\infty) = -\pi/2 .$$

Substituting in (a2.1.6) the values of ψ_1 and ψ_2 given by (a2.1.7), one directly obtains (2.1.15). Similarly, the substitution of (a2.1.8) in (a2.1.6) yields (2.1.16).

Appendix a2.1.2 Planar charge within a circular boundary

The charge element of thickness dr, considered at any arbitrary point N on the circular ring of radius r, which subtends an angle $d\theta$ at the origin, is obtained by multiplying the area of the element $(dr)(r\,d\theta)$ by the surface charge density ρ_s (see Fig. 2.1.3). Treating the charge element $(dr)(r\,d\theta)(\rho_s)$ thus obtained as a point charge, we can find the element of electric field intensity $\mathbf{dE'}$ due to it at P(0, 0, z), with the help of (2.1.14), as

$$\mathbf{dE}' = \frac{\rho_s\, r d\theta\, dr}{4\pi\varepsilon_0\, s^2}\mathbf{a}_{NP} , \tag{a2.1.9}$$

where s (= NP) is the distance of the charge element from P; $\mathbf{a}_{NP}$ is the unit vector directed from N to P, which may be put as

$$\mathbf{a}_{NP} = \frac{\mathbf{NP}}{NP} = \frac{\mathbf{NO}+\mathbf{OP}}{s} = \frac{-r\,\mathbf{a}_r + z\,\mathbf{a}_z}{s} . \tag{a2.1.10}$$

Here, $\mathbf{a}_r$ is the unit vector directed from the origin of the rectangular cartesian coordinates to the charge element at N. Clearly, the direction of $\mathbf{a}_r$ would change from position to position on the circular ring. Substituting (a2.1.10) into (a2.1.9), one obtains

$$\mathbf{dE}' = \frac{(-\rho_s\, r^2 d\theta\, dr)\,\mathbf{a}_r}{4\pi\varepsilon_0\, s^3} + \frac{(\rho_s\, rz\, d\theta\, dr)\,\mathbf{a}_z}{4\pi\varepsilon_0 s^3} . \tag{a2.1.11}$$

Now, let us proceed to find the electric field intensity due to the entire charge ring (of radius r and thickness dr). Choosing to put this field intensity as an element $\mathbf{dE}$, we then have

$$\mathbf{dE} = \int \mathbf{dE}' = \frac{-\rho_S\, r^2 dr}{4\pi\varepsilon_0\, S^3}\int_0^{2\pi} \mathbf{a}_r\, d\theta + \frac{\rho_S\, rz\, dr\, \mathbf{a}_z}{4\pi\varepsilon_0 S^3}\int_0^{2\pi} d\theta . \tag{a2.1.12}$$

In (a2.1.12) we notice that since $\mathbf{a}_r$ changes from the charge element to element it is kept inside the first integral, while since $\mathbf{a}_z$ remains the same for all the charge elements it is kept outside the second integral. Now we see that the first integral vanishes. This is because the contribution to $\mathbf{dE}$ due to any arbitrarily considered charge element, say, at N on the ring will be balanced out by that due to the charge element at N′ considered at a diametrically opposite point, since the values of $\mathbf{a}_r$ corresponding to these two charge elements will be equal but of opposite signs. Thus the entire charge on the ring may be divided into a number of such pairs of charge elements considered at diametrically opposite points, each pair contributing zero to the first integral in (a2.1.12). Thus, putting the first integral as zero and evaluating the second integral, we have from (a2.1.12):

$$\mathbf{dE} = \frac{z\,\rho_s\, rdr}{2\varepsilon_0\, s^3}\mathbf{a}_z . \tag{a2.1.13}$$

Now, the intensity of electric field $\mathbf{E}$ due to the charge on the entire plane bounded by a circle of radius a is obtained by integrating (a2.1.13) as follows:

$$\mathbf{E} = \frac{z\,\rho_s}{2\varepsilon_0}\mathbf{a}_z \int_0^a \frac{rdr}{s^3} . \tag{a2.1.14}$$

We notice that, while integrating (a2.1.13), we have taken $\mathbf{a}_z$ *outside the integration since* $\mathbf{a}_z$ *is the same whatever the radius (r) of the ring* $(0 \le r \le a)$. *The evaluation of the integral becomes easier by introducing a variable* ψ *measuring the half cone angle subtended by the ring of radius r at P (see Fig. 2.1.3). In terms of* ψ,

(a2.1.15) $$r = z \tan\psi, \quad dr = z \sec^2\psi \, d\psi \quad \text{and} \quad s = z \sec\psi \; .$$

Putting (a2.1.15) in (a2.1.14),

$$\mathbf{E} = \frac{z \rho_s}{2\varepsilon_0 z} \mathbf{a}_z \int_0^{\tan^{-1} a/z} \sin\psi \, d\psi$$

which when evaluated passes on to (2.1.17).

Appendix a2.1.3 *Charge distribution over a sphere*

Let us consider an annular belt of infinitesimal thickness $R\,d\theta$ *on the spherical surface at angle* θ *measured from the line directed from the centre of the sphere, 0, to the point P where to find the electric field intensity (see Fig. 2.1.4). If* ρ_s *is the uniform surface charge density, then the charge on an element of area* dA *considered arbitrarily at S is* $\rho_s\, dA$ *and the element of electric field intensity* $\mathbf{dE}'$ *due to this element of charge at P is obtained with the help of (2.1.14) as:*

(a2.1.16) $$\mathbf{dE}' = \frac{\rho_s \, dA}{4\pi\,\varepsilon_0\, s^2} \mathbf{a}_{SP} \; ,$$

where s = (SP) is the distance of the charge element from P, $\mathbf{a}_{SP}$ *is the unit vector directed from S to P. Substituting* $\mathbf{a}_{SP} = \mathbf{s}/s$ *and* $\mathbf{s} = \mathbf{SM} + \mathbf{MP}$ *in (a2.1.16) we may write*

(a2.1.17) $$\mathbf{dE}' = \frac{\rho_S \, dA(\mathbf{SM} + \mathbf{MP})}{4\pi\varepsilon_0 \, s^3} \mathbf{a}_{SP} \; .$$

Here,

(a2.1.18) $$SM = R \sin\theta$$

is the radius of the belt of charge and

(a2.1.19) $$MP = s \cos\alpha \; ,$$

where α *is the half-cone angle subtended by the belt at P. Introducing the unit vectors* $\mathbf{a}_r$ *and* $\mathbf{a}_z$ *as directed from O to P and from M to S, respectively, we can write with the help of (a2.1.18) and (a2.1.19):*

(a2.1.20) $$\mathbf{SM} = -(R \sin\theta)\, \mathbf{a}_z \quad \text{and}$$

$$\mathbf{MP} = (s \cos\alpha)\, \mathbf{a}_r \, .$$

Substituting (a2.1.20) into (a2.1.17),

(a2.1.21) $$\mathbf{dE}' = \frac{\rho_S \, dA}{4\pi\varepsilon_0 \, s^3} [-(R \sin\theta)\mathbf{a}_z + (s \cos\alpha)\mathbf{a}_r] \; .$$

Integrating (a2.1.21) over the entire charge belt,

(a2.1.22) $$\mathbf{dE} = \frac{-\rho_S\, R \sin\theta}{4\pi\varepsilon_0\, s^3}\int \mathbf{a}_z\, dA + \frac{\rho_S\, S\cos\alpha}{4\pi\varepsilon_0\, s^3}\,\mathbf{a}_r\int dA\ .$$

We notice that, in the above integral expression (a2.1.22) for the intensity of electric field due to the charge belt at P, $\mathbf{a}_z$ *is kept inside the integral (the first term) since it varies from element to element (dA), but* $\mathbf{a}_r$ *has been kept outside the integral (the second term) since its does not depend upon where we take the element (dA) on the charge belt. However, the first integral in (a2.1.22) vanishes following the same argument as given while deriving (a2.1.13) from (a2.1.12), thus enabling us to write*

(a2.1.23) $$\mathbf{dE} = \frac{\rho_s\, s\cos\alpha}{4\pi\varepsilon_0\, s^3}\,\mathbf{a}_r\, A\ ,$$

where A *is the area of the belt. Now, the belt has thickness* $R\,d\theta$ *and radius* $R\sin\theta$ *so that*

(a2.1.24) $$A = (2\pi R\ \sin\theta)\,(R\,d\theta)\ .$$

Moreover, we can write (see Fig. 2.1.4)

(a2.1.25) $$\cos\alpha = \frac{r^2+s^2-R^2}{2rs}$$

and $$\cos\theta = \frac{r^2+R^2-s^2}{2rR}$$

whence, by taking the differential, we have

(a2.1.26) $$\sin\theta\, d\theta = \frac{s\,ds}{rR}\ .$$

Substituting (a2.1.24) and (a2.1.25) into (a2.1.23) and taking help of (a2.1.26), one may then express **dE** *as*

$$\mathbf{dE} = \frac{\rho_s\, R}{4\pi\,\varepsilon_0\, r^2}\left[(r^2-R^2)\frac{ds}{s^2}+ds\right]\mathbf{a}_r\ ,$$

which, when integrated between $s = r + R$ *and* $s = r - R$*, covers the contribution due to the entire charge on the spherical surface (see Fig. 2.1.4):*

$$\mathbf{E} = \frac{\rho_s\, R}{4\pi\,\varepsilon_0\, r^2}\left[(r^2-R^2)\int_{r-R}^{r+R}\frac{ds}{s^2}+\int_{r-R}^{r+R} ds\right]\mathbf{a}_r = \frac{\rho_s\, R^2}{\varepsilon_0 r^2}\,\mathbf{a}_r\ ,$$

which, through the relation

$$\rho_s = \frac{Q}{4\pi\, R^2}\ ,$$

passes on to (2.1.19).

Appendix a2.1.4 *Charge distribution over a cylinder*

Let us consider two long line charges arbitrarily considered at S_1 *and* S_2 *on the long cylinder symmetrically placed at an angle* θ *on the two sides of the radial line passing through the point* P *at a radial distance r (see Fig. 2.1.5). The elements of electric field intensity* $\boldsymbol{dE}_1$ *and* $\boldsymbol{dE}_2$ *due to the long line charges at* S_1 *and* S_2 *are given, with the help of (2.1.16), respectively, as*

(a2.1.27) $$\boldsymbol{dE}_1 = \frac{\rho_l'}{2\pi\varepsilon_0 s}\boldsymbol{a}_{S_1P} \quad \text{and}$$

$$\boldsymbol{dE}_2 = \frac{\rho_l'}{2\pi\varepsilon_0 s}\boldsymbol{a}_{S_2P} ,$$

where ρ_l' *and* $R\,d\theta$ *are the line charge density and the width of each of the long line charges at* S_1 *and* S_2 *each at a distance s from P* ($S_1P = S_2P = s$)*.* $\boldsymbol{a}_{S_1P}$ *and* $\boldsymbol{a}_{S_2P}$ *are the unit vector directed from* S_1 *to P and that from* S_2 *to P, respectively:*

(a2.1.28) $$\boldsymbol{a}_{S_1P} = \frac{\boldsymbol{S_1P}}{S_1P} \quad \text{and} \quad \boldsymbol{a}_{S_2P} = \frac{\boldsymbol{S_2P}}{S_2P} .$$

Now, $\boldsymbol{S_1P}$ *and* $\boldsymbol{S_2P}$ *occurring in (a2.1.28) may be expressed as*

(a2.1.29) $$\boldsymbol{S_1P} = \boldsymbol{S_1M} + \boldsymbol{MP} \quad \text{and} \quad \boldsymbol{S_2P} = \boldsymbol{S_2M} + \boldsymbol{MP} .$$

Now, substituting (a2.1.18) through (a2.1.29) in (a2.1.27) and in view of the relation

$$\boldsymbol{S_1M} = -\boldsymbol{S_2M} ,$$

we can write the total contribution to the electric field intensity at P from the pair of line charge elements at S_1 *and* S_2 *as*

(a2.1.30) $$\boldsymbol{dE} = \boldsymbol{dE}_1 + \boldsymbol{dE}_2 = \frac{\rho_l'}{2\pi\varepsilon_0 s^2}(2\boldsymbol{MP}) .$$

Substituting $\boldsymbol{MP} = s\cos\alpha\,\boldsymbol{a}_r$ *in (a2.1.30),*

(a2.1.31) $$\boldsymbol{dE} = \frac{\rho_l'}{\pi\varepsilon_0}\frac{\cos\alpha}{s}\boldsymbol{a}_r ,$$

where α *is the angle made by either of* S_1 *and* S_2 *at the point where the electric field intensity is sought. Now, line charge density* ρ_l' *of the long line of width* $R\,d\theta$ *bears the following relation with line charge density* ρ_l *of the entire cylinder of charge of width* $2\pi R$*, the circumference of the cross section of the cylinder:*

(a2.1.32) $$\rho_l' = \frac{d\theta}{2\pi}\rho_l .$$

Substituting (a2.1.32) into (a2.1.31) and carrying out the integration over the entire cylinder, we obtain the following expression for the electric field intensity due to the charge sprayed over the cylinder:

(a2.1.33) $$\boldsymbol{E} = \int \boldsymbol{dE} = \frac{\rho_l}{2\pi^2\varepsilon_0}\left(\int_0^{\pi}\frac{\cos\alpha}{s}\,d\theta\right)\boldsymbol{a}_r .$$

In (a2.1.33), we have included the contribution from all the line-charge element pairs lying between $\theta = 0$ *and* π *which would cover the entire cylinder, and taken* $\boldsymbol{a}_r$ *essentially outside the integral since it does not depend upon the location of the line charge elements. Now, substituting*

$$\cos\alpha = \frac{r^2 + s^2 - R^2}{2rs}$$

(see Fig. 2.1.5), the integral in (a2.1.33) takes the following form:

$$\boldsymbol{E} = \frac{\rho_l}{4\pi^2 \varepsilon_0 r}\left(\int_0^\pi d\theta + (r^2 - R^2)\int_0^\pi \frac{d\theta}{s^2}\right)\boldsymbol{a}_r \qquad \text{(a2.1.34)}$$

$$= \frac{\rho_l}{4\pi^2 \varepsilon_0 r}\left(\pi + (r^2 - R^2)\ \mathcal{I}\right)\boldsymbol{a}_r \, .$$

where

$$\mathcal{I} = \int_0^\pi \frac{d\theta}{s^2} \, . \qquad \text{(a2.1.35)}$$

Substituting s^2 *obtainable from the relation*

$$\cos\theta = \frac{r^2 + R^2 - s^2}{2Rr}$$

(see Fig. 2.1.5) and introducing

$$A = R^2 + r^2 \quad \text{and} \qquad \text{(a2.1.36)}$$

$$B = -2Rr \, ,$$

the integral $\mathcal{I}$ *given by (a2.1.35) may be expressed as*

$$\mathcal{I} = \int_0^\pi \frac{d\theta}{A + B\cos\theta} \, . \qquad \text{(a2.1.37)}$$

The integral (a2.1.37) may be evaluated as

$$\mathcal{I} = \frac{\pi}{(A^2 - B^2)^{1/2}} \, . \qquad \text{(a2.1.38)}$$

[*Intermediate steps:*

$$\mathcal{I} = \int_0^\pi \frac{d\theta}{A + B\cos\theta} = \int_0^\pi \frac{d\theta}{A(\cos^2\frac{\theta}{2} + \sin^2\frac{\theta}{2}) + B(\cos^2\frac{\theta}{2} - \sin^2\frac{\theta}{2})}$$

$$= \int_0^\pi \frac{d\theta}{(A+B)\cos^2\frac{\theta}{2} + (A-B)\sin^2\frac{\theta}{2}} = \frac{1}{A+B}\int_0^\pi \frac{\sec^2\frac{\theta}{2}\,d\theta}{1 + \frac{A-B}{A+B}\tan^2\frac{\theta}{2}} = \frac{2}{(A^2 - B^2)^{1/2}}\int_0^\infty \frac{du}{1+u^2} \, ,$$

where

$$u = \left(\frac{A-B}{A+B}\right)^{1/2} \tan\frac{\theta}{2}$$

so that

$$du = \frac{1}{2}\left(\frac{A-B}{A+B}\right)^{1/2} \sec^2\frac{\theta}{2}\ d\theta \qquad = \left[\frac{2}{(A^2-B^2)^{1/2}}\left[\tan^{-1} u\right]_0^{\infty} = \frac{\pi}{(A^2-B^2)^{1/2}}\right] J\ .$$

Substituting (a2.1.36) into (a2.1.38),

$$\mathcal{J} = \frac{\pi}{r^2 - R^2}$$

which may now be put in (a2.1.34) to obtain (2.1.21).

2.2 Gauss's Law and Electric Field Intensity

Gauss's law [4] is another concept of electrostatics which is closely related to Coulomb's law. In fact, starting from any one of these laws, the other follows. Gauss's law makes it considerably easy to deal with problems which enjoy some sort of geometrical symmetry. In such problems, the electric field intensity varies, if it does at all, only with one coordinate, and remains constant over a regular surface. According as the constant electric field surfaces, called the Gaussian surfaces, in these problems, may take the planar, cylindrical, spherical shapes, etc., we describe these problems as to have the symmetry of the rectangular, cylindrical, spherical types, etc., respectively. Gauss's law, which has been established, starting from the basic Coulomb's law, in appendix *a2.2.1*, states that the flux of electric displacement vector **D** (= ε**E**) over a closed surface is equal to the charge enclosed by the surface:

$$\oint_S \mathbf{D}.\mathbf{a}_n\ dS = q\ . \qquad (2.2.1)$$

(**D** is also called the electric field flux density). The left hand side of (2.2.1) represents the closed-surface integral of **D** over the surface **S**. dS represents an element of area on the surface over which **D** may be regarded as constant. $\mathbf{a}_n$ is the unit vector drawn outwardly-normal to the surface. The right hand side of (2.2.1) represents the charge enclosed by the surface. This charge may be a single charge or a system or distribution of charges [4] (appendix *a2.2.1*).

While applying Gauss's law (2.2.1) to symmetrical problems, one has to define a surface or a number of surfaces such that the magnitude of the intensity of electric field remains constant over the surface(s) considered, and the direction of the electric field intensity and that of the unit vector $\mathbf{a}_n$ make a known angle with each other. The application of the law may be illustrated with reference to the problems of finding the electric field intensities due to charges uniformly sprayed along a long line, an infinite plane, a spherical surface and a long cylinder. The approaches to these problems using Coulomb's law can be seen to inconveniently involve complex integrations. Take, for instance, the problem of uniform charge distribution over a long cylinder (Example 2.1.2(d)). In principle, the problem involves two integrations. The first integral yields the electric field intensity due to a long line charge of infinitesimal thickness considered on the cylinder (Example 2.1.2(a)), while the second integral gives the electric field intensity due to the entire cylinder. Now, Example 2.2.1 to follow will illustrate how simply the same problem can be dealt with, relatively easily, using Gauss's law without requiring to handle complex integrals.

Example 2.2.1 Evaluate the electric field intensity due to charges sprayed over a long right-circular cylinder (see Example 2.1.2(d)) using Gauss's law.

The problem enjoys the cylindrical symmetry. One may construct a cylindrical surface, called the Gaussian surface, passing through the point where to find the intensity of electric field such that the surface is coaxial with the cylinder on which the charge is sprayed. Obviously, the Gaussian surface itself is right-circular in this case. One may now apply Gauss's law (2.2.1) to a closed volume made out of a finite length of this Gaussian surface (see appendix *a2.2.2*). $\mathbf{D}$ (= $\varepsilon_0\,\mathbf{E}$) is radial to, and has the same value over, the Gaussian cylindrical surface. The unit vector $\mathbf{a}_n$ (Fig. a2.2.1) is also radial. As a result, the evaluation of the closed surface integral of (2.2.1), as applicable to the present problem, becomes extremely simple. The technique leads to the same expression as (2.1.21). (For the details of derivation, see appendix *a2.2.2*). Similarly, the other expressions for symmetrical problems of Example 2.1.2, namely, (2.1.16), (2.1.18) and (2.1.20) can be easily obtained using Gauss's law (see problem 2.7).

Example 2.2.2 Find the electric field intensity inside a long axi-symmetric beam of electrons of uniform volume charge density, using Gauss's law.

Although an electron beam is composed of charges which are in motion, for many useful purposes, their effects may be smeared out and the electron beam regarded as a fluid of charges. Considering the electrons to be all of the same velocity, two snap-shots of the same cross-section of the beam, at two different instants of time, or of two cross sections of the beam at a distance from each other at the same instant of time would be indistinguishable. Consequently, assigning a volume charge density ρ to the electron beam, one can proceed to find the electric field inside the beam in terms of ρ.

For this purpose, a closed Gaussian cylindrical surface of a given axial length and radius is constructed passing through the point where to find the electric field intensity as in Example 2.2.1/appendix *a2.2.2*. In this case, the electric field intensity is radial and, therefore, the contribution to the closed surface integral of the left hand side of Gauss's law (2.2.1) from the end faces is zero, and the contribution is only from the curved cylindrical surface. Further, here $\mathbf{D}$ (= $\varepsilon_0\mathbf{E}$), which remains constant over the cylindrical surface of constant radius, may be taken outside the surface integral. The charge enclosed by the Gaussian surface, as required to evaluate the right hand side of (2.2.1), and as applicable to the present problem, is obtained by multiplying the volume charge density by the volume of the Gaussian enclosure. The method leads to the following expression for the electric field intensity inside the electron beam (for the details of derivation, see appendix *a2.2.3*):

$$\mathbf{E} = \frac{\rho r}{2\varepsilon_0}\mathbf{a}_r . \tag{2.2.2}$$

It is clear from (2.2.2) that the electric field intensity inside the electron beam is minimum, being equal to zero, at the axis of the beam (r = 0) and it increases towards the beam edge.

Appendix a2.2.1 Gauss's law starting from Coulomb's law

Let us consider a surface of arbitrary shape enclosing a point charge q (see Fig. a2.2.1). Let dS be an element of area considered arbitrarily on the surface. Considering the element of area to be infinitesimally small one can take the electric field $\mathbf{E}$ *to be uniform over the element of area. The element of flux* $d\phi_D$ *of the electric displacement* $\mathbf{D}$ $(= \varepsilon_0 \mathbf{E})$ *through the area dS may be defined as*

$$d\phi_D = \mathbf{D}.\mathbf{a}_n\, dS = \varepsilon_o \mathbf{E}.\mathbf{a}_n\, dS \,. \tag{a2.2.1}$$

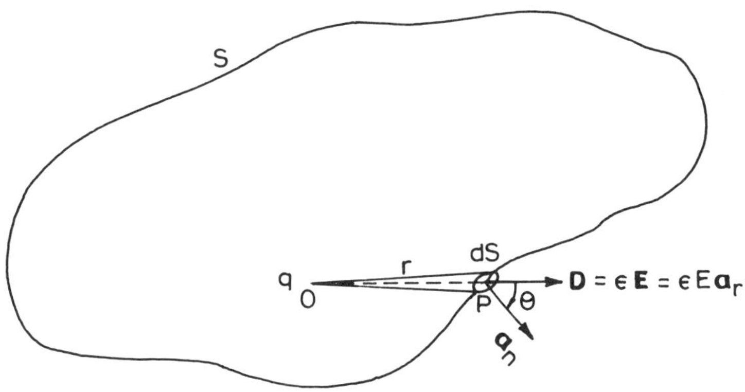

Fig. a2.2.1 The arbitrary "Gaussian" surface enclosing a point charge.

We notice that $\mathbf{E}$ *has the direction from O to P. In other words,*

$$\mathbf{E} = E\, \mathbf{a}_r \,, \tag{a2.2.2}$$

where $\mathbf{a}_r$ *is the unit vector in the direction of* $r(= OP)$ *increasing. Let, in general,* $\mathbf{a}_n$ *and* $\mathbf{a}_r$ *make an angle* θ *with each other. This enables one to express (a2.2.1) through (a2.2.2), noting that* $\mathbf{a}_r.\mathbf{a}_n = \cos\theta$, *as*

$$d\phi_D = \varepsilon_o\, E\, dS \cos\theta \,. \tag{a2.2.3}$$

Now writing

$$E = \frac{q}{4\pi\varepsilon_0 r^2} \,,$$

obtainable from (2.1.14), one may express (a2.2.3) as

$$d\phi_D = \frac{q}{4\pi} d\omega \,, \tag{a2.2.4}$$

where $d\omega$ $(= (dS\cos\theta)/r^2)$ *is the element of solid angle subtended by the element of area dS at the point charge q (at 0).*

Thus, the total flux ϕ_D *through the entire closed surface may be written with the help of (a2.2.1) and (a2.2.4) and by integration over the closed area as follows:*

(a2.2.5) $$\oint d\phi_D = \oint \boldsymbol{D}.\boldsymbol{a}_n \, dS = \frac{q}{4\pi}\oint d\omega \, .$$

Now, since the solid angle subtended by a closed surface at any point inside is 4π, one may substitute

$$\oint d\omega = 4\pi$$

in (a2.2.5) to obtain

(a2.2.6) $$\oint \boldsymbol{D}.\boldsymbol{a}_n \, dS = q$$

which is the mathematical form of Gauss's law as stated by (2.2.1).

We notice that it is easy to extend (a2.2.6) to the case in which the Gaussian surface encloses a charge distribution instead of a single point charge. Thus if the distribution consists of point charges q_1, q_2, q_3 . . . q_n, (a2.2.5) takes the following form

$$\oint \boldsymbol{D}.\boldsymbol{a}_n \, dS = \frac{q_1}{4\pi}\oint d\omega_1 + \frac{q_2}{4\pi}\oint d\omega_2 + \cdots + \frac{q_n}{4\pi}\oint d\omega_n$$

which, noting that $$\oint d\omega_1 = \oint d\omega_2 = \ldots = \oint d\omega_n = 4\pi \, ,$$

simplifies to

(a2.2.7) $$\oint \boldsymbol{D}.\boldsymbol{a}_n \, dS = q_1 + q_2 + \ldots + q_n \, .$$

The right hand side of (a2.2.7) is identified as the net charge enclosed. This makes the proof of Gauss's law more generalised.

Appendix a2.2.2 *Gauss's law for the problem of charge over a cylinder*

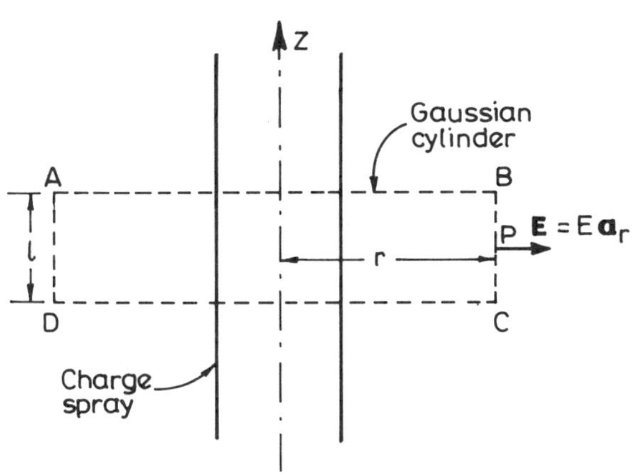

Fig. a2.2.2 Construction of the Gaussian cylinder through a point outside and co-axial with a cylinder over which the charge is uniformly sprayed.

Let us draw a long Gaussian cylinder of length l through the point where to find the electric field intensity. The Gaussian cylinder is symmetrically placed with respect to the long cylindrical surface over which the charge is uniformly distributed. Therefore, the electric field intensity has no azimuthal variation ($\partial/\partial\theta = 0$). Since, the charge cylinder is long, the electric field intensity would not change from point to point in a direction parallel to the length of the cylinder ($\partial/\partial z = 0$). Thus the electric field intensity would have only the radial variation ($\partial/\partial r \neq 0$). Further, in this problem of symmetry, the electric field intensity would be purely radial. This would mean that the portion of the end faces (AB, DC) of the Gaussian cylinder outside the charge cylinder would not contribute to the integrand of the surface integral (2.2.1), since $\mathbf{a}_n$ on each of these surfaces is normal to the radial direction which, here, is also the direction of $\mathbf{D}$ ($= \varepsilon_o \mathbf{E}$) outside the charge distribution. The portion of the end faces inside the charge cylinder would not similarly contribute to the surface integral since the value of $\mathbf{D}$ inside is zero (This can be proved again using Gauss's law easily. The proof is left as an exercise to students).

Thus, in this problem, the contribution to the surface integral of (2.2.1) would come only from the cylindrical surface (constant r) where $\mathbf{a}_n$ and $\mathbf{D}$ ($= \varepsilon_o \mathbf{E}$) have the same direction. In other word, one has, on the Gaussian cylindrical surface (constant r),

$$\mathbf{D} = D\,\mathbf{a}_r$$

and

$$\mathbf{a}_n = \mathbf{a}_r$$

so that (2.2.1), in this case, takes the following form

$$\oint D\,\mathbf{a}_r . \mathbf{a}_r\, dS = q \;, \qquad \text{(a2.2.8)}$$

where q is the charge enclosed by the Gaussian surface which is given by

$$q = \rho_l\, l\,, \qquad \text{(a2.2.9)}$$

where ρ_l is the charge per unit length on the cylinder. Substituting (a2.2.9) into (a2.2.8) and noting that D ($= \varepsilon_o E$) remains the same all over the Gaussian cylinder (constant r) so that it can be taken outside the integral, one has

$$\varepsilon_0 E \oint dS = \rho_l\, l\,. \qquad \text{(a2.2.10)}$$

Now, the integral in (a2.2.10) is simply equal to $2\pi r l$, being the area of the cylindrical surface of radius r and length l, so that (a2.2.10) would lead to

$$E = \frac{\rho_l}{2\pi\varepsilon_o r}\,. \qquad \text{(a2.2.11)}$$

Now that $\mathbf{E}$ is radial, the expression for $\mathbf{E}$ ($= E\mathbf{a}_r$) may be seen to be identical with (2.1.21) as mentioned in Example 2.2.1 (see also Example 2.1.2(d)).

Appendix a2.2.3 *Field intensity inside a cylindrical electron beam*

Let us draw a Gaussian cylinder of length l and radius r, coaxial with and inside a long cylindrical axi-symmetric beam of electrons of constant volume charge density (see Fig. a2.2.3). As in the problem treated in appendix a2.2.2, in this problem, too, there are no contributions to the surface integral of

(2.2.1) from the end faces. The contribution is only from the cylindrical surface of constant r. Thus, the left hand side of Gauss's law (2.2.1) may be written, putting $\mathbf{D} = \varepsilon_o E\, \mathbf{a}_r$ *and* $\mathbf{a}_n = \mathbf{a}_r$, *as*

(a2.2.12) $$\oint \mathbf{D}.\mathbf{a}_n\, dS = \varepsilon_0 E \oint dS = \varepsilon_0 E\, 2\pi\, rl\, .$$

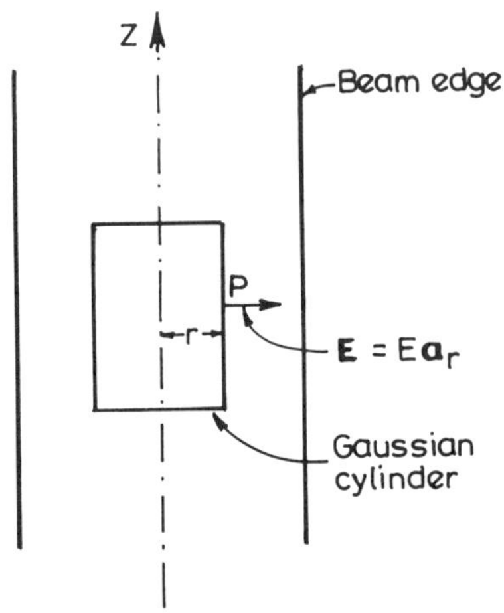

Fig. a2.2.3 Construction of the Gaussian surface through a point (P) inside and coaxial with the axi-symmetric electron beam.

The right hand side of (2.2.1) is:

(a2.2.13) *charge enclosed by the Gaussian cylinder* $= \pi r^2 l\rho$,

being equal to the volume ($\pi r^2 l$) *times the charge density* ρ. *Thus, (2.2.1) through (a2.2.12) and (a2.2.13) yields* E *whence* $\mathbf{E}$ (= $E\, \mathbf{a}_r$) *as (2.2.2).*

2.3 Differential Form of Gauss's Law : Poisson's Equation

The surface integral form (2.2.1) of Gauss's law takes the point form when the surface integral is taken over an element of volume which, in the limit, shrinks to a point (zero volume) [5].

Let us consider an element of volume $d\tau$ around a point where the volume charge density is ρ, say. If q is the charge enclosed by the volume element, then we can write

$$\rho = \mathop{\mathrm{Lt}}_{d\tau \to 0} \frac{q}{d\tau}\, . \tag{2.3.1}$$

Now, applying Gauss's law (2.2.1) to this closed element of volume and using (2.3.1), we have

$$\underset{d\tau \to 0}{\text{Lt}} \frac{\oint \mathbf{D}.\mathbf{a}_n \, ds}{d\tau} = \rho \; . \tag{2.3.2}$$

The left hand side of (2.3.2), that is, the flux of electric displacement vector **D** over a closed element of volume divided by the element of volume, in the limit the volume element shrinking to zero at a point, is defined as the divergence of **D**, abbreviated as div **D** and symbolised as $\nabla.\mathbf{D}$, at the point [6]. Thus (2.3.2), called Poisson's equation, may be written as

$$\text{div}\,\mathbf{D} = \nabla.\mathbf{D} = \rho \quad \text{(Poisson's equation)} \; . \tag{2.3.3}$$

The expressions for

$$\text{div}\,\mathbf{D} = \nabla.\mathbf{D}\left(\underline{\underline{\Delta}} \; \underset{d\tau \to 0}{\text{Lt}} \frac{\oint \mathbf{D}.\mathbf{a}_n \, dS}{d\tau} \right) \tag{2.3.4}$$

in the rectangular, cylindrical and spherical-polar systems of coordinates are (for the derivation of these expressions see appendix *a2.3.1*):

$$\text{div}\,\mathbf{D} = \nabla.\mathbf{D} = \frac{\partial D_x}{\partial x} + \frac{\partial D_y}{\partial y} + \frac{\partial D_z}{\partial z} \qquad \text{(a)}$$

(rectangular system)

$$= \frac{1}{r}\frac{\partial}{\partial r}(r\,D_r) + \frac{1}{r}\frac{\partial D_\theta}{\partial \theta} + \frac{\partial D_z}{\partial z} \qquad \text{(b)}$$

(cylindrical system)

$$= \frac{1}{r^2}\frac{\partial}{\partial r}(r^2 D_r) + \frac{1}{r\sin\theta}\frac{\partial}{\partial \theta}(D_\theta \sin\theta) + \frac{1}{r\sin\theta}\frac{\partial D_\phi}{\partial \phi} \; . \qquad \text{(c)} \tag{2.3.5}$$

(spherical-polar system)

<u>*Example 2.3.1*</u> In an axisymmetric long electron beam of uniform cross-section, supposing that the volume charge density ρ varies as

$$\rho = \frac{\alpha \exp(-pr)}{r} \qquad (r > 0),$$

where α and p are constants, find the space-charge electric field intensity of the beam using Poisson's equation.

In this problem $\partial/\partial z = 0$ (long beam), $\partial/\partial\theta = 0$ (axi-symmetric beam), and we may write with the help of (2.3.3) and (2.3.5(b)):

$$\frac{1}{r}\frac{d}{dr}(rD_r) \;\; (= \rho) \; = \frac{\alpha \exp(-pr)}{r} \; ,$$

which upon integration, and remembering that $D_r = \varepsilon_0 E_r$, yields the space-charge field intensity E_r which can be expressed in the vectorial form as

$$\mathbf{E} = \frac{\alpha}{\varepsilon_0 p}\left(\frac{1-\exp(-pr)}{r}\right)\mathbf{a}_r \quad .$$

Example 2.3.2 Evaluate the electric field intensity inside an electron beam (see Example 2.2.2) using Poisson's equation.

The problem was treated in Example 2.2.2 using Gauss's law. As in Example 2.3.1, the present problem also enjoys the axial symmetry $\partial/\partial z = 0$ (long beam) as well as the azimuthal symmetry ($\partial/\partial\theta = 0$) (axi-symmetric beam of circular cross section). This enables one to write, with the help of (2.3.3) and (2.3.5(c))

$$\frac{1}{r}\frac{d}{dr}(rD_r) = \rho$$

which, upon integration, yields

$$rD_r = \frac{\rho r^2}{2} + A \ ,$$

where A is the integration constant, whence, remembering that $D_r = \varepsilon_0 E_r$, we have,

$$E_r = \frac{\rho r}{2\varepsilon_0} + \frac{A}{r} \quad . \tag{2.3.6}$$

On examining (2.3.6) it may be appreciated that $A = 0$. This is because if $A \neq 0$, $E_r = \infty$ at $r = 0$ which is not physically acceptable. Thus by putting $A = 0$ in (2.3.6) we obtain the expression for the intensity of electric field $\mathbf{E}$ ($= E\,\mathbf{a}_r$) as

$$\mathbf{E} = \frac{\rho r}{2\varepsilon_0}\mathbf{a}_r \ . \tag{2.3.7}$$

We notice that (2.2.2) and (2.3.7) are identical. Thus we have shown, with reference to a particular problem, that Gauss's law in the integral form and Poisson's equation in the point form both, as expected, yield one and the same expression for the electric field intensity due to a given charge distribution.

Appendix a2.3.1 *Divergence in different systems of coordinates*

The expression for the divergence of a vector with particular reference to electric displacement ***D****, that is, for* $\nabla.\mathbf{D}$ *in the rectangular, cylindrical and spherical-polar systems may be deduced as follows. Let us represent the point P in the three systems of coordinates and consider around the point a closed element of volume* $d\tau$ *(see Fig. a2.3.1). Let us define the following quantities as relevant to deducing the expressions for div****D*** *at P in the rectangular, cylindrical and spherical-polar systems of coordinates, stating them under the heads (a), (b) and (c) respectively. Thus the coordinates of P are: (a)* x, y, z *(b)* r, θ, z *and (c)* r, θ, ϕ*. The three orthogonal components of* ***D*** *at P are: (a)* D_x, D_y, D_z*, (b)* D_r, D_θ, D_z *and (c)* D_r, D_θ, D_ϕ*. The volume element* $d\tau$ *= (AE) (AB) (AD) is (a)* $dx\,dy\,dz$ *(b)* $dr\,rd\theta\,dz$ *and (c)* $dr\,rd\theta\,r\sin\theta\,d\phi$ *(see Fig. a2.3.1). IIn Fig. a2.3.1(c), N is the foot of the perpendicular drawn from P on the XY-plane).*

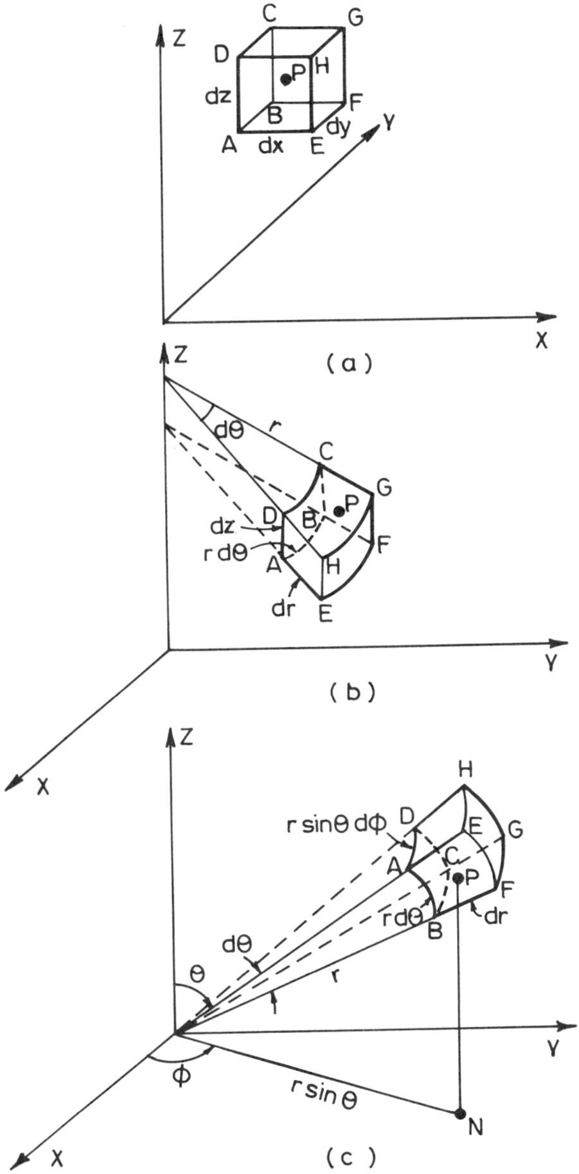

Fig. a2.3.1 Element of volume around a point where the divergence of **D** (= ∇.**D**) is sought in the rectangular **(a)**, cylindrical **(b)** and spherical-polar **(c)** systems of coordinates.

Further, the relevant quantities applicable to the area element (AB)(AD) under the above heads (a) (b) and (c) are:

(i) area value (dS): (a) dy dz (x = constant at x) (b) $rd\theta$ dz (r = constant at r) and (c) $rd\theta\ r\sin\theta\ d\phi$ (r = constant at x);

(ii) the unit vector outwardly normal to the area element: (a) $-\boldsymbol{a}_x$ (b) $-\boldsymbol{a}_r$ and (c) $-\boldsymbol{a}_r$;

(iii) the component of **D** *over the area element: $D_j - (\partial D_j/\partial x)(dx/2)$ [(a) j = x, y, z (b) j = r, θ, z and (c) j = r, θ, φ].*

The corresponding quantities (i)-(iii) for the remaining five area elements (EF)(GH), (AE)(AD), (BF)(BC), (AE)(AB) and (DH)(DC), respectively, are:

(i) (a) dy dz, dx dz, dx dz, dx dy and dx dy (b) $(r + dr)\ d\theta\ dz$, dr dz, dr dz, $dr\ rd\theta$ and $dr\ r\ d\theta$ and (c) $(r + dr)\ d\theta\,(r + dr)\sin\theta\ d\phi$, $dr\ r\sin\theta\, d\phi$, $dr\ r\sin(\theta + d\theta)\ d\theta$, $dr\ r\ d\theta$ and $dr\ r\ d\theta$;

(ii) (a) $\boldsymbol{a}_x$, $-\boldsymbol{a}_y$, $\boldsymbol{a}_y$, $-\boldsymbol{a}_z$ and $\boldsymbol{a}_z$ (b) $\boldsymbol{a}_r$, $-\boldsymbol{a}_\theta$, $\boldsymbol{a}_\theta$, $-\boldsymbol{a}_z$ and $\boldsymbol{a}_z$ (c) $\boldsymbol{a}_r$, $-\boldsymbol{a}_\theta$, $\boldsymbol{a}_\theta$, $-\boldsymbol{a}_\phi$ and $\boldsymbol{a}_\phi$; and

(iii) (a) $D_j + (\partial D_j/\partial x)(dx/2)$, $D_j - (\partial D_j/\partial y)(dy/2)$, $D_j + (\partial D_j/\partial y)(dy/2)$, $D_j - (\partial D_j/\partial z)(dz/2)$ and $D_j + (\partial D_j/\partial z)(dz/2)$, [j = x, y, z] (b) same as (a) [j = r, θ, z] and (c) same as (a) [j = r, θ, φ].

Thus the values of $\boldsymbol{D}.\boldsymbol{a}_n$ *over the six area elements considered are, respectively:*

(a) $-(D_x - (\partial D_x/\partial x)(dx/2))$, $D_x + (\partial D_x/\partial x)(dx/2)$, $-(D_y - (\partial D_y/\partial x)(dy/2))$, $D_y + (\partial D_y/\partial y)(dy/2)$, $-(D_z - (\partial D_z/\partial z)(dz/2))$, $D_z + (\partial D_z/\partial z)(dz/2)$; *(b)* $-(D_r - (\partial D_r/\partial r)(dr/2))$, $D_r + (\partial D_r/\partial r)(dr/2)$, $-(D_\theta - (\partial D_\theta/\partial\theta)(d\theta/2))$, $D_\theta + (\partial D_\theta/\partial\theta)(d\theta/2)$, $-(D_z - (\partial D_z/\partial z)(dz/2))$, $D_z + (\partial D_z/\partial z)(dz/2)$; *and (c)* $-(D_r - (\partial D_r/\partial r)(dr/2))$, $D_r + (\partial D_r/\partial r)(dr/2)$, $-(D_\theta - (\partial D_\theta/\partial\theta)(d\theta/2))$, $D_\theta + (\partial D_\theta/\partial\theta)(d\theta/2)$, $-(D_\phi - (\partial D_\phi/\partial\phi)(d\phi/2))$, $D_\phi + (\partial D_\phi/\partial\phi)(d\phi/2)$.

Now that we know the value of the element of area dS as well as that of $\boldsymbol{D}.\boldsymbol{a}_n$ *over each of the six faces of the element of volume* $d\tau$, *we know the value of* $\boldsymbol{D}.\boldsymbol{a}_n\ dS$ *also over each face. Adding the values of* $\boldsymbol{D}.\boldsymbol{a}_n\ dS$ *over all the six faces of the elements of the closed volume* $d\tau$, *we have*

(a2.3.1) (a)

$$\oint \boldsymbol{D}.\boldsymbol{a}_n\, dS \cong -(D_x - (\partial D_x/\partial x)(dx/2))\,dy\,dz + (D_y + (\partial D_y/\partial x)(dx/2))\,dy\,dz$$
$$- (D_y - (\partial D_y/\partial y)(dy/2))\,dx\,dz + (D_y + (\partial D_y/\partial y)(dy/2))\,dx\,dz$$
$$- (D_z - (\partial D_z/\partial z)(dz/2))\,dx\,dy + (D_z + (\partial D_z/\partial z)(dz/2))\,dx\,dy$$
$$\cong (\partial D_x/\partial x + \partial D_y/\partial y + \partial D_z/\partial z)(dx\,dy\,dz)$$

(b)

$$\oint \boldsymbol{D}.\boldsymbol{a}_n\, dS \cong -\left(D_r - \frac{\partial D_r}{\partial r}\frac{dr}{2}\right) rd\theta\,dz + \left(D_r + \frac{\partial D_r}{\partial r}\frac{dr}{2}\right)(r+dr)\,d\theta\,dz$$
$$-\left(D_\theta - \frac{\partial D_\theta}{\partial\theta}\frac{d\theta}{2}\right) dr\,dz + \left(D_\theta + \frac{\partial D_\theta}{\partial\theta}\frac{d\theta}{2}\right) dr\,dz$$
$$-\left(D_z - \frac{\partial D_z}{\partial z}\frac{dz}{2}\right) dr\,r\,d\theta + \left(D_z + \frac{\partial D_z}{\partial z}\frac{dz}{2}\right) dr\,r\,d\theta$$
$$\cong \left(\frac{\partial D_r}{\partial r} + \frac{1}{r}D_r + \frac{1}{r}\frac{\partial D_\theta}{\partial\theta} + \frac{\partial D_z}{\partial z}\right) dr\,r\,d\theta\,dz$$

(ignoring the higher order term $\frac{\partial D_r}{\partial r}\frac{dr}{2}\,dr\,d\theta\,dz$*)*

$$\cong \left(\frac{1}{r}\frac{\partial}{\partial r}(r\,D_r) + \frac{1}{r}\frac{\partial D_\theta}{\partial \theta} + \frac{\partial D_z}{\partial z} \right) dr\, r\, d\theta\, dz$$

(c) $$\oint \mathbf{D.a}_n\, dS \cong -\left(D_r - \frac{\partial D_r}{\partial r}\frac{dr}{2} \right) r d\theta\; r \sin\theta\, d\phi + \left(D_r + \frac{\partial D_r}{\partial r}\frac{dr}{2} \right)(r + dr) d\theta\, (r + dr)\sin\theta\, d\phi$$

$$-\left(D_\theta - \frac{\partial D_\theta}{\partial \theta}\frac{d\theta}{2} \right) dr\, r \sin\theta\, d\phi + \left(D_\theta + \frac{\partial D_\theta}{\partial \theta}\frac{d\theta}{2} \right) dr\, r \sin(\theta + d\theta)\, d\phi$$

$$-\left(D_\phi - \frac{\partial D_\phi}{\partial \phi}\frac{d\phi}{2} \right) dr\, r\, d\theta + \left(D_\phi + \frac{\partial D_\phi}{\partial \phi}\frac{d\phi}{2} \right) dr\, r\, d\theta$$

$$\cong \frac{\partial D_r}{\partial r} dr\, r\, d\theta\, r \sin\theta\, d\phi + 2D_r\, r\, dr\, d\theta \sin\theta\, d\phi$$

$$+ \frac{\partial D_\theta}{\partial \theta} d\theta\, dr\, r \sin\theta\, d\phi + D_\theta\, dr\, r \cos\theta\, d\theta\, d\phi + \frac{\partial D_\phi}{\partial \phi} d\phi\, dr\, r\, d\theta$$

(putting $\sin(\theta + d\theta) = \sin\theta \cos d\theta + \cos\theta \sin d\theta \approx \sin\theta + \cos\theta\, d\theta$*, and ignoring higher order terms)*

$$\cong \left(\frac{1}{r}\frac{\partial}{\partial r}(r^2 D_r) + \frac{1}{r \sin\theta}\frac{\partial}{\partial \theta}(D_\theta \sin\theta) + \frac{1}{r \sin\theta}\frac{\partial D_\phi}{\partial \phi} \right) dr\, r d\theta\, r \sin\theta\, d\phi\ .$$

We notice that, in each of the three equations (a2.3.1), we have essentially put a sign of approximation. This is because while evaluating the closed surface integral we have assumed the value of **D** *to be the same everywhere over any face of the element of the closed volume considered. This is more and more a valid assumption for smaller and smaller area elements which in turn would mean a smaller and smaller volume element* $d\tau$ *enclosing the point P. In the limit when the volume element shrinks to zero at P* ($d\tau \to 0$)*, then we have from (a2.3.1), remembering the values of* $d\tau$ *in the three systems of coordinates considered, the following exact relations:*

$$\underset{d\tau \to 0}{Lt} \frac{\oint \mathbf{D.a}_n\, dS}{d\tau} = \text{div}\, \mathbf{D} \qquad \textit{(see (2.3.4))}$$

$$= \frac{\partial D_x}{\partial x} + \frac{\partial D_y}{\partial y} + \frac{\partial D_z}{\partial z}$$

(rectangular system)

$$= \frac{1}{r}\frac{\partial}{\partial r}(r\,D_r) + \frac{1}{r}\frac{\partial D_\theta}{\partial \theta} + \frac{\partial D_z}{\partial z}$$

(cylindrical system)

$$= \frac{1}{r}\frac{\partial}{\partial r}(r^2 D_r) + \frac{1}{r \sin\theta}\frac{\partial}{\partial \theta}(D_\theta \sin\theta) + \frac{1}{r \sin\theta}\frac{\partial D_\phi}{\partial \phi}$$

(spherical-polar system)

which are identical with (2.3.5).

2.4 Electrostatic Potential and Electric Field Intensity

The electrostatic potential is analogous to the gravitational potential. When we lift a particle to a height from the ground level, we have to do so by doing some work against the gravitational force of attraction on the particle. This work is stored in the form of potential energy of the particle. Thus, if the particle is let loose from the height it has been elevated to, it will fall back to the ground thereby losing the potential energy it had acquired. The particle elevated to a height is said to be at some gravitational *potential* with respect to the ground which may be considered to be at the zero reference potential. In an analogous way, the electrostatic potential energy is stored in a charged particle when it is moved against the force on the particle due to a static electric field. In the electrostatic case, the zero reference potential is considered to be at infinity. Thus the electrostatic potential at a point is the work done in bringing a point charge from infinity upto the point per unit charge. Quantitatively, if Q is the point charge which is imagined to be brought from infinity upto a point, and W is the work done in the process, then the electrostatic potential V at the point is given by

$$V = W/Q \quad . \tag{2.4.1}$$

The unit of potential, as can be appreciated from (2.4.1), is Joul/Coulomb which is put as volt as a practical unit.

Example 2.4.1 Find the electrostatic potential due to a point charge at a distance from it.

The force on a point charge Q at a distance r from the point charge q can be found using Coulomb's law (2.1.4). The Coulomb force may be regarded as constant over an infinitesimal distance. Then it is easy to find an expression for the element of work done in moving the point charge Q, against this force, through an element of distance. Integrating this expression one can then find the total work done W in moving the point charge Q all the way from infinity upto the point where the electrostatic potential V due to the point charge q is sought. Subsequently, V due to point charge q at a distance r can be found using (2.4.1) as (for the details of deduction, see appendix *a2.4.1*):

$$V = \frac{q}{4\pi\varepsilon_0 r} \quad . \tag{2.4.2}$$

Example 2.4.2 Find the electrostatic potential due to a circular charge disc.

Let the charge be sprayed uniformly over a plane with a circular boundary. The electric field due to this charge distribution at a perpendicular distance measured from the centre of the circle was found in Example 2.1.2 (see (2.1.17)).

For the purpose of evaluating the electrostatic potential due to the given charge distribution, the charge on the plane within the circular boundary is divided into elementary charge rings of infinitesimal thickness as in Example 2.1.2 (see appendix *a2.1.2*, Fig. 2.1.3). The method is to find the element of electrostatic potential due to one such ring and integrate it to obtain the electrostatics potential V due to the entire charge distribution:

$$V = \frac{\rho_s}{2\varepsilon_0}\left((a^2 + z^2)^{1/2} - 1\right), \tag{2.4.3}$$

where ρ_s is the surface charge density, a is the radius of the circular plane charge, and z is the perpendicular distance of the point, where the electrostatic potential is found, measured from the centre of the circle. The meaning of these symbols is already explained in Example (2.1.2). (For the details of derivation of (2.4.3), see appendix *a2.4.2*).

Appendix a2.4.1 *Potential due to a point charge*

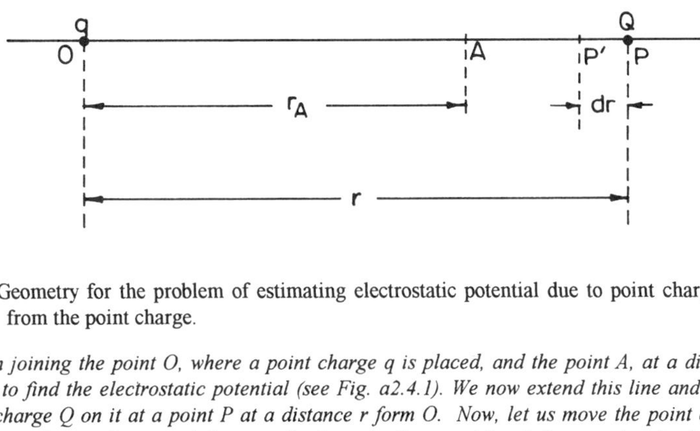

Fig. a2.4.1 Geometry for the problem of estimating electrostatic potential due to point charge q at a point distant r from the point charge.

A line is drawn joining the point O, where a point charge q is placed, and the point A, at a distance r_A from O, where to find the electrostatic potential (see Fig. a2.4.1). We now extend this line and consider another point charge Q on it at a point P at a distance r form O. Now, let us move the point charge Q from P to P' through an infinitesimal distance dr along the line considered. Let us take, for the sake of understanding, that q and Q both have a positive sign. Then the Coulomb force between them is that of repulsion (see Section 2.1). The magnitude of this force F, obtainable from (2.1.4), is

$$F = \frac{qQ}{4\pi\varepsilon_0 r^2} . \tag{a2.4.1}$$

F may be regarded as constant over an infinitesimal distance dr between P and P'. The element of work done dW in moving the point charge Q against this Coulomb force from P to P', through the infinitesimal distance dr, is

$$dW = \frac{qQ}{4\pi\varepsilon_0 r^2} dr . \tag{a2.4.2}$$

The work done W in moving the point charge from infinity ($r = \infty$) upto A ($r = r_A$) is obtained by integrating (a2.4.2) as follows:

(a2.4.3) $$W = \int dW = \int_{r_A}^{\infty} \frac{qQ}{4\pi\varepsilon_0 r^2} dr = \frac{qQ}{4\pi\varepsilon_0}\left[-\frac{1}{r}\right]_{r_A}^{\infty} = \frac{qQ}{4\pi\varepsilon_0 r_A} .$$

With the help of the definition of electrostatic potential, (2.4.1), and (a2.4.3), the electrostatic potential V_A at A is found as

(a2.4.4) $$V_A = \frac{W}{Q} = \frac{q}{4\pi\varepsilon_0 r_A} .$$

Dropping the subscript A in (a2.4.4), the electrostatic potential V at a distance r from the point charge q may be written as

$$V = \frac{q}{4\pi\varepsilon_0 r} ,$$

which is identical with (2.4.2).

Appendix a2.4.2 Potential due to a planar charge within a circular boundary

The geometry and the pertinent conditions of the problem are identical with those of Example 2.1.2. However, here, we are interested in finding the electrostatic potential instead of static electric field intensity. Referring to Fig. 2.1.3, let us first find the element of electrostatic potential dV′ at P (0, 0, z) due to the charge element of thickness dr considered arbitrarily at N on the circular ring of radius r, the element subtending an angle $d\theta$ at the origin. Thus the charge element at N has a value $(dr)(rd\theta)(\rho_s)$ and the element of potential dV′ due to it may be expressed, with the help of (2.4.2) (treating the charge element essentially as a point charge), as

(a2.4.5) $$dV' = \frac{\rho_s r d\theta\, dr}{4\pi\varepsilon_0 s} .$$

Integrating (a2.4.5) we obtain the electrostatic potential due to the charge ring. Denoting this potential as dV, we have

(a2.4.6) $$dV = \int dV' = \frac{\rho_s r\, dr}{4\pi\varepsilon_0 s}\int_0^{2\pi} d\theta = \frac{\rho_s r\, dr}{2\varepsilon_0 s} .$$

The values of r, s and ρ_s remaining the same all over the circular ring, these quantities have been taken outside the integral of (a2.4.6). The integration of (a2.4.6) yields the electrostatic potential V due to the entire circular charge disc as (see Fig. 2.1.3):

(a2.4.7) $$V = \int_0^a \frac{\rho_s r\, dr}{2\varepsilon_0 s} = \frac{\rho_s}{2\varepsilon_0}\int_0^a \frac{r\, dr}{s} .$$

Substituting (a2.1.15) into (a2.4.7),

$$V = \frac{\rho_s z}{2\varepsilon_0}\int_0^{\tan^{-1} a/z} \tan\psi \sec\psi \, d\psi = \frac{\rho_s z}{2\varepsilon_0}\int_0^{\tan^{-1} a/z} d(\sec\psi) = \frac{\rho_s}{2\varepsilon_0}\left[\sec\psi\right]_0^{\tan^{-1} a/z}$$

$$= \frac{\rho_s}{2\varepsilon_0}\left((a^2+z^2)^{1/2} - 1\right),$$

which is identical with (2.4.3).

2.5 Static Electric Field Intensity as a Negative Potential Gradient

The variation of the electrostatic potential with space coordinates in a region around a point may be interpreted to find the static electric field intensity at the point. Let us consider two close by points A and B where the potentials are $V_A = V$ and $V_B = V + dV$, respectively. We may draw surfaces through A and B on which the potentials remain constant at values V and V +dV, respectively. Such surfaces are called the equipotentials V and V + dV (see Fig. 2.5.1), respectively. Let the static electric field intensity **E** present in the region make an angle θ with the line AB. Consequently, if a point charge Q is moved through an infinitesimal distance from B to A, some element of work dW has to be done against the force **F** = Q**E** (see (2.1.1)). The element of work dW may be thus put as

$$dW = \mathbf{F.dR} = Q\mathbf{E.dR}, \tag{2.5.1}$$

where **dR** is the distance vector whose magnitude is dR = AB, and which is directed from A to B.

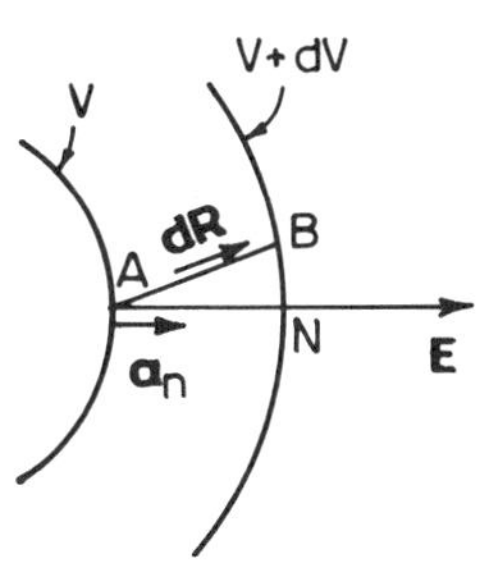

Fig. 2.5.1 Two equipotentials V and V+dV through two nearby points and the electric field intensity **E** normal to the equipotentials.

Now, as a corollary to the definition of electrostatic potential given by (2.4.1) one may deduce the potential difference between A and B from the work done in moving the point charge Q from B to A, that is, as

$$dW = Q(V_A - V_B) = Q[V - (V + dV)] = -Q\,dV. \tag{2.5.2}$$

Combining (2.5.1) and (2.5.2),

$$dV = -\mathbf{E}.d\mathbf{R} . \tag{2.5.3}$$

Supposing that we choose, in the limit, B to lie on the equipotential V itself, that is on the equipotential passing through A, then we have

$$V_A - V_B = V - (V + dV) = -dV = 0$$

which substituted in (2.5.3) makes

$$\mathbf{E}.d\mathbf{R} = 0 \quad (B \to A), \tag{2.5.4}$$

for two infinitesimally close-by points lying on the same equipotential. For such close-by points, in the limit, **dR** obviously becomes tangential to the equipotential. This, in conjunction with (2.5.4), means that **E** and the tangential direction to the equipotential are at right angles to each other or, in other words, **E** is normal to the equipotential. Denoting by $\mathbf{a}_n$ a unit vector normal to the equipotential, **E** may be represented as

$$\mathbf{E} = E\,\mathbf{a}_n . \tag{2.5.5}$$

In another limit, it is of interest to consider the point B lie on the equipotential V + dV and such that AB is along the direction of $\mathbf{a}_n$. In other words, B now lies on N, and (2.5.3) takes the following form

$$dV = -\mathbf{E}.d\mathbf{R} = -(E\,\mathbf{a}_n).(dn\,\mathbf{a}_n) \quad (B \to N) \tag{2.5.6}$$

with the help of (2.5.5) and making the interpretation of **dR** as

$$d\mathbf{R} = dn\,\mathbf{a}_n$$

for B lying on N. Evaluating the dot product, (2.5.6) may be expressed as

$$dV = -E\,dn . \tag{2.5.7}$$

Expressing dV as

$$dV = \frac{\partial V}{\partial n} dn , \tag{2.5.8}$$

and combining (2.5.7) and (2.5.8), we have

$$E = -\frac{\partial V}{\partial n} ,$$

which substituted in (2.5.5), yields

$$\mathbf{E} = -\frac{\partial V}{\partial n}\mathbf{a}_n . \tag{2.5.9}$$

$\partial V/\partial n$ occurring in (2.5.9) has the physical significance of being the maximum rate of variation of potential with distance (at A). The rate is maximum because dn represents the minimum distance between the equipotential V and V + dV. The said maximum is realised in the direction of $\mathbf{a}_n$. Consequently, the quantity $\partial V/\partial n\ \mathbf{a}_n$ not only signifies the maximum rate of variation of potential but also indicates the direction in which this rate is maximum. The quantity is called the gradient of scalar potential V and abbreviated as grad V which is also symbolised as $\nabla\mathbf{V}$ [6]:

$$\frac{\partial V}{\partial n}\mathbf{a}_n = \text{grad } V = \nabla V \,. \tag{2.5.10}$$

Thus, using the notation of (2.5.10), one may express (2.5.9) as

$$\mathbf{E} = -\nabla \mathbf{V} \,. \tag{2.5.11}$$

The expression (2.5.11) is quite useful and provides one with a method to find the static electric field intensity as the negative gradient of electrostatic potential. At this stage, therefore, it is useful to express the gradient of electrostatic potential V in various systems of coordinates (see appendix *a2.5.1*), as:

$$\nabla V = \frac{\partial V}{\partial x}\mathbf{a}_x + \frac{\partial V}{\partial y}\mathbf{a}_y + \frac{\partial V}{\partial z}\mathbf{a}_z \qquad \text{(a)}$$

(rectangular system)

$$= \frac{\partial V}{\partial r}\mathbf{a}_r + \frac{1}{r}\frac{\partial V}{\partial \theta}\mathbf{a}_\theta + \frac{\partial V}{\partial z}\mathbf{a}_z \qquad \text{(b)}$$

(cylindrical system)

$$= \frac{\partial V}{\partial r}\mathbf{a}_r + \frac{1}{r}\frac{\partial V}{\partial \theta}\mathbf{a}_\theta + \frac{1}{r\sin\theta}\frac{\partial V}{\partial \phi}\mathbf{a}_\phi \,. \qquad \text{(c)} \qquad (2.5.12)$$

(spherical-polar system)

The expression for electric field intensity, namely, (2.5.11) in conjunction with (2.5.12) finds applications in problems in which it is relatively easy to find the potential at a point.

Example 2.5.1 Find the electric field intensity due to a short dipole which consists of equal and opposite point charges separated by a distance supposedly much less than the distance of the point where the electric field intensity is sought, the distance being measured from the centre of the line joining the point charges of the dipole.

The problem enjoys the azimuthal symmetry ($\partial/\partial\phi = 0$), the point (P) where the electric field is to be found being represented by the spherical-polar coordinates (r, θ, ϕ). The centre of the dipole is taken at the origin of the coordinates, and the alignment of the dipole considered along the z-axis ($\theta = 0$). Let the negative point charge of the dipole be $-q$ located at $z = -l/2$ while the positive point charge of the dipole be $+q$ located at $z = l/2$ (Fig. 2.5.2).

The method is to first find the electrostatic potential due to the individual point charge $-q$ and q by applying (2.4.2), add these potentials to get the potential at P due to the dipole, and then apply (2.5.11) in conjunction with (2.5.12(c)) to obtain the following expression for the intensity of electric field $\mathbf{E}$ (for the details of derivation, see appendix *a2.5.2*):

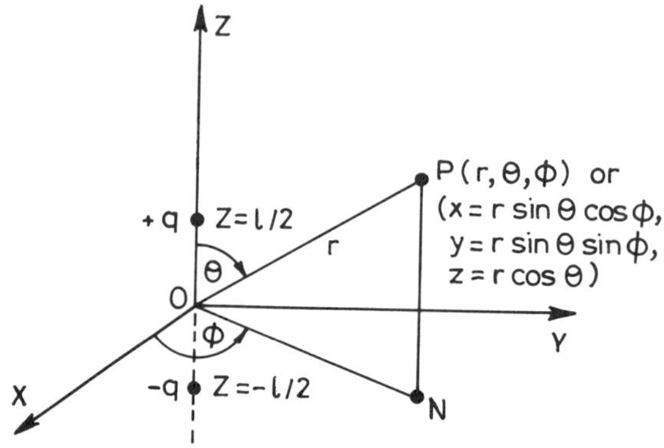

Fig. 2.5.2 The short electric dipole aligned on the z-axis and the point P (r, θ, ϕ) where to find the electric field intensity. (N is the foot of the perpendicular drawn from P on the XY-plane).

$$\mathbf{E} = \frac{p}{4\pi\varepsilon_0 r^3}(2\cos\theta\,\mathbf{a}_r + \sin\theta\,\mathbf{a}_\theta)\,, \tag{2.5.13}$$

where p = ql is the dipole moment, l being the length of the dipole.

Example 2.5.2 Find the intensity of electric field due to a circular charge disc for which the electrostatic potential is already obtained (in Example 2.4.2).

The problem enjoys the planar symmetry, the potential on the plane for a constant value of z being the same everywhere (see (2.4.3)). The procedure of finding **E** is exactly the same as that followed in Example 2.5.1, except that now one has to use (2.5.12(a)) instead of (2.5.12(c)). This yields

$$\mathbf{E} = \frac{\rho_s}{2\varepsilon_0}\left[1 - \frac{1}{\left(1+\frac{a^2}{z^2}\right)^{1/2}}\right]\mathbf{a}_z\,. \tag{2.5.14}$$

If we take $z/a << 1$, the problem resolves down to finding the electric field intensity due to an infinite planar sheet of charge, and (2.5.14) gives

$$\mathbf{E} = \frac{\rho_s}{2\varepsilon_0}\mathbf{a}_z\,. \tag{2.5.15}$$

We notice that (2.5.14) and (2.5.15) are identical, respectively, with (2.1.17) and (2.1.18) which were previously obtained starting from Coulomb's law. Also, it is worth comparing the method presented in this section with the one using Gauss's theorem which, as discussed in Example 2.2.1, also yields an expression identical with (2.5.15) for a problem which enjoys some sort of geometrical symmetry.

Appendix a2.5.1 *Gradient in different systems of coordinates*

The expression for the gradient of a scalar with particular reference to electrostatic potential, that is, for ∇V in the rectangular, cylindrical and spherical-polar systems of coordinates may be found as follows. The incremental potential dV may be written with the help of (2.5.3) and (2.5.11) as

(a2.5.1) $$dV = \nabla V \,.\, d\boldsymbol{R} \,.$$

Denoting the coordinates of A as x, y, z and those B, apart by an infinitesimal distance, as x+dx, y+dy, z+dz (see Fig. 2.5.1), one can express dR as

$$d\boldsymbol{R} = dx\, \boldsymbol{a}_x + dy\, \boldsymbol{a}_y + dz\, \boldsymbol{a}_z$$

which when substituted into (a2.5.1) yields

(a2.5.2) $$dV = \nabla V \,.\, (dx\, \boldsymbol{a}_x + dy\, \boldsymbol{a}_y + dz\, \boldsymbol{a}_z) \,.$$

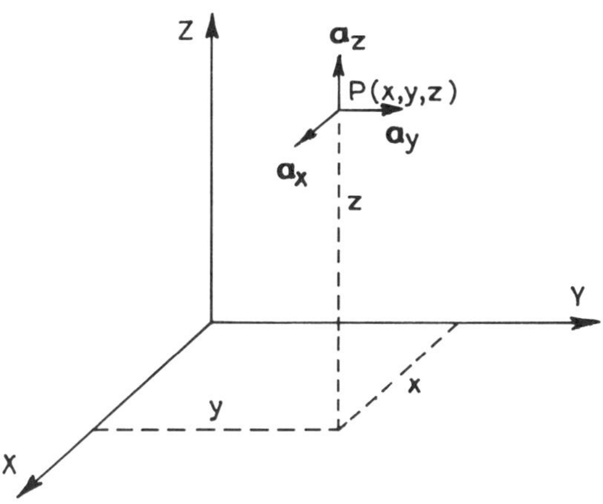

Fig. a2.5.1 Description of the point P (x, y, z) and the unit vectors in the rectangular system of coordinates.

∇V may be expressed in terms of three orthogonal components in the three systems of coordinates of interest as follows:

(a2.5.3) $$\nabla V = (\nabla V)_x\, \boldsymbol{a}_x + (\nabla V)_y\, \boldsymbol{a}_y + (\nabla V)_z\, \boldsymbol{a}_z \quad \textit{(rectangular system)}$$

$$= (\nabla V)_r\, \boldsymbol{a}_r + (\nabla V)_\theta\, \boldsymbol{a}_\theta + (\nabla V)_z\, \boldsymbol{a}_z \quad \textit{(cylindrical system)}$$

$$= (\nabla V)_r\, \boldsymbol{a}_r + (\nabla V)_\theta\, \boldsymbol{a}_\theta + (\nabla V)_\phi\, \boldsymbol{a}_\phi \,. \quad \textit{(spherical-polar system)}$$

Substituting (a2.5.3) into (a2.5.2), one has

(a2.5.4) (a) $dV = [(\nabla V)_x \, \boldsymbol{a}_x + (\nabla V)_y \, \boldsymbol{a}_y + (\nabla V)_z \, \boldsymbol{a}_z] \, . \, (dx \, \boldsymbol{a}_x + dy \, \boldsymbol{a}_y + dz \, \boldsymbol{a}_z)$
(rectangular system)

(b) $dV = [(\nabla V)_r \, \boldsymbol{a}_r + (\nabla V)_\theta \, \boldsymbol{a}_\theta + (\nabla V)_z \, \boldsymbol{a}_z] \, . \, (dx \, \boldsymbol{a}_x + dy \, \boldsymbol{a}_y + dz \, \boldsymbol{a}_z)$
(cylindrical system)

(c) $dV = [(\nabla V)_r \, \boldsymbol{a}_r + (\nabla V)_\theta \, \boldsymbol{a}_\theta + (\nabla V)_\phi \, \boldsymbol{a}_\phi] \, . \, (dx \, \boldsymbol{a}_x + dy \, \boldsymbol{a}_y + dz \, \boldsymbol{a}_z) \, .$
(spherical-polar system)

It is possible to simplify expression (a2.5.4) further by evaluating the dot products involved in the expression. In the rectangular system of coordinates (Fig. a2.5.1), the relevant dot products of unit vectors are:

(a2.5.5)
$$\boldsymbol{a}_x . \boldsymbol{a}_x = 1, \quad \boldsymbol{a}_x . \boldsymbol{a}_y = 0, \quad \boldsymbol{a}_x . \boldsymbol{a}_z = 0 ,$$
$$\boldsymbol{a}_y . \boldsymbol{a}_x = 0, \quad \boldsymbol{a}_y . \boldsymbol{a}_y = 1, \quad \boldsymbol{a}_y . \boldsymbol{a}_z = 0 ,$$
$$\boldsymbol{a}_z . \boldsymbol{a}_x = 0, \quad \boldsymbol{a}_z . \boldsymbol{a}_y = 0, \quad \boldsymbol{a}_z . \boldsymbol{a}_z = 1 .$$
(rectangular system)

In the cylindrical system of coordinates, it is quite easy to appreciate with the help of Fig. a2.5.2 that

(a2.5.6)
$$\boldsymbol{a}_r . \boldsymbol{a}_x = \cos\theta, \quad \boldsymbol{a}_r . \boldsymbol{a}_y = \cos(\pi/2 - \theta) = \sin\theta, \quad \boldsymbol{a}_r . \boldsymbol{a}_z = 0 ,$$
$$\boldsymbol{a}_\theta . \boldsymbol{a}_x = -\cos(\pi/2 - \theta) = -\sin\theta, \quad \boldsymbol{a}_\theta . \boldsymbol{a}_y = \cos\theta, \quad \boldsymbol{a}_\theta . \boldsymbol{a}_z = 0 ,$$
$$\boldsymbol{a}_z . \boldsymbol{a}_x = 0, \quad \boldsymbol{a}_z . \boldsymbol{a}_y = 0, \quad \boldsymbol{a}_z . \boldsymbol{a}_z = 1 .$$
(cylindrical system)

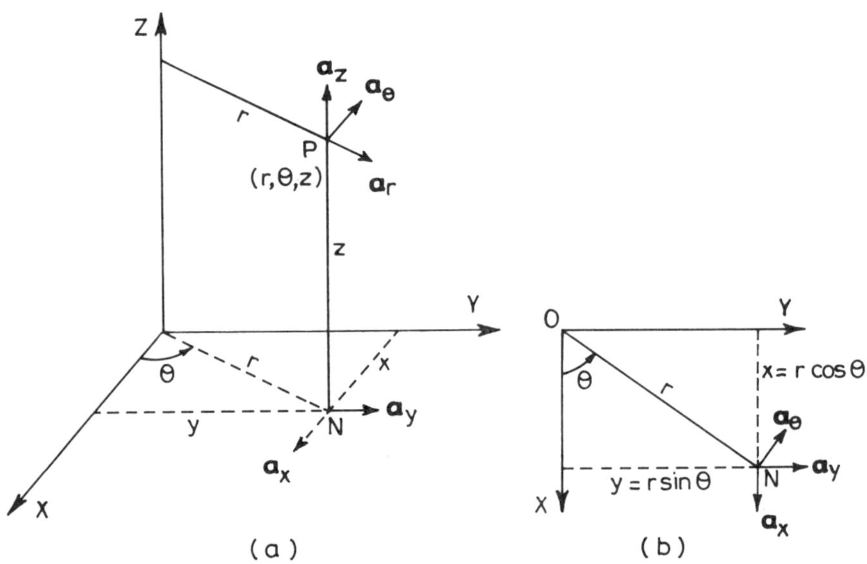

Fig. a2.5.2 Description of the point P (r, θ, z) and the unit vectors in the cylindrical system of coordinates **(a)** and the unit vectors drawn at N, the foot of the perpendicular dropped from P to the XY-plane **(b)** .

Also, one may write in the cylindrical system (Fig. a2.5.2):

(a2.5.7) $$x = r\cos\theta,\quad y = r\sin\theta,\quad z = z;$$
$$dx = -r\sin\theta\, d\theta + \cos\theta\, dr,\quad dy = r\cos\theta\, d\theta + \sin\theta\, dr,\quad dz = dz.$$

Similarly, taking resort to Fig. a2.5.3 one may express in the spherical-polar system of coordinates:

(a2.5.8) $$\mathbf{a}_r.\mathbf{a}_x = \sin\theta\cos\phi,\quad \mathbf{a}_r.\mathbf{a}_y = \sin\theta\sin\phi,\quad \mathbf{a}_r.\mathbf{a}_z = \cos\theta;$$
$$\mathbf{a}_\theta.\mathbf{a}_x = \cos\theta\cos\phi,\quad \mathbf{a}_\theta.\mathbf{a}_y = \sin\phi\cos\theta,\quad \mathbf{a}_\theta.\mathbf{a}_z = -\sin\theta;$$
$$\mathbf{a}_\phi.\mathbf{a}_x = -\sin\phi,\quad \mathbf{a}_\phi.\mathbf{a}_y = \cos\phi,\quad \mathbf{a}_\phi.\mathbf{a}_z = 0.$$
(spherical-polar system)

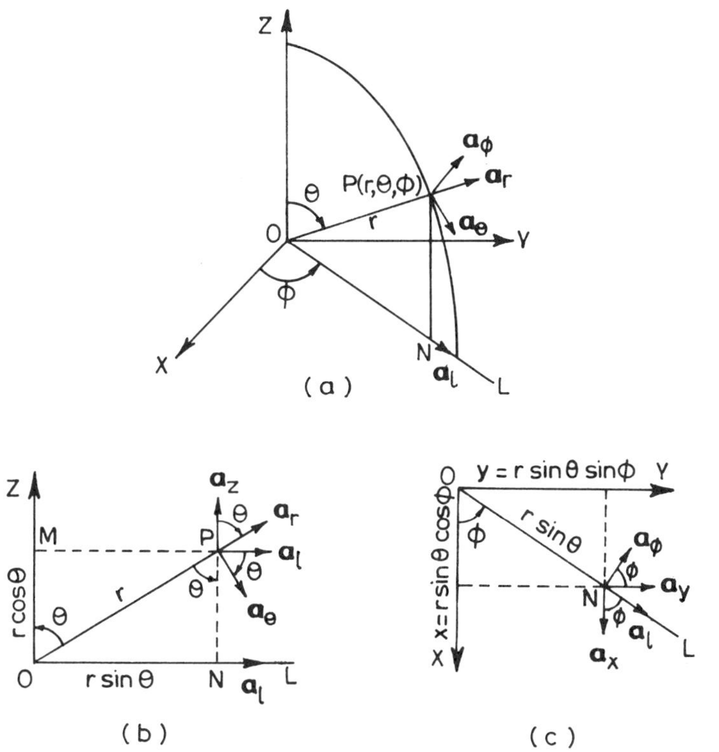

Fig. a2.5.3 Description of the point P (r, θ, ϕ) **(a)** and the unit vectors drawn at P on the ϕ = constant-plane **(b)** as well as those drawn at N, the foot of the perpendicular dropped from P to the XY-plane **(c)**.

In deriving (a2.5.8), it is convenient to introduce an additional unit vector $\mathbf{a}_l$ *(see Fig. a2.5.3), in the direction of OL, the line passing through the origin O and N, the foot of perpendicular from P. It is now easy to appreciate (see Fig. a2.5.3(b)) that* $\mathbf{a}_r$ *may be divided into two components, one parallel to OL and the other along the positive z direction thus enabling one to write* (a2.5.9) $$\boldsymbol{a}_r = \sin\theta\ \boldsymbol{a}_l + \cos\theta\ \boldsymbol{a}_z \quad .$$

Similarly, $\boldsymbol{a}_\theta$ may be divided into two components, one parallel to OL and the other in the negative z direction. This yields

(a2.5.10) $$\boldsymbol{a}_\theta = \cos\theta\ \boldsymbol{a}_l - \sin\theta\ \boldsymbol{a}_z \quad .$$

With the help of (a2.5.9) and (a2.5.10) it now becomes relatively easy to obtain the dot products given by (a2.5.8). For instance, one can use (a2.5.9) to obtain (see Fig. a2.5.3(c)):

$$\boldsymbol{a}_r \cdot \boldsymbol{a}_x = \sin\theta\ \boldsymbol{a}_l \cdot \boldsymbol{a}_x + \cos\theta\ \boldsymbol{a}_z \cdot \boldsymbol{a}_x = \sin\theta\ \cos\phi + (\cos\theta)(0) = \sin\theta\ \cos\phi \ ,$$

which is the first of the products given in (a2.5.8).

For the spherical-polar system (Fig. a2.5.3), the relations between the coordinates (corresponding to (a 2.5.7) for the cylindrical system) are:

(a2.5.11)
$$\begin{aligned} x &= r\sin\theta\cos\phi, \quad y = r\sin\theta\sin\phi, \quad z = r\cos\theta; \\ dx &= r\,(-\sin\theta\ \sin\phi\ d\phi + \cos\phi\cos\theta\ d\theta) + \sin\theta\ \cos\phi\ dr\,, \\ dy &= r\,(\sin\theta\ \cos\phi\ d\phi + \sin\phi\ \cos\theta\ d\theta) + \sin\theta\ \sin\phi\ dr\,, \\ dz &= -\ r\sin\theta\ d\theta + \cos\theta\ dr \ . \end{aligned}$$
(spherical-polar system)

At this stage, now it becomes relatively easy to evaluate the right hand side of (a2.5.4) with the help of (a2.5.5) - (a2.5.8) and (a2.5.11). This gives

(a2.5.12)
$$dV = (\nabla V)_x\ dx + (\nabla V)_y\ dy + (\nabla V)_z\ dz$$
(rectangular system)

$$= (\nabla V)_r\ dr + (\nabla V)_\theta\ r\ d\theta + (\nabla V)_z\ dz$$
(cylindrical system)

$$= (\nabla V)_r\ dr + (\nabla V)_\theta\ r\ d\theta + (\nabla V)_\phi\ r\sin\theta\ d\phi\,.$$
(spherical-polar system)

Also, one can alternatively express dV as

(a2.5.13)
$$dV = \frac{\partial V}{\partial x}dx + \frac{\partial V}{\partial y}dy + \frac{\partial V}{\partial z}dz$$
(rectangular system)

$$= \frac{\partial V}{\partial r}dr + \frac{\partial V}{\partial \theta}d\theta + \frac{\partial V}{\partial z}dz$$
(cylindrical system)

$$= \frac{\partial V}{\partial r}dr + \frac{\partial V}{\partial \theta}d\theta + \frac{\partial V}{\partial \phi}d\phi \ .$$
(spherical-polar system)

Comparing the right hand sides of (a2.5.11) and (a2.5.12), we then have

(a2.5.14)
$$(\nabla V)_x = \frac{\partial V}{\partial x}, \quad (\nabla V)_y = \frac{\partial V}{\partial y}, \quad (\nabla V)_z = \frac{\partial V}{\partial z},$$
(rectangular system)

$$(\nabla V)_r = \frac{\partial V}{\partial r}, \quad (\nabla V)_\theta = \frac{1}{r}\frac{\partial V}{\partial \theta}, \quad (\nabla V)_z = \frac{\partial V}{\partial z},$$

(cylindrical system)

$$(\nabla V)_r = \frac{\partial V}{\partial r}, \quad (\nabla V)_\theta = \frac{1}{r}\frac{\partial V}{\partial \theta}, \quad (\nabla V)_\phi = \frac{1}{r \sin\theta}\frac{\partial V}{\partial \phi}.$$

(spherical-polar system)

On substituting (a2.5.14) in (a2.5.3), one directly obtains (2.5.12), the expression for the gradient of electrostatic potential.

Appendix a2.5.2 *Field intensity due to a short dipole*

The problem enjoys the azimuthal symmetry ($\partial/\partial\phi = 0$), or in other words in this problem it would turn out that the electric field intensity is independent of the ϕ-coordinates of the point where the field is sought. Here we consider the distance between the point charges ($-q$ and $+q$) of the dipole, that is, the length of the dipole l to be much less than the radial distance r of the point where the electric field intensity is sought, the distance r being measured from the centre of the dipole located at the origin O of the coordinate system (see Fig. 2.5.2). If we denote by r^+ the distance of the point charge $+q$, located at $x = 0$, $y = 0$, $z = l/2$, from the point $P(r, \theta, \phi)$ which can be considered to be located at $x = r\sin\theta\cos\phi$, $y = r\sin\theta\sin\theta$, $z = r\cos\theta$ (see Fig. 2.5.2 and also Fig. a2.5.3), then r^+ may be written as (see (2.1.6)):

$$r^+ = [(r\sin\theta\cos\phi)^2 + (r\sin\theta\sin\phi)^2 + (r\cos\theta - l/2)^2]^{1/2}$$

which, upon simplification, takes the following form

(a2.5.15) $$r^+ = r\left(1 + \frac{l^2}{4r^2} - \frac{l\cos\theta}{r}\right)^{1/2}.$$

For a short dipole ($l \ll r$), one may ignore the second term under the square root of the right hand side of (a2.5.15) and expand the latter binomially ignoring higher order terms in view of the condition $l \ll r$ (for a short dipole) to obtain:

(a2.5.16) $$r^+ = r\left(1 - \frac{l\cos\theta}{2r}\right) = r - \frac{l}{2}\cos\theta.$$

Similarly, one can show that r^-, the distance of the point charge $-q$ from P, is given by

(a2.5.17) $$r^- = r + l/2\cos\theta.$$

Making use of (2.4.2), one may now write, for the potential V due to the dipole, as:

(a2.5.18) $$V = \frac{q}{4\pi\varepsilon_0}\left(\frac{1}{r^+} + \frac{1}{r^-}\right).$$

On substituting (a2.5.16) and (a2.5.17) into (a2.5.18), one has

$$V = \frac{q\,l\cos\theta}{4\pi\varepsilon_0(r^2 - l^2/4\cos\theta)},$$

which, for a short dipole ($l \ll r$), takes the following form:

(a2.5.19) $$V = \frac{p\cos\theta}{4\pi\varepsilon_0 r^2},$$

where p (= ql) is called the moment of the dipole. By taking the partial derivatives of (a2.5.19) one can evaluate:

(a2.5.20) $$\frac{\partial V}{\partial r} = \frac{-2p\cos\theta}{4\pi\varepsilon_0 r^3}, \quad \frac{\partial V}{\partial\theta} = \frac{-p\sin\theta}{4\pi\varepsilon_0 r^2}, \text{ and } \frac{\partial V}{\partial\phi} = 0 \text{ (azimuthal symmetry)}.$$

With the help of (2.5.11), (2.5.12(c)) and (a2.5.20), one then obtains the expression for **E** *as identical with (2.5.13).*

2.6 Laplacian Forms of Poisson's and Laplace's Equations

Substituting $\mathbf{D} = \varepsilon_0 \mathbf{E}$ in (2.3.3), we may express Poisson's equation as

$$\nabla.\mathbf{E} = \frac{\rho}{\varepsilon_0} \tag{2.6.1}$$

into which further one may substitute (2.5.11) to obtain

$$\nabla^2 V = -\frac{\rho}{\varepsilon_0} \quad \text{(Poisson's equation)}, \tag{2.6.2}$$

where ∇^2 (*del* squared) is called the Laplacian operator which operates on a scalar quantity, in this case, the electrostatic potential. $\nabla^2 V$ has the following meaning

$$\nabla^2 V = \nabla.\nabla V \quad (= \text{the divergence of the gradient of V}) \tag{2.6.3}$$

which can be further suitably expressed with the help of (2.3.5) and (2.5.12). Here, it may be noted that the expression (2.3.5) is applicable in general for the divergence of any vector. Here, one has to take the divergence of the vector ∇V, the latter in turn being given by (2.5.12). Thus one may expand the Laplacian form of (2.6.3) as

$$\nabla^2 V = \frac{\partial^2 V}{\partial x^2} + \frac{\partial^2 V}{\partial y^2} + \frac{\partial^2 V}{\partial z^2} \tag{a}$$

(rectangular system)

$$= \frac{\partial^2 V}{\partial r^2} + \frac{1}{r}\frac{\partial V}{\partial r} + \frac{1}{r^2}\frac{\partial^2 V}{\partial\theta^2} + \frac{\partial^2 V}{\partial z^2}$$

$$= \frac{1}{r}\frac{\partial}{\partial r}\left(r\frac{\partial V}{\partial r}\right) + \frac{1}{r^2}\frac{\partial^2 V}{\partial\theta^2} + \frac{\partial^2 V}{\partial z^2} \tag{b}$$

(cylindrical system)

$$= \frac{\partial^2 V}{\partial r^2} + \frac{2}{r}\frac{\partial V}{\partial r} + \frac{1}{r^2}\frac{\partial^2 V}{\partial \theta^2} + \frac{\cot\theta}{r^2}\frac{\partial V}{\partial \theta} + \frac{1}{r^2 \sin^2\theta}\frac{\partial^2 V}{\partial \phi^2}$$

$$= \frac{1}{r^2}\frac{\partial}{\partial r}\left(r^2 \frac{\partial V}{\partial r}\right) + \frac{1}{r^2 \sin\theta}\frac{\partial}{\partial \theta}\left(\sin\theta \frac{\partial V}{\partial \theta}\right) + \frac{1}{r^2 \sin^2\theta}\frac{\partial^2 V}{\partial \phi^2} .$$

(spherical-polar system) (c) (2.6.4)

As a corollary to (2.6.2) it follows that, at a point where the charge density is zero, the potential V would satisfy the following equation, named after Laplace [1]:

$$\nabla^2 V = 0 \quad \text{(Laplace's equation)} . \tag{2.6.5}$$

Poisson's and Laplace's equations in the Laplacian form, that is (2.6.2) and (2.6.5) read with the help of (2.6.4) are very much useful in many physical problems [7], [8]. These equations may be solved for the potential V at a point. If required, the electric field intensity may be found from potential making use of (2.5.11). One interesting application of this approach is found in the treatment of electron gun to form a beam of electrons, as discussed in chapter *6*. However, the principles discussed in this section are amply illustrated in the following simple examples.

Example 2.6.1 Show how the solution to Poisson's equation in the Laplacian form would yield the same results for a long axi-symmetric electron beam as obtained in Example 2.2.2. (In fact, the problem has been treated using Poisson's equation in Example 2.3.2. But it is sometimes convenient to start from the Laplacian form of Poisson's equation, namely, (2.6.2)).

In this problem, $\partial/\partial\theta = \partial/\partial z = 0$, and consequently (2.6.2) with the help of (2.6.4(b)) takes the following form

$$\frac{1}{r}\frac{d}{dr}\left(r\frac{dV}{dr}\right) = -\frac{\rho}{\varepsilon_0} , \tag{2.6.6}$$

taking care to change the partial derivative to complete derivative, since V now varies only with r. Further (2.6.6), may expressed as

$$d\left(r\frac{dV}{dr}\right) = -\frac{\rho}{\varepsilon_0} r dr$$

before the latter is integrated to yield

$$r\frac{dV}{dr} = -\frac{\rho}{\varepsilon_0}\frac{r^2}{2} + B, \tag{2.6.7}$$

where B is an integration constant.

The intensity of electric field **E** inside the beam is obtainable from (2.5.11) through (2.5.12(b)), taking $\partial/\partial\theta = \partial/\partial z = 0$, as:

$$\mathbf{E} = -\frac{dV}{dr}\mathbf{a}_r$$

(again changing the partial derivative to complete derivative in (2.5.12(b))) which, with the help of (2.6.7) may be expressed as

$$\mathbf{E} = \frac{\rho r}{2\varepsilon_0}\mathbf{a}_r , \tag{2.6.8}$$

the integration constant B being put equal to zero following the same argument as given in Example 2.3.2 following (2.3.6). (2.6.8) is identical with (2.2.2) or (2.3.7).

<u>*Example 2.6.2*</u> Solve Laplace's equation in the Laplacian form to solve for potential and hence electric field intensity at a point located between a pair of conducting surfaces kept at different potential values. In separate situations, consider that these conducting surfaces are formed by (a) two large parallel plates separated by a finite distance (b) two coaxial long cylinders (c) two concentric spheres (d) two large plates making a wedge separated by an infinitesimal gap at the hinge (e) two long identical coaxial cones placed tip-to-tip with an infinitesimal gap between them in a bi-conical geometry.

(a) Between parallel plates there are no charges and the problem enjoys a rectangular symmetry, the potential varying only along z, say, and the equipotentials being parallel to the XY-plane. This makes $\partial/\partial x = \partial/\partial y = 0$, and enables one to write with the help of (2.6.4(a)):

$$\frac{\partial^2 V}{\partial z^2} = 0$$

which upon integration twice would yield

$$V = A_1 z + B_1 , \tag{2.6.9}$$

where A_1 and B_1 are integration constants which can be evaluated applying boundary conditions. Thus considering the two plates to be located at z = 0 and z = d and taking their potentials as V_0 and zero, respectively, the integration constants are found as:

$$A_1 = -\frac{V_0}{d} \text{ and } B_1 = V_0 . \tag{2.6.10}$$

On substituting (2.6.10) in (2.6.9), the potential V at a point between plates is obtained as

$$V = V_0\left(1 - \frac{z}{d}\right) . \tag{2.6.11}$$

Substituting (2.6.11) in (2.5.12(a)) and noting that $\partial/\partial x = \partial/\partial y = 0$, ∇V can easily be evaluated and put in (2.5.11) to obtain

$$\mathbf{E} = -\frac{dV}{dz}\mathbf{a}_z = \frac{V_0}{d}\mathbf{a}_z . \tag{2.6.12}$$

(b) The problem of long coaxial cylinders continues to enjoy symmetry where one can put $\partial/\partial z = \partial/\partial\theta = 0$, in (2.6.4(b)) enabling one to write

$$\frac{d}{dr}\left(r\frac{dV}{dr}\right) = 0 \tag{2.6.13}$$

taking care of change the partial derivative to complete derivative.

Integrating twice (2.6.13) yields the potential value between cylindrical conductors ($a \le r \le b$):

$$V = A_2 \ln r + B_2 \tag{2.6.14}$$

where A_2 and B_2 are integration constants which can be easily evaluated considering the inner and the outer cylinders to be of radii a and b and at potentials V_0 and zero, respectively :

$$A_2 = \frac{V_0}{\ln a/b} \quad \text{and } B_2 = -\frac{V_0}{\ln a/b}\ln b\,. \tag{2.6.15}$$

On substituting (2.6.15) into (2.6.14) we obtain

$$V = V_0\frac{\ln r/b}{\ln a/b}\,. \tag{2.6.16}$$

Now ∇V can be evaluated substituting (2.6.16) into (2.5.12(b)) whence **E** can be found with the help of (2.5.11) putting $\partial/\partial\theta = \partial/\partial z = 0$ and changing the partial derivative to complete derivative.

$$\mathbf{E} = -\frac{dV}{dr}\mathbf{a}_r = \frac{V_0}{\ln b/a}\frac{1}{r}\mathbf{a}_r\,. \tag{2.6.17}$$

(c) In the case of two concentric spherical conductors, the problem continues to be one-dimensional as in (a) and (b), and now one has $\partial/\partial\theta = \partial/\partial\phi = 0$. In this case one may write with the help of (2.6.5) and (2.6.4(c)):

$$\frac{1}{r^2}\frac{d}{dr}\left(r^2\frac{dV}{dr}\right) = 0$$

which, upon integration, as in (a) and (b), yields the potential between spherical conductors ($a \le r \le b$):

$$V = \left(\frac{r-b}{a-b}\right)\left(\frac{a}{r}\right)V_0\,, \tag{2.6.18}$$

where a and b are the inner and the outer radii of the spherical conductors and V_0 and zero their potentials, respectively. The electric field intensity between conductors is obtained with the help of (2.6.18), (2.5.11) and (2.5.12(c)) as:

$$\mathbf{E} = -\frac{dV}{dr}\mathbf{a}_r = \frac{ab}{b-a}\frac{1}{r^2}V_0\,\mathbf{a}_r\,. \tag{2.6.19}$$

(d) In this problem of two large plates hinged at an angle θ_0, we have the axial and radial symmetry ($\partial/\partial z = \partial/\partial r = 0$), the potential varying only with θ (Fig. 2.6.1). This enables one to write, with the help of (2.6.5) and (2.6.4(b)),

$$\frac{1}{r^2}\frac{d^2v}{d\theta^2} = 0$$

which (as in problems (a)-(c)), upon integration, yields

$$V = \frac{V_0}{-\theta_0}\theta + \frac{-V_0\theta_0}{0-\theta_0} = -V_0\frac{\theta}{\theta_0} + V_0 = V_0\left(1-\frac{\theta}{\theta_0}\right). \tag{2.6.20}$$

Now, one may substitute (2.6.20) into (2.5.11) in which (2.5.12(b)) is plugged, to obtain the expression for the electric field intensity between the hinged plates as:

$$\mathbf{E} = -\frac{1}{r}\frac{dV}{d\theta}\,\mathbf{a}_\theta = -\frac{1}{r}\frac{V_0}{\theta_0}\,\mathbf{a}_\theta \,. \tag{2.6.21}$$

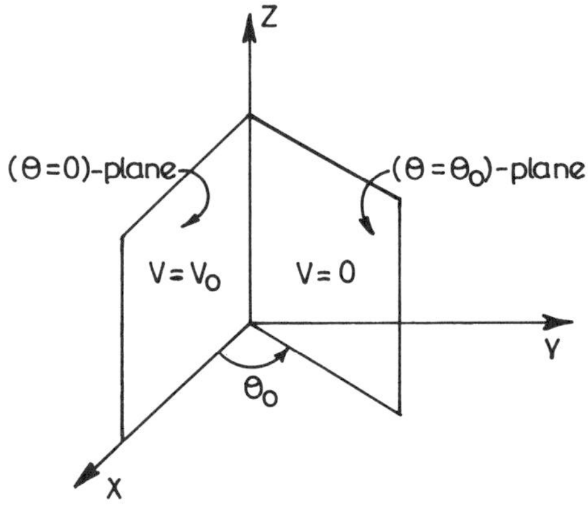

Fig. 2.6.1 The problem of two large hinged plates kept at a difference of potentials.

Here, the potential difference between the hinged plates is taken as V_0, the potential of the plate coinciding with the XZ-plane being V_0 with respect to the other plate at zero potential, the latter being at an angle of θ_0 with respect to the former (Fig. 2.6.1).

(e) The problem of bi-conical geometry in which long identical coaxial cones are placed tip-to-tip with an infinitesimal gap between them enjoys the symmetry in which the potential varying only with θ ($\partial/\partial r = \partial/\partial\phi = 0$) (Fig. 2.6.2). This enables one to write with the help of (2.6.5) and (2.6.4(c))

$$\frac{1}{r^2\sin\theta}\frac{d}{d\theta}\left(\sin\theta\frac{dv}{d\theta}\right) = 0 \tag{2.6.22}$$

which upon integration yields the following expression for potential V at a point between the conical conductors:

$$V = \frac{V_0}{2}\left(1+\frac{\ln\tan\theta/2}{\ln\tan\theta_0/2}\right). \tag{2.6.23}$$

In deriving (2.6.23) from (2.6.22), the integration constants have been found by taking $V = V_0$ at $\theta = \theta_0$, on one of the cones, while $V = 0$, at $\theta = \pi - \theta_0$, on the other (Fig. 2.6.2). The electric field intensity is found with the help of (2.6.23), (2.5.11) and (2.5.12(c)) as:

$$\mathbf{E} = -\frac{1}{r}\frac{dv}{d\theta}\,\mathbf{a}_\theta = \frac{-V_0}{2\sin\theta\ln(\tan\frac{\theta_0}{2})}\,\frac{1}{r}\,\mathbf{a}_\theta \,. \tag{2.6.24}$$

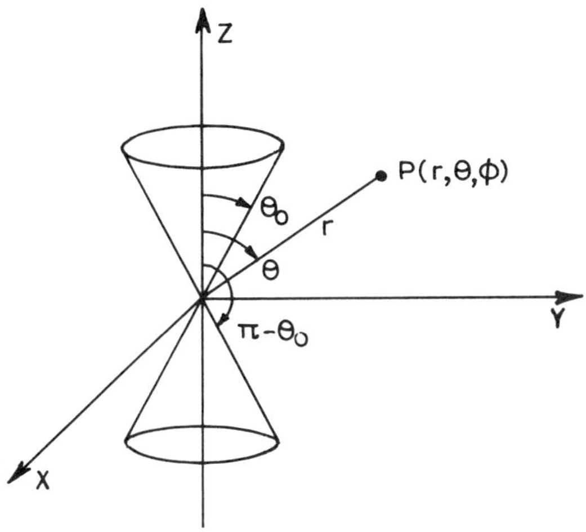

Fig. 2.6.2 The problem of two identical conical conductors placed tip-to-tip with an infinitesimal gap between them.

It is of interest to note that, in problem (a), the potential at a point between the conductors depends upon the location of the point (z) but the electric field intensity is constant in the region between the conductors (being independent of z) (see (2.6.11) and (2.6.12)). In problems (b) and (c), both the potential and the intensity of electric field depend upon the location of the point defined by a single coordinate r (see (2.6.16)-(2.6.19)). In problems (d) and (e), the potential depends upon the location of the point, defined by a single coordinate, namely, θ, which is the azimuthal coordinate in the case of problem (d) and the zenithal coordinate in the case of problem (e), and the electric field intensity in (d) depends only on r and that in (e) on both θ and r (see (2.6.20), (2.6.21), (2.6.23) and (2.6.24)).

2.7 Electrostatic Energy

Energy is stored in static electric field. One has to do some work in order to place a charge at a point by moving it against the field. Consequently, a charge subject to this field is capable of doing work spending this energy. This is first described here with reference to a particular example of a parallel-plate capacitor. A more generalized treatment of the problem will also subsequently follow.

In Example 2.6.2(a), we have considered two large parallel plates at a difference of potential and separated by a distance (Fig. a2.7.1 in appendix *a2.7.1*). The electric field intensity at a point considered in the region between the two plates is given by (2.6.12) in terms of V_0, the potential difference between the plates. It is clear from (2.6.12) that the electric field intensity is constant in the region and is directed as normal to the plates. At this stage, it is also useful to know the expression for the electric field intensity at a point in between plates in terms of the charges on the plates. The expression can be easily obtained using Gauss's law following the approach given in Example 2.2.1 and appendix *a2.2.2*. Thus, the magnitude of the electric field intensity (E) may be found as (for the details, see appendix *a2.7.1*; see also problem 2.4):

$$E = \frac{\rho_s}{\varepsilon_0} . \tag{2.7.1}$$

It also follows from (2.6.12) that the magnitude of the electric field intensity E is

$$E = \frac{V_0}{d} . \tag{2.7.2}$$

Equating the right hand sides of (2.7.1) and (2.7.2), we have

$$\frac{\rho_s}{\varepsilon_0} = \frac{V_0}{d} \tag{2.7.3}$$

If Q_0 be the magnitude of the charge on either of the plates (+ Q_0 on the plate M and - Q_0 on the plate N; see Fig. a2.7.1), and S is the area of each of these plates, then one can write

$$\rho_s = \frac{Q_0}{S}$$

which substituted in (2.7.3) yields

$$\frac{Q_0}{V_0} = \frac{\varepsilon_0 S}{d} . \tag{2.7.4}$$

The quantity in the left hand side of (2.7.4) is the charge per unit potential difference and called the capacitance of the capacitor [9], said to be formed by the pair of parallel plates in the present case under study. It signifies the capability of the system of holding charges for a given potential difference established between the plates. Denoting the capacitance by C, (2.7.4) may be expressed as:

$$C = \frac{\varepsilon_0 S}{d} . \tag{2.7.5}$$

Now, let us proceed to find the work done in charging the conductors of the capacitor. This is practically done by connecting the plates by a source of potential (Fig. a2.7.1). In the process the charge is built up from a zero value to Q_0, and correspondingly also the potential difference is built up between the plates from a zero value to V_0. Let in this process the charge on the capacitor be Q at any instant ($0 < Q < Q_0$), and correspondingly the potential difference between the plates be V (= Q/C). Then the element of work done dW in adding an extra element of charge dQ to the system is

$$dW = VdQ = \frac{Q}{C}dQ \tag{2.7.6}$$

remembering that C = Q/V is a measure of holding charges by the capacitor. Clearly, then the total work done W in charging the capacitor to its full value Q_0 is obtained by integrating (2.7.6) as:

$$W = \int dW = \int_0^{Q_0} \frac{Q}{C} dQ = \frac{Q_0^2}{2C} = 1/2\, C V_0^2 \tag{2.7.7}$$

using the relation $C = Q_0 / V_0$. Substituting (2.7.5) in (2.7.7) and using (2.7.2), now W may be expressed as

$$W = \frac{1}{2}\, \varepsilon_0 E^2 (\, S\, d\,), \tag{2.7.8}$$

where the quantity in parenthesis may be identified as the volume of the capacitor. The energy density U_E, being equal to the energy per unit volume stored in the electric field established between the plates of the capacitor, is

$$U_E\left(= \frac{W}{Ad}\right) = \frac{1}{2}\varepsilon_0 E^2 = \frac{1}{2}(\varepsilon_0 E)(E) = \frac{1}{2}(D)(E) = \frac{1}{2}\, \mathbf{D.E} = \frac{1}{2}\, \mathbf{E\,.D}\,. \tag{2.7.9}$$

Although, the above expression for U_E has been derived here with particular reference to a parallel-plate capacitor, the expression is rather general. A more rigorous proof in the general case is given in appendix *a2.7.2*.

<u>*Example 2.7.1*</u> Verify the validity of the energy expression (2.7.9) with reference to Example 2.6.2(b).

The problem (b) in Example 2.6.2 is that of a system of long coaxial cylinders at different potentials. The problem may be identified as the problem of a coaxial-cable transmission line or a cylindrical capacitor.

The electric field intensity at a point between the coaxial cylinders may be found using Gauss's law following the procedure given in Example 2.2.1, and in more details, in appendix *a 2.2.2*.

It can be shown that in this problem the electric field intensity continues to be given by (2.1.21). Thus the magnitude of the electric field intensity E is given by

$$E = \frac{\rho_l}{2\pi\varepsilon_0 r}, \tag{2.7.10}$$

where ρ_l is the charge per unit length on the positive conductor. E is also obtainable from (2.6.17) as

$$E = \frac{V_0}{\ln(b/a)}\frac{1}{r}, \tag{2.7.11}$$

where a and b are the radii of the inner and the outer conductors respectively. Equating the right hand sides of (2.7.10) and (2.7.11) one obtains

$$\frac{\rho_l}{V_0} = \frac{2\pi\varepsilon_0}{\ln(b/a)}\,. \tag{2.7.12}$$

The left hand side of (2.7.12) represents the charge per unit length per unit potential difference between the conductors and hence represents the capacitance per unit length of the cylindrical capacitor (coaxial cable). For a length 'l' of the capacitor, the capacitance C is given by

$$C\left(=\frac{\rho_l l}{V_0}\right) = \frac{2\pi\varepsilon_0 l}{\ln(b/a)} . \tag{2.7.13}$$

In order to show the validity of (2.7.9) as envisaged we have to integrate U_E, as given by (2.7.9), in the entire volume space τ, say, occupied between the conductors of length l, and show that the resulting quantity is identical with (2.7.7). In other words we have to show the equality

$$\int_\tau U_E \, d\tau = \frac{1}{2} C V_0^2 , \tag{2.7.14}$$

where $d\tau$ is the element of volume of the cylinder of length l and of the thickness dr considered at r $(a < r < b)$. $d\tau$ is given by

$$d\tau = 2\pi r l \, dr . \tag{2.7.15}$$

Substituting (2.7.9) and (2.7.15) into (2.7.14) one obtains:

$$\text{left hand side of (2.7.14)} = \int_\tau \frac{1}{2}\varepsilon_0 E^2 \, 2\pi r l \, dr . \tag{2.7.16}$$

Substituting (2.7.11) into (2.7.16) and carrying out the integration, one has:

$$\text{left hand side of (2.7.14)} = \frac{1}{2}\varepsilon_0 \frac{V_0^2 \, 2\pi l}{(\ln(b/a))^2} \int_a^b \frac{dr}{r} = \frac{1}{2} \frac{2\pi\varepsilon_0 l}{\ln(b/a)} V_0^2 . \tag{2.7.17}$$

Now, the right hand side of (2.7.14), with the help of (2.7.13), is:

$$\text{right hand side of (2.7.14)} = \frac{1}{2} \frac{2\pi\varepsilon_0 l}{\ln(b/a)} V_0^2 . \tag{2.7.18}$$

On examining (2.7.17) and (2.7.18) we easily see that the left and the right-hand side quantities of (2.7.14) are identical thus verifying the equality of (2.7.14) and hence, in a way, testifying to the validity of (2.7.9) in this particular example. However, as mentioned earlier, for a general proof of (2.7.9), the reader may refer to appendix *a2.7.2*.

Appendix a2.7.1 *Field intensity between parallel plates*

Let us take the case of a parallel plate capacitor and construct a rectangular parallelepiped ABCD as the Gaussian surface such that the face AD is inside the positive plate conductor (M), as shown in Fig. a2.7.1. AD is parallel to these plates. The electric field intensity is normal to the plates (see (2.6.12)), and, consequently, it is perpendicular to the face BC and parallel to each of the faces AB and DC.

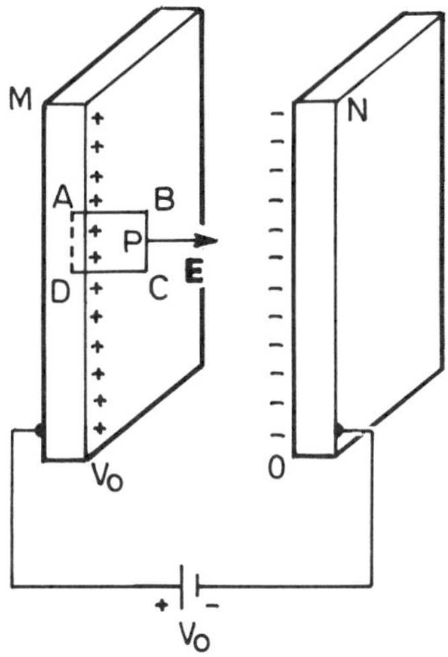

Fig. a2.7.1 Construction of the Gaussian surface, a rectangular parallelopiped, passing though the point (P), where to find the electric field intensity, inside the parallel plate capacitor.

Now, one can easily apply Gauss's law (2.2.1) to the problem. It can be easily appreciated that, since the face AD is inside the conductor, the value of the electric field and hence that of the electric displacement is zero on the face AD (see Example 4.1.2). Consequently, the face AD does not contribute to the surface integral given by the left hand side of (2.2.1). Similarly, none of the faces AB and AD contribute to the surface integral since the electric field intensity is parallel to each of these faces. The only contribution to the surface integral comes from the face BC and becomes equal to $D\alpha = \varepsilon_0 E\alpha$, *where* α *is the area of the face BC. The charge enclosed by the Gaussian surface is the charge on the positive plate on the area which is equal to the area of each of the faces AD and BC. This charge is then* $\rho_s\alpha$, *where* ρ_s *is the surface density of charge on the positive plate, and constitutes the right hand side of Gauss's law (2.2.1). This enables us to write from (2.2.1):*

$$D\alpha = \rho_s \alpha$$

whence the magnitude of the electric field, $E\ (= D/\varepsilon_0)$, *is obtained as*

$$E = \frac{\rho_s}{\varepsilon_0},$$

which is identical with (2.7.1).

Appendix a2.7.2 Energy storage in establishing a charge distribution

In this treatment it will turn out that the amount of work done in establishing a distribution of charges in space can be suitably interpreted to find an expression for the energy stored in static electric field [5]. For this purpose, let us proceed to estimate the work done in establishing a given distribution, say, in positioning the point charges $Q_1, Q_2, Q_3, \ldots$ *at the points* $P_1, P_2, P_3, \ldots$ *where the potentials are* $V_1, V_2, V_3, \ldots$, *respectively. Let these point charges be positioned one by one, starting from* Q_1. *Let* *the point charge Q_1 be brought from infinity upto the point P_1, considering initially that all of the remaining charges are absent from the region. Since the point charge Q_1 is thus brought in the field-free region, no work will be done in the process. If we denote this work by W_1, we may write*

$$W_1 = 0 . \tag{a2.7.1}$$

Once the point charge Q_1 has been positioned at P_1, some amount of work W_2 has to be done in positioning the point charge Q_2, as this has to be done now in the field due to Q_1. W_2 may be expressed with the help of (2.4.1) as

$$W_2 = Q_2 V_{21} , \tag{a2.7.2}$$

where V_{21} is the potential at P_2 due to Q_1. Similarly, one can now find W_3, the work done in positioning the point charge Q_3 at P_3, now in the presence of Q_1 and Q_2 already positioned at P_1 and P_2, respectively. W_3 is given by

$$W_3 = Q_3 (V_{31} + V_{32}) , \tag{a2.7.3}$$

where V_{31} and V_{32} are the potentials at P_3 due to Q_1 and Q_2, respectively. Similarly, one can express the work done, W_4, in positioning the point charge Q_4 at P_4 in the presence of Q_1, Q_2 and Q_3 already positioned at P_1, P_2 and P_3, respectively. The procedure can be extended to find the work done in positioning the remaining charges in the given distribution. Summing up all these quantities of work done, we obtain the total work done W_{total} in establishing the charge distribution as

$$\begin{aligned} W_{total} &= W_1 + W_2 + W_3 + W_4 + \ldots \\ &= Q_2 V_{21} + Q_3 (V_{31} + V_{32}) + Q_4 (V_{41} + V_{42} + V_{43}) \\ &\quad + Q_5 (V_{51} + V_{52} + V_{53} + V_{54}) + \ldots \end{aligned}$$

which may be alternatively put as

$$\begin{aligned} W_{total} &= Q_2 V_{21} + Q_3 V_{31} + Q_3 V_{32} + Q_4 V_{41} + Q_4 V_{42} + Q_4 V_{43} \\ &\quad + Q_5 V_{51} + Q_5 V_{52} + Q_5 V_{53} + Q_5 V_{54} + \ldots \end{aligned} \tag{a2.7.4}$$

where V_{mn} ($m \neq n$, m and n being integers) is the potential at P_m due to the charge at P_n.

The individual terms in the right hand side of (a2.7.4) are of the form $Q_m V_{mn}$. Further, we notice that in view of (2.4.2):

$$Q_m V_{mn} = Q_m \frac{Q_n}{4\pi\varepsilon_0 r_{mn}} , \quad \text{and} \tag{a2.7.5}$$

$$Q_n V_{nm} = Q_n \frac{Q_m}{4\pi\varepsilon_0 r_{mn}} , \tag{a2.7.6}$$

where r_{mn} is the distance between the point P_m and P_n. Thus it follows from (a2.7.5) and (a2.7.6) that

$$Q_m V_{mn} = Q_n V_{nm} \quad (m \neq n) . \tag{a2.7.7}$$

Making use of the relation (a2.7.7) one can express (a2.7.4) alternatively as:

$$\begin{aligned} W_{total} &= Q_1 V_{12} + Q_1 V_{13} + Q_2 V_{23} + Q_1 V_{14} + Q_2 V_{24} + Q_3 V_{34} + Q_1 V_{15} \\ &\quad + Q_2 V_{25} + Q_3 V_{35} + Q_4 V_{45} + \ldots \end{aligned} \tag{a2.7.8}$$

Adding (a2.7.4) and (a2.7.8)

(a2.7.9) $$2\,W_{total} = Q_1 (V_{12} + V_{13} + V_{14} + V_{15} + \ldots) + Q_2 (V_{21} + V_{23} + V_{24} + V_{25} + \ldots) + Q_3 (V_{31} + V_{32} + V_{34} + V_{35} + \ldots) + Q_4 (V_{41} + V_{42} + V_{43} + V_{45} + \ldots) + Q_5 (V_{51} + V_{52} + V_{53} + V_{54} + \ldots)\ .$$

Denoting by V_1, V_2, V_3, V_4, V_5, . . . the potentials at P_1, P_2, P_3, P_4, P_5, . . ., respectively, we may write V_1 as the sum of the potentials at P_1 as contributed by the charges at P_2, P_3, P_4, P_5, . . . , and similarly V_2 as the sum of the potentials at P_2 as contributed by the charges at P_1, P_3, P_4, P_5, . . . , and so on, or in other words:

$$\begin{aligned} V_1 &= V_{12} + V_{13} + V_{14} + V_{15} + \ldots \\ V_2 &= V_{21} + V_{23} + V_{24} + V_{25} + \ldots \\ V_3 &= V_{31} + V_{32} + V_{34} + V_{35} + \ldots \\ V_4 &= V_{41} + V_{42} + V_{43} + V_{45} + \ldots \\ V_5 &= V_{51} + V_{52} + V_{53} + V_{54} + \ldots\ , \end{aligned}$$

and so on, which in turn yields the following simplified version of the (a2.7.9)

(a2.7.10) $$W_{total} = (1/2)(Q_1 V_1 + Q_2 V_2 + Q_3 V_3 + Q_4 V_4 + Q_5 V_5 + \ldots) = (1/2) \sum_{m = 1,\, 2,\, 3,\, \ldots} Q_m V_m\ .$$

The above summation may be changed into an integral form if the distribution of point charges is replaced by a volume charge distribution enabling one to express (a2.7.10) as

(a2.7.11) $$W_E = \frac{1}{2} \oint_\tau V \rho \, d\tau\ ,$$

where ρ is the volume charge density at a point where the potential is V and the integral is extended over a volume in which the charge distribution is enclosed. It may be noted that in (a2.7.11) we have chosen to write W_E in place of W_{total} to indicate that the energy is stored in the electric field.

Substituting ρ from (2.3.3) into (a2.7.11),

$$W_E = \frac{1}{2} \oint_\tau V \nabla . \mathbf{D} \, d\tau$$

which, with the help of the vector identity

$$\nabla.(V\mathbf{D}) = V\,(\nabla.\mathbf{D}) + \mathbf{D}.\,(\nabla V)$$

or

$$V \nabla.\mathbf{D} = \nabla.\,(V\mathbf{D}) - (\mathbf{D}.\,\nabla V)\ ,$$

takes the following form

(a2.7.12) $$W_E = \frac{1}{2} \left(\oint_\tau \nabla.(V\mathbf{D})\, d\tau - \oint_\tau (\mathbf{D}.\nabla V)\, d\tau \right).$$

The first of the volume integrals in (a2.7.12) may be converted into a surface integral by a theorem called the Gauss's divergence theorem in vector calculus which in this context may be written as

(a2.7.13) $$\oint_\tau \nabla.(V\mathbf{D})\, d\tau = \oint_S (V\mathbf{D}).\boldsymbol{a}_n \, dS\ ,$$

(closed-volume integral) *(closed-surface integral)*

where S is the area of the closed surface and dS is an element of area considered on it, and $\boldsymbol{a}_n$ *is a unit vector normal to the element of area. [In (a2.7.13), the theorem refers to a particular vector quantity, namely,* $V\boldsymbol{D}$*. But the theorem is true for any other vector quantity as well. In fact, the validity of the theorem can be easily checked for the vector* $\boldsymbol{D}$ *with the help of the mathematical form of Gauss's law (2.2.1) and Poisson's equation (2.3.3) and working out a few steps. Thus putting*

$$q = \oint_\tau \rho \, d\tau$$

in the right hand side of (2.2.1), we have

$$\oint_S \boldsymbol{D}.\boldsymbol{a}_n \, dS = \oint_\tau \rho \, d\tau ,$$

into which one can now substitute (2.3.3) to get

$$\oint_S \boldsymbol{D}.\boldsymbol{a}_n \, dS = \oint_\tau \nabla.\boldsymbol{D} \, d\tau ,$$

which is the divergence theorem stated with particular reference to the vector $\boldsymbol{D}$*].*

Making use of (a2.7.13), we can now express (a2.7.12) as

$$W_E = \frac{1}{2}\left(\oint_S (V\boldsymbol{D}).\boldsymbol{a}_n \, dS - \oint_\tau (\boldsymbol{D}.\nabla V) \, d\tau \right) . \tag{a2.7.14}$$

It can be shown that the first integral of (a2.7.14) is zero. This can be appreciated by taking the closed surface to be located far from the charge distribution. From such large distances on the surface, the distribution of charges may be regarded as point charges, and the closed surface itself a huge sphere, with the charges located at its centre. It follows from (2.1.14) and (2.4.2) that the quantity ($V\boldsymbol{D}$) *occurring in the integrand of the first integral of (a2.7.14) decreases as* $1/r^2$ *and the element of area dS* ($=2\pi r^2 \sin\theta \, d\theta$*; see appendix a2.3.1), also occurring in the same integrand, increases as* r^2 *so that, over all, the integrand decreases as 1/r, in the limit, approaching zero when the surface, in principle, is taken at an infinite distance from the charge distribution. Consequently, in the limit, the first integral of (a2.7.14) would also tend to zero enabling one to write*

$$W_E = \frac{1}{2}\left(-\oint_\tau (\boldsymbol{D}.\nabla V) \, d\tau \right) ,$$

which, with the help of (2.5.11), takes the following form

$$W_E = \frac{1}{2}\oint_\tau \boldsymbol{D}.\boldsymbol{E} \, d\tau . \tag{a2.7.15}$$

Further, remembering that $\boldsymbol{D} = \varepsilon_0 \boldsymbol{E}$*, one may simplify (a2.7.15) as*

$$W_E = \frac{1}{2}\oint_\tau \varepsilon_0 E^2 \, d\tau . \tag{a2.7.16}$$

In terms of the energy density, we can also express W_E *as*

$$W_E = \oint_\tau U_E \, d\tau . \tag{a2.7.17}$$

Now it is to easy to see by comparing (a2.7.15) or (a2.7.16) with (a2.7.17) that

$$U_E = \frac{1}{2}\,\mathbf{D.E} \quad or, \quad = \frac{1}{2}\,\varepsilon_0 E^2 ,$$

which is identical with (2.7.9).

2.8 Summary

The most fundamental law in electrostatics, namely, Coulomb's law may be expressed in a vectorial form. The expression for the Coulomb force (vector) between two point-charges is given in terms of the coordinates of the points. The result is also expressed in terms of the magnitude of the force and a unit vector.

The concept of electric field intensity due to a point charge is developed from Coulomb's law and the same extended to find the electric field intensity due to a uniform charge distribution.

Gauss's law can be used to find static electric field intensity in problems which enjoy geometrical symmetry. The law may be derived starting from Coulomb's law. The law as applied to a volume element in the limit leads to the point form of Gauss's law known as Poisson's equation which equates the volume charge density at a point to the divergence of the electric displacement vector. The expressions for the divergence in the rectangular, cylindrical and spherical systems of coordinates are given. The deduction of these expressions is, however, given in an appendix. Poisson's equation is illustrated by giving the examples of an electron beam treating the latter as a charge fluid.

The concept of electrostatic potential is developed with the help of a few illustrative examples. The static field intensity is then seen as the negative gradient of potential. The expressions for the gradient in the rectangular, cylindrical and spherical systems of coordinates are given. The derivation of these expressions is given in an appendix. Laplace's equation is a special case of Poisson's equation which is applicable to a charge-free region. It is useful in many electrostatic problems to express these two equations in the Laplacian form. Such an approach is discussed with reference to the problems of rectangular, cylindrical and spherical symmetry. The method gives the electric field intensity in terms of potential distribution on the conductors in such problems.

A simple approach to finding an expression for the energy stored in static electric field through the concept of the capacitance is given in the text. However, a more rigorous and general approach to deriving this expression is also given in an appendix.

The basic concepts of electrostatics are developed in this chapter with the help of a large number of illustrative examples. More specific and practical examples of these concepts will be found in the problems of the formation and the confinement of an electron beam dealt with in chapters *6* and *7*, respectively, in the second part of the treatise.

Problems

2.1 Find the ratio of a practical unit of charge, namely, a Coulomb to a unit of charge in the electrostatic unit (esu). (Note that a unit point charge of 1esu repels an identical charge 1cm away by a force of 1dyne).

Answer: 3×10^9.

2.2 Four equal point charges are distributed, one each of amount q, at the four corners of a square of side a. What amount of charge concentrated at the centre of the square would keep in equilibrium the system of charges?

Answer: $-(1+2\sqrt{2})/4\ q$.

2.3 Starting from (2.1.16), the expression for the electric field intensity due to an infinitely long line charge, deduce (2.1.18), the expression for the electric field intensity due to a sheet of charge sprayed over an infinite plane.

2.4 Show how the electric field intensity due to an infinite planar charged conductor may be shown as equal to that due to an infinite sheet of charge plus that due to another sheet of charge of equal and opposite charge density.

2.5 A charge is distributed uniformly over a planar rectangle 2a×2b. If ρ_s is the surface charge density of the distribution, find the expression for **E** as a function of distance along the normal to the plane drawn at the centre of the rectangle.

Answer: $\mathbf{E} = (\rho_s/\pi\varepsilon_0)\ [\tan^{-1}\{z^2/a^2+z^2/b^2+(z^2/a^2)\ (z^2/b^2)\}^{-1/2}]\mathbf{a}_z$.

Hint: Let the charge be distributed over the XY plane and be bounded by $x = \pm a$ and $y = \pm b$. Let the point where to find the electric field intensity be located on the z-axis, the origin $x = y = z = 0$ being considered at the centre of the rectangle. The electric field intensity due to two symmetrically placed line charges, each parallel to the x-axis and of width dy, one considered at +y and the other at −y, may be found using (2.1.15). The components of electric field intensity due to these line charges will add up in a direction normal to the plane, that is, along z, and will cancel out in the transverse direction. Thus the element of electric field intensity due to these two line charges together are found as:

$$\mathbf{dE} = 4\rho_l\, az\,/[(4\pi\varepsilon_0)\ (y^2+z^2)\ (y^2+z^2+a^2)^{1/2}]\ ,$$

where ρ_l is the charge per unit length of the two line charges. Interpreting ρ_l as ρ_s dy and integrating the foregoing expression for **dE**, between y = 0 and b, one may obtain the desired result.

2.6 Appreciate that, as a special case: a, b >> z, the expression for the electric field intensity obtained in problem 2.5 would pass on to (2.1.18), the expression for the electric field intensity due to infinite planar charge distribution.

2.7 Apply Gauss's law for the symmetric problems of Example 2.1.2 (a), (b) and (c).

Hint: See Example 2.2.1.

2.8 Verify Gauss's law with particular reference to a point charge q placed at the centre of a rectangular parallelepiped 2a×2b×2c as the Gaussian surface.

Hint: Evaluate the surface integral over each of the two faces, of area 2a×2c, perpendicular to the y axis as: $(q/\pi)\tan^{-1}(ca/b\,(a^2+b^2+c^2)^{1/2})$. Similarly, find the surface integrals over the other two pairs of faces and see that all such integrals when added equal q.

2.9 A charge is uniformly distributed inside a sphere of radius R. If the volume charge density is ρ, find the electric field intensity **E** and hence the potential V at $r \le R$.

<u>Answer</u>: $\mathbf{E} = (\rho r/3\varepsilon_0)\mathbf{a}_r$ and $V = \rho\,(3R^2 - r^2)/6\varepsilon_0$.

2.10 The permittivity of a dielectric which completely fills the region between the conductors of a parallel-plate capacitor increases linearly from ε_1, at one plate to ε_2 at the other. If A is the area of the plates and d the distance between them, find the expression for the capacitance C of the capacitor.

<u>Answer</u>: $C = A(\varepsilon_2 - \varepsilon_1)/(d\,\ln(\varepsilon_2/\varepsilon_1))$.

2.11 A capacitor consists of two long coaxial cylindrical conductors of radii a and c, respectively ($a < c$). The region between the two conductors is completely filled by two discrete dielectrics of permittivities ε_1 and ε_2, respectively, the former touching the inner cylinder and the latter touching the outer cylinder, while the two making a common interface at $r = b$ ($a < b < c$). If the potential of the outer cylinder is V_0 with respect to that of the inner one, find (a)the potentials V_1 and V_2 in the regions of permittivities ε_1 and ε_2, respectively, and (b)the capacitance of the capacitor.

<u>Answer</u>: (a)

$$V_1 = \left(\frac{V_0}{(\varepsilon_1/\varepsilon_2)(\ln c/b) + \ln b/a}\right)\ln r/a \qquad (a \le r \le b)$$

$$V_2 = \left(\frac{V_0}{(\varepsilon_1/\varepsilon_2)(\ln c/b) + \ln b/a}\right)((\varepsilon_1/\varepsilon_2)\,(\ln r/b)+\ln b/a) \qquad (b \le r \le c)$$

(b)

$$C = \frac{2\pi\varepsilon_1}{\ln b/a + (\varepsilon_1/\varepsilon_2)(\ln c/b)} .$$

2.12 A capacitor consists of three long coaxial conducting cylinders of radii a, b and c, respectively ($a < b < c$). The central conductor is kept at the ground potential, while the innermost and the outermost cylinders are kept at positive and negative potentials with respect to the ground to such extents that a charge per unit length ρ_{l1} is developed on the former and a charge per unit length $-\rho_{l2}$ is developed on the latter. Taking ε_1 as the permittivity of the region $a \le r \le b$, between the innermost and the central conductors, and ε_2 as the permittivity of the region $b \le r \le c$, between the central and the outermost conductors, obtain the potentials V_1 and V_2 in the regions of permittivities ε_1 and ε_2, respectively.

<u>Answer</u>: $V_1 = (\rho_{l1}/2\pi\varepsilon_1)\ln b/r \quad (a \le r \le b)$, $V_2 = (-\rho_{l2}/2\pi\varepsilon_2)\ln r/b \quad (b \le r \le c)$.

2.13 Finding the force of attraction between the plates of a parallel plate capacitor, show that the pressure on each plate is equal to the energy density stored in electric field between the plates.

<u>Hint</u>: Work done in increasing the distance between the plates by an infinitesimal extent, against the force between them, is equal to the increase in energy of the capacitor due to the increase in the volume of the capacitor.

References

[1] S. Ramo, J.R. Whinnery, and T. Van Duzer, *Fields and Waves in Communication Electtronics* (John Wiley, New York, 1962).

[2] Coulomb's Law Committee, "The teaching of electricity and magnetism at the college level", *American J. Phys.*, **18** (1950) 1-25.

[3] Giorgi, *Elettricita(Milan)***20** (1901) 787-788.

[4] R. Plonsey and R.E. Collin, *Principles and Applications of Electromagnetic Fields* (Tata Mc Graw-Hill, New-York, 1973).

[5] W.H. Hayt, *Engineering Electromagnetics* (Mc Graw-Hill, Kogakusha, 1974).

[6] L.A. Pipes and L.R. Harvill, *Applied Mathematics for Engineers and Physicists* (Mc Graw-Hill Kogakusha, Tokyo, 1970).

[7] E. Weber, *Electromagnetic Fields* (John Wiley, New York, 1950).

[8] E.M. Pugh and E.W. Pugh, *Principles of Electricity and Magnetism* (Addison-Wesley, Reading, 1970).

[9] D. Halliday and R. Resnick, *Physics, Part II* (John Wiley, New York, 1962).

CHAPTER 3

STEADY MAGNETIC FIELDS

3.0 Introduction

Electrostatic field is said to exist near electric charges. Similarly, magnetostatic field exists near a permanent magnet or near a conductor carrying a steady current. The magnetic properties of a long piece of a permanent magnet or a magnetised material are found to be concentrated near its ends known as the poles of the magnets. We can assign a pole strength or *magnetic charge* to each of these poles. These poles of a magnet have equal and opposite signs and they are said to constitute a magnetic dipole just as do equal and opposite electric charges form an electric dipole. It turns out that the simplest magnetic structure in magnetostatics is the magnetic dipole. We do not have here a free or isolated magnetic pole, unlike in the electrostatic case where a free electric charge does exist to form the simplest structure. Further, we notice that like magnetic poles repel, while unlike magnetic poles attract each other just as the like and unlike electric charges do. It may be mentioned that like poles of two permanent magnets (or magnetic dipoles) are situated near the *homologous* ends of the magnets, that is, the ends that pointed the same way during the magnetisation process. The two poles of a permanent magnet are conventionally called the north-seeking or simply north pole and the south-seeking or simply south pole since when suspended freely the magnet is so aligned that these poles point towards the geographical north and south, respectively.

In an alternative formulation the magnetic effects are studied as the effects of direct electric currents or charges in uniform motion. Such effects can be observed plenty in nature. Interestingly, even the effects due to a permanent magnet can be ascribed to currents in the atomic scale. On the contrary, a current loop may be considered as a magnetic dipole and the magnetic effects due to it predicted following the same approach as used in electrostatics to study the effects of an electric dipole.

The electric field is experienced as a force on an electric charge whether the latter is at rest or in motion (see (2.1.1)) and the discussion following it). Similarly, the magnetic field is experienced as a force on an electric charge which is in motion in a direction other than that in which the magnetic field is directed; a stationary electric charge, however, does not experience such a force at all.

3.1 Magnetostatics in Analogy with Electrostatics

The force **F** between two poles (or magnetic charges) of strengths q_{m1} and q_{m2} separated by a distance r in a unbounded, homogeneous medium is given by Coulomb's law of magnetostatics (see (2.1.4)) [1], [2]:

$$\mathbf{F} = \frac{\mu}{4\pi} \frac{q_{m1}\, q_{m2}}{r^2} \mathbf{a}_r , \tag{3.1.1}$$

where μ is a medium property called the permeability. For a free space as well as for a *non-magnetic* medium, the permeability is: $\mu = \mu_0 = 4\pi \times 10^{-7}$ henry per meter. That the unit of permeability is so would be clear from a typical expression for the inductance of an inductor, for instance, see Example 3.2.5. (Here, we may recall that, in section *2.1*, we had put the unit of ε as Coulomb^2/(Newton-meter^2). We could have equally well set a practical unit of ε as farad/meter which is evident from the expression for the capacitance of a capacitor, for instance, expression (2.7.5)). The unit of pole strength or magnetic charge is amp-meter which would be clear, at a later stage, from the considerations of the magnetic moment of a current loop. (See the discussions following (3.2.12)). This will also be evident from the comparison of (3.1.1) with the expression for the force between two current elements to be developed in section *3.3*.

A magnetic flux density **B** is said to exist at a point if a magnetic pole or charge of a magnet experiences a force at the point. Mathematically, the force **F** on the magnetic charge q_m may be stated as

$$\mathbf{F} = q_m \mathbf{B} , \tag{3.1.2}$$

an equation which is analogous to (2.1.1). An equation analogous to (2.1.14) is then easily obtained with the help of (3.1.1) and (3.1.2) and following the approach given in Section *2.1.1*, as:

$$\mathbf{B} = \frac{\mu\, q_m}{4\pi r^2} \mathbf{a}_r , \tag{3.1.3}$$

an expression for the magnetic flux density at a distance r from a pole strength or magnetic charge q_m. We may notice that **E** (see (2.1.14)) and **B** (see (3.1.3)) are the two analogous quantities, both depending on the medium properties, ε and μ, respectively. In this formulation, the magnetic field intensity **H** (= **B** / μ) and the electric displacement **D** (= ε **E**), being independent of μ and ε, as can be seen with the help of (2.1.1) and (3.1.3), respectively, may be considered to be two other analogous quantities.

In (3.1.2) or (3.1.3) it is essentially implied that the magnetic charge q_m is not *free* (and that it is bound to an equal and opposite charge in the simplest structure of a dipole). The absence of a free magnetic charge enables one to write the equation analogous to Gauss's law of electrostatics, namely,

$$\oint \mathbf{D}.\mathbf{a}_n \, dS = q \qquad \text{((2.2.1) re-written)}$$

or, $$\oint \mathbf{E.a}_n \, dS = q/\varepsilon \quad \text{(in view of the relation } \mathbf{E} = \mathbf{D}/\varepsilon)$$

as:

$$\oint \mathbf{H.a}_n \, dS = 0 \qquad \text{(a)}$$

or, $$\oint \mathbf{B.a}_n \, dS = 0 \quad \text{(in view of the relation } \mathbf{B} = \mu\mathbf{H}) \text{ . (b)} \qquad (3.1.4)$$

The point form of (3.1.4), which is easily obtained following the method given in section *2.3*, is:

$$\nabla.\mathbf{H} = 0 \qquad \text{(a)}$$

or, $$\nabla.\mathbf{B} = 0 \qquad \text{(b)} \qquad (3.1.5)$$

which is analogous to (2.3.3) in the electrostatic case. Both (3.1.4) and (3.1.5) have the physical meaning that there are no free magnetic charges. This would also mean that magnetic flux lines are continuous unlike in the electrostatic case. (For further elaboration on this point, see appendix *a3.1.1*).

Finally, it is useful to introduce the expression for energy stored in a static magnetic field analogous to (2.7.9), of the electrostatic case as [3]

$$U_B = \frac{1}{2}\mu_0 H^2 = \frac{1}{2}(\mu_0 H)(H) = \frac{1}{2}\mathbf{B.H} = \frac{1}{2}\mathbf{H.B} \, . \qquad (3.1.6)$$

A comparison of (3.1.6) with (2.7.9) suggests that the analogy between **E** and **B** and that between **D** and **H** as seen earlier in this section continues to be valid. It may be mentioned that as we have verified (2.7.9) with particular reference to the example of a capacitor, we may take up the example of an inductor to check the validity of (3.1.6). Also, it is possible to give a rigorous proof of (3.1.6) (which, however, we would not choose to do here), as has been done in the electrostatic case by interpreting the amount of work done in establishing a distribution of electric charges (appendix *a2.7.2*). In the present case, however, since no free magnetic charges exist, one has to find the work done in establishing a distribution of current loops to obtain an expression for energy stored in a magnetic field [3].

Appendix a3.1.1 *Continuity of magnetic flux lines*

*Flux lines are hypothetical lines which are associated with a vector quantity. For instance, flux lines may refer to **D** and **E** in electrostatics and to **H** and **B** in magnetostatics. The direction of the tangent to the flux line at a point represents the direction of the vector quantity under consideration at the point. The density of flux lines, that is, number of such lines per unit cross-sectional area gives a measure of the magnitude of the concerned vector quantity [4].*

*Let us now consider an electric dipole constituted by point charges + q and - q (see Fig. a3.1.1). We may consider flux lines of **E** (or **D**) to start from the point charge +q and terminate on the point charge - q. With the help of (2.2.1), we can write*

(a3.1.1) (a) $$\oint_{S_1} \boldsymbol{D}.\boldsymbol{a}_n \, dS = q \, ,$$

(b) $$\oint_{S_2} \mathbf{D}.\mathbf{a}_n \, dS = -q \quad \text{and}$$

(c) $$\oint_{S_3} \mathbf{D}.\mathbf{a}_n \, dS = 0 \ ,$$

where S_1 and S_2 are the Gaussian surfaces enclosing the point charges q_1 and q_2, respectively, and S_3 is the Gaussian surface enclosing any other point considered in the region where no charge is present. In terms of lines of flux, we may state that there is a net number of lines of flux leaving the surface S_1, since the right hand of (a3.1.1(a)) is positive; that there is a net number of lines of flux entering the surface S_2, since the right hand side of (a3.1.1(b)) is negative; and that the number of flux lines entering the surface S_3 is same as the number of flux lines leaving the surface S_3 since the right hand side of (a3.1.1(c)) is zero (the surface S_3 enclosing no charges) (Fig. a3.1.1(a)).

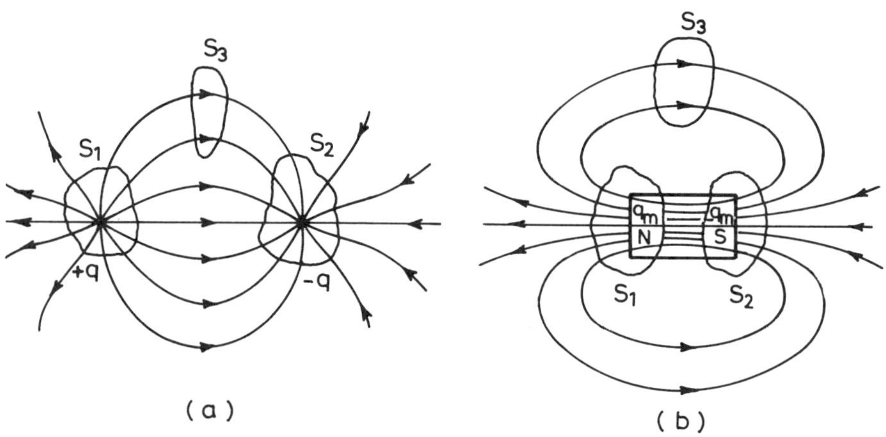

Fig. a3.1.1 Gaussian surfaces S_1 and S_2 enclosing the point charges + q and - q, respectively, in the electrostatic case **(a)** , and similar surfaces enclosing the north and south poles where the magnetic point charges $+q_m$ and $-q_m$ are considered to be concentrated, in the magnetostatic case **(b)**, Gaussian surface S_3 enclosing a charge-free point in both the cases.

In the corresponding magnetic dipole case, we notice that (3.1.4) continues to be valid, irrespective of whether the surface integral is taken over S_1, S_2 or S_3 (see Fig. a3.1.1(b)) which, in terms of flux lines, would mean that the magnetic flux lines are continuous; they, unlike in the electrostatic case, do not start from a positive charge and terminate on a negative charge (see Fig. 3.1.1(a)).

3.2 Magnetic Field of a Current-Carrying Conductor

The two important laws of magnetostatics which are useful in finding the magnetic field caused by a steady current through a conductor are Biot-Savart's and Ampere's circuital laws which take the roles played in electrostatics by Coulomb's and Gauss's laws, respectively.

3.2.1 Biot-Savart's Law

Biot-Savart's law gives a measure of the steady magnetic field intensity due to a current element. The element of current, here, is defined as a vector quantity the magnitude of which is equal to the product of the current through a filamentary conductor (the solid-cylindrical conductor of circular cross-section of radius approaching zero) and the length element. The direction of the current-element vector is given by the direction of the current. Denoting the current as i and the length element as dl, the current-element vector may thus be put as **idl** (Fig. 3.2.1). The magnetic field intensity due to this current element, **idl**, at a distance r from the element is given by Biot-Savart's law as follows [3], [4]:

$$\mathbf{dH} = \frac{i\,\mathbf{dl}\times\mathbf{r}}{4\pi r^3} \,. \tag{3.2.1}$$

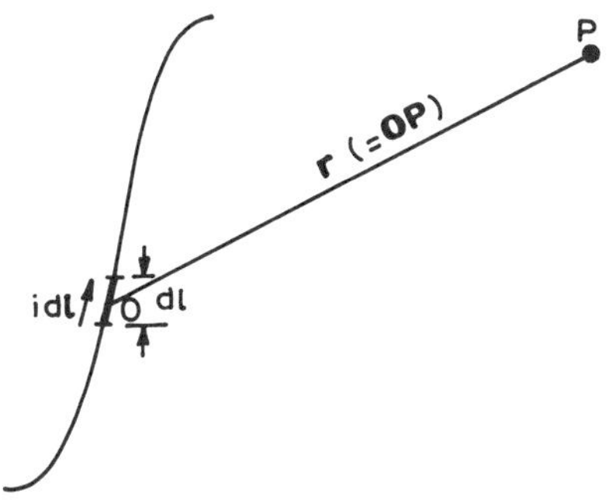

Fig. 3.2.1 The geometry for finding the magnetic field intensity due to a current element using Biot-Sarvart's law.

Here, **r** is a distance vector whose magnitude is equal to the distance between the current element and the point where to find the magnetic field intensity, and which is directed from the current element to the point (Fig. 3.2.1). It is worth comparing (3.2.1) with the following equation in the analogous electrostatic case for an element of line charge ρ_l dl (see (a2.1.1) through (a2.1.2)):

$$\mathbf{dD} = (\varepsilon_0 \mathbf{dE}) = \frac{\rho_l \, dl\, \mathbf{r}}{4\pi r^3} \,.$$

It becomes then obvious that the quantity $i\mathbf{dl}\times\mathbf{r}$ for a current element, in the steady magnetic field case, is analogous to the quantity ρ_l dl **r** for a charge element in the electrostatic case.

Example 3.2.1 Find the steady magnetic field intensity due to a straight wire of a finite length carrying a direct current.

The problem is analogous to that of finding the electric field displacement **D** (= ε_0 **E**) due to a line charge of a finite length. The answer to the problem is found by integrating (3.2.1) for the given length of the straight wire following the same approach as in Example 2.1.2(a). For this purpose, a reference can be made to Fig. 2.1.2, where the filamentary straight conductor may be considered as to coincide with the z-axis. The length of the conductor considered is between the points B and A, the current being directed along the z-axis i.e., from B to A.

Making use of Biot-Savart's law, mathematically stated by (3.2.1), one can find the magnetic field intensity due to the entire length of the conductor as (Fig. 2.1.2)

$$\mathbf{H} = \int \frac{i\, \mathbf{dl} \times \mathbf{R}}{4\pi R^3}, \tag{3.2.2}$$

where the integration is extended over the length of the conductor. Here, **R** is chosen as the distance vector instead of **r** in (3.2.1), the vector being directed from the current element to the point (r, θ, z) where the magnetic field intensity is sought. Interpreting **dl** as dz′ $\mathbf{a}_z$ and **R** as $-(z'-z)\,\mathbf{a}_z + r\,\mathbf{a}_r$ (see appendix *a2.1.1*), one may write (3.2.2) as

$$\mathbf{H} = \int \frac{i\, \mathbf{a}_z \times (-(z'-z)\mathbf{a}_z + r\, \mathbf{a}_r)}{4\pi R^3} dz' = \frac{i\, r\, \mathbf{a}_\theta}{4\pi} \int \frac{dz'}{R^3}, \tag{3.2.3}$$

the reason for keeping r $\mathbf{a}_\theta$ outside the integration being exactly similar to what is stated following (a2.1.4). The integral of (3.2.3) is identical with the second integral of (a2.1.4) which has already been evaluated in appendix *a2.1.1*. Substituting the value of the integral in (3.2.3), we have [5]

$$\mathbf{H} = \frac{i}{4\pi r}(\sin\psi_1 - \sin\psi_2)\mathbf{a}_\theta = \frac{i}{4\pi r}[\sin\tan^{-1}(z'_1 - z)/r - \sin\tan^{-1}(z'_2 - z)/r]\mathbf{a}_\theta. \tag{3.2.4}$$

The meaning of ψ_1, ψ_2 and, correspondingly, that of z'_1 and z'_2 in (3.2.4) are already given in appendix *a2.1.1*. For an infinitely long conductor, one may put $z_1 = \infty$ ($\psi_1 = \pi/2$) and $z'_2 = -\infty$ ($\psi_1 = -\pi/2$). Substituting these values in (3.2.4), the magnetic field intensity due to an infinitely long straight wire is found as:

$$\mathbf{H} = \frac{i}{2\pi r}\mathbf{a}_\theta . \tag{3.2.5}$$

It is interesting to compare (3.2.5) with

$$\mathbf{D} = \varepsilon_0 \mathbf{E} = \frac{\rho_l}{2\pi r}\mathbf{a}_r ,$$

the expression for the electric displacement due to an infinitely long line charge (obtainable from (2.1.16), in the corresponding electrostatic case. In this example, the direct current "i" is found to be analogous to the line charge density "ρ_l". In both the cases, the inverse proportionality of the relevant quantities, namely, **H** and **D** with r is common. The change is, however, in the direction of the field intensity — in the electrostatic case, it is radial and, in the case of the steady magnetic field, it is azimuthal.

Example 3.2.2 Study the special case of Example 3.2.1 in which the point where to find the magnetic field intensity is equidistant from the ends of the straight wire carrying a direct current.

Let the wire carrying a direct current be coincident with the z-axis such that the centre of the wire falls on the origin, and the point P where to find the magnetic field intensity lies on the Y-axis (Fig. 3.2.2). The z-coordinates of the end points of the wire are $z'_1 = l/2$ and $z'_2 = -l/2$, respectively, the length of the wire being l. The rectangular cartesian coordinates of P are (0, r, 0) and the corresponding cylindrical coordinates are (r, $\pi/2$, 0). Then the magnetic field intensity at P can be shown with the help of (3.2.4) to be:

$$\mathbf{H} = \frac{i}{2\pi r}\frac{1}{[1+(\frac{r}{l/2})^2]^{1/2}}\mathbf{a}_\theta \ . \tag{3.2.6}$$

$\mathbf{a}_\theta$, in this geometry, may be seen to be identical with $-\mathbf{a}_x$.

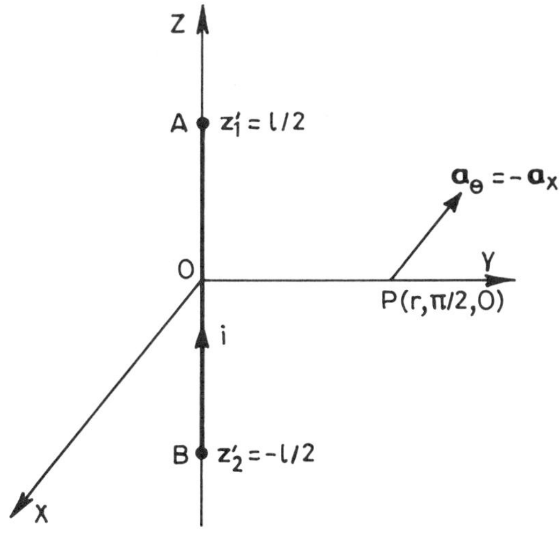

Fig. 3.2.2 Geometry for finding the magnetic field intensity at a point equidistant from the ends of a straight wire of finite length carrying a direct current.

Example 3.2.3 Evaluate the magnetic field intensity due to a closed loop or turn of wire making a regular polygon carrying a direct current, the point where to find the magnetic field intensity being the centre of the polygon. Take up the case of a circular turn as a special case.

Let us choose the regular polygon-turn of wire to lie on the XY-plane (Fig. 3.2.3), and find using (3.2.6) the magnetic field intensity due to the individual sides of the polygon wire and then sum them up. The magnetic field intensity at the origin O found thus is

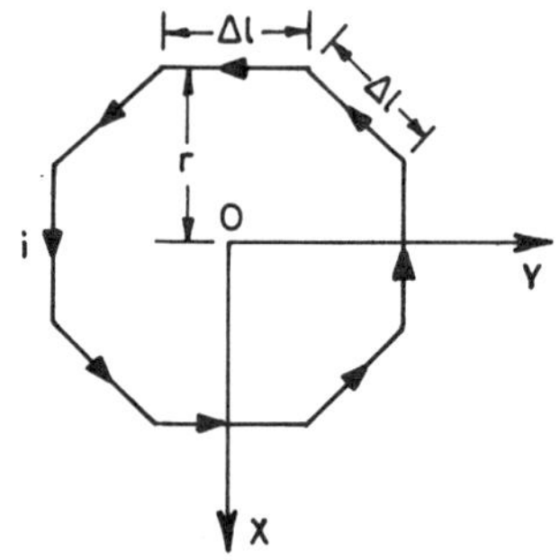

Fig. 3.2.3 A regular polygon-turn of wire carrying a direct current shown to lie on the XY-plane with its centre at the origin (O) where to find the magnetic field intensity.

$$\mathbf{H} = \sum \frac{i}{2\pi r} \frac{1}{[1+(\frac{r}{\Delta l/2})^2]^{1/2}} \mathbf{a}_z, \tag{3.2.7}$$

where Δl is the length of each side of the polygon turn (loop), r is the perpendicular distance of each side from the centre of the polygon where we intend to find the magnetic field intensity, and the summation is extended over all the sides of the polygon. In Fig. 3.2.3, we have shown the polygon to have six sides, typically. If we increase the number of sides of the polygon, $\Delta l \to 0$, and the polygon becomes a circle of radius r. For such a circular turn, we have from (3.2.7), taking $r/(\Delta l/2) >> 1$:

$$\mathbf{H} = \frac{i}{2\pi r} \sum \frac{\Delta l/2}{r} \mathbf{a}_z = \frac{i}{4\pi r^2} \sum \Delta l \mathbf{a}_z, \tag{3.2.8}$$

where the summation has to be taken over the circular turn. Putting $\Sigma\, \Delta l = 2\pi r$, the circumference of the circular turn, in (3.2.8) we then have

$$\mathbf{H} = \frac{i}{2r} \mathbf{a}_z \tag{3.2.9}$$

which shows that the magnetic field intensity is directed perpendicular to the plane of the turn. It can be easily appreciated that if the sense of the current is reversed, the direction of the magnetic field intensity due to the loop would also reverse.

Example 3.2.4 Find applying Biot-Savart's law the magnetic field intensity due to a circular turn of direct current at a point lying on the axis of the turn, that is, on a line which passes through the centre of the turn and is normal to the plane containing the circular turn, the point not necessarily lying at the centre of the turn.

Let us consider the circular turn of radius a, carrying a direct current i, the turn lying on the XY-plane, and find the magnetic field intensity due to it at a point P ($z = d$), distant d from the origin, lying on the z-axis. The magnetic field intensity at P may be found using Biot-Savart's law as (for the details of derivation see appendix *a3.2.1*):

$$\mathbf{H} = \frac{i\alpha}{2\pi(a^2 + d^2)^{3/2}} \mathbf{a}_z , \tag{3.2.10}$$

where α (= πa^2) is the area enclosed by the circular loop. At a distant point, for which one can put $a << d$,

$$\mathbf{H} = \frac{i\alpha}{2\pi d^3} \mathbf{a}_z \quad . \tag{3.2.11}$$

It is of interest to compare (3.2.11) with the expression for the magnetic field intensity at a point on the z-axis due to a short magnetic dipole, also lying on the z-axis with its centre at the origin. At a point located at $z = d >> l$, the length of the dipole, the magnetic field intensity due to the dipole is (for the details of derivation see appendix *a3.2.1*):

$$\mathbf{H} = \frac{q_m \, l}{2\pi d^3} \mathbf{a}_z \, . \tag{3.2.12}$$

Thus it becomes evident from the present example that, as far as the magnetic effects of a circular current loop are concerned, the loop may be considered to behave as a magnetic dipole. It follows from the comparison of (3.2.11) and (3.2.12) that the magnetic moment of the current loop is

$$q_m \, l \;=\; i\,\alpha \, .$$

The above relation also suggests that the unit of the pole strength or magnetic charge is amp-meter as stated in the discussion following (3.1.1). In passing it may be noted that (3.2.10), as a special case of $d = 0$, would pass on to (3.2.9).

<u>*Example 3.2.5*</u> Apply Biot-Savart's law to find the magnetic field intensity at a point lying on the axis of a long solenoid. Show also that the self inductance of such a solenoid is proportional to the volume of the solenoid and to the square of the number of turns per unit length.

The method is to first find the magnetic field intensity due to a single turn and then add contributions to the magnetic field intensity due to all such individual turns of the solenoid. The first part of the problem has already been worked out in Example 3.2.4 (For details, see also appendix *a3.2.1*).

One may consider the length of the solenoid to be very large compared to the distance between two consecutive turns of the solenoid. One may now find the element of magnetic field intensity contributed by the turns occupied in length dz of the solenoid using (3.2.10) and find the magnetic field intensity due to the entire solenoid by integration (see appendix *a3.2.2*, for the details of deduction):

$$\mathbf{H} = n\, i\, \mathbf{a}_z \tag{3.2.13}$$

where i is the direct current through the solenoid, n is the number of turns per unit length, and the axis of the solenoid is along the z-axis, the direction of the magnetic field intensity being given by the screw rule, that is, by the direction of the advancement of a screw turned in the direction of the current through the solenoid.

The self-inductance of the solenoid, L, say, is defined as the magnetic flux per unit current through the solenoid, ϕ_B/i, where $\phi_B = (B\alpha)(nl) = (\mu_0 H\alpha)(nl)$ is the magnetic flux $B\alpha$ (= $\mu_0 H\alpha$), associated with each turn multiplied by the number of turns nl of the solenoid where l and α are its length and cross-sectional area, respectively. Thus, using (3.2.13), L is found as:

$$L \; (= \phi_B / i\,) \;=\; \mu_0 \, n^2 \, (l\,\alpha) \;,$$

which shows that the self-inductance of a long solenoid is proportional to the square of the number of turns per unit length (n^2) and to the volume ($l\alpha$) of the solenoid.

3.2.2 Ampere's Circuital Law in the Integral Form

We have seen in section *3.2.1* that Biot-Savart's law gives an estimate of the steady magnetic field intensity **H** (and hence $\mathbf{B} = \mu_0\mathbf{H}$) due to a direct-current element just as it is possible in electrostatics to obtain electric field intensity **E** (and hence $\mathbf{D} = \varepsilon\mathbf{E}$) due to a static charge element, starting from the basic Coulomb's law. In this section, we would like to take up Ampere's circuital law which has the same role as played by Gauss's law in electrostatics in working out relatively simpler problems which enjoy some sort of symmetry.

At this stage, it is worth recalling Gauss's law (2.2.1) in which we may choose to express the charge enclosed in terms of the volume charge density ρ:

$$\oint \mathbf{D}.\mathbf{a}_n \, dS = \int \rho d\tau \ , \qquad (3.2.14)$$

where the integration is extended over the volume enclosed by the closed Gaussian surface. We have earlier stated Gauss's law for the magnetostatic case as (3.1.4) from the considerations of magnetic charges which do not exist "freely". For the steady magnetic field caused by a direct current, a law which is as useful as, and which corresponds to, Gauss's law (3.2.14) in electrostatics is the Ampere's circuital law which can be mathematically stated as

$$\oint \mathbf{H}.\mathbf{dl} = \oint \mathbf{J}.\mathbf{a}_n \, dS = i \ , \qquad (3.2.15)$$

where the left hand side represents the closed line integral of magnetic field intensity round a closed path which encloses a direct current i which in turn may be expressed in terms of flux of current density **J** through the cross-section of the conductor carrying the current. It is appropriate to mention here that the magnetic field intensity may, in principle, be caused as well by a convection current, for instance, by an electron beam, though in most of the practical situations their effects are found to be negligible.

At this stage, without going into the rigorous proof of (3.2.15) let us restrict ourselves to verifying it with particular reference to the problem of an infinitely long straight wire carrying a direct current. For this purpose, let us consider a circular path of radius r on a plane perpendicular to the straight wire. The magnetic field intensity H at any point on this path is given by (3.2.5). The element of length vector **dl** is given by

$$\mathbf{dl} = r \, d\theta \, \mathbf{a}_\theta \ . \qquad (3.2.16)$$

Substituting (3.2.5) and (3.2.16) into the left hand side of (3.2.15) we have

$$\oint \mathbf{H}.\mathbf{dl} = \oint \frac{i}{2\pi r} \mathbf{a}_\theta . (r d\theta \, \mathbf{a}_\theta) = \frac{i}{2\pi} \oint d\theta = \frac{i}{2\pi}(2\pi) = i \ .$$

which verifies the validity of (3.2.15).

<u>*Example 3.2.6*</u> Work out the problem of Example 3.2.5 using Ampere's Circuital Law.

It is of interest to see in this example the case with which the problem of the solenoid can be dealt with using Ampere's circuital law. We recall that the magnetic field intensity at a point on the axis of a long solenoid was found in Example 3.2.5 (appendices *a3.2.1* and *a3.2.2*) using a method which involves two mathematical integrations, starting from the basic Biot-Savart's law.

The method in this example, and, in fact, in all other similar examples, is based on the choice of a suitable closed path on which to take the line integral of the magnetic field intensity (see (3.2.15)). The closed path is analogous to the closed Gaussian surface in electrostatic problems studied in Section *2.2*.

The closed path, in the present example, is taken on a plane lying on a cross-sectional plane passing through the axis of the solenoid. The application of Ampere's circuital law along suitably chosen paths shows that, while the magnetic field intensity outside the solenoid is zero, that inside the solenoid is given by (3.2.13). (For the details of the present method of deduction, see appendix *a3.2.3*) [4].

3.2.3 Ampere's Circuital Law in the Differential Form

Just as the integral forms of Ampere's circuital law (3.2.15) and Gauss's law (3.2.14) correspond to each other, so should also their differential or point forms do. In fact, just as we derived Poisson's equation in the point form (2.3.3) from Gauss's law in the integral form (2.2.1) or (3.2.14), we may expect to obtain a point form of Ampere's circuital law starting from its integral form (3.2.15) [5].

The method followed, for this purpose, is to consider the elements of closed area normal to the direction of the coordinate axes, that is, to $\mathbf{a}_x$, $\mathbf{a}_y$, $\mathbf{a}_z$ in the rectangular system, to $\mathbf{a}_r$, $\mathbf{a}_\theta$, $\mathbf{a}_z$, in the cylindrical system, and to $\mathbf{a}_r$, $\mathbf{a}_\theta$, $\mathbf{a}_\phi$ in the spherical-polar system, such elements of area being considered around the point where the point form of Ampere's circuital law is desired. The line integral of $\mathbf{H}$ along a closed path enclosing an element of area, divided by the area element, in the limit, the latter shrinking to zero to coincide with the point yields a component of the quantity called to curl of $\mathbf{H}$, also denoted as $\nabla\times\mathbf{H}$, at the point, the component being directed normally to the area element considered, with a sense determined by the linear motion of a screw turned in the direction in which the line integral is taken. The quantity is further found, with the help of (3.2.15), to be equal to a component of current density $\mathbf{J}$, the component being in the same direction as that of curl $\mathbf{H}$ (= $\nabla\times\mathbf{H}$). Further, on combining the three orthogonal components in the three coordinates axes, respectively (for the details of derivation, see appendix *a3.2.4*), we get:

$$\text{curl}\,\mathbf{H} = \nabla\times\mathbf{H} = \mathbf{J}\,. \tag{3.2.17}$$

The expressions for curl $\mathbf{H}$, in the three systems of coordinates, are (see appendix *a3.2.4*):

$$\text{curl}\,\mathbf{H} = \nabla\times\mathbf{H} = \left(\frac{\partial H_z}{\partial y} - \frac{\partial H_y}{\partial z}\right)\mathbf{a}_x + \left(\frac{\partial H_x}{\partial z} - \frac{\partial H_z}{\partial x}\right)\mathbf{a}_y + \left(\frac{\partial H_y}{\partial x} - \frac{\partial H_x}{\partial y}\right)\mathbf{a}_z \quad \text{(a)}$$

(rectangular system)

$$= \left(\frac{1}{r}\frac{\partial H_z}{\partial\theta} - \frac{\partial H_\theta}{\partial z}\right)\mathbf{a}_r + \left(\frac{\partial H_r}{\partial z} - \frac{\partial H_z}{\partial r}\right)\mathbf{a}_\theta + \frac{1}{r}\left(\frac{\partial (r\,H_\theta)}{\partial r} - \frac{\partial H_r}{\partial\theta}\right)\mathbf{a}_z \quad \text{(b)}$$

(cylindrical system)

$$= \frac{1}{r\sin\theta}\left(\frac{\partial(H_\phi \sin\theta)}{\partial\theta} - \frac{\partial H_\theta}{\partial\phi}\right)\mathbf{a}_r + \frac{1}{r}\left(\frac{1}{\sin\theta}\frac{\partial H_r}{\partial\phi} - \frac{\partial(rH_\phi)}{\partial r}\right)\mathbf{a}_\theta + \frac{1}{r}\left(\frac{\partial(rH_\theta)}{\partial r} - \frac{\partial H_r}{\partial\theta}\right)\mathbf{a}_\phi$$

(spherical-polar system)

(c) (3.2.18)

3.2.4 Magnetic Vector Potential

Just as the problem of finding static electric field is simplified through an intermediate step of determining the electrostatic potential, so is it useful to introduce the magnetic vector potentials in steady magnetic field problems. The magnetic vector potential can be related to the steady current element and the distance of such element from the point where to find the potential, just as it is possible to relate the electrostatic potential with the charge element and distance (see Section *2.5*).

For this purpose let us start from Biot-Savart's law which gives the magnetic flux density **dB** (= μ_0 **dH**) due to a current element **idl** (see (3.2.1)) as:

$$\mathbf{dB} = (\mu_0 \mathbf{dH}) = \frac{\mu_0}{4\pi}\frac{i\,\mathbf{dl}\times\mathbf{r}}{r^3}. \qquad (3.2.19)$$

Let the current element be located at the source point (x′, y′, z′) and the point where the magnetic flux density is observed be (x, y, z). One may then write

$$\mathbf{r} = (x - x')\,\mathbf{a}_x + (y - y')\,\mathbf{a}_y + (z - z')\,\mathbf{a}_z \qquad (a)$$

and

$$r = [(x - x')^2 + (y - y')^2 + (z - z')^2]^{1/2}. \qquad (b) \qquad (3.2.20)$$

Now, with the help of (2.5.12a) and (3.2.20) one may show (see appendix *a3.2.5*):

$$\nabla\left(\frac{1}{r}\right) = -\frac{1}{r^3}\,\mathbf{r}, \qquad (3.2.21)$$

where the gradient (∇) operation is taken with respect to the field point variables (x, y, z). Making use of (3.2.21) one may then express (3.2.19) as

$$\mathbf{dB} = -\frac{\mu_0}{4\pi}\,i\mathbf{dl}\times\nabla\left(\frac{1}{r}\right). \qquad (3.2.22)$$

Now using the vector identity

$$\nabla\times(\psi\mathbf{G}) = \psi\nabla\times\mathbf{G} - \mathbf{G}\times\nabla\psi,$$

where ψ and **G** are the two scalar and vector quantities, respectively, one has by taking ψ = 1/r and **G** = i **dl**,

$$-\mathbf{idl} \times \nabla\left(\frac{1}{r}\right) = \nabla \times \left(\frac{1}{r}\mathbf{idl}\right) - \frac{1}{r}\nabla \times \mathbf{idl}$$

which when substituted in (3.2.22) yields

$$\mathbf{dB} = \frac{\mu_0}{4\pi}\left(\nabla \times (\frac{1}{r}\mathbf{idl}) - \frac{1}{r}\nabla \times \mathbf{idl}\right). \tag{3.2.23}$$

Now, since the curl operation in the second term of (3.2.23) is taken with respect to the field-point variables (x, y, z) while the current element **idl** is represented by the source point variables (x′, y′, z′), the second term in the right hand side of (3.2.23) will be zero. Thus retaining only the first term in the right hand side of (3.2.23), one may write:

$$\mathbf{dB} = \frac{\mu_0}{4\pi}\nabla \times (\frac{1}{r}\mathbf{idl}) = \nabla \times \left(\frac{\mu_0}{4\pi}\frac{\mathbf{idl}}{r}\right) . \tag{3.2.24}$$

Choosing to write the elemental vector quantity under the curl operation in (3.2.24) as

$$\mathbf{dA} = \left(\frac{\mu_0}{4\pi}\right)\left(\frac{\mathbf{idl}}{r}\right), \tag{3.2.25}$$

one may express (3.2.24) as

$$\mathbf{dB} = \nabla \times \mathbf{dA} . \tag{3.2.26}$$

The magnetic flux density due to a given current distribution is obtained by summing up the contributions from the current elements as

$$\mathbf{B} = \int \mathbf{dB} = \int \nabla \times \mathbf{dA} = \nabla \times \int \mathbf{dA} = \nabla \times \mathbf{A}, \tag{3.2.27}$$

where the integration is extended over the current elements, and

$$\mathbf{A} = \frac{\mu_0}{4\pi}\int \frac{\mathbf{idl}}{r} \tag{3.2.28}$$

is called the magnetic vector potential. Now, that the divergence of the curl of a vector quantity is zero makes in this case

$$\nabla . \nabla \times \mathbf{A} = 0$$

which, with the help of (3.2.27), yields

$$\nabla . \mathbf{B} = 0,$$

a relation which is in conformity with (3.1.5).

In the following two Examples 3.2.7 and 3.2.8, the magnetic flux density is found through the magnetic vector potential. In the first of these, one finds **dA** with the help of (3.2.25) and then **dB** (= ∇×**dA**) with the help of (3.2.26), the latter on integration resulting into **B** (= ∫ **dB**). In the second one, one first integrates **dA** to obtain **A** (= ∫ **dA**) (see (3.2.28)). Subsequently, taking curl of **A** one may obtain **B** (= ∇×**A**) (see (3.2.27)).

<u>*Example 3.2.7*</u> Consider the problem of a straight wire carrying a steady current of Example 3.2.1, and find the magnetic field intensity using the vector potential approach.

With the help of (3.2.25) we may find **dA** at P (r, θ, z) due to a current element $i\,\mathbf{dl} = i\,dz'\,\mathbf{a}_z$ considered arbitrarily at S on the wire as (see Fig. 2.1.2 with reference to Example 3.2.1):

$$\mathbf{dA} = \frac{\mu_0\, i\, dz'}{4\pi R}\mathbf{a}_z = \frac{\mu_0\, i\, dz'\,\mathbf{a}_z}{4\pi[r^2 + (z'-z)^2]^{1/2}} \tag{3.2.29}$$

whence **dB** (= ∇×**dA**) may be found with the help of (3.2.18(b)). In this example, $(\mathbf{dA})_\theta = (\mathbf{dA})_r = 0$ (see (3.2.29)), and also $\partial/\partial\theta = 0$ (azimuthal symmetry). Thus,

$$\mathbf{dB} = -\left[\frac{\partial}{\partial r}\left(\frac{\mu_0 i}{4\pi}\right)(dz')(r^2 + (z'-z)^2)^{-1/2}\right]\mathbf{a}_\theta$$

$$= \left(\frac{\mu_0 i}{4\pi}\right)(r)\left(\frac{dz'\,\mathbf{a}_\theta}{(r^2 + (z'-z)^2)^{3/2}}\right) = \frac{\mu_0\, i\, r\, dz'}{4\pi R^3}\mathbf{a}_\theta$$

which upon integration yields an expression for **H** (= ∫ $\mathbf{dB}/\mu_0$) that becomes identical with (3.2.3). Finally, from (3.2.3) one obtains an expression identical with (3.2.4) as in Example 3.2.1.

<u>*Example 3.2.8*</u> Apply the concept of vector potential to find the expression for the magnetic field intensity due to a circular turn of current at a distant point. Also, as a special case, verify the result with that obtained in Example 3.2.4.

As in Example 3.2.4, let the circular turn now lie on the XY-plane but the point where the magnetic flux density or intensity is sought be situated off the z-axis at P represented by r, θ, ϕ in spherical-polar coordinates (see Fig. a3.2.8). The problem enjoys spherical symmetry ($\partial/\partial\phi = 0$), and no generality is lost if one takes the point P at (r, θ, 0). The element of vector potential **dA** at P due to a current element i**dl** considered arbitrarily at ϕ = ϕ′ (x = a cos ϕ′, y = a sin ϕ′, z = 0), taking P far from the turn (r >> a), is given by (see appendix *a3.2.6*) [6], [7]:

$$\mathbf{dA} = \left(\frac{\mu_0}{4\pi r}\right)(i\, a\, d\phi'\cos\phi')\left(1 + \frac{a}{r}\sin\theta\cos\phi'\right)\mathbf{a}_\phi \tag{3.2.30}$$

which integrated over the entire circular turn (between ϕ′ = 0 and 2π) yields (see appendix *a3.2.6*):

$$\mathbf{A} = \frac{\mu_0 i\,\pi a^2 \sin\theta}{4\pi r^2}\mathbf{a}_\phi \tag{3.2.31}$$

whence the magnetic flux **B** (= ∇×**A**) is obtained as (using (3.2.18(b)), however, taking **A** instead of **H** therein):

$$\mathbf{B} = \mu_0 i\,\pi a^2\left(\frac{\cos\theta}{2\pi r^3}\mathbf{a}_r + \frac{\sin\theta}{4\pi r^3}\mathbf{a}_\theta\right). \tag{3.2.32}$$

Interpreting r as d and putting $\pi a^2 = \alpha$, the area of the turn, and taking a particular case θ = 0, one may obtain **H** (= $\mathbf{B}/\mu_0$) from (3.2.32) that directly passes on to (3.2.11) noting that, for θ = 0, $\mathbf{a}_r$ becomes identical with $\mathbf{a}_z$.

Appendix a3.2.1 *Circular loop as a magnetic dipole*

*The problem is to find the magnetic field intensity due to the circular current loop at a point ($z \neq 0$, in general) on the z-axis (Fig. a3.2.1). The element of magnetic field intensity due to a current element considered arbitrarily at M is first found using Biot-Savart's law (3.2.1) whence the magnetic field intensity **H** due to the entire current loop is found by integration (Fig. a3.2.1) as:*

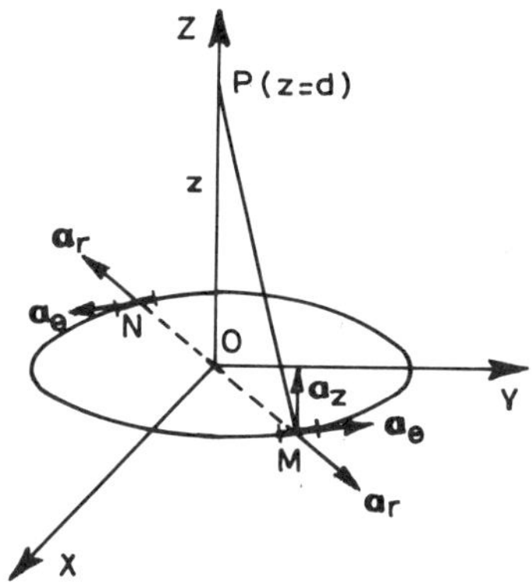

Fig. a3.2.1 A circular current loop is considered on the XY-plane for the prediction of the magnetic field intensity due to it at a point lying on the z-axis, at z = d.

(a3.2.1)
$$\boldsymbol{H} = \int \frac{i\, dl\, \boldsymbol{a}_\theta \times \boldsymbol{MP}}{4\pi (MP)^3} = \frac{i\, dl\, \boldsymbol{a}_\theta \times \left(-a\, \boldsymbol{a}_r + d\, \boldsymbol{a}_z\right)}{4\pi\, (a^2 + d^2)^{3/2}}$$

$$= \frac{i\, a\, \mathbf{a}_z}{4\pi \left(a^2 + d^2\right)^{3/2}} \int dl + \frac{i\, d}{4\pi \left(a^2 + d^2\right)^{3/2}} \int \mathbf{a}_r\, dl\ ,$$

*where a is the radius of the circular turn. d is the distance of the point P where to find the magnetic field intensity from the centre of the circle, the point lying on a line normal to the plane of the turn, passing through the centre of the turn. In other words, the cartesian coordinates of P are (0, 0, d). dl is a length element arbitrarily considered at M, the current element there being i **dl** = i dl $\boldsymbol{a}_\theta$ (Fig. a3.2.1).*

The right hand side of (a3.2.1) consists of two integrals. $\boldsymbol{a}_z$ does not change from element to element and hence it is kept outside the first of the two integrals. However, $\boldsymbol{a}_r$ although it has the same magnitude (namely, unity) its direction changes from element to element. The direction of $\boldsymbol{a}_r$ at M is opposite to that at a diametrically opposite point N. Therefore, the contributions due to these diametrically opposite elements cancel out in the second integral. Considering the entire circular loop as to consist of such diametrically opposite current elements, we then appreciate that the value of the second integral of the right hand side of (a3.2.1) is zero. This simplifies (a3.2.1) to:

(a3.2.2)
$$\boldsymbol{H} = \frac{i\, a\, \boldsymbol{a}_z}{4\pi\, (a^2 + d^2)^{3/2}} \int dl = \frac{i\, a\, \boldsymbol{a}_z\, 2\pi a}{4\pi\, (a^2 + d^2)^{3/2}}$$

noting that $\int dl = 2\pi a$. *Putting* $\alpha = \pi a^2$, *(a3.2.2) becomes identical with (3.2.10) which further directly passes on to (3.2.11), on putting* $d = 0$.

As for the derivation of (3.2.11), one may refer to the analogous problem of a dipole already dealt with in electrostatics. Let the magnetic dipole lie on the z-axis, with its centre located at the origin, and the point (P) where to find the magnetic field intensity be also located on the Z-axis ($z = d$; $d >> l$) *(see Fig. a3.2.2).*

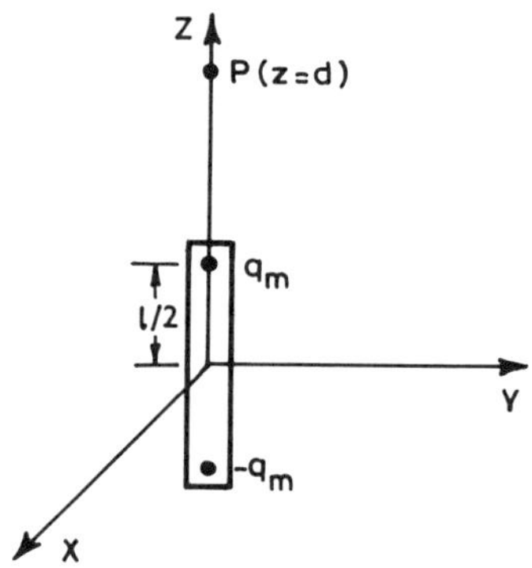

Fig. a3.2.2 Geometry of the problem of finding magnetic field intensity due to a magnetic dipole-equivalent of a circular-current loop.

In an analogous situation of electrostatics we have, with the help of (2.5.13), putting $\theta = 0$ *at the point P,*

$$D\,(=\varepsilon_0 E) = \frac{q\,l}{2\pi d^3}\,\boldsymbol{a}_z\,. \tag{a3.2.3}$$

We then draw the following analogy for the magnetic dipole case (see section 3.1):

$$H = \frac{q_m\,l}{2\pi d^3}\,\boldsymbol{a}_z$$

which is identical with (3.2.12).

Appendix a3.2.2 *Application of Biot-Savart's law to a long solenoid*

Let us first find the magnetic field intensity due to turns lying on the plane $z = 0$ *and occupying an elemental distance range dz, the point P where to find the magnetic field intensity being located at a perpendicular distance z from the plane of the turns.*

There are ndz number of turns located within the range dz, where n is the number of turns per unit length regarded as constant throughout the solenoid. Putting $d = z$ *in (3.2.10), we may write the element of magnetic field intensity* ***dH*** *caused by such turns as follows:*

$$dH = \frac{\alpha\, i\, n\, dz}{2\pi\,(a^2 + z^2)^{3/2}}\,\boldsymbol{a}_z \tag{a3.2.4}$$

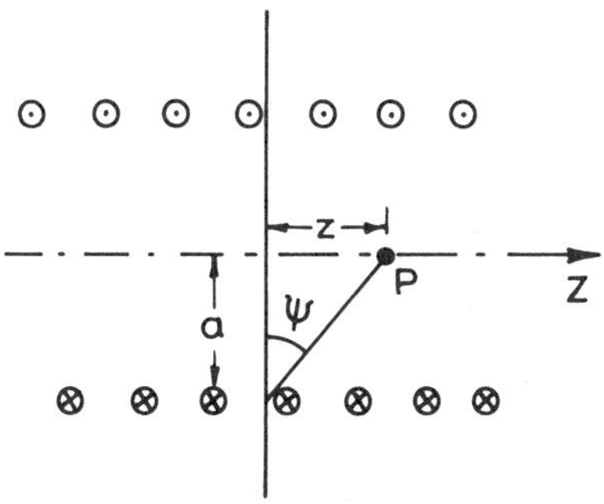

Fig. a3.2.3 Geometry of the problem of finding magnetic field intensity at a point on the axis of a long solenoid.

where a is the radius of the solenoid, and α *(* $=\pi a^2$*) is its area of cross-section. Integration of (a3.2.4) yields:*

$$H = \int_{-\infty}^{\infty} \frac{\alpha \, i \, n \, dz}{2\pi (a^2 + z^2)^{3/2}} \, a_z \, , \tag{a3.2.5}$$

where the integration limits are taken as $\pm\infty$ *taking an infinitely long solenoid. Substitution of*

$$\tan\psi = z/a, \quad \text{whence also } dz = a \sec^2\psi \, d\psi \, ,$$

into (a3.2.5) gives

$$H = \frac{ni}{2}\left(\int_{-\pi/2}^{\pi/2} \cos\psi \, d\psi\right) a_z = \frac{ni}{2}\left[\sin\psi\right]_{-\pi/2}^{\pi/2} a_z = ni \, a_z$$

which is identical with (3.2.13).

Appendix a3.2.3 Application of Ampere's circuital law to a long solenoid

Let us consider a long solenoid with its turns relatively closely packed. We notice that, between the turns of the solenoid, the fields would cancel out, that the field at P outside the solenoid due to the upper part of the solenoid would tend to be canceled out by that due to the lower part of the solenoid, and that the field due to all the turns, contributed by both the upper and lower parts, would add up along the positive z-direction inside the solenoid (Fig. a3.2.4).

Let us first consider a closed rectangular path abcd passing through P, a point arbitrarily considered outside the solenoid (Fig. a3.2.4). We choose the sides da and bc to be essentially at different distances from the axis of the solenoid. From physical considerations (see magnetic flux lines) it may be appreciated that the direction of the magnetic field intensity would be perpendicular to the sides ab and cd. Therefore, the line integral $\int \mathbf{H}.d\mathbf{l}$ *would be each zero along the paths ab and cd.*

$$\int_{ab} \mathbf{H}.d\mathbf{l} = \int_{cd} \mathbf{H}.d\mathbf{l} = 0 \; . \tag{a3.2.6}$$

Now, since the solenoid is considered to be long, it enjoys the symmetry $\partial/\partial z = 0$ *(besides the azimuthal symmetry* $(\partial/\partial\theta = 0)$*). For arguments sake, let us consider that a finite value of* $\mathbf{H}$ *say,* $\mathbf{H}_{01}$ *exists at P outside the solenoid, assuming that the field due to the upper part of the solenoid is not exactly canceled out by that due to its lower part. It can be easily seen that, if it happens so (Fig. a3.2.4), we may choose to write the field intensity at P as*

$$\boldsymbol{H}_{01} = -H_{01}\ \boldsymbol{a}_z$$

which would also be the values at all the points along the path ad. Similarly, let us write the magnetic field intensity $\boldsymbol{H}_{02}$ *at points along bc as*

$$\boldsymbol{H}_{02} = -H_{02}\ \boldsymbol{a}_z\ .$$

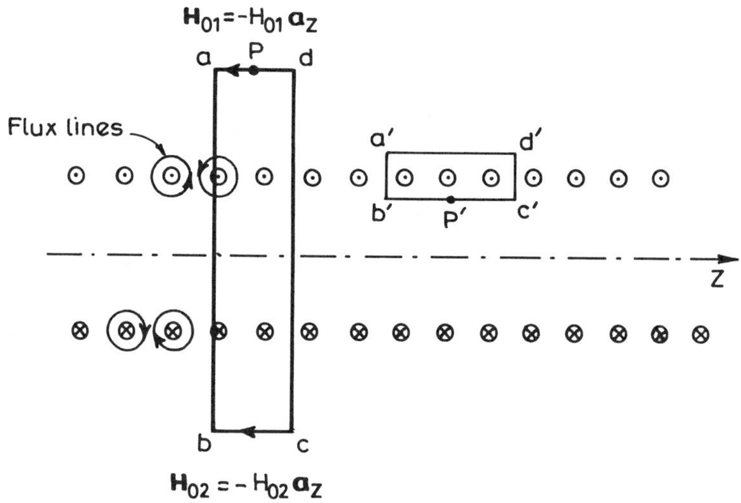

Fig. a3.2.4 Closed paths abcd and a'b'c'd' for the calculation of magnetic field intensity due to a long solenoid using Ampere's circuital law.

Since the paths ad and bc are at different distances from the axis of the solenoid,

(a3.2.7) $$H_{01} \neq H_{02}\ ,\ \text{unless}\ H_{01} = H_{02} = 0\ .$$

The line integrals along the paths da and bc may be then written as

(a3.2.8) $$\int_{da} \boldsymbol{H}.d\boldsymbol{l} = \int_{da} \boldsymbol{H}_{01}.d\boldsymbol{l} = \int_{da} -H_{01}\boldsymbol{a}_z.(-dz\,\boldsymbol{a}_z) = H_{01}\int_{da} dz = H_{01}\, l$$

and

(a3.2.9) $$\int_{bc} \boldsymbol{H}.d\boldsymbol{l} = \int_{bc} \boldsymbol{H}_{02}.d\boldsymbol{l} = \int_{bc} -H_{02}\boldsymbol{a}_z.(dz\,\boldsymbol{a}_z) = -H_{02}\int_{bc} dz = -H_{02}\, l\ ,$$

where l *(= da = bc) is the length of each of the paths bc and da. Using (a3.2.6), (a3.2.8) and (3.2.9), we then have*

(a3.2.10) $$\oint_{abcd} \boldsymbol{H}.d\boldsymbol{l} = (H_{01} - H_{02})\, l\ .$$

Now, it can be seen that the current enclosed by the area abcd is zero, there being as many turns contributing current out of the paper due to the upper part of the solenoid as these are turns contributing current into the paper due to the lower part of the solenoid. This makes the right hand side of Ampere's circuital law (3.2.15), as applied to the closed path abcd, zero and, therefore, enables one to write, with the help of (a3.2.10):

$$(H_{01} - H_{02})\, l = 0 .$$

which yields, since $l \neq 0$,

$$H_{01} = H_{02} . \tag{a3.2.11}$$

In view of (a3.2.7) and (a3.2.11), one may then easily infer that

$$H_{01} = H_{02} = 0 .$$

Thus the magnetic field intensity at any point outside the solenoid is zero. This also agrees with physical expectations since the magnetic flux lines which are closely packed inside a long cylinder is distributed throughout the region outside the solenoid extending upto infinity.

Once we are satisfied in appreciating that the magnetic field intensity outside a long solenoid is zero, it becomes a relatively simpler task for us to find an expression for the magnetic field intensity inside the solenoid again by using Ampere's circuital law. For this purpose, let us now consider a closed rectangular path $a'b'c'd'$, *passing through* P', *a point inside the solenoid where to find the magnetic field intensity, the point* P' *not necessarily lying on the axis of the solenoid, and the sides* $a'b'$ *and* $c'd'$ *being each perpendicular to the axis(Fig. a3.2.4). As in (a3.2.6), we have:*

$$\int_{a'b'} \mathbf{H}.\mathbf{dl} = \int_{c'd'} \mathbf{H}.\mathbf{dl} = 0 \tag{a3.2.12}$$

and, since the magnetic field intensity outside the solenoid is already shown to be zero,

$$\int_{d'a'} \mathbf{H}.\mathbf{dl} = 0 . \tag{a3.2.13}$$

Further, since the magnetic field intensity $\mathbf{H}$ *inside the solenoid, which has been seen to be parallel with the axis along the positive z-direction, may be considered to be constant along the path* $b'c'$, *as the length of the solenoid considered is essentially very large, we may write* $\mathbf{H} = H\,\mathbf{a}_z$, *and*

$$\int_{b'c'} \mathbf{H}.\mathbf{dl} = \int_{b'c'} H\,\mathbf{a}_z.dz\,\mathbf{a}_z = H\int_{b'c'} dz = H\,l' , \tag{a3.2.14}$$

where $l' (= b'c')$ *is the length of the path* $b'c'$. *Combining (a3.2.12) - (a3.2.14),*

$$\oint_{a'b'c'd'} \mathbf{H}.\mathbf{dl} = H\,l' . \tag{a3.2.15}$$

Now, the current enclosed by the area $a'b'c'd'$ *is equal to the number of turns enclosed nl times the current of the solenoid i. This makes the right hand side of Ampere's circuital law (3.2.15), as applied to the closed path* $a'b'c'd'$, *equal to* $(nl')(i)$, *where n is the number of turns per unit length. Also, as applied to the present problem, the left hand side of Ampere's circuital law (3.2.15) is given by (a3.2.15). This enables one to write*

$$H\,l' = n\,l'\,i$$

whence

$$\mathbf{H} = H\,\mathbf{a}_z = n\,i\,\mathbf{a}_z$$

which is identical with (3.2.13).

Appendix a3.2.4 Expansion of curl in different systems of coordinates

*Let us consider a closed path ABCD enclosing an element of area dS_x (= dy dz) (x = constant), surrounding the point P (x, y, z) where we wish to find curl **H**, first in the rectangular system of coordinates (Fig. a3.2.5). The closed area element to start with is considered normal to the x-direction, with x = constant, and may be put as (Fig. a3.2.5):*

(a3.2.16) $$\mathbf{dS}_x = dS_x \, \mathbf{a}_x = dy \; dz \; \mathbf{a}_x \; .$$

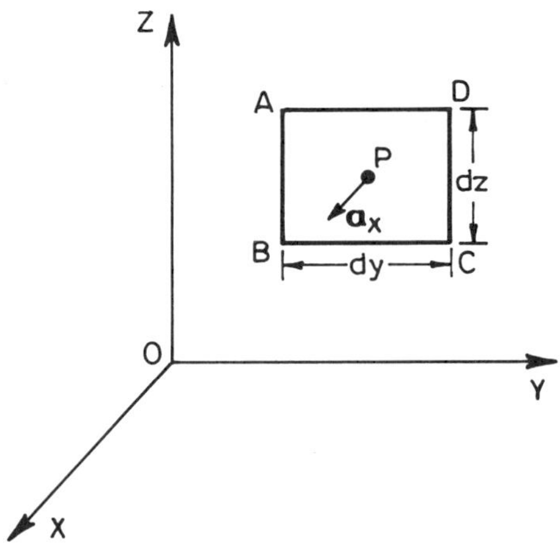

Fig. a3.2.5 A closed path ABCD surrounding the point P where to find a component of curl **H** (= ∇×**H**), in the rectangular system (here, the component being considered typically, along x).

*Let the magnetic field intensity at P be denoted by **H**. The z-component of magnetic field intensity along the path BA is then $H_z - (\partial H_z/\partial y)(dy/2)$ and that along the path CD is $H_z + (\partial H_z/\partial y)(dy/2)$. Similarly, the y-components of magnetic field intensity along the paths BC and AD are $H_y - (\partial H_y/\partial z)(dz/2)$ and $H_y + (\partial H_y/\partial z)(dz/2)$, respectively. The magnetic field intensity component along each of these paths is regarded as "approximately" uniform considering the path length to be extremely small. The line integral of magnetic field intensity along the closed path ABCD enclosing the area element dS (= dy dz) is then approximately obtained as:*

(a3.2.17) $$\oint_{\substack{dS_x \\ (\text{path } ABCD)}} \mathbf{H}.\mathbf{dl} \cong \int_{(AB)} \mathbf{H}.\mathbf{dl} + \int_{(BC)} \mathbf{H}.\mathbf{dl} + \int_{(CD)} \mathbf{H}.\mathbf{dl} + \int_{(DA)} \mathbf{H}.\mathbf{dl}$$

$$\cong \underbrace{\left(H_z - \frac{\partial H_z}{\partial y}\frac{dy}{2} \right) \mathbf{a}_z.(-dz\,\mathbf{a}_z)}_{(AB)} + \underbrace{\left(H_y - \frac{\partial H_y}{\partial z}\frac{dz}{2} \right) \mathbf{a}_z.(dy\,\mathbf{a}_y)}_{(BC)}$$

$$+ \underbrace{\left(H_z + \frac{\partial H_z}{\partial y}\frac{dy}{2} \right) \mathbf{a}_z.(dz\,\mathbf{a}_z)}_{(CD)} + \underbrace{\left(H_y + \frac{\partial H_y}{\partial z}\frac{dz}{2} \right) \mathbf{a}_y.(-dy\,\mathbf{a}_y)}_{(AD)}$$

$$\cong \left(\frac{\partial H_z}{\partial y} - \frac{\partial H_y}{\partial z} \right) dy\, dz$$

whence,

(a3.2.18)
$$\underset{\substack{dy \to 0 \\ dz \to 0 \\ ds_x (= dydz) \to 0}}{Lt} \frac{\oint \boldsymbol{H.dl}}{dS_x} = \frac{\partial H_z}{\partial y} - \frac{\partial H_y}{\partial z} .$$

It may be noted that in (a3.2.17) a sign of approximation has been added since the magnetic field intensity along the paths AB, BC, CD and DA have been considered to be uniform. Obviously, the approximation tends to become exact as the path lengths tend to zero ($dy \to 0$, $dz \to 0$) which would in turn make the closed area element shrink to the point P ($dS_x \to 0$) (Fig. a3.2.5). Thus the left hand side of (a3.2.18), which would now become exact by virtue of $dS_x \to 0$, and refer to the point (P), is defined as $(\text{curl } H)_x$, the x-component of curl ***H****. We may then express (a3.2.18) as:*

(a3.2.19)
$$(\text{curl } \boldsymbol{H})_x = (\nabla \times \boldsymbol{H})_x = \frac{\partial H_z}{\partial y} - \frac{\partial H_y}{\partial z} .$$

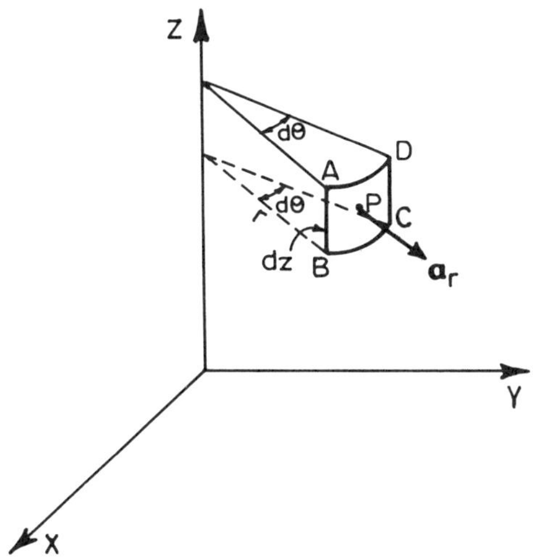

Fig. a3.2.6 A closed path ABCD surrounding the point P where to find a component of curl **H** (= ∇×**H**), in the cylindrical system (here, the component being, typically, the r-component).

Similarly, by considering area elements normal to the y- and z-directions, respectively, we can show the y- and z-components of curl ***H*** *as* (a3.2.20) $$(curl\,\boldsymbol{H})_y = (\nabla\times\boldsymbol{H})_y = \frac{\partial H_x}{\partial z} - \frac{\partial H_z}{\partial x} \quad \text{and}$$

(a3.2.21) $$(curl\,\boldsymbol{H})_z = (\nabla\times\boldsymbol{H})_z = \frac{\partial H_y}{\partial x} - \frac{\partial H_x}{\partial y} \;.$$

*Combining (a3.2.19)-(a3.2.21), the expression of curl **H** at a point is thus obtained as:*

(a3.2.22) $$curl\,\boldsymbol{H} = \nabla\times\boldsymbol{H} = (curl\,\boldsymbol{H})_x\,\boldsymbol{a}_x + (curl\,\boldsymbol{H})_y\,\boldsymbol{a}_y + (curl\,\boldsymbol{H})_z\,\boldsymbol{a}_z$$

$$= \left(\frac{\partial H_z}{\partial y} - \frac{\partial H_y}{\partial z}\right)\boldsymbol{a}_x + \left(\frac{\partial H_x}{\partial z} - \frac{\partial H_z}{\partial x}\right)\boldsymbol{a}_y + \left(\frac{\partial H_y}{\partial x} - \frac{\partial H_x}{\partial y}\right)\boldsymbol{a}_z \;,$$

which is identical with (3.2.18(a)).

*Next, let us proceed to find an expression for curl **H** in the cylindrical system of coordinates. (curl **H**)$_r$, the r-component of curl **H** may be found by considering a closed area element dS_r, with r = constant, the element of area being considered normal to the direction of **a**$_r$ and enclosing the point P where we wish to find curl **H** (Fig. a3.2.6).*

The z-components of magnetic field intensity along BA and CD (Fig. a3.2.6) are $H_z - (\partial H_z/\partial\theta)(d\theta/2)$ and $H_z + (\partial H_z/\partial\theta)(d\theta/2)$, respectively, and the θ-component of magnetic field intensity along BC and AD are $H_\theta - (\partial H_\theta/\partial z)(dz/2)$ and $H_\theta + (\partial H_\theta/\partial z)(dz/2)$, respectively. Hence, one has:

(a3.2.23) $$\underset{\substack{dS_r\\(\text{path } ABCD)}}{\oint} \boldsymbol{H.dl} = \oint_{AB}\boldsymbol{H.dl} + \oint_{BC}\boldsymbol{H.dl} + \oint_{CD}\boldsymbol{H.dl} + \oint_{DA}\boldsymbol{H.dl}$$

$$\cong \underbrace{\left(H_z - \frac{\partial H_z}{\partial\theta}\frac{d\theta}{2}\right)\mathbf{a}_z.(-dz\,\mathbf{a}_z)}_{(AB)} + \underbrace{\left(H_\theta - \frac{\partial H_\theta}{\partial z}\frac{dz}{2}\right)\mathbf{a}_\theta.(r\,d\theta\,\mathbf{a}_\theta)}_{(BC)}$$

$$+ \underbrace{\left(H_z + \frac{\partial H_z}{\partial\theta}\frac{d\theta}{2}\right)\boldsymbol{a}_z.(dz\,\boldsymbol{a}_z)}_{(CD)} + \underbrace{\left(H_\theta + \frac{\partial H_\theta}{\partial z}\frac{dz}{2}\right)\boldsymbol{a}_\theta.(-r d\theta\,\boldsymbol{a}_\theta)}_{(DA)}$$

$$\cong \frac{\partial H_z}{\partial\theta}dz\,d\theta - \frac{\partial H_\theta}{\partial z}r\,d\theta\;dz \;.$$

In obtaining (a3.2.23) we have taken the path lengths AB, BC, CD and DA as dz, rdθ, dz and rdθ, respectively. Also, as in (a3.2.18), a sign of approximation has been used. Dividing (a3.2.23) by dS_r = $dz\;r\;d\theta$ and taking the limit $dS_r \to 0$ to make the relation exact, we may write an equation similar to (a3.2.18) as

(a3.2.24) $$\underset{\substack{dz\to 0\\ d\theta\to 0\\ ds_r=(dz r d\theta)\to 0}}{Lt}\;\frac{\oint_{dS_r} H.dl}{dS_r} = \frac{1}{r}\frac{\partial H_z}{\partial\theta} - \frac{\partial H_\theta}{\partial z}$$

*By definition, the left hand side of (a3.2.24) is (curl **H**)$_r$ = (∇×**H**)$_r$, the r-component or curl **H** = (∇×**H**), expressed in the cylindrical system of coordinates. Then (a3.2.24) may be expressed as* (a3.2.25) $$(\text{curl}\, \boldsymbol{H})_r = (\nabla \times \boldsymbol{H})_r = \frac{1}{r}\frac{\partial H_z}{\partial \theta} - \frac{\partial H_\theta}{\partial z} .$$

Similarly, by considering area elements normal to the θ- and z-directions, one can show

(a3.2.26) $$(\text{curl}\, \boldsymbol{H})_\theta = (\nabla \times \boldsymbol{H})_\theta = \frac{\partial H_r}{\partial z} - \frac{\partial H_z}{\partial r} .$$

(a3.2.27) $$(\text{curl}\, \boldsymbol{H})_z = (\nabla \times \boldsymbol{H})_z = \frac{1}{r}\left(\frac{\partial (r H_\theta)}{\partial r} - \frac{\partial H_r}{\partial \theta}\right) .$$

Combining (a3.2.25)-(a3.2.27), we directly obtain the required expression for curl $\boldsymbol{H} = \nabla\times\boldsymbol{H}$ *as given by (3.2.18(b)), in the cylindrical system of coordinates.*

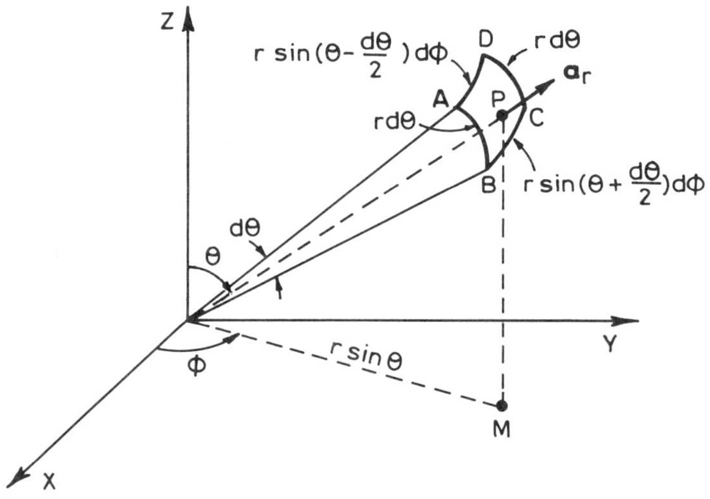

Fig. a3.2.7 A closed path ABCD surrounding the point P where to find a component of curl **H** (= ∇×**H**) in the spherical polar system (here, the component being, typically, the r-component). (M is the foot of the perpendicular drawn from P on the XY-plane).

In order to obtain (3.2.18(c)), the expression for curl $\boldsymbol{H} = \nabla\times \boldsymbol{H}$ *in the spherical-polar system of coordinates, we may proceed in a similar way. Thus in order to obtain the expression for (curl* **H**), *the r-component of* $\nabla\times \boldsymbol{H}$, *one has to consider an element of area perpendicular to the direction of* $\boldsymbol{a}_r$, *around the point P (Fig. a3.2.7).*

The θ - component of magnetic field intensity along AB and DC are $H_\theta - (\partial H_\theta/\partial\phi)(d\phi/2)$ *and* $H_\theta + (\partial H_\theta/\partial\phi)(d\phi/2)$, *respectively, and the components of magnetic field intensity along BC and AD are* $H_\phi - (\partial H_\phi/\partial\theta)(d\theta/2)$ *and* $H_\phi + (\partial H_\phi/\partial\theta)(d\theta/2)$, *respectively. The path lengths AB, BC, CD and DA are* $rd\theta$, $r\sin(\theta + d\theta/2)\, d\phi$, $rd\theta$ *and* $r\sin(\theta - d\theta/2)\, d\phi$, *respectively (Fig. a3.2.7). Accordingly, then one may write:*

(a3.2.28)
$$\oint_{\substack{dS_r \\ \text{(path ABCD)}}} \boldsymbol{H}.\boldsymbol{dl} \cong \oint_{AB} \boldsymbol{H}.\boldsymbol{dl} + \oint_{BC} \boldsymbol{H}.\boldsymbol{dl} + \oint_{CD} \boldsymbol{H}.\boldsymbol{dl} + \oint_{DA} \boldsymbol{H}.\boldsymbol{dl}$$

$$\cong \left(H_\theta - \frac{\partial H_\theta}{\partial \phi}\frac{d\phi}{2}\right)\boldsymbol{a}_\theta .rd\theta\,\boldsymbol{a}_\theta + \left(H_\phi + \frac{\partial H_\phi}{\partial \theta}\frac{d\theta}{2}\right)\boldsymbol{a}_\phi .r(\sin(\theta + \frac{d\theta}{2}))d\phi\,\boldsymbol{a}_\phi$$

$$+ \left(H_\theta + \frac{\partial H_\theta}{\partial \phi}\frac{d\phi}{2}\right)\boldsymbol{a}_\theta .(-rd\theta\,\boldsymbol{a}_\theta) + \left(H_\phi - \frac{\partial H_\phi}{\partial \theta}\frac{d\theta}{2}\right)\boldsymbol{a}_\phi .\left(-r(\sin(\theta - \frac{d\theta}{2}))d\phi\right)\boldsymbol{a}_\phi$$

$$\cong -\frac{\partial H_\theta}{\partial \phi} d\phi\, d\theta\, r + \left(H_\phi + \frac{\partial H_\phi}{\partial \theta}\frac{d\theta}{2}\right) rd\phi\,(\sin\theta\cos\frac{d\theta}{2} + \cos\theta\sin\frac{d\theta}{2})$$

$$-\left(H\phi - \frac{\partial H_\phi}{\partial \theta}\frac{d\theta}{2}\right) rd\phi(\sin\theta\cos\frac{d\theta}{2} - \cos\theta\sin\frac{d\theta}{2})$$

$$\cong -(\partial H_\theta/\partial\phi)\,(d\phi)\,(d\theta)\,(r) + (H_\phi\cos\theta)(d\theta)\,(rd\phi) + (\partial H_\phi/\partial\theta)\,(rd\theta)\,(d\phi)\,(\sin\theta)\,.$$

(using the approximation: $\cos(d\theta/2) \approx 1$, $\sin(d\theta/2) \approx d\theta/2$)

Dividing (a3.2.28) by dS_r ($= rd\theta\, r\sin\theta\, d\phi$) and taking the limit $dS_r \to 0$ to make the relation exact, we may then write an equation similar to (a3.2.18) or (a3.2.24) as:

(a3.2.29)
$$\underset{\substack{d\theta\to 0 \\ d\phi\to 0 \\ ds_r(=rd\theta\, r\sin\theta\, d\phi)\to 0}}{Lt} \frac{\oint_{dS_r} \boldsymbol{H}.\boldsymbol{dl}}{dS_r} = -\frac{1}{r\sin\theta}\frac{\partial H_\theta}{\partial \phi} + \frac{1}{r\sin\theta} H_\phi\cos\theta + \frac{1}{r}\frac{\partial H_\phi}{\partial \theta}$$

$$= \frac{1}{r\sin\theta}\left(\frac{\partial}{\partial\theta}(H_\phi\sin\theta) - \frac{\partial H_\theta}{\partial\phi}\right) .$$

*By definition, the left hand side of (a3.2.29) is (curl **H**)$_r$ =($\nabla\times$ **H**)$_r$, the r-component of curl **H** = $\nabla\times$ **H**, expressed in the spherical-polar system of coordinates. This enables one to write (a3.2.29) as:*

(a3.2.30)
$$(\text{curl}\,\boldsymbol{H})_r = (\nabla\times\boldsymbol{H})_r = \frac{1}{r\sin\theta}\left(\frac{\partial}{\partial\theta}(H_\phi\sin\theta) - \frac{\partial H_\theta}{\partial\phi}\right).$$

Similarly, one may take area elements normal to the θ- and z-directions, respectively, to show:

(a3.2.31)
$$(\text{curl}\,\boldsymbol{H})_\theta = (\nabla\times\boldsymbol{H})_\theta = \frac{1}{r}\left(\frac{1}{\sin\theta}\frac{\partial H_r}{\partial\theta} - \frac{\partial(rH_\phi)}{\partial r}\right) \text{, and}$$

(a3.2.32)
$$(\text{curl}\,\boldsymbol{H})_\phi = (\nabla\times\boldsymbol{H})_\phi = \frac{1}{r}\left(\frac{\partial(r\,H_\theta)}{\partial r} - \frac{\partial H_r}{\partial\theta}\right).$$

Combining (a3.2.30)-(a3.2.32), one then directly obtains the required expression for curl ***H*** *as given by (3.2.18(c)), in the spherical-polar system of coordinates. Now, let us work out a few steps to express curl* ***H*** *=($\nabla\times$* ***H****) as the current density as given by (3.2.17); rather we would proceed to show the expression components-wise and subsequently build (3.2.17) by combining these components. For instance, let us find the r-component of (3.2.17) in the spherical-polar system of coordinates. For this purpose, we may refer back to Fig. a3.2.7. The left hand side of (a3.2.29) may be written with the help of (3.2.15) as:*

$$\underset{\substack{d\theta\to 0\\ d\phi\to 0\\ ds_r(=rd\theta\, r\sin\theta\, d\phi)\to 0}}{Lt}\ \frac{\dfrac{\oint \boldsymbol{H}.\boldsymbol{dl}}{dS_r}}{dS_r} = \underset{ds_r\to 0}{Lt}\ \frac{i}{ds_r}, \qquad \text{(a3.2.33)}$$

where i has to be interpreted as the current enclosed by the closed path ABCD (Fig. a3.2.7) and may be put as:

$$i \cong \boldsymbol{J}.\boldsymbol{a}_r\ dS_r \cong (J_r\ \boldsymbol{a}_r + J_\theta\ \boldsymbol{a}_\theta + J_\phi\ \boldsymbol{a}_\phi).\ \boldsymbol{a}_r\ \ dS_r = J_r\ \ dS_r\ . \qquad \text{(a3.2.34)}$$

The relation (a3.2.34) is approximate in that the current density is considered to be uniform over the element of area ABCD. One may express (a3.2.34) as:

$$\frac{i}{dS_r} \cong J_r\,,$$

and make the relation exact by taking the limit $dS_r \to 0$ as follows:

$$\underset{ds_r\to 0}{Lt}\ \frac{i}{dS_r} = J_r\ . \qquad \text{(a3.2.35)}$$

Making use of (a3.2.35) one then has from (a3.2.33), remembering that the left hand side of (a3.2.33) as stated following (a3.2.29) is equal to (curl ***H****)$_r$ (= ($\nabla\times$* ***H****)$_r$):*

$$(\text{curl}\ \boldsymbol{H})_r = (\nabla\times\boldsymbol{H})_r = J_r\ . \qquad \text{(a3.2.36)}$$

Similarly, one can take area elements perpendicular to the θ- and ϕ-directions in the spherical-polar system of coordinates, and show

$$(\text{curl}\ \boldsymbol{H})_\theta = (\nabla\times\boldsymbol{H})_\theta = J_\theta \quad \text{and} \qquad \text{(a3.2.37)}$$

$$(\text{curl}\ \boldsymbol{H})_\phi = (\nabla\times\boldsymbol{H})_\phi = J_\phi\ . \qquad \text{(a3.2.38)}$$

Combining (a3.2.36) - (a3.2.38), we then have the relation (3.2.17) derived.

It may be noted that we have taken up, for the above derivation, typically, the spherical-polar system of coordinates. We could have as well taken up any other system (for instance, rectangular or cylindrical) and derived the same result, namely, (3.2.17).

Appendix a3.2.5 A mathematical identity for the gradient of 1/r

With the help of (2.5.12(a)),

$$\nabla\left(\frac{1}{r}\right) = \frac{\partial}{\partial x}\left(\frac{1}{r}\right)\boldsymbol{a}_x + \frac{\partial}{\partial y}\left(\frac{1}{r}\right)\boldsymbol{a}_y + \frac{\partial}{\partial z}\left(\frac{1}{r}\right)\boldsymbol{a}_z = -\frac{1}{r^2}\frac{\partial r}{\partial x}\boldsymbol{a}_x - \frac{1}{r^2}\frac{\partial r}{\partial y}\boldsymbol{a}_y - \frac{1}{r^2}\frac{\partial r}{\partial z}\boldsymbol{a}_z\ . \qquad \text{(a3.2.39)}$$

Taking partial derivatives of (3.2.20(b)), one has

$$\frac{\partial r}{\partial x} = \frac{1}{2}((x-x')^2 + (y-y')^2 + (z-z')^2)^{-1/2}\ (2(x-x')) = \frac{1}{r}(x-x'),$$

$$\frac{\partial r}{\partial y} = \frac{1}{r}(y-y') \quad and \quad \frac{\partial r}{\partial z} = \frac{1}{r}(z-z') .$$

which, when substituted in (a3.2.39), gives

$$\nabla\left(\frac{1}{r}\right) = -\frac{1}{r^3}((x-x')\mathbf{a}_x + (y-y')\mathbf{a}_y + (z-z')\mathbf{a}_z) = -\frac{1}{r^3}\mathbf{r},$$

which is identical with (3.2.21).

Appendix a3.2.6 Magnetic field of a circular turn of current.

The element of vector potential ***dA*** *due to a current element, i* ***dl*** *considered arbitrarily on the circular turn, say at S, that is, at $\phi = \phi'$, which is shown in Fig. a3.2.8(a), as to be placed on the XY-plane, is (see 3.2.25):*

(a3.2.40)
$$\mathbf{dA} = \frac{\mu_0 i\, \mathbf{dl}}{4\pi R},$$

where R is the distance of the current element at S from the point P (r, θ, ϕ) where the magnetic flux density is sought. We use here the spherical-polar coordinates and notice that the problem enjoys azimuthal symmetry ($\partial/\partial\phi = 0$). No generality is lost in the result if the x-axis is so chosen that one may take, for P, $\phi = 0$. The projection M of the point P on the XY-plane would then lie on the x-axis itself, the direction of $\mathbf{a}_\phi$ becoming the same as that of $\mathbf{a}_y$ (see Fig. a3.2.8(b)).

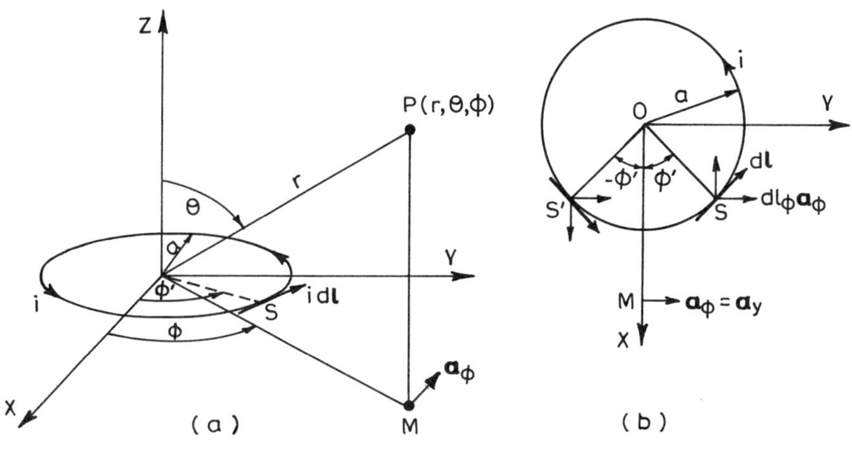

Fig. a3.2.8 Geometry of the problem in Example 3.2.7 **(a)** and the circular turn of current as well as two symmetrically placed current elements on it at s ($\phi = \phi'$) and s′ ($\phi = -\phi'$) shown on the XY-plane **(b)**.

One can express R in terms of the cartesian coordinates of the current element at S, namely, $x = a\cos\phi'$, $y = a\sin\phi'$, $z = 0$, 'a' being the radius of the circular turn and those at P, namely, $x = r\sin\theta\cos\phi = r\sin\theta$, $y = r\sin\theta\sin\phi = 0$, $z = r\cos\theta$ (in view of $\phi = 0$). Thus, (see 2.1.6) one may write,

$$R = ((r\sin\theta - a\cos\phi')^2 + (0 - a\sin\phi')^2 + (r\cos\theta - 0)^2)^{1/2} = (r^2 + a^2 - 2r\,a\sin\theta\cos\phi')^{1/2}$$

$$\cong (r^2 - 2r\,a\sin\theta\cos\phi')^{1/2} \quad (\text{in view of } a^2 << r^2) \quad \cong r\,[1 - 2(a/r)\sin\theta\cos\phi']^{1/2},$$

and substitute the latter in (a3.2.40) to obtain,

$$\text{(a3.2.41)} \qquad \mathbf{dA} = \frac{\mu_0 i\,[1 - 2(a/r)\sin\theta\cos\phi']^{-1/2}\,\mathbf{dl}}{4\pi r} = \frac{\mu_0 i}{4\pi r}\,[1 + (a/r)(\sin\theta)(\cos\phi')]\,\mathbf{dl}\,.$$

(on binomial expansion and ignoring higher powers of $2(a/r)\sin\theta\cos\phi'$, in view of $a << r$ and $\sin\theta\cos\phi' < 1$).

We notice from (a3.2.41), and with reference to Fig. a3.2.8(b), that the contributions to the vector potential from a pair of current elements which are symmetrically placed with respect to $\phi = 0$, say at S ($\phi = \phi'$) and S' ($\phi = -\phi'$), would add up in the y-direction and cancel out in the x-direction. The component of $\mathbf{dA}$ in the y-direction at P can be seen to be the component dA_ϕ, the component in the direction of $\mathbf{a}_\phi$ ($\mathbf{a}_\phi = \mathbf{a}_y$) at M, and hence at P, as indicated in Fig. a3.2.8(b). dA_ϕ with the help of (a3.2.41), may be expressed as:

$$\text{(a3.2.42)} \quad dA_\phi = \frac{\mu_0 i}{4\pi r}\,[1 + (a/r)(\sin\theta)(\cos\phi')]\,dl_\phi = \frac{\mu_0 i}{4\pi r}\,[1 + (a/r)(\sin\theta)(\cos\phi')]\,dl\cos\phi'$$

$$(\text{putting } dl_\phi = dl\cos\phi')$$

$$= \frac{\mu_0 i}{4\pi r}\,[1 + (a/r)(\sin\theta)(\cos\phi')]\,a\,d\phi'\cos\phi'$$

$$(\text{putting } dl = a\,d\phi'),$$

where dl_ϕ is the component of $\mathbf{dl}$ in the y-direction which is the direction of $\mathbf{a}_\phi$ at P. This enables one to express $\mathbf{dA}$ as (3.2.30). A_ϕ, the ϕ-component of $\mathbf{A}$ at P, may then be obtained by integrating (a3.2.42) over the entire circular turn, that is, between $\phi' = 2\pi$ and 0, as follows:

$$\text{(a3.2.43)} \quad A_\phi = \frac{\mu_0 i\,a}{4\pi r}\left(\int_0^{2\pi}\cos\phi'\,d\phi' + \frac{a}{r}\sin\theta\int_0^{2\pi}\cos^2\phi'\,d\phi'\right) = \frac{\mu_0 i\,a}{4\pi r}\left(0 + \frac{a\sin\theta}{r}\pi\right) = \frac{\mu_0 i\pi a^2\sin\theta}{4\pi r^2}.$$

This enables one to express $\mathbf{A}$ as (3.2.31). Then the magnetic flux density $\mathbf{B} = \nabla\times\mathbf{A}$ (see (3.2.27)) is obtained as (with the help of the expression for the curl of a vector quantity in the spherical-polar coordinate, that is, using (3.2.18(c)), taking here $\mathbf{A}$ instead of $\mathbf{H}$):

$$\mathbf{B} = \mu_0 i\,\pi\,a^2\left(\frac{\cos\theta}{2\pi r^3}\mathbf{a}_r + \frac{\sin\theta}{4\pi r^3}\mathbf{a}_\theta\right),$$

an expression which is identical with (3.2.32).

3.3 Lorentz Force

The force on a point charge q, moving with a velocity **v** in a steady magnetic field (or flux density B), called the Lorentz force **F**, is given by [2], [3]:

$$\mathbf{F} = q\,\mathbf{v} \times \mathbf{B}\,. \qquad (3.3.1)$$

It is of interest to compare the Lorentz force given by (3.3.1) with the force due to a static electric field intensity given by (2.1.1). It is worth recalling that (2.1.1) is valid whether the charge is stationary or in the motion. For a moving charge, one may notice that the electrostatic force, and hence the acceleration of the point charge, is independent of the motion of the point charge and may have, in general, a component in the direction of the velocity. Thus it is possible, by electrostatic field, to change the magnitude of velocity and hence the kinetic energy of the point charge. The same is not true with the Lorentz force. It is clear from (3.3.1) that such a force, and hence the acceleration on the point charge caused thereby, is perpendicular to the motion of the point charge. Obviously, then the acceleration caused by the Lorentz force can not alter the magnitude of the velocity of the point charge and hence its kinetic energy. (It is imperative then that for devices based on the principle of conversion of the kinetic energy of charged particles, such as a linear-beam accelerator or a travelling-wave tube, it is the electric field rather than the magnetic field that becomes responsible in the energy-exchange mechanism).

3.3.1 Magnetic Force on a Current-Carrying Conductor

One of the manifestations of the Lorentz force is the force on a conductor carrying a direct current placed in a magnetic field. A case of interest is the one in which the conductor takes the form of a wire. The drifting of charge carriers through the wire constitutes a current. If the wire is placed in a magnetic field, the drifting charge carriers will be subjected to the Lorentz force given by (3.3.1). The force **dF** on a differential current element **idl** on the wire is given by [2], [3]:

$$\mathbf{dF} = \mathbf{idl} \times \mathbf{B}\,. \qquad (3.3.2)$$

(For the derivation of (3.3.2), one may see appendix *a3.3.1*).

Integrating (3.3.2) one may the find the force on the entire length of the conductor as:

$$\mathbf{F} = \int \mathbf{idl} \times \mathbf{B}\,. \qquad (3.3.3)$$

Example 3.3.1 Find the force per unit length on a long straight wire carrying a direct current, exerted by another long wire parallel to the first, also carrying a direct current. Hence also appreciate that "like" currents attract and "unlike" currents repel.

Let the two long straight wires lie on the YZ-plane, be parallel to the z-axis, located at $y = d_1$ and d_2 ($d_2 > d_1$), and carry currents i_1 and i_2 both directed along z (Fig. 3.3.1). The magnetic flux density $\mathbf{B}_{21}$ $(= \mu_0 \mathbf{H}_{21})$ due to the first of these wires at any point on the second one is given with the help of (3.2.5) as:

$$\mathbf{B}_{21} = -\frac{\mu_0 i_1}{2\pi d}\mathbf{a}_x , \qquad (3.3.4)$$

where d $(= d_2 - d_1)$ is the perpendicular distance between the wires. The force $\mathbf{dF}_{21}$ on a current element $i_2\,\mathbf{dl}_2$, considered on the second wire due to the current in the first wire may be written with the help of (3.3.2) as:

$$\mathbf{dF}_{21} = i_2\,\mathbf{dl} \times \mathbf{B}_{21}$$

which, on replacing $\mathbf{dl}_2$ by $dl_2\,\mathbf{a}_z$, may be put as

$$\mathbf{dF}_{21} = i_2\,dl_2\,\mathbf{a}_z \times \mathbf{B}_{21} . \qquad (3.3.5)$$

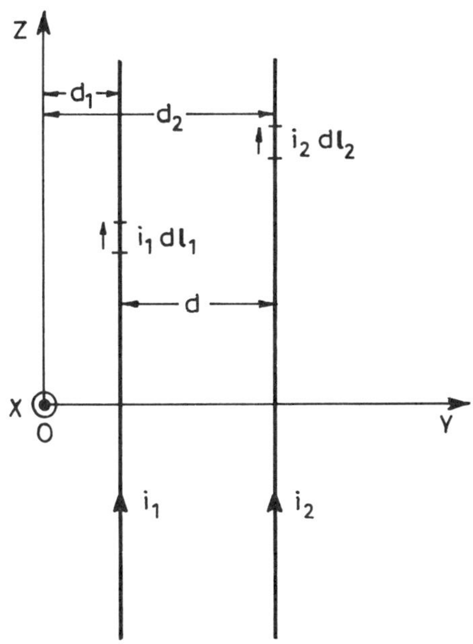

Fig. 3. 3.1 Two long parallel wires carrying direct currents i_1 and i_2 lying on the YZ-plane showing the arbitrary current elements $i_1\,\mathbf{dl}_1$ and $i_2\,\mathbf{dl}_2$, respectively, on them.

Combining (3.3.4) and (3.3.5), one may then express the force per unit length exerted on the second wire, due to current in the first, as:

$$\frac{\mathbf{dF}_{21}}{dl_2} = -\frac{\mu_0 i_1\, i_2\, \mathbf{a}_y}{2\pi d} . \qquad (3.3.6)$$

Similarly, one may write the force per unit length exerted on the first wire, due to current in the second, as:

$$\frac{\mathbf{dF}_{12}}{\mathrm{dl}_1} = -\frac{\mu_0 i_1 i_2 \mathbf{a}_y}{2\pi d}. \tag{3.3.7}$$

We see on examining (3.3.6) and (3.3.7), that $\mathbf{F}_{21}$ and $\mathbf{F}_{12}$ have the same magnitude but are oppositely directed, $\mathbf{F}_{21}$ being in the negative y-direction and $\mathbf{F}_{12}$ in the positive y-direction. We may recall that the currents, which we have taken here, are both along the positive z-direction, constituting "like" currents. Thus we see that "like" currents attract. The currents continue to be "like" and forces exactly given by (3.3.6) and (3.3.7) if the directions of both the currents are reversed. However, if the direction of any one of these currents is reversed, the signs of the right hand sides of both (3.3.6) and (3.3.7) would change thereby rendering $\mathbf{F}_{21}$ be in the positive y-direction while $\mathbf{F}_{12}$ be in the negative y-direction which, in other words, would suggest that "unlike" currents do repel.

Appendix a3.3.1 Magnetic force on a differential current element.

*The force on the current element i**dl** is constituted by sum of the Lorentz forces extorted on the charge carriers present in the element. If α is the cross-sectional area of the current element and dl its length, then the number of charge carriers within the element is n α dl, where n is the number density (that is, number per unit volume), of the charge carriers. The force (**dF**) exerted on these charge carriers each of charge q, say, is equal to the Lorentz force on each charge carrier, given by (3.3.1), multiplied by the number of charge carrier, and which is also equal to the force exerted on the current element. Thus,*

$$\mathbf{dF} = n\,\alpha\,dl\,q\,\mathbf{v}\times\mathbf{B}. \tag{a3.3.1}$$

*We take here all the charge carriers to have the same velocity, namely, **v**.*

It can be shown that the right hand side of (a3.3.1) is related to the current i. For this purpose, let us separately consider a conductor of a uniform unit cross-sectional area and of length numerically equal to v between the faces A and B, say. The number of charge carriers within this length (= v), between the faces A and B, is nv. If the charge carriers are drifted from the face A to B, then all these charge carriers (nv in number) would cross the face B per second. This is because the charges which are present on the face A would transverse the length v, the distance between the faces A and B, in one second, just crossing the face B. All other charges present within the volume between the faces A and B would then cross the face B within the time span of one second. The charge crossing the face B in one second is then obtained by multiplying the charge of each carrier, q, by the number, nv, and constitutes the one-dimensional current density (current per unit area) J. Thus J is given by

$$J = n\,q\,v\ . \tag{a3.3.2}$$

With the help of (a3.3.2), one may express (a3.3.1) as

$$\mathbf{dF} = \alpha\,dl\,\mathbf{J}\times\mathbf{B}, \tag{a3.3.3}$$

***J** and **v** having the same directions. Substituting*

$$\mathbf{J}\,\alpha\,dl = I\,\mathbf{dl},$$

***dl** and **J** having the same direction, in (a3.3.3) we obtain*

$$\mathbf{dF} = i\,\mathbf{dl}\times\mathbf{B},$$

an expression which is identical with (3.3.2).

3.4 Summary

Unlike in electrostatics, free or isolated magnetostatic 'charges' do not exist. This, in other words, means that magnetic flux lines are continuous. However, the force between two magnetic charges (or poles) which belong to two different long magnets, respectively, say, is given, as in electrostatics, by what we may call as Coulomb's law of magnetostatics. Similarly, one may also write, as in electrostatics, Gauss's law and Poisson's equation as appropriate to a steady magnetic field. Thus one finds an analogy between electrostatic and steady-magnetic field problems.

Steady magnetic field may be caused by a direct current. So, in this case, the direct current takes the role of stationary charge in electrostatics. Biot-Savart's law may be used to find steady magnetic field due to a direct current. A number of problems illustrate the use of the law. Biot-Savart's law takes the role of Coulomb's law in this respect.

Ampere's law in steady magnetic field problems plays the same role as does Gauss's law in electrostatics, in working out problems which enjoy geometrical symmetry. The point or differential form of Ampere's law relates the curl of magnetic field intensity to the current density. The expressions for curl in the rectangular, cylindrical and spherical polar system for coordinates are given. The derivation of these expressions is given in an appendix.

Also, the concept of vector potential is developed from first principles and its usefulness illustrated in steady magnetic field problems. Finally, Lorentz force expression on a moving point charge in steady magnetic field is extended to derive an expression for the force on a current-carrying conductor. It is of interest to compare the result that like currents attract and unlike currents repel with an analogous result in electrostatics that like charges repel and unlike charges attract.

The concepts developed in this chapter would be referred to in developing the time-varying field concepts in the following chapter. Also, they will have direct relevance to the problem of confinement of the electron beam as discussed in Chapter 7.

Problems

3.1 A square ($2a\times2a$) turn of wire lying on a plane carries a dc current i. Find the magnetic field intensity **H** at a point lying on the axis normal to the plane and passing through the mid-point of the coil, at a perpendicular distance d.

<u>Answer</u>: $\mathbf{H} = [2\,ia^2/(\pi\,(a^2+d^2)\,(2a^2+d^2)^{1/2}\,)]\mathbf{a}_n$,
where $\mathbf{a}_n$ is the unit vector directed perpendicular to the plane of the turn whose sense is given by the cork-screw rule, that is, in this case, by the linear motion of a screw rotated round the current direction.

3.2 (a) Find the magnetic field intensity at a point (x, y, z) due to a thin infinitely long conducting strip of finite width W lying on the plane y =0 and carrying a current per unit width $\mathbf{J}_S = J_S\mathbf{a}_z$. (b) Study the special case of (a) in which the point where to find the magnetic field intensity is located at a perpendicular distance d from the centre line of the strip. (c) Also appreciate that, as a special case of (b), the result passes on to (3.2.5).

Answer: (a) $\mathbf{H} = (-J_S/2\pi)(\tan^{-1}(x+W/2)/y) - \tan^{-1}((x-W/2)/y)\mathbf{a}_x + (J_S/4\pi)(\ln (y^2+(x+W/2)^2)/(y^2+(x-W/2)^2)$.
(b) $\mathbf{H} = -(J_S/\pi)(\tan^{-1}(W/2d))\mathbf{a}_x$,
(c) $\mathbf{H} = (i/2\pi r)\mathbf{a}_\theta$ (3.2.5), interpreting J_S, $\mathbf{a}_x$ and d as i/W, $-\mathbf{a}_r$ and r, respectively.

3.3 Expressing the right hand side of (3.2.15) in terms of **J** and making use of (3.2.17), verify Stoke's theorem of vector calculus with reference to the magnetic field intensity vector.

3.4 Apply Ampere's circuital law to find the magnitude of magnetic field intensity H inside a solenoid which is bent in the form of a closed circle to make a toroid of circular cross-section of N turns, in terms of i, the current through the toroid and r, the radial distance of the point from the axis of the toroid.

Answer: $H = Ni/(2\pi r)$.

3.5 The inner and the outer radii of a toroid of rectangular cross-section, and of N turns, are a and b, respectively, and its cross-sectional dimension perpendicular to the radial plane is h. Integrate the magnetic flux density over the toroid cross-sectional area and hence find an expression for the inductance L of the toroid.

Answer: $L = (\mu_0/2\pi)\, N^2\, h \ln (b/a)$.

3.6 Obtain an expression for the inductance per unit length L of a coaxial cable in terms of the radius of the central (inner) conductor a and the inner and the outer radii b and c, respectively, of the outer conductor.

Answer: $L = \mu_0/(8\pi) + (\mu_0/(2\pi)) \ln (b/a) + (\mu_0/(2\pi))\,(c^2-b^2)^{-2}\,(c^4 \ln (c/b) - c^2\,(c^2-b^2) + (1/4)\,(c^4-b^4))$.

Hint: Apply Ampere's circuital law (3.2.15) to find the magnetic field intensity, and hence the magnetic flux density, in the three regions: $0 \le r \le a$, $a \le r \le b$ and $b \le r \le c$, respectively. Find the element of flux through an element of a rectangular strip of width dr, lying between r and r+dr on a radial plane, and of unit axial length, in each of these regions. In order to find the magnetic flux and hence the inductance per unit length, one may integrate the elements of flux in the respective regions between the specified limits, that is, between r = 0 and a; r = a and b; and r = b and c; respectively.

3.7 An electron is shot with a dc velocity perpendicular to a uniform dc magnetic field of flux density of magnitude B. Show that the electron executes a circular motion with an angular frequency (cyclotron) $\omega_c = |\eta|\, B$, where η is the charge to mass ratio of an electron.

3.8 An electron of charge e is shot with a dc velocity v_0 in a crossed dc electric and magnetic fields at right angles to both dc electric field intensity of magnitude E_0, and magnetic flux density of magnitude B_0, which themselves are orthogonal to each other. (a) Establish the equations of motion and, (b) subsequently, solve them for the electron trajectory. (c) Also plot the trajectory for six values of v_0, namely, $v_0 = 0$, $0.5\, E_0/B_0$, E_0/B_0, $1.5\, E_0/B_0$, $2\, E_0/B_0$ and $2.5\, E_0/B_0$, respectively.

Answer: Taking the dc magnetic field to be along positive x, the electric field intensity along negative y and the dc velocity v_0 of the electron, at z = 0, in general, along positive z, one has, (a) the equations of motion as:

$$m\frac{d^2x}{dt^2} = 0\ ;\ m\frac{d^2y}{dt^2} = e\left(-E_0 + B_0\frac{dz}{dt}\right);\ \text{and}\ m\frac{d^2z}{dt^2} = -eB_0\frac{dy}{dt}.$$

(b) Subject to the given condition that $dz/dt = v_0$ at t = 0, the solutions to the equations of motion are:

$$x = 0;\ y = -\omega_c^{-1}\,(v_0 - E_0/B_0)\,(1-\cos\omega_c t);\ \text{and}\ z = (v_0 - E_0/B_0)(\sin\omega_c t)/\omega_c + (E_0/B_0)t$$

where ω_c is the electron cyclotron frequency.

(c) The electron trajectory is shown below [8]:

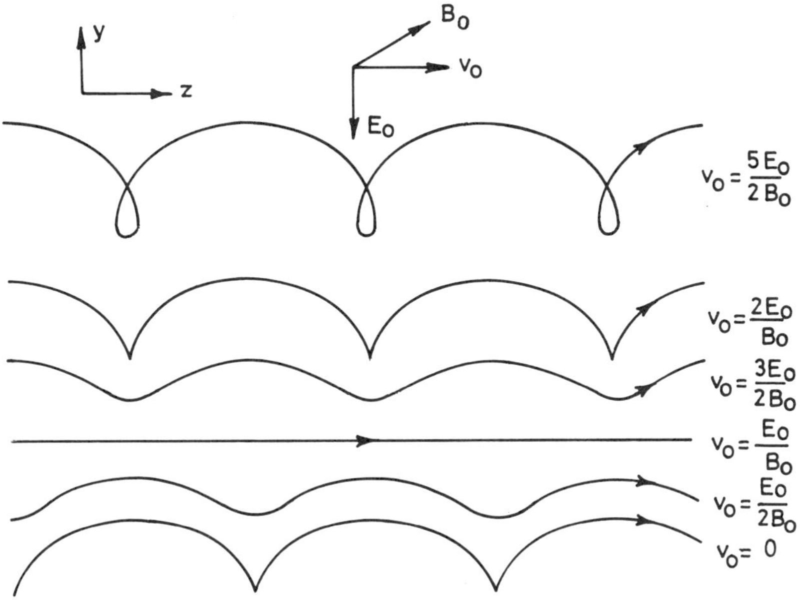

3.9 A current 1mA passes through 1 mm×1 mm cross section of a rectangular block of an n-type semiconductor immersed in a magnetic field of flux density 0.1Tesla perpendicular to the length of the block. The voltage developed across the thickness, perpendicular to both the length of the block and the direction of magnetic field intensity, is 0.3mV. What is the order to the donor density in the semiconductor?

Answer: $2\times10^{21}\ \mathrm{m}^{-3}$.

Hint: In the n-type semiconductor the charges (electrons) are deflected by the Lorentz force and accumulated on one face of the semiconductor. Under the equilibrium, the force on the electrons exerted by the electric field set up within the semiconductor due to deflected charges is balanced by the Lorentz force. The voltage developed across the thickness is the built-in electric field times the thickness.

3.10 With particular reference to a rectangular current loop carrying a current i, which is immersed in a magnetic field of flux density **B**, show that the torque τ developed on the loop is given by:

$$\tau = i\,\alpha \times \mathbf{B},$$

where α is the area vector, the magnitude of which is the product of the length and the breadth of the loop, and the directions of which is given by the cork-screw rule, that is, by the linear motion of a screw which is rotated in the direction of the current.

3.11 A large conducting plate of thickness h and conductivity σ is immersed in a dc magnetic field of flux density B which is directed perpendicular to the plane of the plate. The plate is moved parallel to its plane, with a velocity v. Find the force experienced by the unit area of the plate.

Answer: $B^2\sigma\, hv$.

3.12 A planar transmission line consist of two identical, parallel, thin rectangular ribbons separated by a perpendicular-distance D, each of infinite length and finite width W. If $\pm$ I are the currents parallel to the length of the line, (a) find the force of repulsion per unit length between the two conductors, F, and (b) show that, for close-by ribbons (D << W), and for fixed currents per unit width $\pm J_S$ ($= \pm I/W$) carried by the line, the quantity is proportional to the width of the line.

Answer: (a) $F = (\mu_0 I^2/(\pi W^2))[W \tan^{-1}(W/D) - D \ln(1+(W/D)^2)^{1/2}]$,
(b) $F = (1/2)\,\mu_0 J_S^2\, W$ (for D << W) (proportional to W).

3.13 Two coaxial circular loops of radii a_1 and a_2, carrying azimuthal currents I_1 and I_2, respectively, have their planes separated by a distance d. Find the magnitude of the force F between them.

Answer: $F = (3\pi/2)\,[\mu_0 I_1 I_2 a_1^2 a_2^2 d/(a_1^2 + d^2)^{5/2}]$.

References

[1] H. Semat, *Fundamentals of Physics* (Pitman, New York City, 1951).

[2] M.A. Plonus, *Applied Electromagnetics* (Mc Graw-Hill, New York, 1978).

[3] R. Plonsey and R.E. Collin, *Principles and Applications of Electromagnetic Fields* (Mc Graw-Hill, New York, 1961).

[4] D. Halliday and R. Resnick, *Physics, Pt. II* (John Wiley, New York, 1966).

[5] W.H. Hayt, Jr., *Engineering Electromagnetics* (Mc Graw-Hill, New York, 1989).

[6] S. Ramo, J.R. Whinnery, and T.V. Duzer, *Fields and Waves in Communication Electronics* (John Wiley, New York, 1965).

[7] L. Solymar, *Lectures on Electromagnetic Theory: A Short Course for Engineers* (Oxford Univ. Press, London, 1976).

[8] J.F. Gittins, *Power Travelling-Wave Tubes* (American Elsevier, New York, 1965).

CHAPTER 4

TIME-VARYING ELECTRIC AND MAGNETIC FIELDS

4.0 Introduction

In the preceding two chapters, our concern had been, respectively, with static-electric field and static-magnetic field or steady-current phenomena. In this chapter let us take up situations in which electric and magnetic fields vary with time.

4.1 Continuity Equation

Let us start with the problem of flow of charges. We remarked earlier (see Example 2.2.2) that an electron beam may be considered as a charge fluid; and that, as long as the beam current remains constant, a cross section of the beam at a particular position is indistinguishable from the cross section at the same position at a different instant of time, or at a different position on the axis of the beam at the same instant of time, and as such one can treat such an electron beam essentially in the realm of electrostatics (Example 2.2.2). For such a dc beam-current case, whether the beam is composed of electrons or any other charge species irrespective, if we consider an elementary volume enclosing the beam, the net flux of current density **J** through the surface of the enclosure becomes zero. Mathematically, this may be stated in terms of the following closed surface integral:

$$\oint \mathbf{J}.\mathbf{a}_n \, dS = 0 , \qquad (4.1.1)$$

where dS is an element of area on the enclosing surface and $\mathbf{a}_n$ is the unit vector drawn outwardly, normal to the element of area dS. One may also interpret (4.1.1) by stating that the current or time rate of flow of charge into the enclosure is same as that out of the enclosure and, consequently, that the volume charge density of the beam within the enclosure does not change with time.

Now, let us extend the above case to the time-varying phenomena in which the volume charge density of the beam does change with time (say, due to modulation of the beam in the presence of a time-varying electromagnetic field, see section *8.2*). Correspondingly, let the charge density inside the enclosure, considered to have an element of volume $\Delta\tau$, say, decrease with time due to a positive charge diverging or fluxing out of the element of volume. Thus, in this situation, the current or time rate of charge flow given by the right hand side of (4.1.1) has essentially a non-zero value given by:

$$\oint \mathbf{J}.\mathbf{a}_n \, dS = -\frac{dQ}{dt}, \tag{4.1.2}$$

where Q is the charge within the volume element $\Delta\tau$ at a given instant of time. One may see that, for a positive charge flowing out of the volume, the charge within the volume decreases, and this will make dQ/dt have a negative value. Hence, with a negative sign the right hand side of (4.1.2) becomes positive. This will also make in this case the left hand side of (4.1.2), i.e., the flux of current density **J** through the surface enclosing the volume a positive quantity. Expressing Q in terms of volume charge density ρ as

$$Q = \int_{\Delta\tau} \rho \, d\tau, \tag{4.1.3}$$

and substituting (4.1.3) in (4.1.2), one has

$$\oint \mathbf{J}.\mathbf{a}_n \, dS = -\frac{d}{dt}\int_{\Delta\tau} \rho \, d\tau. \tag{4.1.4}$$

Here, one may mention that though ρ is a function of the space and time coordinates Q is a function only of time; this will be clear from (4.1.3) wherefrom it can be seen that the 'definite' volume integral would make Q independent of the space coordinates. If the surface enclosing the charge is chosen to be constant, and since the integral is convergent, one may replace the complete derivative in the right hand side of (4.1.4) by the partial derivative and put it under the definite integral. This enables one to express (4.1.4) as

$$\oint \mathbf{J}.\mathbf{a}_n \, dS = -\int_{\Delta\tau} \frac{\partial\rho}{\partial\tau} d\tau. \tag{4.1.5}$$

Dividing (4.1.5) by $\Delta\tau$ and taking the limit $\Delta\tau \to 0$ one may write

$$\underset{\Delta\tau\to 0}{\mathrm{Lt}} \frac{\mathbf{J}.\mathbf{a}_n \, dS}{\Delta\tau} = -\underset{\Delta\tau\to 0}{\mathrm{Lt}} \frac{\int_{\Delta\tau} \frac{\partial\rho}{\partial t} d\tau}{\Delta\tau}. \tag{4.1.6}$$

When the enclosed element of volume $\Delta\tau$ is very small ($\Delta\tau \to 0$), the volume charge density and its partial derivative may be considered to be uniform inside the volume $\Delta\tau$. This makes

$$\int_{\Delta\tau} \frac{\partial \rho}{\partial t} d\tau \cong \frac{\partial \rho}{\partial t} \int_{\Delta\tau} d\tau = \frac{\partial \rho}{\partial t} \Delta\tau ,$$

which when substituted in (4.1.6), yields

$$\underset{\Delta\tau \to 0}{\mathrm{Lt}} \frac{\mathbf{J.a}_n dS}{\Delta\tau} = -\frac{\partial \rho}{\partial t} . \tag{4.1.7}$$

But, by definition, the left hand side of (4.1.7) is simply the divergence of **J** (= **∇.J**) (as defined following (2.3.2)), so that (4.1.7) may be written as

$$\nabla.\mathbf{J}_c = -\frac{\partial \rho}{\partial t} . \tag{4.1.8}$$

(4.1.8) is known as the continuity equation, and is extensively used in time-varying field phenomena. The subscript c has been used here to indicate that the current density **J** may be contributed by either the conduction or convection phenomenon of charge transport.

Example 4.1.1 Verify the continuity equation (4.1.8) starting from (4.1.2), while making use of Gauss's law (2.2.1), instead of (4.1.3), and Poisson's equation (2.3.3).

If one chooses to express Q by Gauss's law (2.2.1), instead of (4.1.3), and goes through exactly the same procedural steps as done in arriving at (4.1.7) starting from (4.1.2), one easily obtains the following equation instead of (4.1.7).

$$\underset{\Delta\tau \to 0}{\mathrm{Lt}} \frac{\mathbf{J.a}_n \, dS}{\Delta\tau} = - \underset{\Delta\tau \to 0}{\mathrm{Lt}} \frac{\int \frac{\partial \mathbf{D}}{\partial t} . \mathbf{a}_n \, dS}{\Delta\tau}$$

which, by the definition of divergence (see the discussion following (2.3.2)), takes the following form:

$$\nabla.\mathbf{J} = -\nabla.\frac{\partial \mathbf{D}}{\partial t} . \tag{4.1.9}$$

Now, since the divergence essentially involves partial derivatives, one may choose to interchange the sequence of the divergence and the partial time-derivative in (4.1.9) and write

$$\nabla.\mathbf{J} = -\frac{\partial}{\partial t} \nabla.\mathbf{D} . \tag{4.1.10}$$

Substituting Poisson's equation (2.3.3) in (4.1.10), one then obtains

$$\nabla.\mathbf{J} = -\frac{\partial \rho}{\partial t} \quad \text{(continuity equation)} . \tag{4.1.11}$$

Interpreting **J** as $\mathbf{J}_c$, (4.1.11) is found to be identical with (4.1.8).

<u>*Example 4.1.2*</u> Show that an excess charge placed in the interior of a conductor would die out with a time constant that is inversely proportional to the conductivity of the conductor. As a corollary to this, also appreciate that the electric field inside a conductor is zero.

Let us proceed to find at what rate would the volume charge density at a point inside the conductor change due to the placement of an excess charge [1]-[3]. For this purpose let us substitute

$$\mathbf{J}_c = \sigma \mathbf{E} \quad \text{(Ohm's law)}$$

into (4.1.8) and make use of the relation

$$\nabla.\mathbf{E} = \frac{\rho}{\varepsilon_0}$$

(Poisson's equation; (2.3.3) read with the help of the relation $\mathbf{D} = \varepsilon_0 \mathbf{E}$)

to obtain

$$\frac{\partial \rho}{\partial t} = -\frac{\sigma \rho}{\varepsilon_0} \quad . \tag{4.1.12}$$

Integrating (4.1.12), we obtain

$$\int \frac{\partial \rho}{\rho} = -\int \frac{\sigma}{\varepsilon_0} dt \; ,$$

that is,

$$\ln \rho = -\frac{\sigma}{\varepsilon_0} t + \ln \rho_0 \quad . \tag{4.1.13}$$

Here, we choose to write the integration constant as $\ln \rho_0$. Then (4.1.13) yields

$$\rho = \rho_0 \exp(-t / \tau), \tag{4.1.14}$$

where,

$$\tau = \frac{\varepsilon_0}{\sigma} \tag{4.1.15}$$

is the time constant, called the relaxation time, with which the charge density decays (according to (4.1.14)). τ is inversely proportional to σ as can be seen from (4.1.15).

For numerical appreciation, if we put $\sigma = 5.7 \times 10^7$ mho m^{-1} corresponding to copper, we get $\tau = 1.55 \times 10^{-19}$ s. This calculation suggests that, within a time which is negligible, for a good conductor, the excess charge in the interior of the conductor will disappear (only to reappear on the outer surface of the conductor).

Thus, as a corollary to the above finding, considering a Gaussian surface as close as possible to the outer surface, but still keeping it within the conductor, and applying Gauss's law (2.2.1), one can easily show that the electric field intensity in the interior of the conductor is zero.

4.2 Time-Varying Electric Fields

Let us take a capacitor, and for that matter a simple parallel-plate capacitor [1] (see section *2.7*), and consider it to be filled with a dielectric that does not conduct. It is our experience that, when such a capacitor is connected to an *ac* source of *voltage*, an *ac* current is recorded in the external circuit and, further, that there is no such current, in the steady state, if the source is a *dc* instead of an *ac* one. This suggests that, with the *ac* source of voltage in the external circuit, there is a time-varying electric field established in the dielectric in a form called the *displacement* current. The charge Q{t} on the capacitor at any instant of time is related to instantaneous voltage V{t} across the parallel-plate capacitor as (see (2.7.4)):

$$Q\{t\} = \frac{\varepsilon S}{d} V\{t\} , \tag{4.2.1}$$

where S is the area of each plate and d is the distance between the plates of the capacitor. Differentiation (4.2.1) with respect to time t, and using (2.7.2), one obtains

$$J_d = \frac{i_d}{S} = \frac{dD}{dt} , \tag{4.2.2}$$

where J_d is the magnitude of current per unit area or current density and i_d is the current through the capacitor, in the presence of time-varying electric displacement D (=ε E). The region between the plates being filled up with a dielectric, ε_0 has been replaced by ε, the permittivity of the dielectric, while using (2.7.4). The subscript d stands for 'displacement', the displacement current density J_d being obtained simply as the time differentiation of the quantity, electric displacement D.

Example 4.2.1 A parallel-plate capacitor is filled with a *lossy* dielectric having a finite conductivity. Obtain a phase relation between the voltage across and current through it, and hence power loss in the capacitor.

The current density in the region between the plates is given by

$$J = \sigma_d E + \varepsilon \frac{dE}{dt} , \tag{4.2.3}$$

where the first term, according to Ohm's law, is the conduction current density, σ_d being the conductivity of the dielectric; the second term is the displacement current density given by (4.2.2) (interpreting D as εE). Considering the time dependence of E as exp jωt, one may express (4.2.3) as

$$J = \sigma_d E + j\omega\varepsilon E = j\omega(\varepsilon' - j\varepsilon'')E = j\omega\varepsilon_{complex} E , \tag{4.2.4}$$

where $$\varepsilon_{complex} = \varepsilon' - j\varepsilon'' , \quad \varepsilon' = \varepsilon \text{ and } \varepsilon'' = \sigma_d/\omega . \tag{4.2.5}$$

In the circuit-theoretical description, the current I through the capacitor in terms of the voltage V across the capacitor is obtained by multiplying (4.2.4) by S, and using (2.7.2), as follows (Fig. a4.2.1):

$$I = \frac{V}{R} + \frac{V}{1/j\omega C} = V\left(\frac{1}{R} + \frac{1}{1/j\omega C}\right) \tag{4.2.6}$$

where $$C = \varepsilon' S/d \text{ and } R = d/(\omega\varepsilon'' S). \tag{4.2.7}$$

Equation (4.2.4) suggests that the capacitor filled with a lossy dielectric is essentially a *leaky* capacitor which may be considered as equivalent to an *ideal* capacitor of capacitance C in parallel with the resistance R. For $\sigma_d = 0$, the value of R is infinity (see (4.2.7) through (4.2.5)); or, in other words, we are left with the ideal capacitor alone, the first term of (4.2.6) being zero. In this case, I leads V by a phase angle of $\pi/2$. However, for $\sigma_d \neq 0$, R has a finite value (see (4.2.7) through (4.2.5)), and then it also follows from (4.2.6) that I fails to lead V by the stipulated angle $\pi/2$, and that the angle δ it "falls short" by is given by a quantity called the "loss-tangent" as follows (see appendix *a 4.2.1*):

$$\tan\delta = \frac{1}{\omega CR} = \frac{\varepsilon''}{\varepsilon'} = \frac{\sigma_d}{\omega\varepsilon} . \tag{4.2.8}$$

The "loss-tangent" is a measure of power loss P_L in the capacitor, which may be shown to be (see appendix *a 4.2.1*):

$$P_L = V' I' \sin\delta , \tag{4.2.9}$$

where the primed quantities represent the effective values. One may note that, for $\sigma_d = 0$, $\delta = 0$ (see (4.2.8)), P_L, as can be seen from (4.2.8), has a zero value.

In the above, the concept of displacement current density has been developed with particular reference to a parallel-plate capacitor. A more general, field-analytical approach would be to start from the continuity equation (4.1.8) [2] which is re-written here as:

$$\nabla.\mathbf{J}_c + \frac{\partial\rho}{\partial t} = 0 . \tag{4.2.10}$$

With the help of (2.3.3), one may express (4.2.10) as:

$$\nabla.(\mathbf{J}_c + \frac{\partial \mathbf{D}}{\partial t}) = 0 \tag{4.2.11}$$

Since the divergence of the curl of a vector is zero, one may write, for the vector field **H**,

$$\nabla.(\nabla \times \mathbf{H}) = 0 . \tag{4.2.12}$$

Now, at this stage, let us recall the point form of Ampere's circuital law, namely, (3.2.17), and substitute it in (4.2.12). This yields

$$\nabla.\mathbf{J} = 0 . \tag{4.2.13}$$

The interpretation of **J**, in the present time-varying field case, then immediately follows, by comparing (4.2.13) with (4.2.11), as:

$$\mathbf{J} = \mathbf{J}_c + \frac{\partial \mathbf{D}}{\partial t} . \qquad (4.2.14)$$

The first term of the right hand sides of (4.2.14), as discussed following (4.1.8), is the conduction or convection current density according as the flow of charges is through a conductor or free space. The second term of the right hand side of (4.2.14) is identified as the displacement current density $\mathbf{J}_d$ (see (4.2.2) for such a current, for the particular case of a parallel-plate capacitor):

$$\mathbf{J}_d = \frac{\partial \mathbf{D}}{\partial t} . \qquad (4.2.15)$$

Thus, as for the displacement current density, (4.2.15) should be taken as a more general expression than (4.2.2). Further, for the time-varying phenomena, Ampere's circuital law in the differential or point form (3.2.17) takes the following form with the interpretation of $\mathbf{J}$ as given by (4.2.14):

$$\nabla \times \mathbf{H} = \mathbf{J}_c + \frac{\partial \mathbf{D}}{\partial t} . \qquad (4.2.16)$$

Also, at this stage, it is of interest to go a step backward to obtain the integral form of Ampere's circuital law (3.2.15), for the time-varying field situation, as:

$$\oint \mathbf{H}.d\mathbf{l} = \oint (\mathbf{J}_c + \frac{\partial \mathbf{D}}{\partial t})\mathbf{a}_n.dS . \qquad (4.2.17)$$

At this stage, it is worth recalling how the differential or point form of Ampere's circuital law has been obtained starting from its integral form (see Section *3.2.3*). (4.2.16) or (4.2.17) further suggests that, even in the absence of the conduction or convection current density ($\mathbf{J}_c = 0$), a magnetic field is caused by a time-varying electric field ($\partial \mathbf{D}/\partial t \neq 0$).

Appendix a 4.2.1 *Phasor diagram for a leaky capacitor*

With reference to a leaky capacitor, (4.2.6) may be used to draw a phasor diagram, Fig. a 4.2.1, which shows the angle δ by which the current I falls short of $\pi/2$ in leading the voltage V. Here, V_0 and I_0 represent the peak values of V and I, respectively, which may be expressed as

(a 4.2.1) $$V = V_0 \sin \omega t$$

(a 4.2.2) $$I = I_0 \sin (\omega t + \phi), \quad \text{where}$$

(a 4.2.3) $$\phi = \pi/2 - \delta .$$

The angle δ is given by (see Fig. a 4.2.1):

(a4.2.4) $$\tan \delta = \frac{V_0 / R}{\omega C V_0} = \frac{1}{\omega CR} .$$

Substituting (4.2.7) in (a 4.2.4), one has

(a4.2.5) $$\tan\delta = \frac{\varepsilon''}{\varepsilon'} .$$

Again substituting (4.2.5) in (a 4.2.5), one obtains

(a4.2.6) $$\tan\delta = \frac{\sigma_d}{\omega\varepsilon} .$$

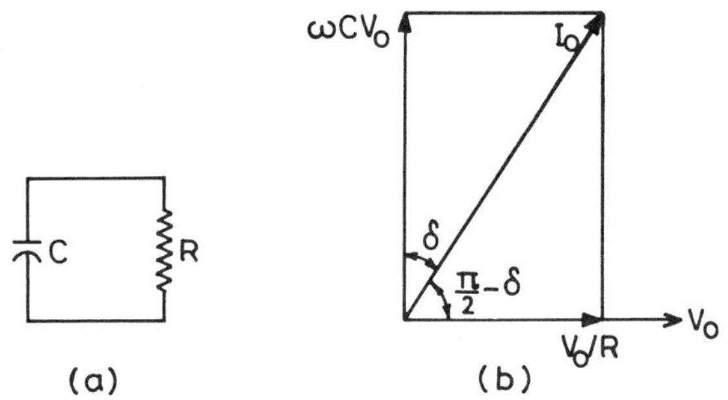

Fig. a 4.2.1 Equivalent circuit of a leaky capacitor **(a)** and phasor diagram to represent the voltage across, and current through, the capacitor **(b)** .

Expressions (a4.2.4)-(a4.2.6) are all put together and expressed in the text as (4.2.8).
Averaging the instantaneous power, VI, over a period, one obtains

(a4.2.7) $$P_L = \frac{\int_0^{2\pi} VI \, d(\omega t)}{\int_0^{2\pi} d(\omega t)} .$$

Substituting V from (a4.2.1) and I from (a4.2.2) into (a4.2.7), and carrying out the integrations, one easily obtains

(a4.2.8) $$P_L = \frac{1}{2} V_0 \, I_0 \, \cos\phi .$$

Substituting ϕ from (a 4.2.3) into (a 4.2.8), P_L may be expressed as:

$$P_L = V'_0 \, I'_0 \, \sin\delta$$

which is found to identical with (4.2.9). (Here, the primes indicate the effective values).

4.3 Time-Varying Magnetic Fields

It is an experimental fact that an emf (electromotive force) is induced in a circuit which is linked with a magnetic flux that changes with time. A quantitative estimate of the emf due to what is known as electromagnetic induction follows from Faraday's law [4], [5].

Faraday's law can be demonstrated with particular reference to a close-circuit setup of Fig. 4.3.1. It consists of a linear part A_1A_2, parallel to x-direction, that can freely slide on a pair of straight rails running parallel to the z-direction, the whole circuit lying on the XZ plane immersed in a magnetic field of flux density **B** which is uniform in space and time and pointed towards positive y direction. Let the current in the circuit be i due to an applied emf $\mathcal{E}$; the direction of the current is the positive x direction through the portion of length l between the pair of rails.

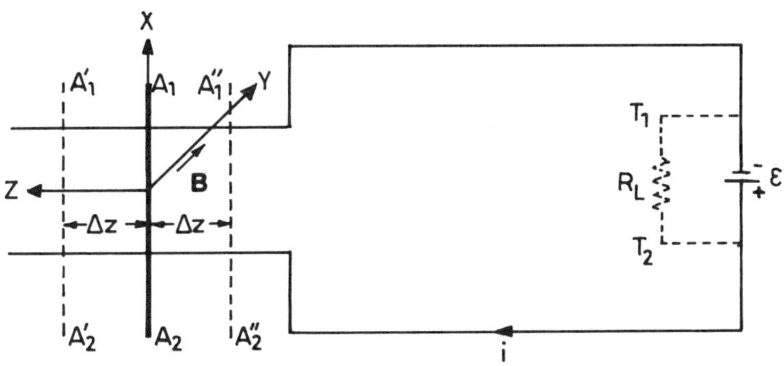

Fig. 4.3.1 A closed-circuit set-up to establish Faraday's law. (The source of emf $\mathcal{E}$ has to be replaced by a load resistance R_L, shown between T_1T_2, with reference to Example 4.3.2).

The part A_1A_2 experiences a force **F**, which according to (3.3.3), is

$$\mathbf{F} = i\,l\,\mathbf{a}_x \times B\mathbf{a}_y = i\,l\,B\mathbf{a}_z\,. \tag{4.3.1}$$

The direction of **F**, as can be seen from (4.3.1), is the positive z-direction, and, as a result, A_1A_2 moves in the positive z direction, thereby increasing the area of the circuit. Let the part A_1A_2 move by in an infinitesimal distance Δz in an infinitesimal time Δt, to its new position $A'_1A'_2$. Now, the work done by the source of emf, in time Δt, in maintaining the current i, is $\mathcal{E}\,i\,\Delta t$. A part of this, $i^2R\,\Delta t$, is used to overcome the resistance, say, R of the circuit and is lost as the *ohmic loss* of the circuit, while the remaining part is spent in doing the work, $\mathbf{F}.\mathbf{a}_z\,\Delta z$, in moving A_1A_2 to $A'_1A'_2$. Hence, by the principle of conservation of energy, one may write

$$\mathcal{E}\,i\,\Delta t = i^2R\,\Delta t + \mathbf{F}.\mathbf{a}_z\,\Delta z\,,$$

which, with help of (4.3.1), takes the form:

$$\mathcal{E}\, i\, \Delta t = i^2 R\, \Delta t + i\, l\, B\, \Delta z\ . \tag{4.3.2}$$

Dividing (4.3.2) by Δt and rearranging terms, one now obtains:

$$i = \frac{\mathcal{E} - l B \dfrac{\Delta z}{\Delta t}}{R}\ . \tag{4.3.3}$$

But it is of interest to note here that when A_1A_2 moves to $A'_1A'_2$, the area of the circuit changes by the elements $\Delta S = l\,\Delta z$, and as a result, the flux linked with the circuit changes by

$$\Delta\phi_B = \mathbf{B}.\mathbf{a}_n\, \Delta S = B\, \mathbf{a}_y.\mathbf{a}_y\, dS = B\, l\, \Delta z\ , \tag{4.3.4}$$

where $\mathbf{a}_n$ ($= \mathbf{a}_y$) is the unit vector normal to the elemental area.
Dividing (4.3.4) by Δt one has

$$\frac{\Delta\phi_B}{\Delta t} = B l \frac{\Delta z}{\Delta t}\ ,$$

which when substituted in (4.3.3), with infinitesimal time $\Delta t \to 0$, gives the instantaneous circuit current i as

$$i = \frac{\mathcal{E} - Blv}{R} = \frac{\mathcal{E} - \dfrac{d\phi_B}{dt}}{R}\ , \tag{4.3.5}$$

where v (= dz/dt) is the magnitude of the velocity of A_1A_2. We notice that, in this set up, left to electromagnetic action alone, the area of the circuit linked with magnetic flux density B increases, and hence ϕ_B increases with time. Here, in this typical set up, $d\phi_B/dt$ has to be interpreted as the time rate of increment of magnetic flux.

It is clear from (4.3.5) that the circuit current is controlled by two terms in the numerator of the right hand side of (4.3.5); the first term is the applied emf $\mathcal{E}$, and the second term is $-d\phi_B/dt$, the induced emf $\mathcal{E}_{ind}$:

$$\mathcal{E}_{ind} = -\frac{d\phi_B}{dt}\ . \tag{4.3.6}$$

In this typical setup, the induced emf $\mathcal{E}_{ind}$ (= $-\, d\phi_B/dt = -\, Blv$) is a negative quantity. The induced emf thus sets up an induced current i_{ind} opposite to the current due to the applied emf $\mathcal{E}$. The direction of i_{ind} is then the negative x direction in A_1A_2.

One may now attribute i_{ind} to the flow of charges caused by an induced electric field **E** exerting a force on these charges. If we choose to move a test charge q, say, around the closed path coinciding with the circuit, then the work done W on the charge will be the closed line integral (see(2.5.1)):

$$W = \oint \mathbf{F}.\mathbf{dl}\ , \tag{4.3.7}$$

where **F** is the force on the test charge exterted by the induced electric field. Substituting **F** from (2.1.1) in (4.3.7),

$$W = q\oint \mathbf{E.dl} \,. \tag{4.3.8}$$

It is also easy to appreciate that (see (2.5.2)):

$$W = q\ \mathcal{E}_{ind} \,. \tag{4.3.9}$$

Therefore, combining (4.3.8) and (4.3.9),

$$\mathcal{E}_{ind} = \oint \mathbf{E.dl} \,. \tag{4.3.10}$$

Now, a very useful form of the law of electromagnetic induction, known as Faraday's law, is obtained by combining (4.3.9) with (4.3.10):

$$\oint \mathbf{E.dl} = -\frac{d\phi_B}{dt} \qquad \text{(Faraday's law).} \tag{4.3.11}$$

Expressing ϕ_B as the close surface integral of magnetic flux density **B** over the area enclosed by the path, one may write Faraday's law (4.3.11) in the following form:

$$\oint \mathbf{E.dl} = -\frac{d}{dt}\oint \mathbf{B.a}_n dS \quad \text{(Faraday's law).} \tag{4.3.12}$$

It may be mentioned here that one may choose to cause a time variation of magnetic flux linked with the closed path in two ways: i)by a time-varying path, and more precisely, area enclosed by the path, while keeping the magnetic field constant, as has been done in the typical hypothetical setup considered above to establish (4.3.11) or (4.3.12) and ii)by a time-varying magnetic field linked with the closed path, the area enclosed by the path remaining unchanged with time, as may be the case in numerous practical situations of interest in electromagnetics. In the latter case, the complete time-derivative (4.3.12) may be replaced by the partial derivative and put under the convergent, definite integral (see discussion following (4.1.4)), enabling one to write another form of Faraday's law as:

$$\oint \mathbf{E.dl} = -\oint \frac{\partial \mathbf{B}}{\partial t}\mathbf{.a}_n dS \quad \text{(Faraday's law).} \tag{4.3.13}$$

4.3.1 Differential Form of Faraday's Law

If we compare Faraday's law, given by (4.3.13), with Ampere's circuital law, given by (3.2.15), we will find a similarly in their forms. In fact, if we replace **H** by **E**, and also **J** by $-\partial\mathbf{B}/\partial t$ in the latter, we shall immediately get the former. It then follows that, by following the same procedure as given in section *3.2.3*, one should arrive at the differential or point form of Faraday's law from its integral form. Thus if we now again replace **H** by **E**, and **J** by - $\partial\mathbf{B}/\partial t$, in (3.2.17), we shall have the following differential or point form:

$$\nabla \times \mathbf{E} = -\frac{\partial \mathbf{B}}{\partial t} \quad \text{(Faraday's law)}. \tag{4.3.14}$$

Thus, Faraday's law, (4.3.13) or (4.3.14), gives a quantitative estimate of the electric field caused by a time-varying magnetic field.

Simultaneously with Faraday's law is named another law called Lenz's law according to which the direction of the current due to electromagnetic induction will be such that the effect of this induced current opposes the cause to which it is due. The minus sign in Faraday's law in various forms: (4.3.6) and (4.3.11)-(4.3.14), in fact, certifies to Lenz's law. This point will be further clear referring back to the setup of Fig. 4.3.1 (and to a few examples to follow). In the setup of Fig. 4.3.1, i_{ind} has a negative x direction (see the discussion following (4.3.6)). Accordingly, the force on A_1A_2 due to i_{ind} will be $- i_{ind}\, l\, \mathbf{a}_x \times B\, \mathbf{a}_y = - i_{ind}\, l\, B\, \mathbf{a}_z$ which has a direction opposite to what is given by (4.3.1). Thus the induced current opposes the causes to which it is due.

<u>*Example 4.3.1*</u> Obtain the direction of the induced current in the circuit of Fig. 4.3.1 (a) when the polarity of the source of emf is reversed; (b) when the polarity of the source of emf remains unchanged, and $A_1\, A_2$ is moved with a constant velocity in the negative z direction by an external mechanical force; (c) when the polarity of the source of emf is reversed and $A_1\, A_2$ is moved with a constant velocity in the positive z direction by an external mechanical force.

(a) Since the polarity of the source of emf $\mathcal{E}$ has been reversed (Fig. 4.3.1), the direction of i in A_1A_2 is the negative x direction. Due to electromagnetic action, $A_1\, A_2$ will now experience a force in the negative z-direction, given by (4.3.1) but with a negative sign in its right hand side. Let $A_1\, A_2$ move through a distance Δz in an infinitesimal time Δt to its new position $A_1''\, A_2''$ (Fig. 4.3.1). The energy of the source of emf is utilised in moving $A_1\, A_2$ to $A_1''A_2''$, and, in this example, (4.3.2) continues to be valid, to satisfy the principle of conservation of energy. Further, the current in the circuit continues to be given by (4.3.5) and the induced emf by (4.3.6). The minus sign in (4.3.6) continues to suggest that the direction of i_{ind} is opposite to that due to the applied emf $\mathcal{E}$, that is, the direction of i_{ind} is the positive x direction, and, therefore, the force due to this on $A_1\, A_2$ will be $i_{ind}\, l\, \mathbf{a}_x{\times}B\, \mathbf{a}_y = i_{ind}\, l\, B\, \mathbf{a}_z$, opposing the force due to electromagnetic action responsible for moving $A_1\, A_2$ in the negative z direction; this is precisely what is also predicted by Lenz's law.

In this example, it may be noted, the effect of the electromagnetic action is to decrease the area of the circuit linked with the magnetic flux with time, and hence $d\phi_B/dt$ represents the time rate of *decrement* of magnetic flux. This is contrary to what is in the original example of Fig. 4.3.1 (with the polarity of the source of emf unchanged), where, it may be recalled, left to electromagnetic action, the area of the circuit linked with the magnetic flux increases with time, and hence $d\phi_B/dt$ represents the time rate of *increment* of magnetic flux.

(b) In this example, an external mechanical force is required to move $A_1\, A_2$ in the negative z-direction, say, to $A_1''A_2''$ through a distance Δz in time Δt; this force has to overcome the force given by (4.3.1). Thus the source of energy to the system comprises both the source of emf and the one doing the mechanical work, and, therefore, the energy balance equation of the type (4.3.2), in this case, takes the following form:

$$\mathcal{E}\, i\, \Delta t \; + \; i\, l\, B\, \Delta z \; = \; i^2 R\, \Delta t \; . \tag{4.3.15}$$

Using (4.3.15), an equation similar to (4.3.5) for the circuit current i may be obtained as:

$$i \; = \; \frac{\mathcal{E} + Blv}{R} \; = \; \frac{\mathcal{E} - \dfrac{d\phi_B}{dt}}{R} \; , \tag{4.3.16}$$

where $d\phi_B/dt$, as in the original configuration of Fig. 4.3.1, represents the time rate of increment of magnetic flux (see the interpretation of $d\phi_B/dt$ as discussed in the concluding paragraph of the solution to (a)). In the present example (b), the area of the circuit linked with the magnetic flux actually decreases with time (as $A_1 A_2$ moves to $A_1''A_2''$ (see Fig. 4.3.1)), and, consequently, $d\phi_B/dt$ has a negative value. This is also indicated by (4.3.16), whence $d\phi_B/dt = - Blv$, which has a minus sign.

It is also clearly suggested by (4.3.16) that the current has been increased by the motion of $A_1 A_2$. In other words, i_{ind} in this case adds to the current due to the applied emf, being essentially in the positive x-direction. The force due to i_{ind} on $A_1 A_2$, $i_{ind}\, l\, \mathbf{a}_x \times B\, \mathbf{a}_y = i_{ind}\, l\, B\, \mathbf{a}_z$, is along z, opposing the external mechanical force responsible for the motion of $A_1 A_2$. Thus the direction of i_{ind}, in this example, conforms with Lenz's law.

(c) In this example, the polarity of the source of emf is reversed, and, as in example (a), left to electromagnetic action above, $A_1 A_2$ would have moved in the negative z-direction. However, $A_1 A_2$, in this case, as a result of an external mechanical force, has been moved in the positive z-direction, say, to $A_1'A_2'$ through a distance Δz in time Δt. Here, it can be easily appreciated, the energy balance equation continues to be given by (4.3.15) and the circuit current by (4.3.16). Here, however, as in example (a) $d\phi_B/dt$ represents the time rate of *decrement* of magnetic flux (see the interpretation of $d\phi_B/dt$ as discussed in the concluding paragraph of the solution to (a)). In the present example (c), the area of the circuit linked with the magnetic flux actually increases with time (as $A_1 A_2$ moves to $A_1'A_2'$ (Fig. 4.3.1)), and, consequently, $d\phi_B/dt$ has a negative value. This is also indicated by (4.3.16) whence $d\phi_B/dt = - Blv$, which has a minus sign.

It is also suggested by (4.3.16), as applicable to the present case, that the circuit current has been increased by the induced current, the latter essentially being in the negative x-direction. Consequently, the force on A_1A_2 due to i_{ind} will be $- i_{ind}\, l\, \mathbf{a}_x \times B\, \mathbf{a}_y = - i_{ind}\, l\, B\, \mathbf{a}_z$ which opposes the positive z-directed mechanical force responsible for the motion of $A_1 A_2$. The direction of the induced current thus agrees to the prediction of Lenz's law.

As for the magnitude of the induced emf, $\mathcal{E}_{ind}$, for each of the examples (a)-(c) (as also for the original case depicted by Fig. 4.3.1), one may easily note from (4.3.5), (4.3.6) and (4.3.16) [6]:

$$\left| \mathcal{E}_{ind} \right| \; = \; \left| \frac{-d\phi_B}{dt} \right| \; = \; Blv \; . \tag{4.3.17}$$

<u>Example 4.3.2</u> Take the circuit arrangement with pertinent parameters of Fig. 4.3.1, but replace the source of emf by a load resistance R_L, considered much greater than the resistances of the other parts of the circuit. Find the magnitude and sense of the load

current when A_1 A_2 moves with a uniform velocity a) to the right (that is, in the negative z-direction and b) to the left (that is, in the positive z-direction).

(a) Let us treat the problem as a special case of Example 4.3.1 (b) taking $\mathcal{E} = 0$. (4.3.16) then yields the load current i_L :

$$i_L = i_{ind} = \frac{Blv}{R_L} . \tag{4.3.18}$$

The direction of the above current has already been found in Example 4.3.1 (b) as the positive x-direction in A_1 A_2; this will be from T_1 to T_2 through the load resistance (see Fig. 4.3.1).

(b) Let us treat the problem as a special case of Example 4.3.1 (c) taking $\mathcal{E} = 0$. (4.3.16) again yields the load current which is identically given by (4.3.18). The direction of the current, as has already been found in Example 4.3.1 (c), is the negative x-direction; this will correspond to a current from T_2 to T_1 through the load resistance.

Example 4.3.3 (a) A straight vertical conductor, immersed in a magnetic field which is constant in space and time and which is directed towards east, moves perpendicular to its length with a constant velocity. What is the magnitude and polarity of the motional emf developed when the conductor moves i) towards south ii) towards north? (b) If the above conductor now rotates in the vertical plane with a uniform angular velocity about a horizontal axis which passes through one of the ends of the conductor, what is the magnitude and polarity of the motional emf developed when the rotation is i) clockwise ii) anticlockwise, viewing the conductor down the magnetic field?

(a) The problem is same as Example 4.3.2 except that now we have no circuit through which the induced current can pass (see Fig. a4.3.1). The emf $\mathcal{E}_{ind}$ developed between the ends of the straight conductor is given by (4.3.17) which may be rewritten as [6]:

$$\mathcal{E}_{ind} = B\,l\,v\,, \tag{4.3.19}$$

where l has to be interpreted as the total length of the conductor (A_1 A_2). In the framework of Example 4.3.2 (Fig. a4.3.2), the conductor is aligned in the x-direction, and the magnetic field along y, the conductor moving in the negative and positive z-directions corresponding respectively to the southward (i) and northward (ii) movement in the present example.

Referring back to Example 4.3.2 (a), one may recall that the induced load current was from T_1 to T_2 through the load resistor. This immediately suggests that the upper end of the conductor has developed the positive polarity for the case i) of the present example. Similarly, referring back to Example 4.3.2 (b), one may easily see that, for the case ii) the polarity of the upper end of the conductor is negative.

An alternative method of finding the motional emf in this problem is to consider the separation of charges inside the conductor due to their motion in the magnetic field and then find a condition of equilibrium corresponding to a balance between the electrostatic and Lorentz forces acting on a free charge in the conductor. For the details, one may see (appendix *a4.3.1*).

(b) As in (a), the conductor moves perpendicular to its length in this problem, too (see Fig. a4.3.2). However, the speeds of the various portions of the conductor, unlike in (a), are different here. Now, if we divide the entire length of the conductor into length elements, we can find the elemental motional emf $d\mathcal{E}_{ind}$ developed between the ends of each such element. Summing such elements of emf, one can find the emf, $\mathcal{E}_{ind}$, between the ends of the conductor as follows (for the details of derivation, see appendix *a4.3.2*)):

$$\mathcal{E}_{ind} = \frac{1}{2} B \omega l^2 , \qquad (4.3.20)$$

where ω is the angular velocity of rotation of the conductor. The end of the conductor at the axis of rotation will develop the negative or positive polarity of the induced emf according as the rotation of the conductor is clockwise or anticlockwise, viewing along the direction in which the magnetic field points.

Example 4.3.4 Workout the problem of Example 4.3.3(b) when the straight conductor is replaced by a circular conducting disc with its centre located at the axis of rotation to make a homopolar generator.

The induced emf between any peripheral point and the centre of the disc continues to be given by (4.3.20). The peripheral points of the disc acquires a positive or negative polarity, with respect to the centre of the disc, according as the rotation of the disc is clockwise or anticlockwise. In a practical set-up, two brushes make contacts with the disc at the centre (axle) and along the periphery (see Fig. 4.3.2, for the other details). Such a generator is called *homopolar* because the induced emf has the same sense irrespective of time. For instance, the terminal T_1 develops a positive polarity, for the clockwise rotation of the disc. The polarity reverses for the anticlock rotation.

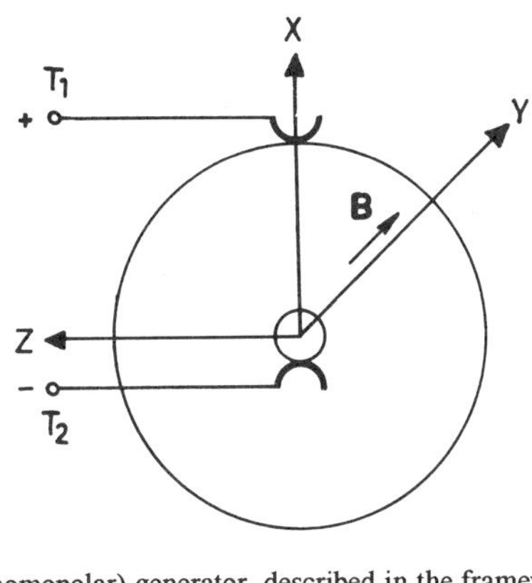

Fig. 4.3.2 Faraday disc (homopolar) generator, described in the framework of Example 4.3.3(b) in which the disc lies on the xz-plane; the brushes lie on the x-axis; and the magnetic field is directed along y.

Example 4.3.5 A loop of a conducting wire is placed in a magnetic flux which in uniform in space but decreases uniformly at the rate of 3 webers per second. If the loop lies on the plane of the paper, and the magnetic field intensity is perpendicular to and into this plane, and, further, a load resistance, 2 ohms is connected to the turn (as shown in Fig. 4.3.3), find the value of the current and its direction through the resistor.

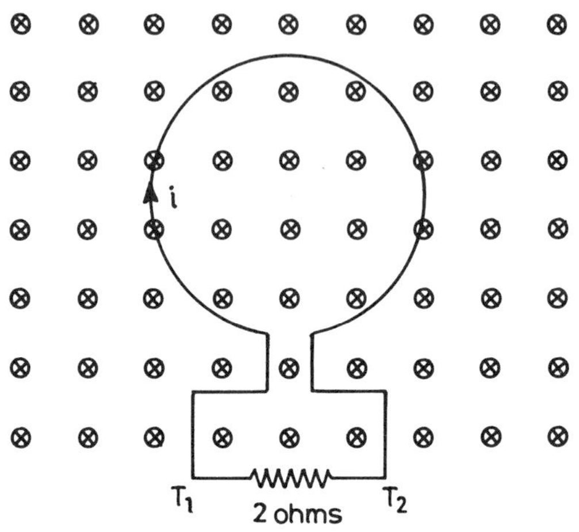

Fig. 4.3.3 The circuit arrangement of Example 4.3.5. The magnetic field direction, perpendicular to the plane of the current loop, is shown by crosses (away from the reader).

The value of the emf according to (4.3.6) is simply 3 volts. Since the resistance is 2 ohms, the current is 1.5 (= 3/2) amps. This current, according to Lenz's law, should give rise to a magnetic field such that it adds to the already existing magnetic field which is decreasing and thus opposes the cause of electromagnetic induction (in this case, the decrement of magnetic flux). This obviously necessitates that the current in the loop has to be as shown in Fig. 4.3.3; this current thus will be from the terminal T_2 to T_1 through the resistor.

Example 4.3.6 A solenoid yields an axial magnetic flux density which remains constant over its cross section but whose magnitude B increases with time t as:

$$B\{t\} = 0.1 + 0.2t + 0.3t^2 \text{ Tesla} .$$

What is the electric field intensity **E** inside the solenoid at a distance 1 cm from its axis, at t = 2 sec? Use both the integral and point forms of Faraday's law, separately, to the problem.

A cross section of the solenoid is shown in Fig. 4.3.4 described in the cylindrical system of coordinates. The magnetic flux density, which is constant over the cross section, is considered to be in the z-direction:

$$\mathbf{B} = B\,\mathbf{a}_z = (0.1 + 0.2t + 0.3t^2)\,\mathbf{a}_z \text{ Tesla} . \qquad (4.3.21)$$

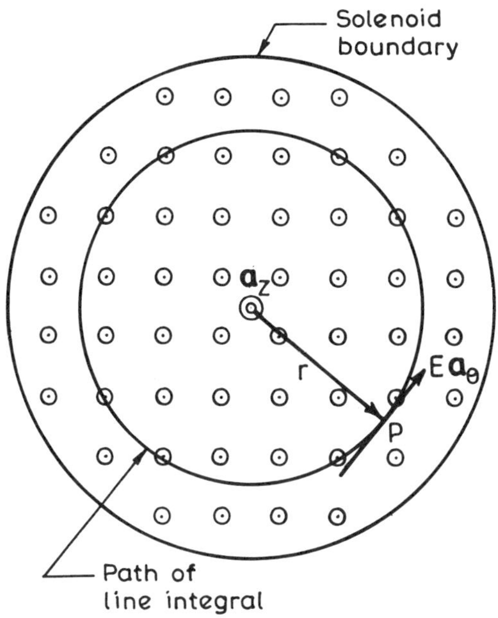

Fig. 4.3.4 Circular cross-section of the solenoid over which the axial magnetic flux density remains constant at a given instant of time, showing the point P defined by the radial coordinate r where to find the electric field intensity **E**. The sense of circulation (azimuthal direction) of **E** (clockwise or anti-clockwise) will depend on how the magnetic flux density varies with time. The magnetic field direction, perpendicular to the cross section of the solenoid, is shown by dots (towards the reader).

The magnetic field is restricted to be within the circular boundary. The electric field intensity **E**, here, essentially points in the azimuthal direction, that is, $\mathbf{E} = E\,\mathbf{a}_\theta$. The problem is similar to that of finding the magnetic field intensity which points in the azimuthal direction at any point inside a straight current-carrying conductor of circular cross section. Just as one may apply Ampere's circuital law to the latter, one may use Faraday's law to solve the present problem.

Using the integral form (4.3.13), one has

$$\oint E_\theta \mathbf{a}_\theta . dl\, \mathbf{a}_\theta = -\oint \frac{\partial B}{\partial t} \mathbf{a}_z . \mathbf{a}_z\, dS\,, \tag{4.3.22}$$

where one has to take the surface integral over the cross-sectional area within the circle passing through P where to find the electric field intensity. Since B is constant with space coordinates (see (4.3.21)), and also since E_θ, because of the symmetrical nature of the problem, remains constant over the circular periphery passing through P, one may take these quantities (E_θ and $\partial B/\partial t$) outside the integrations of (4.3.22) giving

$$E_\theta \oint dl = -\frac{\partial B}{\partial t} \oint dS\,,$$

whence

$$E_\theta \, 2\pi r = -\frac{\partial B}{\partial t}\pi r^2 ,$$

(**B** depending only on t) which yields

$$\mathbf{E} = E_\theta \mathbf{a}_\theta = -\frac{1}{2}\frac{\partial B}{\partial t} r \, \mathbf{a}_\theta . \tag{4.3.23}$$

If one chooses to start from the point form (4.3.14), one may write in terms of the z-component:

$$(\nabla \times \mathbf{E})_z = -\frac{\partial B}{\partial t} = -\frac{dB}{dt} , \tag{4.3.24}$$

(since **B** depends only on t). Using the expression for $\nabla\times\mathbf{E}$ in the cylindrical system of coordinates (see (3.2.18(b))), where the vector **H**, in this case, has to be interpreted as **E**), and remembering that $E_r = 0$, and/or $\partial/\partial\theta = 0$ (azimuthal symmetry), we may write (4.3.24) as:

$$\frac{1}{r}\frac{d}{dr}(r E_\theta) = -\frac{dB}{dt} . \tag{4.3.25}$$

One may note that, now since r and E_θ both depend only on r, and B depends only on t, one may replace the partial derivatives by the complete derivatives.

From (4.3.25),

$$d(r E_\theta) = -\frac{dB}{dt} r \, dr$$

which upon integration gives

$$E_\theta = -\frac{dB}{dt}\frac{r}{2} + \frac{A}{r} , \tag{4.3.26}$$

where A is an integration constant. By putting r = 0 (which describes the axis of the solenoid), E_θ becomes ∞ unless one puts A = 0. The finiteness of E_θ at r = 0, therefore, demands that A = 0. (4.3.26) thus becomes

$$E_\theta = -\frac{dB}{dt}\frac{r}{2} . \tag{4.3.27}$$

Then substituting (4.3.27) in

$$\mathbf{E} = E \, \mathbf{a}_\theta$$

we have

$$E = -\frac{dB}{dt}\frac{r}{2}\mathbf{a}_\theta . \tag{4.3.28}$$

It is of interest to note that (4.3.23) and (4.3.28) which are obtained independently starting from the integral and point forms of Faraday's law, respectively, are, as expected, identical.

It is clear from (4.3.28) that, when B increases with time, as it does in the present example, then, according to (4.3.21), **E** will have the direction of - $\mathbf{a}_\theta$; in other words, **E** will run anticlockwise down the magnetic field, that is, clockwise, as viewed from above the plane of the cross section of the solenoid considered (see Fig. 4.3.4 where, however, the general direction of **E** is shown). The electric field lines, in this case, are clearly continuous forming closed loops. Obtaining B from (4.3.21), and putting t = 2 sec and r = 1 cm (= 10^{-2} meter), in this problem, one has

$$\mathbf{E} = -7 \, \mathbf{a}_\theta \text{ mV/m} .$$

Appendix a4.3.1 *Motional emf across a moving straight conductor*

*Let us consider here an alternative approach to finding the motional emf developed in the problem of Example 4.3.3(a). The free charges — the electrons inside the conductor, while being carried away with the linear conductor in the negative z-direction ($\mathbf{v} = -v\,\mathbf{a}_z$), each experience a Lorentz force **F** due to the magnetic field applied in the y-direction:*

(a 4.3.1) $$\mathbf{F} = |e|\, v\, \mathbf{a}_z \times B\, \mathbf{a}_y = -|e|\, v\, B\, \mathbf{a}_x \,,$$

*where $-|e|$ is the negative electronics charge. As a result of this Lorentz force which has a negative x-direction, the negative charges (electrons) pile up in the vicinity of A_2, leaving a net positive charge in the vicinity of A_1 (see Fig. a4.3.1). The piling up of charges continue until an equilibrium is reached when the Lorentz force **F** on an electron is cancelled out by the electrostatic field **E** developed due to the separation of charges. Thus, under equilibrium:*

(a 4.3.2) $$-|e|\,\mathbf{E} + \mathbf{F} = 0 \,.$$

Substituting (a 4.3.1) in (a 4.3.2)

(a 4.3.3) $$\mathbf{E} = -v\, B\, \mathbf{a}_x \,.$$

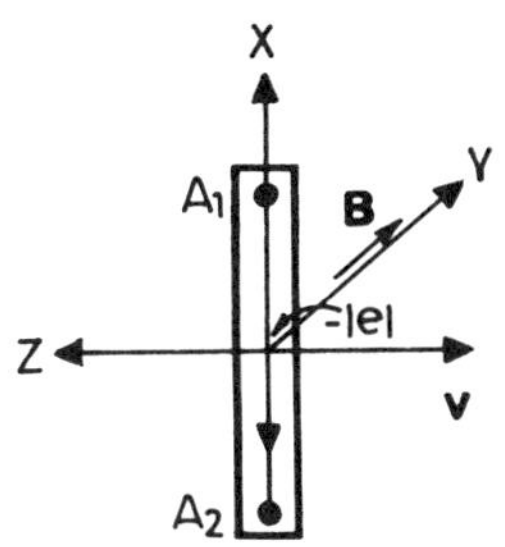

Fig. a4.3.1 The separation of charges inside a straight conductor due to the motion of the conductor in a magnetic field.

***E** can also be expressed in terms of the potential gradient with the help of (2.5.11) and (2.5.12), and noting that, in this case, the charges are separated only along x thereby making $\partial V/\partial y = \partial V/\partial z = 0$:*

(a 4.3.4) $$\mathbf{E} = -\frac{\partial V}{\partial x}\mathbf{a}_x \,.$$

Combining (a 4.3.3) and (a 4.3.4) and integrating, one then has the potential difference, $V_{A_1A_2}$, between A_1 and A_2 as follows:

$$V_{A_1A_2} = \int dV = \int_{x=0}^{x=l} vB\, dx = Blv \,,$$

which is identical with the expression for the induced emf given by (4.3.19).

Appendix a 4.3.2 *Motional emf across a rotating straight conductor*

Let us, for the sake of convenience, treat the problem in the framework of case (i)of Example 4.3.3(a) (see Figs. a4.3.1 and a4.3.2). Thus, at any instant of time, one may choose the length of the conductor to be coincident with the x-axis. Then for the clockwise rotation, the linear velocity of the conductor will be in the negative z-direction. However, the magnitude of this velocity is now not the same, at all the points of the conductor. The element of length dr located at a distance r from the axis of rotation has a velocity of magnitude $v = \omega r$, *where* ω *is the angular velocity of rotation. One can treat the problem as identical with the case (i) of Example 4.3.3 (a) as applied to this element. The element of emf* $d\,\mathcal{E}_{ind}$ *induced between the ends of the element (of length dr), according to (4.3.19), is:*

$$d\,\mathcal{E}_{ind} = (B)(dr)(v) = (B)(dr)(\omega r)\,, \quad \text{(a 4.3.5)}$$

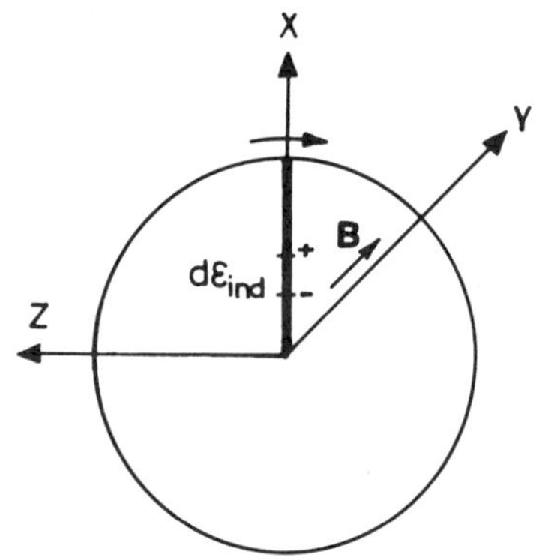

Fig. a4.3.2 Motional emf across a conductor rotating in the xz plane in a uniform, y-directed magnetic field, the rotation being clockwise viewed down the magnetic field.

with the end of the element farther from the axis of rotation having a positive polarity. Therefore, the total emf $\mathcal{E}_{ind}$ *is obtained by integrating as (a 4.3.5) as:*

$$\mathcal{E}_{ind} = \int d\,\mathcal{E}_{ind} = \int_0^l B\omega r\,dr = \frac{1}{2}B\omega l^2\,, \quad \text{(a4.3.6)}$$

now the positive polarity of the entire emf being located at the end of conductor, at the periphery of the path of rotation. For the anticlockwise rotation of the conductor, one has similarly to refer to the case (ii) of Example 4.3.3(a). The induced emf continues to be given by (a4.3.6), however, the end of the conductor at the axis of rotation having now a positive polarity. One may notice that (a 4.3.6) is identical with (4.3.20).

4.4 Maxwell's Equations

The following set of four self-consistent equations are known as Maxwell's equations [7]-[9] expressed in two forms — integral and differential or point forms which have one-to-one correspondence:

<u>Integral Form</u> <u>Differential or Point Form</u>

$$\oint \mathbf{D}.\mathbf{a}_n \, dS = \oint \rho d\tau \quad (a) \qquad \nabla.\mathbf{D} = \rho \quad (b) \qquad (4.4.1)$$

$$\oint \mathbf{B}.\mathbf{a}_n \, dS = 0 \quad (a) \qquad \nabla.\mathbf{B} = 0 \quad (b) \qquad (4.4.2)$$

$$\oint \mathbf{E}.\mathbf{dl} = -\oint \frac{\partial \mathbf{B}}{\partial t}.\mathbf{a}_n \, dS \quad (a) \qquad \nabla \times \mathbf{E} = -\frac{\partial \mathbf{B}}{\partial t} \quad (b) \qquad (4.4.3)$$

$$\oint \mathbf{H}.\mathbf{dl} = \oint (\mathbf{J}_c + \frac{\partial \mathbf{D}}{\partial t}).\mathbf{a}_n \, dS \quad (a) \qquad \nabla \times \mathbf{H} = \mathbf{J}_c + \frac{\partial \mathbf{D}}{\partial t} \quad (b) \qquad (4.4.4)$$

Maxwell's equations (4.4.1)-(4.4.4), in general, refer to time-varying situations. We notice that (4.4.1) and (4.4.2) have not changed from their corresponding static/steady forms. Thus, (4.4.1(a)) and (4.4.1(b)) are identical with (2.2.1) and (2.2.3), respectively; and similarly, (4.4.2(a)) and (4.4.2(b)) are identically same as (3.1.4(b)) and (3.1.5(b)), respectively. In writing (4.4.1(a)), however, we have replaced the charge enclosed q occurring in the right hand side of (2.2.1) by the relevant closed volume integral (see (4.1.3)). Further, (4.4.1) states that the electric flux lines originate from, or terminate, on an electric charge, if present, and also that the electric field/displacement is the *effect* or result while the charge density is the *cause*. Similarly, (4.4.2) states that the magnetic flux lines are continuous forming closed loops and also that there are no free magnetic charges (or poles). (In this context, the discussion in appendix *a3.1.1* continues to be of relevance). The other Maxwell's equations, namely, (4.4.3(a)) or (4.4.3(b)) and (4.4.4(a)) or (4.4.4(b)) are nothing but (4.3.13) or (4.3.14) and (4.2.17) or (4.2.16) rewritten, respectively. Further, (4.4.3) suggests that the time-varying magnetic field gives rise to continuous electric flux lines forming closed loops. (For instance, see Example 4.3.6). Also (4.4.4) suggests that the magnetic field is the result of the convection/conduction current or the time-varying electric field as the cause. Again, it is also suggested by (4.4.4) that closed loops of magnetic flux lines form due to time-varying electric field (see the following example).

<u>*Example 4.4.1*</u> A parallel-plate capacitor with circular plates each of radius 'a' yields an axial electric field perpendicular to the plates which remains constant over its cross section, its magnitude increasing at the rate of $1 \mathrm{Vcm^{-1}s^{-1}}$. Find the magnetic field intensity at a radial distance 1 cm (< a) from the axis of the capacitor.

The problem may be treated in parallel with that of Example 4.3.6. The axis of the capacitor is the z-axis, in the cylindrical system of coordinates, passing through the centre of, and perpendicular to, the plates (see Fig. 4.4.1). Following exactly the same procedure as in Example 4.3.6, but now starting from either the integral form (4.4.4(a)) (which is (4.2.17) re-written) or the differential or point form (4.4.4(b)) (which is (4.2.16) re-written), one may easily show:

$$\mathbf{H} = \frac{dD}{dt}\frac{r}{2}\mathbf{a}_\theta \quad . \qquad (4.4.5)$$

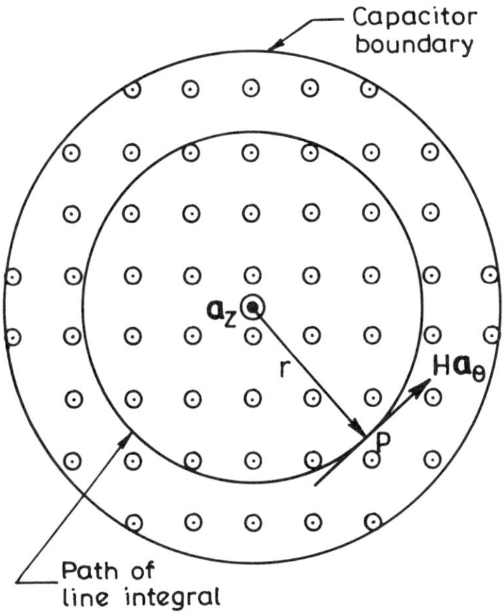

Fig. 4.4.1 A circular cross section perpendicular to the axis of a parallel-plate capacitor made of circular plates, showing the point P, defined by the radial coordinates r, where to find the magnetic field intensity. The sense of circulation of **H** (clockwise or anti-clockwise) will depend on how the electric field intensity varies with time. The electric field direction, perpendicular to the cross section of the capacitor, is shown by dots (towards the reader).

It may be noted that, here, $\mathbf{J}_c = 0$, and that the magnetic field is contributed only by the displacement current density (dD/dt). We notice that (4.4.5) is analogous to (4.3.28); by replacing **E** and **-B** by **H** and **D**, respectively, one obtains the former from the latter. And, in fact, exactly the same analogy exists between the starting equations, namely, (4.4.3) and (4.4.4) from which (4.3.28) and (4.4.5) are derived, respectively, in an analogous manner.

In the present problem $dE/dt = 1$ volt cm^{-1} $s^{-1}=10^2$ volt m^{-1} s^{-1} , $r = 1$ cm $= 10^{-2}$ meter. Substituting these values, while remembering that $\mathbf{D} = \varepsilon\,\mathbf{E}$, in (4.4.5), **H** may be calculated as:

$$\mathbf{H} = 4.43\,\mathbf{a}_\theta \text{ pA/m} .$$

Thus, in this problem, as E considered along z, increases with t, **H** will have the same direction as $\mathbf{a}_\theta$ (instead of $-\mathbf{a}_\theta$ as in Example 4.3.6); in other words, **H** will run clockwise down the magnetic field, that is, anticlockwise as viewed from above the plane of the cross section of the capacitor considered, the magnetic field lines forming closed loops (see Fig. 4.4.1).

4.5 Wave Equation

A time-varying electric field is accompanied by a magnetic field and, similarly, a time-varying magnetic field gives rise to an electric field. This is amply made clear with the help of Examples 4.3.6 and 4.4.1 which actually exemplify the two Maxwell's equations (4.4.3) and (4.4.4), respectively, which are essentially coupled equations in **E** and **H**. It becomes quite useful to decouple these two equations, and that is what we are going to do now with the help of the remaining two Maxwell's equations, namely, (4.4.1) and (4.4.2), to obtain two separate equations in individual field quantities **E** and **H**, respectively.

Let us restrict ourselves to only isotropic and homogeneous unbounded media. In an isotropic medium, ε, μ and σ are essentially scalar quantities so that $\mathbf{D}$ ($= \varepsilon\mathbf{E}$), $\mathbf{B}$ ($= \mu\mathbf{H}$) and $\mathbf{J}_c$ ($= \sigma\mathbf{E}$) (considering that $\mathbf{J}_c$ is due to conduction) have the same directions as **E**, **H** and **E**, respectively. Further, in a homogenous medium, ε, μ and σ do not change with space coordinates. Now, taking the curl of both sides (4.4.3(b)), and making use of i)the relations: $\mathbf{B}$ ($= \mu\mathbf{H}$) and $\mathbf{D}$ ($= \varepsilon\mathbf{E}$), ii)the equations: (4.4.1(b)) and (4.4.4(b)), and iii)the vector identity: $\nabla\times\nabla\times\mathbf{E} = \nabla(\nabla.\mathbf{E}) - \nabla^2\mathbf{E}$, one easily obtains,

$$\Box^2 \mathbf{E} = \mu\frac{\partial \mathbf{J}_c}{\partial t} + \nabla\left(\frac{\rho}{\varepsilon}\right) \tag{4.5.1}$$

where the differential operator

$$\Box^2 = \nabla^2 - \mu\varepsilon\frac{\partial^2}{\partial t^2} \tag{4.5.2}$$

is called the d'Alembertian.

The vector equation (4.5.1) is essentially an inhomogeneous linear differential equation, composed of three separate scalar equations in the three 'scalar' components of **E**, respectively. The Laplacian operator ∇^2 on a particular component is given by (2.6.4) where the quantity V has to be replaced by the desired field component, in the appropriate system of coordinates. The equation (4.5.1) is known as the wave equation in **E**, since the solution to such an equation yields periodic and propagating nature of **E**, characteristic of a wave. The inhomogeneous wave equation in **H** may similarly be obtained. For this purpose, one has now to take the curl of both sides of (4.4.4(b)), and make use of i) the relations: $\mathbf{B}$ ($= \mu\mathbf{H}$) and $\mathbf{D}$ ($= \varepsilon\mathbf{E}$), as before, ii)the equations: (4.4.2(b)) and (4.4.3(b)) and iii)the vector identity: $\nabla\times\nabla\times\mathbf{H} = \nabla(\nabla.\mathbf{H}) - \nabla^2\mathbf{H}$. Thus one obtains

$$\Box^2 \mathbf{H} = -\nabla\times\mathbf{J}_c\ . \tag{4.5.3}$$

It may be noted that the right-hand sides of the wave equations are contributed by the source quantities — the current and the electric charge, for the electric field (see (4.5.1)), and the current alone, for the magnetic field (see (4.5.3)).

4.5.1 Propagation through a Free Space

Let us now turn towards a specific application of wave equations in studying the behaviour of propagating electromagnetic waves. One may associate with a wave propagating through a medium at least one physical quantity which varies periodically in space and with time. Further the quantity should exhibit the propagative nature of waves. Proper functions of space and time coordinates to represent such a quantity are given in appendix *a 4.5.1*.

A plane wave is one for which the wavefront is a plane, the wavefront being a surface over which the phase of the varying physical quantity is the same. The wave is called a uniform plane wave if over the equiphase surface or wavefront the amplitude of the quantity also remains the same. Usually, uniform plane waves are obtained at large distances from the source of such waves. If the propagation of a uniform plane wave is chosen as to be along z, then for such a wave, one may take that there is no variation of field quantities along x and y, that is,

$$\partial/\partial x = \partial/\partial y = 0 . \qquad (4.5.4)$$

The wave equation in a source-free, free-space region ($\mathbf{J}_c = \rho = 0$), may be written from (4.5.1) through (4.5.2) as the following homogeneous equation:

$$\nabla^2 \mathbf{E} = \mu_0 \varepsilon_0 \frac{\partial^2 \mathbf{E}}{\partial t^2}$$

which may be further expanded in terms of individual field components as follows:

$$(\nabla^2 E_x)\mathbf{a}_x + (\nabla^2 E_y)\mathbf{a}_y + (\nabla^2 E_z)\mathbf{a}_z = \mu_0 \varepsilon_0 \frac{\partial^2}{\partial t^2}(E_x \mathbf{a}_x + E_y \mathbf{a}_y + E_z \mathbf{a}_z)$$

which further, with the help of (2.6.4(a)), invoking (4.5.4), may be read as,

$$\nabla^2 E_{x,y,z} = \frac{\partial^2 E_{x,y,z}}{\partial z^2} = \mu_0 \varepsilon_0 \frac{\partial^2 E_{x,y,z}}{\partial t^2} . \qquad (4.5.5)$$

Further, here, since $\rho = 0$, equation (4.4.1(b)), read with the help of (2.3.5(a)), remembering that $\mathbf{D} = \varepsilon_0 \mathbf{E}$, yields, again, in view of (4.5.4), $\partial E_z/\partial z = 0$ whence

$$\frac{\partial^2 E_z}{\partial z^2} = 0 . \qquad (4.5.6)$$

Substituting (4.5.6) in (4.5.5),

$$\frac{\partial^2 E_z}{\partial t^2} = 0 . \qquad (4.5.7)$$

The solution to (4.5.7) is:

$$E_z = a\,t + b\,, \tag{4.5.8}$$

where a and b are constants. If a and b are non-zero constants ($a \neq 0$, $b \neq 0$), then the axial electric field intensity E_z varies linearly with t, having an initial value ($E_z = b$) at $t = 0$; if $b = 0$, but $a \neq 0$, then E_z is directly proportional to t, having zero initial value ($E_z = 0$) at $t = 0$; and if $a = 0$, but $b \neq 0$, then E_z has a non-zero constant value irrespective of time ($E_z = b$). However, since E_z is associated with a wave, the former can neither have a value linearly varying with, nor be directly proportional to, time, nor have a value which remains constant with time. This immediately suggests that one, in this case, has to take $a = b = 0$ which yields (see (4.5.8)):

$$E_z = 0 \text{ (uniform plane wave)}\ . \tag{4.5.9}$$

Thus, for a uniform plane wave, there exists no electric field component perpendicular to the equiphase, equi-amplitude surface or wave front, that is, parallel to the direction of propagation. Imposing (4.5.9), one may write (4.5.5) as

$$\frac{\partial^2 E_x}{\partial z^2} = \mu_0\varepsilon_0 \frac{\partial^2 E_x}{\partial t^2} \qquad \text{(a)}$$

$$\frac{\partial^2 E_y}{\partial z^2} = \mu_0\varepsilon_0 \frac{\partial^2 E_y}{\partial t^2}\,. \qquad \text{(b)} \tag{4.5.10}$$

Considering the basic form of periodic variation of the field quantities with time as sinusoidal, and using the phasor notation to represent this variation, that is, taking the field quantities to vary as $\exp j\omega t$, one may express (4.5.10) as

$$\frac{\partial^2 E_{x,y}}{\partial z^2} = -\,\omega^2 \mu_0\varepsilon_0 E_{x,y}$$

which has the simple solution:

$$E_{x,y} = E_{x_1,y_1} \exp(-j\beta z) + E_{x_2,y_2} \exp(j\beta z)\,,$$

where the time-dependence factor $\exp(j\omega t)$ is understood; allowing this factor to appear, the solution becomes:

$$E_{x,y} = E_{x_1,y_1} \exp j(\omega t - \beta z) + E_{x_2,y_2} \exp j(\omega t + \beta z)\,. \tag{4.5.11}$$

The first and the second terms of (4.5.11) represent the forward and the backward waves, respectively (see appendix *a4.5.1*); the subscripted quantities represent the amplitudes. β ($= \omega/v_p$) is the phase propagation constant, v_p being the phase velocity given by:

$$v_p = (\mu_0\,\varepsilon_0)^{-1/2} = c \tag{4.5.12}$$

(remembering that c is the symbol used for velocity of light in free space).

One could as well have started from the wave equation (4.5.3) instead of (4.5.1), with $\mathbf{J}_c = 0$; used (4.4.2(b)) instead of (4.4.1(b)) and followed the same procedure as underlined here. This would lead one to:

$$H_z = 0 \text{ (uniform plane waves)} \tag{4.5.13}$$

and,

$$H_{x,y} = H_{x_1,y_1} \exp j(\omega t - \beta z) + H_{x_2,y_2} \exp j(\omega t + \beta z), \tag{4.5.14}$$

the subscripted quantities again representing the amplitudes.

Equations (4.5.9), (4.5.11), (4.5.13) and (4.5.14) all go to suggest that transverse electromagnetic (TEM) waves, with electric and magnetic fields each transverse to the direction of propagation, with no component along the latter direction, are supported in an unbounded free-space region.

Example 4.5.1 Show that the refractive index of an isotropic, homogenous medium is equal to the square root of the dielectric constant of the medium.

The treatment presented in this sub-section (section *4.5.1*) continues to be valid also for a homogeneous, isotropic, dielectric medium, for which, however, one has to replace ε_0 by ε, the permitivity of the dielectric in (4.5.10). The refractive index n, which is the ratio of c, the velocity of light in free-space, to the velocity of light in the dielectric, may then be written with the help of (4.5.12) as:

$$n = \frac{c}{v} = \left(\frac{\varepsilon}{\varepsilon_0}\right)^{1/2} = \varepsilon_r^{\,1/2},$$

where ε_r ($= \varepsilon/\varepsilon_0$) is the relative permittivity or dielectric constant of the medium.

Example 4.5.2 Show that, for a uniform plane TEM wave, in an unbounded free-space region, the electric and magnetic field intensity vectors are at right angles to each other, and that the ratio of their magnitudes is a constant of the region.

Putting $J_c = 0$ (free-space region) in (4.4.4(b)), where the curl operation is spelt out using (3.2.18(a)), and remembering (4.5.4), (4.5.9) and (4.5.13), one has

$$-\frac{\partial H_y}{\partial z}\mathbf{a}_x + \frac{\partial H_x}{\partial z}\mathbf{a}_y = \varepsilon_0 \frac{\partial E_x}{\partial t}\mathbf{a}_x + \varepsilon_0 \frac{\partial E_y}{\partial t}\mathbf{a}_y$$

which may be written component-wise as:

$$-\frac{\partial H_y}{\partial z} = \varepsilon_0 \frac{\partial E_x}{\partial t} \qquad \text{(a)}$$

$$\frac{\partial H_x}{\partial z} = \varepsilon_0 \frac{\partial E_y}{\partial t}. \qquad \text{(b)} \qquad (4.5.15)$$

Substituting (4.5.11) and (4.5.14), however, retaining only the first term in each of them, which is tantamount to considering a wave travelling in the positive z direction, into (4.5.15) and considering that the field quantities vary as exp ($j\omega t$), one may easily obtain

$$\beta H_y = \omega \varepsilon_0 E_x$$

and

$$-\beta H_x = \omega \varepsilon_0 E_y$$

whence

$$\frac{E_x}{H_y} = -\frac{E_y}{H_x} = \frac{\beta}{\omega\varepsilon_0} = \frac{1}{v_p\varepsilon_0} = \frac{1}{c\varepsilon_0} = (\frac{\mu_0}{\varepsilon_0})^{1/2} = \eta \text{ , say .} \qquad (4.5.16)$$

In deducing (4.5.16), use has been made of (4.5.12). Now **E** and **H** may be written in terms of their components E_x, E_y and H_x, H_y, respectively (remembering that $E_z = H_z = 0$ (see (4.5.9) and (4.5.13)) as follows:

$$\mathbf{E} = E_x\,\mathbf{a}_x + E_y\,\mathbf{a}_y \text{ ;} \qquad \text{(a)}$$

$$E = (E_x^2 + E_y^2)^{1/2} \qquad \text{(b)} \qquad (4.5.17)$$

and

$$\mathbf{H} = H_x\,\mathbf{a}_x + H_y\,\mathbf{a}_y \text{ ;} \qquad \text{(a)}$$

$$H = (H_x^2 + H_y^2)^{1/2} \text{ .} \qquad \text{(b)} \qquad (4.5.18)$$

It is easy to see from (4.5.16) that

$$E_x H_x + E_y H_y = 0; \qquad (4.5.19)$$

and, from (4.5.17b) and (4.5.18b), that

$$\frac{E}{H} = \left(\frac{E_x^2 + E_y^2}{H_x^2 + H_y^2}\right)^{1/2} \text{ .} \qquad (4.5.20)$$

Also, one may see, using (4.5.17(a)) and (4.5.18(a)), that

$$\mathbf{E.H} = (E_x\,\mathbf{a}_x + E_y\,\mathbf{a}_y)\mathbf{.}(H_x\,\mathbf{a}_x + H_y\,\mathbf{a}_y) = E_x H_x + E_y H_y \text{ .} \qquad (4.5.21)$$

Thus, combining (4.5.19) and (4.5.21) we have,

$$\mathbf{E\,.\,H} = 0 \text{ ,}$$

a result that says that the electric and magnetic field intensity vectors are orthogonal to each other. Substituting $E_x = \eta\, H_y$ and $E_y = -\,\eta\, H_x$ from (4.5.16) into (4.5.20), one has

$$E \,/\, H = \eta \text{ .} \qquad (4.5.22)$$

E/H has the dimension of an impedance and η (= $\mu_0/\varepsilon_0)^{1/2}$ (see (4.5.16)) is a constant of the medium called the intrinsic impedance. Here, the medium is considered to be a free space. It is easy to appreciate that (4.5.22) is also valid for a medium other than a free space by interpreting η as the intrinsic impedance $(\mu/\varepsilon)^{1/2}$ of the concerned medium, μ and ε being the permeability and permittivity of the medium, respectively.

4.5.2 Propagation through a Conducting Medium

The wave equation (4.5.1) through (4.5.2), in this case ($\rho = 0$, $\mathbf{J}_c = \sigma\,\mathbf{E}$), may be written as:

$$\nabla^2\mathbf{E} = \mu\varepsilon\frac{\partial^2\mathbf{E}}{\partial t^2} + \mu\sigma\frac{\partial\mathbf{E}}{\partial t}\;. \tag{4.5.23}$$

The solution to (4.5.23) may be obtained following exactly the same procedure as outlined in section *4.5.1*. In this case, the following equation, analogous to (4.5.7), is obtained:

$$\frac{\partial^2 E_z}{\partial t^2} + \frac{\sigma}{\varepsilon}\frac{\partial E_z}{\partial t} = 0\;. \tag{4.5.24}$$

Further, an equation analogous to (4.5.8) is obtained by solving (4.5.21) as follows:

$$E_z = a + b\exp(-\,t/\tau)\,, \tag{4.5.25}$$

where a and b are constants and τ ($= \sigma/\varepsilon_0$) is the time-constant with which the second term of the right hand side of (4.5.25) decays. For a good conductor, the value of τ is negligibly small (for instance, $\tau = 1.55\times10^{-19}$ s; see Example 4.1.2), so that one can ignore the second term of the right hand side of (4.5.25) and write the latter as:

$$E_z = a\;. \tag{4.5.26}$$

As argued following (4.5.8), one has to put $a = 0$ suggesting, according to (4.5.26), that, here, E_z should vanish:

$$E_z = 0\;. \tag{4.5.27}$$

Now, taking (4.5.27) and (4.5.4) into account, the equations analogous to (4.5.10), in the present case, are:

$$\frac{\partial^2 E_x}{\partial z^2} = \mu\varepsilon\frac{\partial^2 E_x}{\partial t^2} + \mu\sigma\frac{\partial E_x}{\partial t} \qquad \text{(a)}$$

$$\frac{\partial^2 E_y}{\partial z^2} = \mu\varepsilon\frac{\partial^2 E_y}{\partial t^2} + \mu\sigma\frac{\partial E_y}{\partial t}\;. \qquad \text{(b)} \tag{4.5.28}$$

Considering the field quantities to vary as $\exp(j\omega t)$, one may express (4.5.28) as

$$\frac{\partial^2 E_{x,y}}{\partial z^2} = j\omega\mu(\sigma + j\omega\varepsilon)E_{x,y}\;,$$

which (remembering the time-dependence factor of $\exp(j\omega t)$), has a solution analogous to (4.5.11):

$$E_{x,y} = E_{x_1,y_1}\exp\,(j\omega t - \Upsilon z) + E_{x_2,y_2}\exp\,(j\omega t + \Upsilon z) \tag{4.5.29}$$

where

$$\Upsilon = [j\omega\mu(\sigma + j\omega\varepsilon)]^{1/2} \quad . \tag{4.5.30}$$

For a highly conducting medium $\sigma >> \omega\varepsilon$, Υ, from (4.5.30), is given by:

$$\Upsilon = (j\,\omega\mu\sigma)^{1/2} , \tag{4.5.31}$$

which may be separated into a real and an imaginary part, α and β, respectively (see appendix *a4.5.2*), as follows:

$$\Upsilon = \alpha + j\,\beta \ , \tag{4.5.32}$$

where

$$\alpha = \beta = (\omega\,\mu\,\sigma / 2)^{1/2} . \tag{4.5.33}$$

Now, substituting (4.5.32) in (4.5.29), one has

$$E_{x,y} = E_{x_1,y_1} \exp(-\alpha z) \exp j(\omega t - \beta z) + E_{x_2,y_2} \exp(\alpha z) \exp j(\omega t + \beta z) . \tag{4.5.34}$$

The first and the second terms of (4.5.34), as in (4.5.11), represent the forward and the backward waves, respectively. We may notice that the coefficients of the factors $\exp j(\omega t \pm \beta z)$ in (4.5.34) represent the amplitudes. Thus the amplitudes of the forward and the backward waves, as they progress, each attenuate exponentially. Here, z has to be interpresented as positive, for the forward wave, that is, for the first term, and as negative, for the backward wave, that is, for the second term, of (4.5.34). Now, considering the forward wave only, one may choose to express the amplitude of the electric field intensity as:

$$E_{x_1,y_1} \exp(-\alpha z) = E_{x_1,y_1} \exp(- z/\delta) \tag{4.5.35}$$

where

$$\delta = 1/\alpha \ (=1/\beta)$$

which, with the help of (4.5.33), is given by:

$$\delta = \frac{1}{(\pi\,f\,\mu\sigma)^{1/2}} , \tag{4.5.36}$$

where $f\,(=\omega/2\pi)$ is the frequency of the wave. α is called the attenuation constant (It may be noted that β here, as in section *4.5.1*, continues to be called as the phase propagation constant). δ is called the skin depth of the conducting medium. It is so called because, if we define $z = 0$ as the plane of interface between a free space and a semi-infinite conducting medium, then the depth of penetration of the electric field into the conducting medium ($\infty > z > 0$), beyond the interface ($z = 0$), the *skin* of the conductor, before the amplitude of the field is reduced to (1/e) times its value at the interface or skin, as can be seen with the help of (4.5.35), is $z = \delta$.

The value of the skin depth, given by (4.5.36), decreases with the increase of the frequency of the wave and conductivity of the medium. Thus, for a highly conducting medium and or at high frequencies, the amplitude of the field significantly attenuates even with small distances into a conducting medium; in other words, the field is confined near the surface or skin of the conductor ($\delta \to 0$ as σ, $f \to \infty$). The phenomenon is called the skin effect.

Further, as in section *4.5.1*, one may, starting from the wave equation (4.5.3), with $\mathbf{J}_c = \sigma\, \mathbf{E}$, and using (4.4.2(b)), easily show:

$$H_z = 0 \tag{4.5.37}$$

and

$$H_{x,y} = H_{x_1,y_1} \exp(j\omega t - \Upsilon z) + H_{x_2,y_2} \exp(j\omega t + \Upsilon z)\,. \tag{4.5.38}$$

It is obvious from (4.5.27), (4.5.29), (4.5.37) and (4.5.38) that, in the conducting medium also, as in free space or in a homogeneous isotropic dielectric, the electric and magnetic fields of a uniform plane wave will be each transverse to the direction of propagation, having no component along the latter direction. In other words, an isotropic, homogeneous conductor, like a free space or dielectric, supports the TEM mode of a uniform plane wave.

<u>*Example 4.5.3*</u> For a TEM mode supported by a conducting medium, obtain an expression for the intrinsic impedance of the medium, and show that the electric and magnetic fields are orthogonal.

Following exactly the same method as in Example 4.5.2, one may easily obtain, starting from (4.4.4b), an equation analogous to (4.5.15) as follows:

$$-\frac{\partial H_y}{\partial z} = \sigma E_x + \varepsilon_0 \frac{\partial E_x}{\partial t} \qquad \text{(a)}$$

$$\frac{\partial H_x}{\partial z} = \sigma E_y + \varepsilon_0 \frac{\partial E_y}{\partial t}\,. \qquad \text{(b)} \tag{4.5.39}$$

Substituting (4.5.29) and (4.5.38), however, taking in each of them the first term to consider only the forward wave as in Example 4.5.2, into (4.5.39), we obtain:

$$\Upsilon H_y = (\sigma + j\omega\varepsilon_0)\, E_x \qquad \text{(a)}$$

$$-\Upsilon H_x = (\sigma + j\omega\varepsilon_0)\, E_y\,. \qquad \text{(b)} \tag{4.5.40}$$

From (4.5.40), one may write using (4.5.30):

$$\eta = \frac{E_x}{H_y} = -\frac{E_y}{H_x} = \left(\frac{j\omega\mu}{\sigma + j\omega\varepsilon}\right)^{1/2}. \tag{4.5.41}$$

It is of interest to check that, as a special case $\sigma \ll \omega\varepsilon$, (4.5.41) passes on to (4.5.16) in which μ_0 and ε_0 have to be taken as μ and ε, respectively, as appropriate to the medium taken. Starting from (4.5.41), and proceeding as discussed following (4.5.16) in Example 4.5.2, one may again show that

$$\mathbf{E} \cdot \mathbf{H} = 0 ,$$

that is, the electric and magnetic field vectors are orthogonal; and also that $\frac{E}{H} = \eta = (\frac{j\omega\mu}{\sigma + j\omega\varepsilon})^{1/2}$, the intrinsic impedance of the conducting medium.

<u>*Example 4.5.4*</u> Obtain an expression for the surface resistance of a semi-infinite planar conductor.

Let a uniform plane wave, travelling from a free-space region, be incident on a semi-infinite planar conductor occupying a region: $0 < z < \infty$, the direction of propagation being z (see Fig. 4.5.1) [10]. Let the coordinate axis be so chosen that the electric field is directed along y. The electric field intensity **E** inside the conductor associated with a wave travelling in the positive z direction into the conductor is then given by the first term of (4.5.34):

$$\mathbf{E} = E_{y1} \exp(-\alpha z) \exp j(\omega t - \beta z)\, \mathbf{a}_y . \tag{4.5.42}$$

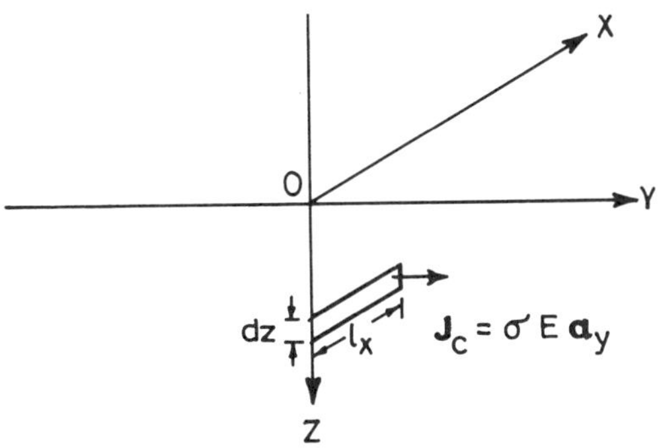

Fig. 4.5.1 A rectangular strip of width l_x, along x, perpendicular to the length of the conductor along y, and thickness Δz, along z, considered at a depth z, inside the conductor, measured from the surface (XY-plane) of a semi-infinite conductor, occupying the region: $o < z < \infty$.

The conduction current density $\mathbf{J}_d$ inside the conductor at a depth z below the free-space-conductor interface is given by (Ohm's law)

$$\mathbf{J}_c = \sigma \mathbf{E} . \tag{4.5.43}$$

With the help of (4.5.42) and (4.5.43) the *amplitude* (time-independent part) of the elemental conduction current, ΔI, through an elemental strip of width l_x and thickness dz (Fig. 4.5.1) may be written as:

$$\Delta I = \mathbf{J}_c \cdot \mathbf{dS} = [\sigma E_{y1} \exp\{-(\alpha + j\beta)z\}]\, \mathbf{a}_y \cdot (l_x \Delta z)\, \mathbf{a}_y , \tag{4.5.44}$$

where the second term of the dot product, namely **ds** (= $l_x \Delta z\, \mathbf{a}_y$) is the area vector of the elemental strip. The total current per *unit width* , also known as the linear current density, J_s is obtained by first evaluating the dot-product involved in (4.5.44), then integrating the latter from z = 0 to ∞, and finally dividing it by the width of the strip l_x, as follows:

$$J_s = \int_0^\infty \sigma E_{y1} \exp[-(\alpha + j\beta)z]\, dz \,. \tag{4.5.45}$$

The surface impedance z_s is defined as:

$$Z_s = \frac{E_{y1}}{J_s} \,, \tag{4.5.46}$$

being the ratio of the potential drop per *unit length* at the free-space-conductor interface (z = 0) (Fig. 4.5.1) to the current *per unit width* . Thus the surface impedance may also be called the impedance per unit length per unit width. The unit of J_s is amp/m and that of Z_s is ohms per square (some times expressed as ohms/□) or, simply, ohms. Evaluating the integral of (4.5.45), and substituting it in (4.5.46), one then obtains

$$Z_s = \frac{\alpha}{\sigma} + j\frac{\beta}{\sigma} \,. \tag{4.5.47}$$

Making use of the relation α = β = 1/δ (see the relation following (*4.5.35*), one may write (4.5.47) as:

$$Z_s = \frac{1}{\sigma\delta} + j\frac{1}{\sigma\delta} \,. \tag{4.5.48}$$

The real part of Z_s is called the surface resistance of the conductor R_s which is thus given by:

$$R_s = \frac{1}{\sigma\delta} \,. \tag{4.5.49}$$

Making use of (4.5.36), one may also express (4.5.49) as:

$$R_s = \left(\frac{\pi f \mu}{\sigma}\right)^{1/2} \,. \tag{4.5.50}$$

<u>*Example 4.5.5*</u> Show that the surface impedance of a thick planar conductor is equal to the intrinsic impedance of the conducting medium.

The intrinsic impedance of a conductor η has already been found as (see Example 4.5.3):

$$\eta = \left(\frac{j\omega\mu}{\sigma + j\omega\varepsilon}\right)^{1/2}$$

which, for a good conductor (σ >> ωε), may be written as

$$\eta = \frac{1}{\sigma}(j\omega\mu\sigma)^{1/2}$$

which may also be expressed as (see (4.5.31)-(4.5.33)):

$$\eta = \frac{\alpha}{\sigma} + j\frac{\beta}{\sigma} . \qquad (4.5.51)$$

Comparing (4.5.47) with (4.5.51),

$$Z_s = \eta \quad (\sigma >> \omega\varepsilon) ,$$

showing that, in the present case, the surface impedance and the intrinsic impedance are equal.

Example 4.5.6 For a solid cylindrical conductor of high conductivity, show that the high-frequency ac-to-dc resistance ratio is equal to the ratio of the radius of cross-section-to- twice the skin depth.

At high frequencies, and for a high-conductivity conductor, the value of the skin depth as given by (4.5.36) would be much smaller than the radius of cross section of the conductor ($\delta << a$). Consequently, one may ignore the curvature of the conductor and *develop* the latter as a planar conductor of width equal to the circumference of the conductor, $2\pi a$, where a is the radius of the cross-section of the conductor. If l is the length of the cylindrical conductor considered, then the ac resistance, R_{ac}, of the conductor may be written as:

$$R_{ac} = R_s \frac{1}{2\pi a} , \qquad (4.5.52)$$

while its dc resistance, R_{dc} , as:

$$R_{dc} = \frac{1}{\sigma}\frac{1}{\pi a^2} . \qquad (4.5.53)$$

Dividing (4.5.52) by (4.5.53), one obtains

$$\frac{R_{ac}}{R_{dc}} = \frac{R_s \sigma a}{2} . \qquad (4.5.54)$$

Substituting (4.5.49) into (4.5.54), the ratio of the two resistances becomes:

$$\frac{R_{ac}}{R_{dc}} = \frac{a}{2\delta} .$$

Clearly, in view of $\delta << a$, one has $R_{ac} >> R_{dc}$.

Example 4.5.7 For sub-marine communication through sea-water ($\sigma = 4$ mho/m; $\varepsilon_r = 80$), show that relatively low frequencies should be preferred, from the standpoint of attenuation.

For sea-water, $\sigma = 4$ mho/m; $\varepsilon_r = 80$,

$$\frac{\omega\varepsilon}{\sigma} = \frac{2\pi f \varepsilon_0 \varepsilon_r}{\sigma} = 1.1 \times 10^{-9} \quad \text{(sea-water)} . \qquad (4.5.55)$$

For relatively 'low' frequencies, say, $f = \omega/2\pi = 10$ kHz ($= 10^4$ Hz), typically, we see from (4.5.55) that $\omega\varepsilon/\sigma \sim 1.1\times10^{-5}$, a value much less than unity. For $\omega\varepsilon/\sigma \ll 1$, that is, for such low frequencies, the expression for the attenuation constant α_{low} is given by (see (a4.5.15)):

$$\alpha_{low} = \left(\frac{\omega\mu\sigma}{2}\right)^{1/2} . \tag{4.5.56}$$

Similarly, for relatively 'high' frequencies, say, $f = 10$ GHz (10^{10} Hz), typically, (4.5.54) yields $\omega\varepsilon/\sigma \sim 11$. For such a case, taking $\omega\varepsilon/\sigma \gg 1$, one has the expression for the attenuation constant, α_{high}, as (see (a4.5.14)):

$$\alpha_{high} = \frac{\sigma}{2}\left(\frac{\mu}{\varepsilon}\right)^{1/2} . \tag{4.5.57}$$

By division, (4.5.56) and (4.5.57) give

$$\frac{\alpha_{low}}{\alpha_{high}} = \left(\frac{2\omega\varepsilon}{\sigma}\right)^{1/2} , \tag{4.5.58}$$

where ω refers to the low frequency value. For $f = (\omega/2\pi) = 10$ kHz, the ratio of the two attenuation constant values from (4.5.55) comes out to be: $\alpha_{low}/\alpha_{high} \sim 4.6\times10^{-3}$, showing that the low-frequency attenuation constant is much less than the high frequency attenuation constant. Therefore, for submarine communication, through sea-water, relatively lower frequencies are preferred [11].

As for the size of the radiating system (antenna), the use of low values of frequency would lead to relatively larger sizes of the radiating elements. However, one point of relief is that the size of the radiating element, which increases with the wavelength in the medium, is less in sea-water than in free space. This becomes clear from the expression for the wavelength in the conducting medium which is given, with the help of (a 4.5.15) and (4.5.36), by

$$\lambda_{medium} = \frac{2\pi}{\beta} = 2\pi\frac{1}{(\pi f \mu_0 \sigma)^{1/2}} = 2\pi\delta . \tag{4.5.59}$$

It is of interest to compare λ_{medium} with the wavelength in free space given by

$$\lambda_{freespace} = \frac{c}{f} . \tag{4.5.60}$$

For $f = 10^4$ Hz, typically, we thus have, from (4.5.60), the free space wavelength as 3×10^4 m (= 30 km !), suggesting antenna sizes ($\sim \lambda$) of fantastic dimensions. However, for the same frequency, (4.5.59) yields the value of $\lambda_{medium} \sim 15$ m which makes the antenna sizes quite reasonable.

Appendix a4.5.1 *Representation of a propagating wave*

Let us try the following function to represent a physical quantity associated with a propagating wave:

$$f\{z, t\} = f\{t - z/v\} , \tag{a4.5.1}$$

where v is a constant having the dimension of a velocity. We may see from (a 4.5.1) that

$$f\{z + \Delta z, t + \Delta t\} = f\{t + \Delta t - (z + \Delta z)/v\} = f\{t - z/v + \Delta t - \Delta z/v\} . \tag{a4.5.2}$$

If we now choose to impose the condition

(a4.5.3) $$\Delta t - \Delta z / v = 0$$

in (a4.5.2), we obtain

(a4.5.4) $$f\{z + \Delta z, t + \Delta t\} = f\{t - z/v\} .$$

Combining (a 4.5.1) and (a 4.5.4) it is easy to see that

(a 4.5.5) $$f\{z, t\} = f\{z+\Delta z, \ t+\Delta t\} .$$

It is stated by (a 4.5.5) that, subject to (a 4.5.3) which, in the limit $\Delta t \to 0$, may be represented as

(a 4.5.6) $$v = dz / dt \ ,$$

the physical quantity represented by the chosen function at a point z at a time t is realised also at a point Δz apart in the direction of propagation of the wave after a time Δt, the velocity of the wave being v, as suggested by (a 4.5.6).

If we now choose to take the variation of the physical quantity of interest to be periodic, then in phasor notation the quantity, for instance, the magnitude of the electric field intensity of a forward wave may be written as:

(a 4.5.7) $$E_1 \ \{z, t\} = f\{t - z/v\} = E_{01} \ \exp j\omega (t - z/v) ,$$

where E_{01} is the amplitude of the electric field intensity E_1 of the forward wave.Similarly, it is easy to appreciate that a function

$$f\{z, t\} = f\{t + z/v\}$$

represents a quantity associated with a wave travelling in the negative z direction. Subject to (a 4.5.6), one may see, in this case, that

$$f\{z, t\} = f\{z-\Delta z, \ t+\Delta t\} .$$

Thus, for a wave travelling in the negative z direction, the expression for the magnitude of the electric field intensity (analogous to (a 4.5.7)) may be written as:

(a 4.5.8) $$E_2 \ \{z, t\} = f\{t + z/v\} = E_{02} \ \exp j\omega (t + z/v) ,$$

where E_{02} is the amplitude of the electric field intensity E_2 of the backward waves.

The phase velocity of the wave travelling in the positive z-direction may be obtained by putting the phase part of the electric field intensity given by the right hand side of (a4.5.7) as a constant:

$$\omega (t - z/v) = a \text{ constant},$$

which, upon differentiation, yields

(a 4.5.9) $$dz / dt = - v .$$

It is to be noted that, for a wave travelling in the positive z-direction, (a4.5.6) and (a4.5.8) are identical. Now, writing β ($=\omega / v$) as the phase propagation constant, one may write expressions, alternative to (a4.5.7) and (a4.5.8), as:

(a4.5.10) $$E_1 = E_{01} \ \exp \ j(\omega t - \beta z) \ \text{(forward wave)} , \text{ and}$$

$$E_2 = E_{02} \ \exp \ j(\omega t + \beta z) \ \text{(backward wave)} .$$

(a4.5.10) has been used to interpret the first and second terms of (4.5.11).

Appendix a4.5.2 Complex propagation constant for a conducting medium

Combining (4.5.31) with (4.5.32), one has

$$\alpha + j\beta = (j\omega\mu\sigma)^{1/2}$$

which on squaring yields

$$\alpha^2 - \beta^2 + 2j\alpha\beta = j\omega\mu\sigma \; . \qquad \text{(a4.5.11)}$$

Equating the real and imaginary parts of (a4.5.11), we have

$$\alpha = \beta$$

and

$$\alpha\beta = \omega\mu\sigma/2$$

whence we have then

$$\alpha = \beta = (\omega\mu\sigma/2)^{1/2} , \qquad \text{(a 4.5.12)}$$

which is identical with (4.5.33). Similarly, starting from (4.5.30), instead of (4.5.31), it is easy to show

$$\alpha, \beta = \left(\frac{\omega\mu(\sigma^2 + \omega^2\varepsilon^2)^{1/2} \mp \omega^2\mu\varepsilon}{2} \right)^{1/2} . \qquad \text{(a 4.5.13)}$$

For $\sigma << \omega\varepsilon$, one may choose to write (a4.5.13) as

$$\alpha, \beta = \left(\frac{\omega^2\mu\varepsilon[(1 + \sigma^2/\omega^2\varepsilon^2)^{1/2} \mp 1]}{2} \right)^{1/2}$$

and expand the latter binomially, while remembering that $\sigma/\omega\varepsilon << 1$, and show

$$\alpha = \frac{\sigma}{2}(\mu/\varepsilon)^{1/2}; \quad \beta = \omega(\mu\varepsilon)^{1/2} \quad (\sigma << \omega\varepsilon) . \qquad \text{(a 4.5.14)}$$

For $\sigma >> \omega\varepsilon$, similarly, one may by expressing (a 4.5.13) as:

$$\alpha, \beta = \left(\frac{\omega\mu\sigma(1 + \omega^2\varepsilon^2/\sigma^2)^{1/2} \mp \omega^2\mu\varepsilon}{2} \right)^{1/2}$$

and expanding the latter binomially, while remembering that $\omega\varepsilon << \sigma$, show

$$\alpha, \beta = \left(\frac{\omega\mu\sigma}{2}\right)^{1/2} \quad (\sigma >> \omega\varepsilon) \qquad \text{(a 4.5.15)}$$

which is identical with (a 4.5.12).

4.6 Scalar and Vector Potentials in Time-Varying Fields

We have seen in the preceding chapters that the static electric field intensity may be conveniently obtained from the scalar electrostatic potential by taking its negative gradient (see (2.5.11)) and the magnetic flux density in the non-time varying case from the magnetic vector potential by taking its curl (see (3.2.27)). Let us now see if one can extend the concepts of scalar and vector potentials to the time-varying situations, too.

One of the Maxwell's equations, (4.4.3(b)), may be re-written as:

$$\nabla \times \mathbf{E} = -\frac{\partial \mathbf{B}}{\partial t} . \tag{4.6.1}$$

Expression **B** in terms of vector potential **A** as (see (3.2.27)):

$$\mathbf{B} = \nabla \times \mathbf{A} , \tag{4.6.2}$$

and substituting (4.6.2) in (4.6.1), one has

$$\nabla \times \mathbf{E} = -\frac{\partial}{\partial t}(\nabla \times \mathbf{A}). \tag{4.6.3}$$

The curl operation essentially involves partial derivatives. Therefore, one may choose to interchange the positions of the partial derivatives in the right hand side of (4.6.3). This enables one to express (4.6.3) as:

$$\nabla \times \left(\mathbf{E} + \frac{\partial \mathbf{A}}{\partial t} \right) = 0 . \tag{4.6.4}$$

Now, as we know, the curl of the gradient of a scalar vanishes. Here, as can be seen from (4.6.4), the curl of the vector that has vanished is $\mathbf{E} + \partial \mathbf{A}/\partial t$. Therefore, one may put the latter as equal to the gradient of a scalar ψ, say. In other words, one may write:

$$\mathbf{E} + \frac{\partial \mathbf{A}}{\partial t} = \nabla \psi$$

whence we have the electric field intensity in the time-varying case as

$$\mathbf{E} = \nabla \psi - \frac{\partial \mathbf{A}}{\partial t} . \tag{4.6.5}$$

In the non-time-varying case, $\partial/\partial t = 0$, (4.6.5) should pass on to the static field expression (2.5.11) which may be re-written here as:

$$\mathbf{E} = -\nabla V . \tag{4.6.6}$$

Therefore, putting $\partial/\partial t = 0$ in (4.6.5) and comparing the latter with (4.6.6), we immediately identify

$$\psi = -V . \tag{4.6.7}$$

In view of (4.6.7), the expression for the electric field intensity, (4.6.5), takes the following form:

$$\mathbf{E} = -\nabla V - \frac{\partial \mathbf{A}}{\partial t}, \tag{4.6.8}$$

in terms of a scalar and a vector potential, V and **A**, respectively.

Let us now express, in terms of **A**, one of the Maxwell's equations, namely, (4.4.4(b)), which may be written in terms of **B** (= μ**H**) and **E** (= **D**/ε) as:

$$\nabla \times \mathbf{B} = \mu \mathbf{J} + \mu\varepsilon \frac{\partial \mathbf{E}}{\partial t}. \tag{4.6.9}$$

Substituting (4.6.2) and (4.6.8) in (4.6.9), and remembering the vector identity $\nabla\times\nabla\times\mathbf{A} = \nabla(\nabla.\mathbf{A}) - \nabla^2\mathbf{A}$, one obtains the following equation in terms of the two potentials **A** and V:

$$\Box^2 \mathbf{A} = \nabla^2 \mathbf{A} - \mu\varepsilon \frac{\partial^2 \mathbf{A}}{\partial t^2} = -\mu \mathbf{J} + \nabla(\nabla.\mathbf{A} + \mu\varepsilon \frac{\partial V}{\partial t}). \tag{4.6.10}$$

We may obtain another equation in **A** and V by taking the divergence of (4.6.8), as follows:

$$\nabla.\mathbf{E} = -\nabla^2 V - \nabla.\frac{\partial \mathbf{A}}{\partial t}. \tag{4.6.11}$$

Now, since the divergence essentially involves partial derivatives, one may interchange the partial derivatives in the right hand side of (4.6.11). Also, one may substitute (4.4.1) in the left hand side of (4.6.11). Thus one may easily obtain from (4.6.11):

$$\nabla^2 V + \frac{\partial}{\partial t}(\nabla.\mathbf{A}) = -\frac{\rho}{\varepsilon}. \tag{4.6.12}$$

Equation (4.6.10) and (4.6.12) each involve, V and **A**, and each of them may be decoupled, subject to the following condition known as Lorentz condition:

$$\nabla.\mathbf{A} + \mu\varepsilon \frac{\partial V}{\partial t} = 0. \tag{4.6.13}$$

Thus, if we substitute (4.6.13) in (4.6.10) and (4.6.12), we obtain the following symmetric inhomogeneous wave equations separately in **A** and V, respectively [12], (compare them with (4.5.1)):

$$\Box^2 \mathbf{A} = \nabla^2 \mathbf{A} - \mu\varepsilon \frac{\partial^2 \mathbf{A}}{\partial t^2} = -\mu \mathbf{J} \tag{4.6.14}$$

and

$$\Box^2 V = \nabla^2 V - \mu\varepsilon \frac{\partial^2 V}{\partial t^2} = -\frac{\rho}{\varepsilon}. \tag{4.6.15}$$

Thus, in (4.6.14) and (4.6.15), we find **A** and V, respectively, both decoupled in the form of two wave equations in these quantities.

Next let us proceed to find the solutions of (4.6.14) and (4.6.15) for **A** and V, respectively. For this purpose let us recall (section 2.5) how in the electrostatic case one obtains the electric field intensity from the scalar potential which, in turn, may be expressed in terms of a static charge distribution, with the help of (2.4.2), as:

$$V = \frac{1}{4\pi\varepsilon}\int_{\tau}\frac{\rho d\tau}{r}, \tag{4.6.16}$$

where ρ is the volume charge density at a point called the source point where we consider a volume element $d\tau$, and r is the distance of the latter from the point of observation where the electrostatic potential is sought. Here, we have considered the element of charge $\rho d\tau$ in the volume element to be essentially a point change thereby making it possible to use (2.4.2) to find the electrostatic potential due to the charge element considered. Integrating such potentials over the entire volume of charge distribution, one obtains (4.6.16). Further, by putting $\partial/\partial t = 0$, and hence $\partial^2/\partial t^2 = 0$ (for the static case) in (4.6.15), we have

$$\nabla^2 V = -\frac{\rho}{\varepsilon_0}, \tag{4.6.17}$$

which, as expected becomes identical with Poisson's equation (2.6.2).

Now, if we move on to the time-varying case ($\partial/\partial t \neq 0$) but let $\rho = 0$ in (4.6.15), we have

$$\nabla^2 V = \mu\varepsilon\frac{\partial^2 V}{\partial t^2}. \tag{4.6.18}$$

In form, (4.6.18) is seen to be identical with the homogeneous wave equation (4.5.5). Therefore, the solution of (4.6.18) could be identical in form with that of (4.5.5). In other words, the solution of (4.6.18) for V will be given by (4.5.11) and hence contain a phase factor of $\exp j(\omega t - \beta r) = \exp j\omega(t - r/v_p)$ (in view of the relation $\beta = \omega/v_p$, where $v_p = (\mu\varepsilon)^{-1/2}$, see the discussion following (4.5.9) and Example 4.5.1), corresponding to a forward wave directed from the source point to the observation point. Now, if we combine i)the foregoing result for the time varying case ($\partial/\partial t \neq 0$) and $\rho = 0$ with ii) the result for the static case ($\partial/\partial t = 0$) and $\rho \neq 0$, that is with (4.6.16), we can easily guess that the solution of (4.6.15), in the general case for $\partial/\partial t \neq 0$ and $\rho \neq 0$, would be:

$$V = \frac{1}{4\pi\varepsilon}\int_{\tau}\frac{\rho \exp j\omega(t - r/v_p)}{r}d\tau. \tag{4.6.19}$$

Similarly, starting from the expression (3.2.28) for the magnetic vector potential in terms of a given current distribution in steady magnetic field case, one may guess the solution of (4.6.14) as (for the details, see appendix *a4.6.1*):

$$\mathbf{A} = \frac{\mu}{4\pi}\int_{\tau} \frac{\mathbf{J} \exp j\omega(t - r / v_p)}{r} d\tau . \tag{4.6.20}$$

It is obvious from the expression for the scalar and vector magnetic potentials given by (4.6.19) and (4.6.20) that as the source quantities, ρ and **J**, vary as exp($j\omega t$), at an instant of time t, the potentials, V and **A**, vary as exp $j\omega t$, where $t' = t - r/v_p$ is a time which is earlier with respect to t. In other words, these potentials are *retarded* to the extent of $t - t' = r/v_p$, the disturbances in the source taking a time r/v_p to reach the observation point at a distance r, while travelling with the velocity v_p. One may also choose to write (4.6.19) and (4.6.20), the expression for the retarded potentials, at an instant of time t, $(V)_t$ and $(\mathbf{A})_t$, in terms of the source quantities $(\rho)_{t-r/v_p}$ and $(\mathbf{J})_{t-r/v_p}$ at an earlier instant of time $t - r/v_p$, by keeping the time dependent part as understood, as follows:

$$(V)_t = \frac{1}{4\pi\varepsilon}\int \frac{(\rho)_{t-r/v_p}}{r} d\tau , \tag{4.6.21}$$

and

$$(\mathbf{A})_t = \frac{\mu}{4\pi}\int \frac{(\mathbf{J})_{t-r/v_p}}{r} d\tau . \tag{4.6.22}$$

Now that the potentials are obtainable with the help of (4.6.19) and (4.6.20), one can find the electric and magnetic fields with the help of (4.6.8) and (4.6.2). It appears that, whereas only the vector potential **A** is required to find **B** (see (4.6.2)), both the scalar potential V and the vector potential **A** are required to find **E**(see (4.6.8)). However, writing by phasor notation, $\partial V/\partial t$ as $j\omega V$(see (4.6.19)) and $\partial \mathbf{A}/\partial t$ as $j\omega\mathbf{A}$(see (4.6.20)), we can write **E**, in terms of only **A**, with the help of (4.6.8) and (4.6.13), as follows:

$$\mathbf{E} = - j\omega\left(\mathbf{A} + \frac{\nabla(\nabla.\mathbf{A})}{\omega^2\mu\varepsilon}\right). \tag{4.6.23}$$

The relation (4.6.2) and (4.6.23) find extensive applications in the study of antennas which have, however, not received any special attention in this treatise, though one may see these relations applied in the problem of an infinitesimal dipole in Example 4.6.1 to follow, which may be considered as the first step to entering the area of antennas. However, we intend to apply, in the following chapter (*5*), some of the concepts developed in this section in deriving an equivalent-circuit model of a structure which supports slow electromagnetic waves; such a structure finds applications, as we shall see (in chapter *8*), in a growing-wave electron beam device — the travelling-wave tube.

Example 4.6.1 Deduce the expressions for the electric and magnetic field intensities due to an infinitesimal oscillating dipole of a given current amplitude placed in free space. Hence show also that, at very large distances from the dipole, the wave impedance is equal to the intrinsic impedance of the medium.

Let us place the infinitesimally small dipole as coincident with the z-axis, and at the origin, of a rectangular system of coordinates (Fig. 4.6.1) [2]. The length of the dipole is so small that one may take the amplitude of the current over it to be the same everywhere. For such a one-dimensional current element one may define A_Z , the z-component of the vector potential at a point P, where the field quantities are sought, at a distance r from the origin, as (see (4.6.20)):

$$A_z = \frac{\mu_0}{2\pi} J\, d\tau \exp(-j\beta r)\ , \tag{4.6.24}$$

where $\exp(j\omega t)$ is understood; J is the amplitude of the one-dimensional current density along z. It is implied that the filamentary current element is of uniform cross section α, say. Then one may express J as I/α and $d\tau$ as αdl, where dl is the infinitesimal length of the current element. This enables one to write

$$J\, d\tau = I dl$$

which when substituted in (4.6.24) makes

$$A_z = \frac{\mu_0}{2\pi} I dl \exp(-j\beta r)\ . \tag{4.6.25}$$

It is useful here to represent the point P, where the field quantities are sought, by the coordinates r, θ, ϕ in the spherical-polar system, the origin of which is considered to be the same as that of the rectangular system (Fig. 4.6.1). (The spherical-polar coordinates of P are related to its rectangular coordinates x, y, z as: $x = r \sin\theta \cos\phi$, $y = r \sin\theta \sin\phi$ and $z = r \cos\theta$).

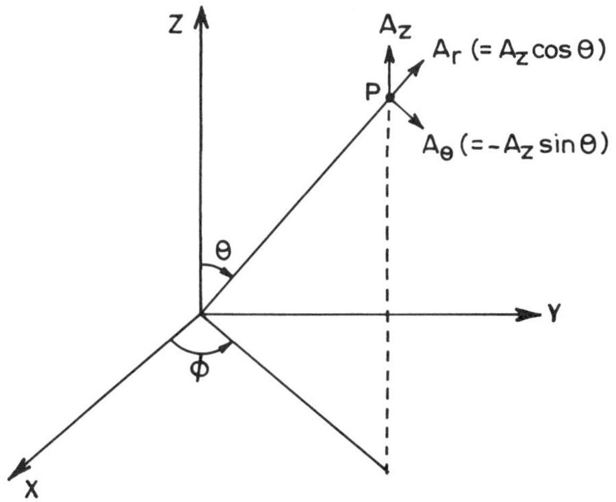

Fig. 4.6.1 The components of vector potential due to an infinitesimal oscillating dipole.

The radial (r –), zenithal (θ –) and azimuthal (ϕ –) components of the vector potential may be written, in terms of A_Z, the z-component of the vector potential, as:

$$A_r = A_z \cos\theta, \quad A_\theta = -A_z \sin\theta \quad \text{and} \quad A_\phi = 0\,. \tag{4.6.26}$$

One may now first find **∇.A** with the help of (2.3.5(c)), (4.6.25) and (4.6.26) as:

$$\nabla.\mathbf{A} = \frac{-\mu_0 Idl}{4\pi}\left(\frac{1}{r^2}+\frac{j\beta}{r}\right)\cos\theta\exp(-j\beta r)\;; \tag{4.6.27}$$

and then obtain the gradient of (4.6.27), with the help of (2.5.12(c)), as:

$$\nabla(\nabla.\mathbf{A}) = -\frac{\mu_0 Idl}{4\pi}\left(-\frac{2}{r^3}-\frac{2j\beta}{r^2}+\frac{\beta^2}{r}\right)\cos\theta\exp(-j\beta r)\mathbf{a}_r$$

$$+\frac{\mu_0 Idl}{4\pi}\left(\frac{1}{r^3}+\frac{j\beta}{r^2}\right)\sin\theta\exp(-j\beta r)\mathbf{a}_\theta\,. \tag{4.6.28}$$

The expression for ∇(**∇.A**), (4.6.28), thus obtained, and that for **A**, which may be written with the help of (4.6.26) and (4.6.25), as

$$\mathbf{A} = A_z\cos\theta\,\mathbf{a}_r - A_z\sin\theta\,\mathbf{a}_\theta$$

$$= \frac{\mu_0 Idl}{4\pi}\cos\theta\exp(-j\beta r)\mathbf{a}_r - \frac{\mu_0 Idl}{4\pi}\sin\theta\exp(-j\beta r)\mathbf{a}_\theta, \tag{4.6.29}$$

may now substituted in (4.6.23) to obtain an expression for the electric field intensity **E** which component-wise reads as:

$$E_r = -\frac{j\,Idl}{2\pi}\frac{\eta}{\beta}\cos\theta\left(\frac{1}{r^3}+\frac{j\beta}{r^2}\right)\exp(-j\beta r)$$

$$E_\theta = -\frac{j\,Idl}{4\pi}\frac{\eta}{\beta}\sin\theta\left(\frac{-\beta^2}{r}+\frac{1}{r^3}+\frac{j\beta}{r^2}\right)\exp(-j\beta r)$$

$$E_\phi = 0\,, \tag{4.6.30}$$

where $\eta = (\mu_0/\varepsilon_0)^{1/2}$ and $\beta = \omega(\mu_0\varepsilon_0)^{1/2}$. η and β ($= \omega/v_p$) may be identified as the intrinsic impedance and the propagation constant, (v_p(= c) being the phase velocity), respectively, as relevant to propagation of electromagnetic waves in free space (see (4.5.12) and (4.5.16)).

Similarly, the magnetic flux density **B** (= ∇×**A**; see (4.6.2)) may be obtained by taking the curl of (4.6.29) with the help of (3.2.18(c)), whence the expression for **H** (= **B**/μ_0) follows which may be written, component-wise, as:

$$H_r = 0, \quad H_\theta = 0, \quad H_\phi = \frac{Idl}{4\pi}\sin\theta\left(\frac{j\beta}{r}+\frac{1}{r^2}\right)\exp(-j\beta r)\,. \tag{4.6.31}$$

At large distances from the dipole, one may ignore, the terms in (4.6.30) and (4.6.31) that vary inversely as r^2 and r^3 enabling one to write:

$$E_r = 0, \quad E_\theta = \frac{j\,\text{idl}\,\eta\beta}{4\pi r}\sin\theta\exp(-j\beta r), \quad E_\phi = 0\,;$$

$$H_r = 0, \quad H_\theta = 0, \qquad H_\phi = \frac{j\text{Idl}\beta}{4\pi r}\sin\theta\exp(-j\beta r). \tag{4.6.32}$$

The wave impedance E_θ/H_ϕ then can easily be seen, from (4.6.32), to be equal to η, the intrinsic impedance of the medium, here, a free space.

Appendix a4.6.1 *Solution of the inhomogeneous wave equation in vector potential*

Replacing the current element $I\,\mathbf{dl}$ by $\mathbf{J}\,d\tau$ (and writing μ in place of μ_0, in general) in (3.2.8), one may write

(a4.6.1) $$\mathbf{A} = \frac{\mu}{4\pi}\int_\tau \frac{\mathbf{J}\,d\tau}{r},$$

for the steady magnetic field case. Further, component-wise, in the rectangular system of coordinates, say, one may write (a 4.6.1) as:

(a4.6.2) $$A_{x,y,z} = \frac{\mu}{4\pi}\int_\tau \frac{J_{x,y,z}\,d\tau}{r}.$$

Equation (a 4.6.2), for each component, is similar to the integral expression for electrostatic potential in terms of static charge distribution, namely, (4.6.16). Here μ and $J_{x,y,z}$ of (a4.6.2) are analogous to $1/\varepsilon$ and ρ, respectively, of (a4.6.2). Extending the analogy to the point form of scalar potential (4.6.17), that is, to Poisson's equation, one may then write,

(a4.6.3) $$\nabla^2 A_{x,y,z} = -\mu J_{x,y,z}\,.$$

Summing up the components of (a4.6.3) vectorialy, one may thus form the following vector Poisson's equation in the steady magnetic field case:

(a4.6.4) $$\nabla^2\mathbf{A} = -\mu\mathbf{J}\,.$$

In the time-varying situation, but with $\mathbf{J} = 0$, we have, from (4.6.14), the following equation:

(a4.6.5) $$\nabla^2\mathbf{A} = \mu\varepsilon\frac{\partial^2\mathbf{A}}{\partial t^2}.$$

Now, in section (4.6.1), the solution of the inhomogeneous equation (4.6.15) was obtained, with the help of (4.6.16), (4.6.17) and (4.6.18), as (4.6.19). Following the same approach, one may similarly obtain the solution of the inhomogeneous equation (4.6.14), now with the help of (a4.6.1), (a4.6.4) and (a4.6.5), as (4.6.20).

4.7 Electromagnetic Boundary Conditions

The electric and magnetic fields obey Maxwell's equations (4.4.1(b))-(4.4.4(b)), in their differential or point form, at every point in a medium. Now, if the medium has an interface with another medium, then the medium properties abruptly change at the interface, and, consequently, the field quantities on the two sides of the interface [12], in these two different media, would, in general, be also different. However, the quantities would pass through a common set of boundary conditions at the interface; these condition can be obtained using the integral form of Maxwell's equations (4.4.1(a))-(4.4.4(a)) by a method to be discussed in this section. Subject to these boundary conditions, the field quantities in either of the media may be determined by solving the wave equations (4.5.1) and (4.5.3). Further, these boundary conditions can be used in predicting the characteristics of a physical structure which supports an electromagnetic wave; one such use will be made in the next chapter while studying a practical wave-guiding structure.

Practical situations which quite often interest us are the interface between one dielectric and another dielectric and that between a dielectric and a conductor; further, for each of these situations, we may be interested in static and time-varying phenomena. The general boundary conditions as well as those valid in these individual situations and phenomena which may be derived as special cases therefrom are listed in Table 4.7.1 (see Example 4.7.1). All these boundary conditions concern with either the normal components of electric displacement and magnetic flux density or the tangential components of electric and magnetic field intensities at the interface between two media. In Table 4.7.1, the subscripts 1 and 2 refer to the quantities in the two regions 1 and 2, say, respectively, and the values of these quantities are taken at a point on the interface; $\mathbf{a}_n$ is the unit vector directed from the region 1 to 2, at the point. ρ_s and $\mathbf{J}_s$ are the surface-charge and current densities, respectively, yet to be defined.

Table 4.7.1 Boundary Conditions in General and Special Cases

dielectric (1)- dielectric (2) interface (static as well as time-varying)		general boundary conditions		conductor (1)- dielectric (2) interface (static)	(time-varying)
$(\mathbf{D}_2 - \mathbf{D}_1).\mathbf{a}_n = 0$	$\Leftarrow$	$(\mathbf{D}_2 - \mathbf{D}_1).\mathbf{a}_n = \rho_s$	$\Rightarrow$	$\mathbf{D}_2.\mathbf{a}_n = \rho_s$	$\mathbf{D}_2.\mathbf{a}_n = \rho_s$
$(\mathbf{B}_2 - \mathbf{B}_1).\mathbf{a}_n = 0$	$\Leftarrow$	$(\mathbf{B}_2 - \mathbf{B}_1).\mathbf{a}_n = 0$	$\Rightarrow$	$(\mathbf{B}_2 - \mathbf{B}_1).\mathbf{a}_n = 0$	$\mathbf{B}_2.\mathbf{a}_n = 0$
$\mathbf{a}_n\times(\mathbf{H}_2 - \mathbf{H}_1) = 0$	$\Leftarrow$	$\mathbf{a}_n\times(\mathbf{H}_2 - \mathbf{H}_1) = \mathbf{J}_s$	$\Rightarrow$	$\mathbf{a}_n\times(\mathbf{H}_2 - \mathbf{H}_1) = 0$	$\mathbf{a}_n\times\mathbf{H}_2 = \mathbf{J}_s$
$\mathbf{a}_n\times(\mathbf{E}_2 - \mathbf{E}_1) = 0$	$\Leftarrow$	$\mathbf{a}_n\times(\mathbf{E}_2 - \mathbf{E}_1) = 0$	$\Rightarrow$	$\mathbf{a}_n\times\mathbf{E}_2 = 0$	$\mathbf{a}_n\times\mathbf{E}_2 = 0$

In order to obtain the general boundary conditions, let us consider a point P on the interface separating two regions, one of permittivity ε_1, permeability μ_1, and conductivity σ_1 and the other of corresponding quantities ε_1, μ_2 and σ_2, respectively (Fig. 4.7.1).

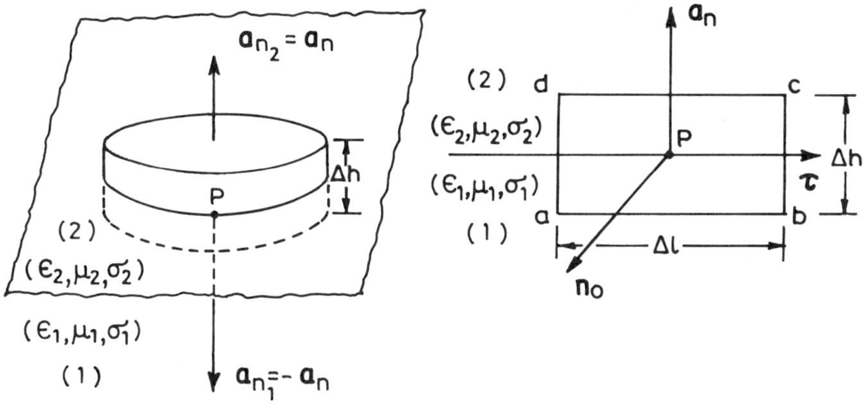

(a) (b)

Fig. 4.7.1 A small pill-box type Gaussian enclosure(**a**) and a small rectangle(**b**) across the interface between two regions of dissimilar physical properties.

Now, the boundary condition for the normal components of electric displacement is obtained as follows. Let us construct a small pill-box-type *Gaussian* enclosure (see section *2.2*), enclosing the point on the interface where the condition is sought. Let the box be of infinitesimal height Δh such that the two end faces of the box, each of infinitesimal area Δs, lie in the two regions 1 and 2, respectively (Fig. 4.7.1(a)). Applying (4.4.1(a)) to the Gaussian enclosure, and following the method outlined in section 2.2, one may obtain the desired boundary condition as (see appendix *a 4.7.1*, for the details of deduction):

$$(\mathbf{D}_2 - \mathbf{D}_1).\mathbf{a}_n = \underset{\Delta h \to 0}{\mathrm{Lt}} \rho \Delta h, \tag{4.7.1}$$

which, with the help of the following definition of surface density ρ_s, in terms of volume charge density ρ at P:

$$\rho_s = \underset{\Delta h \to 0}{\mathrm{Lt}} \rho \Delta h , \tag{4.7.2}$$

reads as

$$(\mathbf{D}_2 - \mathbf{D}_1).\mathbf{a}_n = \rho_s . \tag{4.7.3}$$

It is implied in the definition of (4.7.3) that $\rho \to \infty$ as $\Delta h \to 0$. This, in otherwords, implies that the charge is sprayed over the interface in the form of a *surface charge* in a thin layer. For situations where ρ is finite, the right hand side of (4.7.1) becomes null, and also ρ_s , from (4.7.2), becomes zero, and, consequently, the boundary condition, in this case, is obtained from either (4.7.1) or (4.7.3) as:

$$(\mathbf{D}_2 - \mathbf{D}_1).\mathbf{a}_n = 0 \; . \tag{4.7.4}$$

A typical situation where (4.7.4) would be valid is the dielectric-dielectric interface, for both time-dependent and independent cases, considered later in Example 4.7.1.

Following the same method as used in deducing (4.7.3), one may easily obtain the boundary condition for the normal component of magnetic flux density, but now applying (4.4.2(a)), instead of (4.4.1(a)) to the Gaussian pill box enclosure (Fig. 4.7.1(a)), as (appendix *a4.7.1*):

$$(\mathbf{B}_2 - \mathbf{B}_1).\mathbf{a}_n = 0 \; . \tag{4.7.5}$$

Next, let us find the boundary condition for the tangential components of magnetic field intensity. For this purpose, let us construct a small rectangle of length Δl and width Δh, with its lengths lying parallel to the interface in the regions 1 and 2, respectively (Fig. 4.7.1(b)). Applying (4.4.4(a)) to the elementary rectangle, and following the method outlined in section *3.2.2*, the following boundary condition is obtained (sec appendix *a4.7.1*, for the details of deduction):

$$\mathbf{a}_n \times (\mathbf{H}_2 - \mathbf{H}_1) = \operatorname*{Lt}_{\Delta h \to 0} \mathbf{J}_c \, \Delta h \; , \tag{4.7.6}$$

where $\mathbf{J}_c$ is the conduction current density at P (Fig. 4.7.1(b)). Defining a surface current density $\mathbf{J}_s$ as (of (4.7.2)):

$$\mathbf{J}_s = \operatorname*{Lt}_{\Delta h \to 0} \mathbf{J}_c \, \Delta h \; , \tag{4.7.7}$$

the boundary condition (4.7.6) may be expressed as:

$$\mathbf{a}_n \times (\mathbf{H}_2 - \mathbf{H}_1) = \mathbf{J}_s \; . \tag{4.7.8}$$

It is implied in the definition (4.7.7) that $\mathbf{J}_c \to \infty$ as $\Delta h \to 0$ and that there exists a *sheet current* on the interface in a thin layer. For a situation where $\mathbf{J}_c$ is finite, for instance, in Example 4.7.1, for the dielectric-dielectric interface, in both time-dependent and independent situations or for the conductor-dielectric interface, in a time-independent situation, one has $\mathbf{J}_s = 0$ from (4.7.7); and as such the boundary condition may be read from either (4.7.6) or (4.7.8) as:

$$\mathbf{a}_n \times (\mathbf{H}_2 - \mathbf{H}_1) = 0 \; . \tag{4.7.9}$$

Similarly, the boundary condition for the tangential components of electric field intensity may be obtained, following the same method as used in deducing (4.7.8), however, now applying (4.4.3(a)), instead of (4.4.4(a)), to the elementary rectangle (Fig. 4.7.1(b)), as follows (for the details of deduction, see appendix *a 4.7.1*):

$$\mathbf{a}_n \times (\mathbf{E}_2 - \mathbf{E}_1) = 0 \; . \tag{4.7.10}$$

It may be mentioned that the two sets of boundary conditions, namely (4.7.3)-(4.7.5) and (4.7.8)-(4.7.10) involve terms of the types $\mathbf{M}.\mathbf{a}_n$ and $\mathbf{a}_n\times\mathbf{M}$, respectively, where **M** is a relevant quantity (**E**, **H**, **D**, **E**) at the interface point. Now, since $\mathbf{M}.\mathbf{a}_n$ (= M cosθ) and $|\mathbf{a}_n\times\mathbf{M}|$ (= M sinθ), θ being the angle between **M** and $\mathbf{a}_n$, represent the components of **M**, normal and tangential to the interface, these two seats of boundary conditions refer to the normal and tangential components of the relevant quantity, respectively, as stated earlier while introducing these boundary conditions. Therefore, the general boundary conditions developed here, in the mathematical form, may be stated in physical terms as follows.

(i) The normal components of electric displacement are discontinuous at the interface between two regions if there is a layer of charge on the interface, the amount of discontinuity being equal to the surface charge density (as stated by (4.7.3)). At the charge-free interface, the normal components of electric displacement are continuous (as stated by (4.7.4)).

(ii) The normal components of magnetic flux density are continuous, at the interface between two regions (as stated by (4.7.5)).

(iii) The tangential components of magnetic field intensity are discontinuous or continuous at the interface between two regions, according as there is a sheet current present or absent, at the interface, the amount of discontinuity, if present, being equal to the surface current density (as stated by (4.7.8) and (4.7.9)), respectively.

(iv) The tangential components of electric field intensity are continuous at the interface between two regions (as stated by (4.7.10)).

One may note here that the amount of discontinuity of a physical quantity (normal component of electric displacement or tangential component of electric field intensity) at the interface may be a *cause* while the surface quantity (surface-charge or current density) an *effect*, and vice-versa, in various situations.

Example 4.7.1 Obtain the boundary conditions at the dielectric-dielectric and conductor-dielectric interfaces, for both time independent (static) and time-varying phenomena.

An excess charge placed in the interior of a conductor decays or disappears within no time only to re-appear at the surface of the conductor, while, in the interior of a dielectric such a charge would take a long time to decay, rather it would be distributed in the bulk of the dielectric with a finite volume charge density (see Example 4.1.2). Consequently, in view of (4.7.2), one may take $\rho_s = 0$ at a dielectric-dielectric interface, though $\rho_s \neq 0$ ($\rho \to \infty$) at a conductor-dielectric interface. In either of the dielectric regions forming an interface, further, **E**, **D**, **H** and **B** all have non-zero values, for both time-dependent and-independent situations. Also, for both these situations, **E** and **D** each will be zero inside a conductor region that forms an interface with a dielectric region; they have non-zero values in the latter region. For such an interface, further, in static cases, **H** and **B** have non-zero values inside the conductor but in time-varying cases, since the electric and magnetic fields are coupled (see sections *4.2-4.4*), **H** and **B** would have zero values inside the conductor in view of **E** and **D** having zero values. As for the surface current, $\mathbf{J}_s$ has zero values at the dielectric-dielectric interface since the dielectric will not conduct any current ($\mathbf{J}_c = 0$), static and time-varying phenomena irrespective. At the conductor-dielectric interface, however, $\mathbf{J}_c$ has a finite value in the static case, the conduction current being through the bulk of the conductor. This makes $\mathbf{J}_s = 0$ at the conductor-dielectric interface, in time-independent situations. However, in time-varying situations, the current is present in a thin layer at the conductor-dielectric interface which makes $\mathbf{J}_s \neq 0$ ($\mathbf{J}_c \rightarrow \infty$; see (4.7.7)).

The foregoing discussions on some special cases of the general boundary conditions (4.7.3), (4.7.5), (4.7.8) and (4.7.10) may be summarised as follows. In the case of a dielectric forming an interface with another dielectric, these boundary conditions are interpreted with $\mathbf{E}_{1,2} \neq 0$, $\mathbf{D}_{1,2} \neq 0$, $\mathbf{H}_{1,2} \neq 0$, $\mathbf{B}_{1,2} \neq 0$; and $\mathbf{J}_s = \rho_s = 0$, where the subscripts 1 and 2 refer to the quantities in the conductor region and the subscript 2 to those in the dielectric region. The general boundary conditions so interpreted for special cases of dielectric-dielectric and conductor-dielectric interfaces for both static and time-varying phenomena have been presented in Table 4.7.1.

<u>*Example 4.7.2*</u> Two regions (1) $z > 0$ and (2) $z < 0$ of relative permeabilities μ_{r1} and μ_{r2}, respectively, form an interface at $z = 0$, with a surface current density $\mathbf{J}_s = J_{sx}\mathbf{a}_x + J_{sy}\mathbf{a}_y$. Obtain an expression for $\mathbf{B}_2$ at the interface in terms of $\mathbf{B}_1 = B_{1x}\,\mathbf{a}_x + B_{1y}\,\mathbf{a}_y + B_{1z}\,\mathbf{a}_z$. The subscripts refer to the regions.

$\mathbf{B}_1$ is given as

$$\mathbf{B}_1 = B_{1x}\,\mathbf{a}_x + B_{1y}\,\mathbf{a}_y + B_{1z}\,\mathbf{a}_z \tag{4.7.11}$$

whence $\mathbf{H}_1$ $(= \mathbf{B}_1/\mu_1 = \mathbf{B}_1/(\mu_0\,\mu_{r1}))$ may be found as

$$\mathbf{H}_1 = \frac{B_{1x}}{\mu_0\,\mu_{r1}}\mathbf{a}_x + \frac{B_{1y}}{\mu_0\,\mu_{r1}}\mathbf{a}_y + \frac{B_{1z}}{\mu_0\,\mu_{r1}}\mathbf{a}_z\,. \tag{4.7.12}$$

Let $\mathbf{B}_2$, which is to be found, and $\mathbf{H}_2 = \mathbf{B}_2/\mu_2 = \mathbf{B}_2/\mu_0\,\mu_{r2}$ be expressed as

$$\mathbf{B}_2 = B_{2x}\,\mathbf{a}_x + B_{2y}\,\mathbf{a}_y + B_{2z}\,\mathbf{a}_z \tag{4.7.13}$$

and

$$\mathbf{H}_2 = H_{2x}\,\mathbf{a}_x + H_{2y}\,\mathbf{a}_y + H_{2z}\,\mathbf{a}_z\,. \tag{4.7.14}$$

Substituting (4.7.11) and (4.7.13) in (4.7.5) and remembering that, in this problem, the unit vector $\mathbf{a}_n$ directed from the region 1 ($z > 0$) to 2 ($z < 0$), is:

$$\mathbf{a}_n = -\,\mathbf{a}_z\,, \tag{4.7.15}$$

one may show

$$(B_{1z} - B_{2z})\,\mathbf{a}_z \cdot \mathbf{a}_z = 0$$

whence one has

$$B_{2z} = B_{1z}\,. \tag{4.7.16}$$

Similarly, substituting (4.7.12), (4.7.14) and (4.7.15) in (4.7.8) and noting that $\mathbf{J}_s$ in this problem is given as

$$\mathbf{J}_S = J_{Sx}\,\mathbf{a}_x + J_{Sy}\,\mathbf{a}_y\,,$$

one may easily obtain,

$$-(H_{2x} - H_{1x})\,\mathbf{a}_y + (H_{2y} - H_{1y})\,\mathbf{a}_x = J_{Sx}\,\mathbf{a}_x + J_{Sy}\,\mathbf{a}_y$$

whence one has

$$H_{2x} - H_{1x} = -J_{Sy} \tag{4.7.17}$$

and $$H_{2y} - H_{1y} = J_{Sx} . \tag{4.7.18}$$

Putting, $H_{1(2)x(y)} = B_{1(2)x(y)}/(\mu_0\, \mu_{r\,1(2)})$ in (4.7.17) and (4.7.18), one then has

$$B_{2x} = (\mu_{r2}/\mu_{r1})\, B_{1x} - \mu_0\, \mu_{r2}\, J_{Sy} \tag{4.7.19}$$

and $$B_{2y} = (\mu_{r2}/\mu_{r1})\, B_{1y} + \mu_0\, \mu_{r2}\, J_{Sx} . \tag{4.7.20}$$

Substituting (4.7.16), (4.7.19) and (4.7.20) in (4.7.13), $\mathbf{B}_2$ is then obtained as:

$$\mathbf{B}_2 = ((\mu_{r2}/\mu_{r1})\, B_{1x} - \mu_0\, \mu_{r2}\, J_{sy})\mathbf{a}_x + (\mu_0\, \mu_{r2}\, J_{sx} + (\mu_{r2}/\mu_{r1})\, B_{1y})\mathbf{a}_y + B_{1z}\, \mathbf{a}_z .$$

Example 4.7.3 Deduce an expression for the surface charge density on the spherical capacitor consisting of two concentric spherical conductors having the geometry of Example 2.6.2(c), in terms of potential difference between the conductors, and hence deduce an expression for the capacitance of the capacitor.

The following boundary condition for the conductor-dielectric interface (see Example 4.7.1 and Table 4.7.1) may be applied to the outer surface of the inner conductor or to the inner surface of the outer conductor:

$$\mathbf{D}.\mathbf{a}_n = \rho_s , \tag{4.7.21}$$

where the value of $\mathbf{D}$ (= ε $\mathbf{E}$) at a point between the conductors is given, with the help of (2.6.19), by

$$\mathbf{D} = \frac{\varepsilon\, a\, b}{b-a}\, \frac{1}{r^2}\, V_0 \mathbf{a}_r , \tag{4.7.22}$$

where a and b are the radii of the inner and the outer spheres, respectively; r is the distance of the point from the common centre of the conducting spheres; V_0 is the potential of the inner conductor; and ε is the permittivity of the dielectric between the conductors. $\mathbf{a}_n$ may be interpreted as $\mathbf{a}_r$ and -$\mathbf{a}_r$ for the inner and the outer conductors, respectively. With this interpretation of $\mathbf{a}_n$, one may now substitute (4.7.22), with r = a and b, in (4.7.21) to obtain the following expression for the surface charge densities (ρ_s) for the inner and outer conductors, respectively:

$$\rho_s = \frac{\varepsilon\, a\, b}{b-a}\, \frac{1}{a^2}\, V_0 \quad \text{(inner conductor)} ,$$

$$= -\frac{\varepsilon\, a\, b}{b-a}\, \frac{1}{b^2}\, V_0 \quad \text{(outer conductor)} . \tag{4.7.23}$$

The charges on the inner and the outer conductors may be found by multiplying ρ_s given by (4.7.23) by $4\pi a^2$, the area of the outer surface of the inner conductor, and $4\pi b^2$, the area of the inner surface of the outer conductor, respectively. Dividing the magnitude of charge thus obtained on either of the conductors by V_0, the potential difference between the conductors, one obtains the following expression for C, the capacitance of the capacitor:

$$C = \frac{4\pi\varepsilon}{1/a - a/b} .$$

Appendix a 4.7.1 Derivation of general boundary conditions

If we apply (4.4.1a) to the Gaussian pill-box enclosure (see Fig. 4.7.1a), we have:

$$\mathbf{D}_2 . \mathbf{a}_{n\,2}\ \Delta S + \mathbf{D}_1 . \mathbf{a}_n\ \Delta S + \text{wall contributions} \cong \rho\, \Delta h\, \Delta S\ , \tag{a4.7.1}$$

where $\mathbf{D}_1$ *and* $\mathbf{D}_2$ *are the electric displacements on the upper and lower faces of the pill box, and* $\mathbf{a}_{n1}$ *and* $\mathbf{a}_{n2}$ *are the unit vectors in the outward normal directions to these end faces, respectively. The areas of the end faces, each* ΔS*, being infinitesimally small, it is correct, though not exactly, to assume* $\mathbf{D}_1$ *and* $\mathbf{D}_2$ *to be uniform over the end faces while estimating the surface integral of the left side of (4.4.1(a)) from these faces. The third term of the left hand side of (a4.7.1) is the contribution to this surface integral from the side wall of the pill box. It is easy to appreciate that, for the small pill box, the end faces are parallel (Fig. 4.7.1(a)), and, therefore, one can write*

$$\mathbf{a}_{n\,2} = \mathbf{a}_n \quad \text{and} \quad \mathbf{a}_{n\,1} = -\mathbf{a}_n\ . \tag{a4.7.2}$$

Here $\mathbf{a}_n$ *is the unit vector at P, directed from region 1 to 2. The right hand side of (a4.7.1), which follows from the right hand side (4.4.1(a)), is the charge enclosed by the pill box. Here, the enclosure volume* $\Delta\tau = \Delta h\ \Delta S$ *is considered to be infinitesimally small so that within this volume one may approximately take the volume charge density* ρ *to be uniform. The assumption that* $\mathbf{D}_1$, $\mathbf{D}_2$ *and* ρ *have been considered as uniform leads one to write (a4.7.1) as an approximate relation. The relation becomes an exact relation as* $\Delta S \to 0$ *and* $\Delta h \to 0$*, whence* $\Delta\tau\,(= \Delta h\ \Delta S) \to 0$*, which amounts to making the volume element shrink to the point P on the surface where the boundary condition is sought. Further, if* $\Delta h \to 0$*, one may very well ignore the contribution to the surface integral from the wall of the pill-box (Fig. 4.7.1(a)), and hence ignore the third term of the left hand side of (a4.7.1). This enables one to write the approximate relation (a4.7.1) as the exact relation (4.7.1), which in view of (a4.7.2) passes on to (4.7.3).*

Now, if we apply (4.4.2(a)) instead of (4.4.1(a)) to the Gaussian pill box enclosure (Fig. 4.7.1(a)), we have an expression similar to (a4.7.1), though with $\mathbf{D}_1$ *and* $\mathbf{D}_2$ *replaced by* $\mathbf{B}_1$ *and* $\mathbf{B}_2$*, respectively, and with its right hand side put equal to zero. Starting from the expression thus obtained, and following the same method as outlined above for the derivation of (4.7.3), we easily obtain (4.7.5).*

Now, for the boundary condition involving the tangential components of magnetic field intensity let us consider a closed rectangular path of infinitesimal length and breadth Δh *across the cross section of the interface such that the lengths of the rectangle lie close and parallel to the interface (Fig. 4.7.1(b)). If we apply (4.4.4(a)) to the closed rectangular path abcd enclosing the element of area* $\Delta l\ \Delta h$*, we can show:*

$$\mathbf{H}_1 . \tau\, dl - \mathbf{H}_2 . \tau\, dl + \text{breadth contributions} \cong (\mathbf{J}_c + \frac{\partial \mathbf{D}}{\partial t}).\mathbf{n}_0\, \Delta l\, \Delta h\ , \tag{a4.7.3}$$

where τ *is the unit vector parallel to the length of the rectangle directed from a to b (or d to c),* $\mathbf{n}_0$ *is the unit vector normal to the elementary area, the sense of which follows the cork-screw rule when the angular rotation of the screw follows the paths a to b, b to c, and so, along which the line integral of the left hand side of (4.4.4(a)) is taken to obtain the left hand side of (a4.7.3). The third term of the left hand side of (a4.7.3) is contributed by the line integral along the breadths. It may be noted that (a4.7.3) has been written here as an approximate relation. This is because we have assumed* $\mathbf{J}_c$ *and* $\mathbf{D}$ *to be uniform over the elementary area* $\Delta l \times \Delta h$*, while evaluating the surface integral of the right hand side of (4.4.4a) to obtain the right hand side of (a4.7.3). The relation becomes exact when the elementary area becomes vanishingly small, corresponding to* $\Delta l \to 0$, $\Delta h \to 0$*, shrinking to the point P (see Fig. 4.7.1(b)). Further, for* $\Delta h \to 0$*, one can ignore the breadth contributions. Thus the exact relation is obtained from (a4.7.3) as:*

(a4.7.4) $$(\boldsymbol{H}_1 - \boldsymbol{H}_2).\boldsymbol{\tau} = \underset{\Delta h \to 0}{Lt} (\boldsymbol{J}_c + \frac{\partial \boldsymbol{D}}{\partial t}).\boldsymbol{n}_0 \, \Delta h \, .$$

Substituting $\boldsymbol{a}_n \times \boldsymbol{n}_0$ *for* $\boldsymbol{\tau}$ *(Fig. 4.7.1(b)) in (a4.7.4) one may write*

(a4.7.5) $$(\boldsymbol{H}_1 - \boldsymbol{H}_2).(\boldsymbol{a}_n \times \boldsymbol{n}_0) = \underset{\Delta h \to 0}{Lt} (\boldsymbol{J}_c + \frac{\partial \boldsymbol{D}}{\partial t}).\boldsymbol{n}_0 \, \Delta h \, .$$

Invoking the vector identity, $\boldsymbol{A}.(\boldsymbol{B}\times\boldsymbol{C}) = \boldsymbol{C}.(\boldsymbol{A}\times\boldsymbol{B})$ *in the left hand side of (a4.7.5), one may write*

(a4.7.6) $$\boldsymbol{n}_0.(\boldsymbol{H}_1 - \boldsymbol{H}_2)\times \boldsymbol{a}_n = \underset{\Delta h \to 0}{Lt} (\boldsymbol{J}_c + \frac{\partial \boldsymbol{D}}{\partial t}).\boldsymbol{n}_0 \, \Delta h \, .$$

Using the vector identities $\boldsymbol{A}\times\boldsymbol{B} = -\boldsymbol{B}\times\boldsymbol{A}$ *and* $\boldsymbol{A}.\boldsymbol{B} = \boldsymbol{B}.\boldsymbol{A}$ *in the left and right-hand sides of (a4.7.6), respectively, one may now express the latter, after the re-arrangement of terms, as:*

(a4.7.7) $$\boldsymbol{n}_0.[\boldsymbol{a}_n \times (\boldsymbol{H}_2 - \boldsymbol{H}_1) - \underset{\Delta h \to 0}{Lt} (\boldsymbol{J}_c + \frac{\partial \boldsymbol{D}}{\partial t}) \Delta h] = 0$$

as the dot product of a unit vector, here, $\boldsymbol{n}_0$ *and another vector* **G***, say, which may be identified as*

(a 4.7.8) $$\boldsymbol{G} = \boldsymbol{a}_n \times (\boldsymbol{H}_2 - \boldsymbol{H}_1) - \underset{\Delta h \to 0}{Lt} (\boldsymbol{J}_c + \frac{\partial \boldsymbol{D}}{\partial t}) \Delta h \, .$$

Therefore, from (a4.7.7),

(a4.7.9) $$G \cos\theta = 0 \, ,$$

where θ *is the angle between* **G** *and* $\boldsymbol{n}_0$ *(Fig. 4.7.1(b)), which depends upon the orientation of the elementary rectangular path. Since the orientation of the path can be chosen arbitrarily,* θ *and hence* $\cos\theta$ *may take arbitrary values and, for each of such values, (a4.7.9) is valid. This clearly demands that*

$$G = 0 \, ,$$

which, in other words, from (a 4.7.8) means

(a4.7.10) $$\boldsymbol{a}_n \times (\boldsymbol{H}_2 - \boldsymbol{H}_1) = \underset{\Delta h \to 0}{Lt} (\boldsymbol{J}_c + \frac{\partial \boldsymbol{D}}{\partial t}) \Delta h = \underset{\Delta h \to 0}{Lt} \boldsymbol{J}_c \Delta h + \underset{\Delta h \to 0}{Lt} \frac{\partial \boldsymbol{D}}{\partial t} \Delta h \, .$$

Since **D***, and hence its derivative* $\partial\boldsymbol{D}/\partial t$*, are finite, the second term of (a4.7.10) vanishes, and (a4.7.10) passes on to the desired boundary condition for the tangential components of magnetic field intensity, namely, (4.7.8), in view of (4.7.7).*

Similarly, by applying (4.4.3(a)), instead of (4.4.4(a)), to the closed rectangular path abcd (Fig. 4.7.1(b)), one may easily show

(a4.7.11) $$\boldsymbol{a}_n \times (\boldsymbol{E}_2 - \boldsymbol{E}_1) = - \underset{\Delta h \to 0}{Lt} \frac{\partial \boldsymbol{B}}{\partial t} \Delta h \, .$$

Since **B***, and hence its derivative, are finite, the right hand side of (a4.7.11) becomes null, and one obtains the desired boundary condition for the tangential components of electric field intensity as (4.7.10).*

4.8 Power Flow and Poynting Theorem

Energy may be stored in electromagnetic field, transported by electromagnetic waves and lost in materials such that it remains balanced in the process. The energy balance may be mathematically formulated in terms of the time-rate of flow of energy, that is, power flow as follows.

Let us first recall the following vector identity involving any two vectors, here **E** and **H**, say:

$$\nabla.(\mathbf{E}\times\mathbf{H}) = \mathbf{H}.(\nabla\times\mathbf{E}) - \mathbf{E}.(\nabla\times\mathbf{H}) . \qquad (4.8.1)$$

Substituting Maxwell's equation (4.4.3(b)) and (4.4.4(b)) in (4.8.1), remembering $\mathbf{D} = \varepsilon\mathbf{E}$ and $\mathbf{B} = \mu\mathbf{H}$, and taking μ and ε each as time-independent, one may write:

$$\nabla.(\mathbf{E}\times\mathbf{H}) = -\mu\,\mathbf{H}.\frac{\partial \mathbf{H}}{\partial t} - \varepsilon\mathbf{E}.\frac{\partial \mathbf{E}}{\partial t} - \mathbf{E}.\mathbf{J}_c$$

$$= -\mu H\frac{\partial H}{\partial t} - \varepsilon E\frac{\partial E}{\partial t} - \mathbf{E}.\mathbf{J}_c \ . \qquad (4.8.2)$$

In view of the following simple results of differentiation,

$$\frac{\partial}{\partial t}(\mu H^2) = 2\mu H\frac{\partial H}{\partial t} \quad \text{and} \quad \frac{\partial}{\partial t}(\varepsilon E^2) = 2\varepsilon E\frac{\partial E}{\partial t} ,$$

one may choose to express (4.8.2) as:

$$\nabla.(\mathbf{E}\times\mathbf{H}) = -\frac{\partial}{\partial t}(\frac{1}{2}\mu H^2) - \frac{\partial}{\partial t}(\frac{1}{2}\varepsilon E^2) - \mathbf{E}.\mathbf{J}_c \ . \qquad (4.8.3)$$

Integrating (4.8.3) over a closed volume τ, say, one may then write:

$$\oint_\tau \nabla.(\mathbf{E}\times\mathbf{H})d\tau = -\frac{\partial}{\partial t}\oint_\tau(\frac{1}{2}\mu H^2 + \frac{1}{2}\varepsilon E^2)d\tau - \oint_\tau \mathbf{E}.\mathbf{J}_c\, d\tau \ . \qquad (4.8.4)$$

By Gauss's theorem (see (a2.7.13), as applied to the vector $(\mathbf{E}\times\mathbf{H})$, the left hand side of (4.8.4), which is a volume integral may be converted into the surface integral over the surface area S bounding the volume τ, thus enabling one to express (4.8.4) as:

$$\oint_S (\mathbf{E}\times\mathbf{H}).\mathbf{a}_n\, dS = \oint_S \boldsymbol{\mathcal{P}}.\mathbf{a}_n\, dS = -\frac{\partial}{\partial t}\oint_\tau(\frac{1}{2}\mu H^2 + \frac{1}{2}\varepsilon E^2)d\tau - \oint_\tau \mathbf{E}.\mathbf{J}_c\, d\tau \qquad (4.8.5)$$

or, alternatively, by changing the sign of (4.8.5) as:

$$-\oint_S (\mathbf{E}\times\mathbf{H}).\mathbf{a}_n\, dS = -\oint_S \boldsymbol{\mathcal{P}}.\mathbf{a}_n\, dS = \frac{\partial}{\partial t}\oint_\tau(\frac{1}{2}\mu H^2 + \frac{1}{2}\varepsilon E^2)d\tau + \oint_\tau \mathbf{E}.\mathbf{J}_c\, d\tau \qquad (4.8.6)$$

where $\mathbf{a}_n$ is the unit vector directed outwardly normal to an element of area dS considered on the surface S enclosing the volume τ. $\mathcal{P}$ (= $\mathbf{E}\times\mathbf{H}$) is a power density vector having the dimension of (volt/m)×(amp/m) = watt/m^2, called the Poynting vector, at any instant of time.

It is clear then that the left hand sides of (4.8.5) and (4.8.6) represent the outward and inward fluxes, respectively, of the instantaneous Poynting vector through the surface enclosure. The quantities $(1/2)\varepsilon E^2$ and $(1/2)\mu H^2$ being the volume energy densities stored in electric and magnetic fields given by (2.7.9) and (3.1.6), however, interpreting therein ε_0 and μ_0 as ε and μ, respectively, the first term of the right hand side of (4.8.6) represents the time rate of increase of the total electromagnetic energy (being the sum of the energies stored in the electric and magnetic fields) within the volume τ. If there is a conducting region in the volume, then $\mathbf{J}_c$, interpreted as the conduction current density, has the same direction as $\mathbf{E}$, so that the third term is a positive quantity representing the power dissipated, being the ohmic loss or I^2R *loss* term, at any instant of time. For instance, if a direct current I passes through a uniform-cross-section conductor of conductivity σ (or resistivity $(1/\sigma)$), cross-sectional area α and length l, say; and V (= El) is the potential drop across the length of the conductor (E being the potential drop per unit length) then the third term is identified as the I^2R *loss* as follows:

$$\oint_\tau \mathbf{E}.\mathbf{J}_c\, d\tau = \oint_\tau \mathbf{E}.\sigma\mathbf{E}\, d\tau = \int_\tau \sigma E^2\, d\tau = \sigma E^2 \int_\tau d\tau = \sigma E^2 l\alpha = \sigma(\frac{V^2}{l^2})l\alpha = \frac{V^2}{R} = I^2R,$$

with the help of the Ohm's law relation V =IR, where R ($=1/\sigma)(l/\alpha$) is the resistance of the conductor. If there are free charges present in the volume, $\mathbf{J}_c$ has to be interpreted as the convection current density, and then the third term of (4.8.6) would represent the energy spent in moving the charges in unit time.

In the presence of an external current, Maxwell's equation (4.4.4(b)) would include an additional term, namely, the external current density $\mathbf{J}_{ext}$, say, in its right hand side. Consequently, (4.8.5) would be modified to the following form:

$$\oint_S (\mathbf{E}\times\mathbf{H}).\mathbf{a}_n\, dS = -\frac{\partial}{\partial t}\oint_\tau (\frac{1}{2}\mu H^2 + \frac{1}{2}\varepsilon E^2)d\tau - \oint_\tau \mathbf{E}.\mathbf{J}_c\, d\tau - \oint_\tau \mathbf{E}.\mathbf{J}_{ext}\, d\tau$$

which may be re-arranged to be expressed as:

$$-\oint_\tau \mathbf{E}.\mathbf{J}_{ext}\, d\tau = \oint_S (\mathbf{E}\times\mathbf{H}).\mathbf{a}_n\, dS + \frac{\partial}{\partial t}\oint_\tau (\frac{1}{2}\mu H^2 + \frac{1}{2}\varepsilon E^2)d\tau + \oint_\tau \mathbf{E}.\mathbf{J}_c\, d\tau. \qquad (4.8.7)$$

If there is a power generator present in the volume τ, then $\mathbf{J}_{ext}$ may be interpreted as the generator current density which would be directed oppositely to $\mathbf{E}$ [11] thereby making the left hand side of (4.8.7) a positive quantity which would now stand for the instantaneous power generated in the volume τ. The first term of the right hand side of (4.8.7) would denote the outward flux of Poynting vector over the surface S, while the second and the third term of its right hand side, as discussed following (4.8.6), represent the time rate of increase in the total electromagnetic energy and the instantaneous ohmic loss within the volume, respectively.

Thus, (4.8.6) and (4.8.7) are the integral statements of an energy balance theorem called the Poynting theorem [13]. The theorem, as would be clear from the discussions following (4.8.6) and (4.8.7), has the physical meaning that i) for power sources external to a volume considered, the instantaneous inward flux of Poynting vector across the surface bounding the volume, that is, the instantaneous power entering the volume, is equal to the time rate of increase in electromagnetic energy in the volume plus the instantaneous power dissipated within the volume (as stated by (4.8.6)); and ii) for power sources placed within a volume, the instantaneous power generated within the volume equals the time rate of increase in electromagnetic energy in the volume plus the power dissipated within the volume at any instant, plus the instantaneous power which leaves the volume, the latter being equal to the outward flux of instantaneous Poynting vector over the surface bounding the volume considered.

Example 4.8.1 Find the direction of the Poynting vector over a long, round straight wire carrying a direct current. Identify the power absorbed as the I^2R loss, from the flux of the Poynting vector.

The magnetic field intensity on the surface of a long wire carrying a direct current I may be found using either Biot-Savart's or Ampere's circuital law (for instance, see (3.2.5)) as:

$$\mathbf{H} = \frac{I}{2\pi a}\mathbf{a}_\theta\,, \tag{4.8.8}$$

where a is the wire radius. The current **J** is related to the electric field intensity **E**, within the wire by the Ohm's law,

$$\mathbf{J} = \sigma\,\mathbf{E}$$

whence one may write,

$$\mathbf{E} = \frac{J}{\sigma}\mathbf{a}_z = \frac{I}{\pi a^2}\frac{1}{\sigma}\mathbf{a}_z\,, \tag{4.8.9}$$

where $\mathbf{J} = J\,\mathbf{a}_z$ is the one-dimensional current density, the wire being considered to be aligned parallel to the z-axis, the magnitude, J, of which is given by: $J = I/(\pi a^2)$.

The Poynting vector $\mathbf{E}\times\mathbf{H}$ may be found with the help of (4.8.8) and (4.8.9) as:

$$\mathbf{E}\times\mathbf{H} = \left(\frac{I}{\pi a^2}\frac{1}{\sigma}\mathbf{a}_z\right)\times\left(\frac{I}{2\pi a}\mathbf{a}_\theta\right) = \frac{I}{\pi a^2}\frac{1}{\sigma}\frac{I}{2\pi a}\mathbf{a}_z\times\mathbf{a}_\theta = -\frac{I^2\,\mathbf{a}_r}{(\pi a^2)(\sigma)(2\pi a)}\,. \tag{4.8.10}$$

Thus the Poynting vector in this problem radially points inward, and it has no component along z. The power P entering the wire extending over the length l is obtained from the inward flux of Poynting vector $\mathcal{P}$, given by the left hand side of the integral statement (4.8.6) of the Poynting theorem, as:

$$P = -\int \mathcal{P}.\mathbf{a}_n\, dS = -\oint (\mathbf{E}\times\mathbf{H}).\mathbf{a}_n\, dS\ ,$$

which, with the help of (4.8.10), is found as:

$$P = \frac{I^2\, \mathbf{a}_r.\mathbf{a}_r}{(\pi a^2)(\sigma)(2\pi a)}\oint dS = \frac{I^2}{(\pi a^2)(\sigma)(2\pi a)}(2\pi al)\ , \qquad (4.8.11)$$

where we have taken $\mathbf{a}_n$ as $\mathbf{a}_r$, and the integration of the element of area as $2\pi al$, being the area of the curved cylindrical surface of the wire normal to the radial direction. (One may notice that, if one considers a cylindrical volume enclosing, and touching the wire, over the length l considered, then there would be no contributions to the flux of the vector ($\mathbf{E}\times\mathbf{H}$) over the end faces of the cylinder, since both the vector and the planes of the end faces are radial; hence, the contributions in this case, will be only from the curved cylindrical surface). Using the following expression for the resistance R of the wire:

$$R = \frac{l}{\sigma\pi a^2}\ , \qquad (4.8.12)$$

one may then write (4.8.11) as:

$$P = I^2 R\ . \qquad (4.8.13)$$

The power entering the wire is thus identified as the well-known I^2R loss term given by the right hand side of (4.8.13).

It is of interest to note, while deriving (4.8.13), for this problem, that the power enters the wire from the surrounding electromagnetic field radially inward, instead of axially. Further, one may also identify the I^2R loss term as the right hand side of (4.8.6), which is one of the integral statements of the Poynting theorem, as follows. In the present problem, which concerns with a direct current passing through the wire, the time-derivative part, that is, the first term, of the right hand side of (4.8.6), is zero, while its second term may be written in terms of the potential drop per unit length V/l as:

$$\oint \mathbf{E}.\mathbf{J}_c\, d\tau = \oint (\mathbf{E}.\sigma\mathbf{E})\, d\tau = \sigma E^2 \oint d\tau = \sigma E^2 \pi r^2 l = \sigma\,(V/l)^2\, \pi r^2\, l$$

which, with the help of (4.8.12) and Ohm's law relation V = IR, simplifies to:

$$\oint \mathbf{E}.\mathbf{J}_c\, d\tau = V^2/R = I^2 R$$

the right hand side of which may be seen as identical with the right hand side of (4.8.13). In other words, one may verify in this problem the validity of the Poynting's theorem.

<u>*Example 4.8.2*</u> A long coaxial cable carries a direct current. Show that the power carried by the cable is the product of the potential difference between the conductors and the current carried. Assume the conductors to be of high conductivity.

Since the conductivity of the materials making the inner and outer conductors are high, one may take in this problem the tangential components of electric field at the conducting interfaces to be zero (see Example 4.7.1/Table 4.7.1); this, in other words, means that the electric field intensity **E** is radial in the region between the conductors, given by (2.6.17), at a radial point distant r from the axis of the cable, which may be re-written here as:

$$\mathbf{E} = \frac{V_0}{\ln(b/a)}\frac{1}{r}\mathbf{a}_r, \tag{4.8.14}$$

where V_0 is the potential difference between the conductors, the inner and outer conductors being of radii a and b, respectively, the former being chosen to be at a higher potential. The magnetic field intensity at the point in the region between the conductors can be easily found in terms of the current, I, carried by the conductor, using Ampere's circuital law (see section *3.2.2*) as:

$$\mathbf{H} = \frac{I}{2\pi r}\mathbf{a}_\theta . \tag{4.8.15}$$

Taking the cross product of **E** and **H** given by (4.8.14) and (4.8.15), respectively, we get $\mathcal{P}$ as:

$$\mathcal{P} = \mathbf{E}\times\mathbf{H} = \frac{V_0}{\ln(b/a)}\frac{1}{r}\frac{I}{2\pi r}\mathbf{a}_z . \tag{4.8.16}$$

Thus, in this problem, the Poynting vector has the axial direction. One may now find the element of power dP flowing through an element of area, $2\pi r\,dr$, of an annular zone lying between r and r + dr, on the cross section of the cable, as:

$$dP = \mathcal{P}.(2\pi r\,dr)\,\mathbf{a}_z . \tag{4.8.17}$$

The power propagating down the cable, through the cross section lying between r = a and b, is obtained by substituting (4.8.16) in (4.8.17) and integrating the latter to obtain [14]

$$P = \int dP = \frac{V_0 I}{\ln(b/a)}\int_a^b \frac{dr}{r} = \frac{V_0 I}{\ln(b/a)}\ln\frac{b}{a} = V_0 I ,$$

the desired result.

<u>*Example 4.8.3*</u> Starting from the evaluation of Poynting vectors, show with, reference to Examples 4.3.6 and 4.4.1, that the energy stored in an inductor is $(1/2)LI^2$ and that stored in a capacitor is $(1/2)CV^2$, where L is the inductance of the inductor and I the current built up in it, and C is the capacitance of the capacitor and V the potential to which it is charged.

Let us first take the problem of the inductor of Example 4.3.6. When the current builds up in the inductor, the axial magnetic field intensity inside it grows with time, and the electric field intensity **E** at a radial distance r from the axis of the inductor is obtainable from (4.3.28), and hence the Poynting vector $\mathcal{P}$ at the curved surface of the solenoid (r = a) is given by:

$$\mathcal{P} = \mathbf{E}\times\mathbf{H} = -(\frac{dB}{dt}\frac{a}{2}\mathbf{a}_\theta)\times(H\mathbf{a}_z) = -\mu_0(\frac{dH}{dt})(H)(\frac{a}{2})\mathbf{a}_r , \tag{4.8.18}$$

where a is the radius of the solenoid. Thus it is clear from (4.8.18) that, for a magnetic field increasing with time corresponding to a current building up with time, $\mathcal{P}$ is directed radially inward corresponding to power entering the solenoid in the radial direction. The power P entering the solenoid through the area 2π al of the solenoid of length l, say, is,

$$P = -\oint_S \mathcal{P}.\mathbf{a}_n \, dS . \tag{4.8.19}$$

Substituting (4.8.18) into (4.8.19), and putting $\mathbf{a}_n = \mathbf{a}_r$, one then has

$$P = \mu_0 \frac{dH}{dt} H \frac{a}{2} \int dS = \mu_0 \frac{dH}{dt} H \frac{a}{2}(2\pi a l) = \frac{d}{dt}\left(\frac{1}{2}\mu_0 H^2\right)(\pi a^2 l) . \tag{4.8.20}$$

Integrating (4.8.20) in time one then obtains the energy stored W in the inductor as:

$$W = \frac{1}{2} \mu_0 H^2 (\pi a^2 l) . \tag{4.8.21}$$

Substituting into (4.8.21) the following expression for H, obtainable from (3.2.13):

$$H = n I ,$$

and remembering the expression for the inductance L, given following (3.2.13), as,

$$L = \mu_0 n^2 l \alpha ,$$

where α ($= \pi a^2$) is the cross-sectional area of the solenoid, one then finds W as:

$$W = \frac{1}{2} L I^2 , \tag{4.8.22}$$

the required expression for the energy stored in an inductor.

Similarly, one may proceed to deal with the problem of the circular-parallel-plate capacitor given in Example 4.4.1. As the charge builds up in the capacitor, the axial electric field intensity inside it grows with time, and the resulting magnetic field intensity **H** is given by (4.4.5), and hence the Poynting vector $\mathcal{P}$ (= **E**×**H**) at a peripheral point (r = a) of the capacitor by:

$$\mathcal{P} = \mathbf{E} \times \mathbf{H} = (E \mathbf{a}_z) \times \left(\frac{\partial D}{dt} \frac{a}{2} \mathbf{a}_\theta\right) = -\varepsilon_0 \frac{dE}{dt} E \frac{a}{2} \mathbf{a}_r . \tag{4.8.23}$$

We notice that (4.8.18) and (4.8.23) are identical in nature except that in the capacitor problem ε_0 and E play the same role as μ_0 and H, respectively, have done in the inductor problem. Therefore, following exactly the same approach as in the inductor problem one should be able to write, for the capacitor, the following expression, analogous to (4.8.21):

$$W = \left(\frac{1}{2} \varepsilon_0 E^2\right) (\pi a^2 d) , \tag{4.8.24}$$

where d is the distance of separation between the plates of the capacitor (being the length of the cylinder, the end faces of which are formed by the parallel circular plates of the capacitor). Noting that the electric field intensity E, obtainable from (2.7.2) is $E = V_0/d$, and that the capacitance C, obtainable from (2.7.5) is: $C = \varepsilon_0 \pi a^2/d$, one then obtains from (4.8.24) the following expression for W, the energy stored in the capacitor:

$$W = \frac{1}{2} C V^2 .$$

4.8.1 Complex Poynting Vector for Average and Reactive Power

We come across numerous practical situations in which the field intensities vary in a simple-harmonic manner representable by a sine or a cosine function. (It may be noted that the sine and cosine functions with proper interpretation of their initial phases represent one and the same variation). Any other type of time-periodic variation can also be considered as to be constituted by a number of time-harmonic 'Fourier' 'sinusoids'. For a simple example, let us recall the power flow problem of a long straight wire of Example 4.8.1. Let the current I, unlike in that example, however, now vary as $I = I_0 \cos\omega t$ with I_0 as the peak value. It is then quite easy to appreciate, with the help of (4.8.10), in conjunction with (4.8.8) and (4.8.9), that the direction of **E** and **H** change simultaneously whenever the sign of I changes. Thus, for both positive and negative half cycles of 'sinusoidal' variation of current, the direction of the instantaneous Poynting vector $\mathcal{P}$ is found to be radially inward, corresponding to a power lost in the wire. Taking the average of (4.8.13), one then can find P_{av}, the average power absorbed in the wire-resistor, as:

$$P_{av} = \frac{\left(\int_0^{\pi} I^2 d(\omega t)\right) R}{\int_0^{\pi} d(\omega t)} = \frac{\left(\int_0^{\pi} I_0^2 \cos^2 \omega t \, d(\omega t)\right) R}{\int_0^{\pi} d(\omega t)} = \frac{1}{2} I_0^2 R \text{ , for the first half cycle;}$$

$$= \frac{\left(\int_{\pi}^{2\pi} I^2 d(\omega t)\right) R}{\int_{\pi}^{2\pi} d(\omega t)} = \frac{\left(\int_{\pi}^{2\pi} I_0^2 \cos^2 \omega t \, d(\omega t)\right) R}{\int_{\pi}^{2\pi} d(\omega t)} = \frac{1}{2} I_0^2 R \text{, for the second half cycle;}$$

and

$$= \frac{\left(\int_0^{2\pi} I^2 d(\omega t)\right) R}{\int_0^{2\pi} d(\omega t)} = \frac{\left(\int_0^{2\pi} I_0^2 \cos^2 \omega t \, d(\omega t)\right) R}{\int_0^{2\pi} d(\omega t)} = \frac{1}{2} I_0^2 R \text{, for the full cycle.} \tag{4.8.25}$$

Next, we refer to Example 4.8.3 and examine the Poynting vectors given by (4.8.18) and (4.8.23) for the cases of an inductor and a capacitor, respectively, considering *sinusoidal* variation of fields. For such a variation, it is easy to appreciate that H and dH/dt, and so also E and dE/dt, would have the same (both positive or both negative) and opposite (one positive and other negative) signs in alternate quarter cycles. This further enables one to appreciate, with the help of (4.8.18) and (4.8.23), that the direction of the Poynting vector becomes, alternately, radially inward and outward in quarter cycles, corresponding to a power alternately surging back and forth between the source and the inductor or capacitor as the case be, thereby making zero the average power absorbed in the latter.

Quantitatively, further, one may elaborate these phenomena as follows. For the resistor problem of Example 4.8.1, taking $I = I_0 \cos\omega t$ the electric and magnetic field intensities may be written, with the help of (4.8.9) and (4.8.8), as:

$$\mathbf{E} = E_0 \cos \omega t \, \mathbf{a}_E \quad \text{and} \quad \mathbf{H} = H_0 \cos \omega t \, \mathbf{a}_H , \qquad (4.8.26)$$

$\mathbf{a}_E$ and $\mathbf{a}_H$ being the unit vectors in the directions of $\mathbf{E}$ and $\mathbf{H}$, respectively, and, $E_0 = I_0/(\pi a^2 \sigma)$, $H = I_0/(2\pi a)$; $\mathbf{a}_E = \mathbf{a}_z$, $\mathbf{a}_H = \mathbf{a}_\theta$. The average value of Poynting vector $\mathcal{P}_{av}$, is then obtained using (4.8.26), as,

$$\mathcal{P}_{av} = (\mathbf{E}\times\mathbf{H})_{av} = \frac{\int_0^{2\pi} E_0 H_0 \cos^2 \omega t \, d(\omega t)}{\int_0^{2\pi} d(\omega t)} (\mathbf{a}_E \times \mathbf{a}_H) = \frac{1}{2} E_0 H_0 (\mathbf{a}_E \times \mathbf{a}_H) , \qquad (4.8.27)$$

here, $\mathbf{a}_E \times \mathbf{a}_H = \mathbf{a}_z \times \mathbf{a}_\theta = -\mathbf{a}_r$, giving the direction of the Poynting vector as the radially inward direction.

Similarly, for the inductor problem of Example 4.8.3 one may take (see Fig. 4.3.4):

$$\mathbf{H} = H_0 \cos \omega t \, \mathbf{a}_H , \qquad (4.8.28)$$

with $\mathbf{a}_H$ interpreted as $\mathbf{a}_z$; and hence $\mathbf{E}$, the electric field intensity on the periphery of the solenoid, with the help of (4.3.28) through (4.8.28), may be expressed as:

$$\mathbf{E} = (-\frac{d\dot{B}}{dt}\frac{a}{2})\mathbf{a}_\theta = (\mu_0 \frac{dH}{dt}\frac{a}{2})(-\mathbf{a}_\theta) = -E_0 \sin \omega t \, \mathbf{a}_E = E_0 \cos(\omega t + \frac{\pi}{2}) \, \mathbf{a}_E , \quad (4.8.29)$$

with $\mathbf{a}_E$ interpreted as $-\mathbf{a}_\theta$, where $E_0 = (\mu_0 \, \omega a/2) H_0$.

Now, one may use (4.8.28) and (4.8.29) to average out (4.8.18), over a cycle, to obtain:

$$\mathcal{P}_{av} = (\mathbf{E}\times\mathbf{H})_{av} = \frac{E_0 H_0 \int_0^{2\pi} -\sin\omega t \cos\omega t \, d(\omega t)}{\int_0^{2\pi} d(\omega t)} \mathbf{a}_E \times \mathbf{a}_H = 0 . \qquad (4.8.30)$$

In (4.8.30), $\mathbf{a}_E \times \mathbf{a}_H$ ($= -\mathbf{a}_\theta \times \mathbf{a}_z$) continues to be equal to $-\mathbf{a}_r$ as in (4.8.27). The zero value of the integral occurring in the numerator of (4.8.30), however, makes the average value $\mathcal{P}_{av} = 0$. Similarly, for the capacitor problem of Example 4.8.3, in time-periodic situation, let $\mathbf{E}$ be expressed as (see Fig. 4.4.1):

$$\mathbf{E} = E_0 \cos\omega t \;\mathbf{a}_E\,, \tag{4.8.31}$$

with $\mathbf{a}_E$ interpreted as $\mathbf{a}_z$. Then one may express **H**, the magnetic field intensity on the periphery of the capacitor, with the help of (4.4.5) through (4.8.31), as:

$$\mathbf{H} = \frac{dD}{dt}\frac{a}{2}\mathbf{a}_\theta = \varepsilon_0 \frac{dE}{dt}\frac{a}{2}\mathbf{a}_\theta = -H_0 \sin\omega t\;\mathbf{a}_E = H_0\cos(\omega t + \frac{\pi}{2})\;\mathbf{a}_E\,, \tag{4.8.32}$$

with $\mathbf{a}_E$ interpreted as $\mathbf{a}_\theta$, where $H_0 = (\varepsilon_0\,\omega a/2)\,E_0$.

Substituting (4.8.31) and (4.8.32) in (4.8.23), and taking the average of the latter over a cycle, one then obtains:

$$\mathcal{P}_{av} = (\mathbf{E}\times\mathbf{H})_{av} = \frac{E_0 H_0 \int_0^{2\pi} -\sin\omega t\cos\omega t\, d(\omega t)}{\int_0^{2\pi} d(\omega t)}\mathbf{a}_E \times \mathbf{a}_H = 0 \tag{4.8.33}$$

with $\mathbf{a}_E\times\mathbf{a}_H$ ($= \mathbf{a}_z \times \mathbf{a}_\theta$) continuing to be equal to $-\mathbf{a}_r$ as in(4.8.27) and (4.8.30).

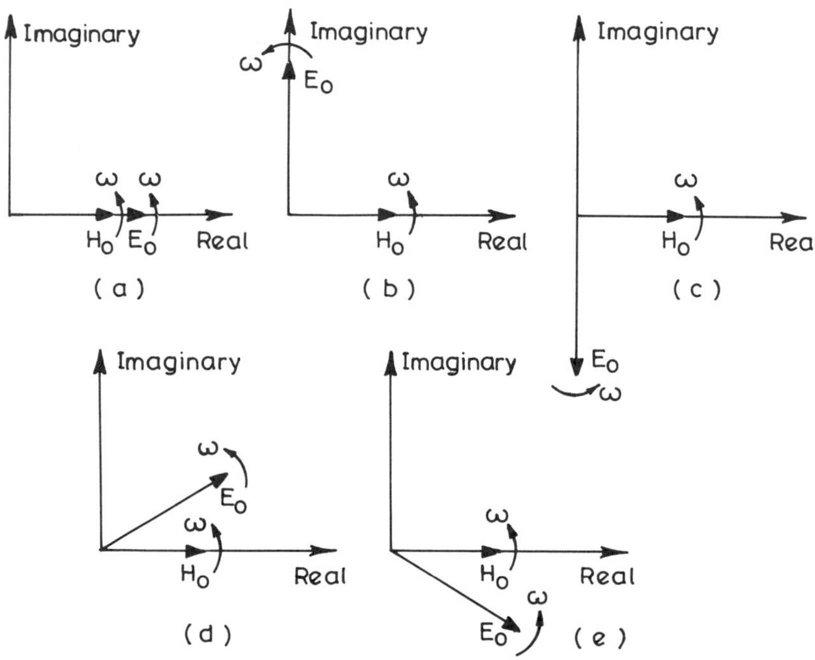

Fig. 4.8.1 Amplitude of electric and magnetic field intensities, represented in the complex plane (Argand diagram), as vectors rotating with an angular velocity ω, for a resistor **(a)**, an inductor **(b)**, a capacitor **(c)**, an inductor in series with a resistor **(d)**, and a capacitor in parallel with a resistor **(e)**.

For the sake of convenience, quite often, the time-periodic electric and magnetic field intensities are represented in a complex plane to obtain what is known as a phasor or Argand diagram. In this diagram, the amplitude of the field quantities may be depicted as vectors rotating with an angular velocity ω, being equal to the angular frequency of the simple-harmonic variation of field quantities. In this description the instantaneous field quantity is obtained as the projection of the amplitude vectors on the real (or imaginary) axis. In the resistor problem of Example 4.8.1, one may notice from (4.8.26), the electric and magnetic field intensities are in phase. The amplitudes of these field quantities, therefore, may be chosen to be represented on the real axis of the phasor diagram (Fig. 4.8.1(a)). Further, for such a case, the Poynting vector is found to average out to a non-zero value given by (4.8.27). In the inductor and capacitor problems of Example 4.8.3, the electric and magnetic field intensities are found to be in phase quadrature, the former leading the latter in the capacitor example (compare (4.8.28) with (4.8.29)), while the former lagging the latter in the inductor example (compare (4.8.31) with (4.8.32)). Hence, in the phasor diagram, if we choose the amplitude of the magnetic field intensity as represented on the real axis, the amplitude of the electric field intensity would appear on the positive imaginary axis of the diagram, for an inductor (Fig. 4.8.1(b)), and on the negative imaginary axis of the diagram, for a capacitor (Fig. 4.8.1(c)). One may remember that, for such field quantities in phase quadrature, the Poynting Vector averages out to zero, as can be seen from (4.8.30) and (4.8.33).

A more general case is one in which the electric and magnetic field intensities differ in phase by an angle θ which is neither zero (corresponding to the fields not being in phase) nor $\pi/2$ (corresponding to the fields not being in phase quadrature). For instance, in the case of an inductor in series with a resistor, or for a lossy inductor, the electric field intensity would lead the magnetic field intensity in phase by an angle θ ($\pi/2 < \theta < 0$)(Fig. 4.8.1(d)). The electric field intensity amplitude E_0 may be resolved, in this case, into the in-phase component, $E_0 \cos\theta$, and the quadrature component, $E_0 \sin\theta$. The average value of the Poynting vector in this example is contributed only by the in-phase component, since the quadrature component may be seen to average out to zero by simply interpreting E_0 as $E_0 \sin\theta$ in (4.8.30). Therefore, for this problem, interpreting E occurring in (4.8.27) as the in-phase component, $E_0 \cos\theta$, one obtains:

$$\mathcal{P}_{av} = \frac{1}{2} E_0 H_0 \cos\theta \, \mathbf{a}_E \times \mathbf{a}_H \, . \tag{4.8.34}$$

It may be noted here that, in the alternative phasor notation, the electric field intensity may be represented as:

$$\mathbf{E} = E_0 \exp(j\theta) \exp(j\omega t) \, \mathbf{a}_E \tag{4.8.35}$$

the magnitude of the real and imaginary parts of the time-independent amplitudes part of which, namely, $E_0 \cos\theta$ and $E_0 \sin\theta$, may be identified as the in-phase and quadrature components, respectively.

Similarly, in the case of a capacitor in parallel with a resistor, or for a lossy capacitor, the electric field intensity would lag the magnetic field intensity by an angle θ ($\pi/2 < \theta < 0$) (Fig. 4.8.1(e)); the amplitude of the electric field intensity, in this case, may be resolved into the in-phase component $E_0 \cos\theta$, and the quadrature component, $- E_0 \sin\theta$. The electric field intensity may, therefore, be represented, in the phasor notation, as in (4.8.35), as:

$$\mathbf{E} = E_0 \exp(-j\theta) \exp(j\omega t)\, \mathbf{a}_E \, . \tag{4.8.36}$$

The in-phase component remaining unchanged, the average value of the Poynting vector, in this case, continues to be given by (4.8.34). Noting that the magnetic field intensity has been chosen in the phasor diagram as to lie in the real axis (Fig. 4.8.1), one may represent it, in the phasor notation, as:

$$\mathbf{H} = H_0 \exp(j\omega t)\, \mathbf{a}_H \, . \tag{4.8.37}$$

Introducing the complex conjugate of $\mathbf{H}$ as:

$$\mathbf{H}^* = H_0 \exp(-j\omega t)\, \mathbf{a}_H \, , \tag{4.8.38}$$

one may define a complex Poynting vector $\mathcal{P}$ (= $\mathbf{E}\times\mathbf{H}^*$), which, with the help of (4.8.35), (4.8.36) and (4.8.38), may be expressed as:

$$\mathcal{P}_{\text{complex}} = \mathbf{E}\times\mathbf{H}^* = E_0H_0 \exp(\pm j\theta)\, \mathbf{a}_E\times\mathbf{a}_H = E_0H_0 (\cos\theta \pm j \sin\theta)\, \mathbf{a}_E\times\mathbf{a}_H \, , \tag{4.8.39}$$

the plus and minus signs corresponding to the leading and lagging phases (Fig. 4.8.1(d) and 4.8.1(e)), respectively. One may now easily see that half the real (*Re*) part of $\mathcal{P}_{\text{complex}}$, given by (4.8.39), is identical with $\mathcal{P}_{\text{av}}$, given by (4.8.34), that is:

$$\mathcal{P}_{\text{av}} = \frac{1}{2}\, Re\, \mathcal{P}_{\text{complex}} = \frac{1}{2}\, Re\, (\mathbf{E}\times\mathbf{H}^*) \, . \tag{4.8.40}$$

The above expression, namely (4.8.40), finds extensive applications in electromagnetic engineering (for instance, in the antenna-gain calculation or in estimating power propagating down a structure supporting electromagnetic waves (see the next chapter), etc.). One may also define a reactive power density $\mathcal{P}_{\text{reactive}}$, which does not contribute to the average value of Poynting vector, as half the imaginary (*Im*) part of $\mathcal{P}_{\text{complex}}$, given by (4.8.39), as follows:

$$\mathcal{P}_{\text{reactive}} = \frac{1}{2}\, Im\, (\mathbf{E}\times\mathbf{H}^*) = \pm \frac{1}{2}\, E_0H_0 \sin\theta\, \mathbf{a}_E\times\mathbf{a}_H \, . \tag{4.8.41}$$

It may be mentioned that the magnitude of $\mathcal{P}_{\text{reactive}}$, that is, $\pm(1/2)E_0H_0\sin\theta$, can be identified as one of the sides of a right angled power-(density) triangle, the hypotenuse of which represents the *virtual* power density magnitude calculated without regard to the phase difference between the electric and magnetic field intensities, that is, the magnitude of $\mathcal{P}_{av}$ given by (4.8.27), while the remaining other side of the triangle representing the magnitude of the average Poynting vector $\mathcal{P}_{av}$ given by (4.8.34).

<u>*Example 4.8.4*</u> Obtain an expression for the average power lost per unit area in a planar, semi-infinite conductor, in terms of the surface resistivity and surface current density, considering a uniform plane electromagnetic wave normally incident, from a free space region, on the conductor.

Let the conductor, called the region 1, extend from z = 0 to ∞, and the wave be incident normally on it from the free-space region, called the region 2, at the planar interface at z = 0 (Fig. 4.8.2). Let the electric and magnetic field intensities at a given instant, along x and y, respectively, be represented, in the phasor notation, as:

$$\mathbf{E} = E_0 \exp(j\omega t)\, \mathbf{a}_x \text{ and } \mathbf{H} = H_0 \exp(j\omega t)\, \mathbf{a}_y ,$$

respectively, so that one may write (4.8.40), in this case, as

$$\mathcal{P}_{av} = \frac{1}{2} Re\, \mathbf{E}\times\mathbf{H}^* = \frac{1}{2} Re(E_0)(H_0^*)\, \mathbf{a}_z , \tag{4.8.42}$$

which shows that the power flows into the conductor along the z-direction.

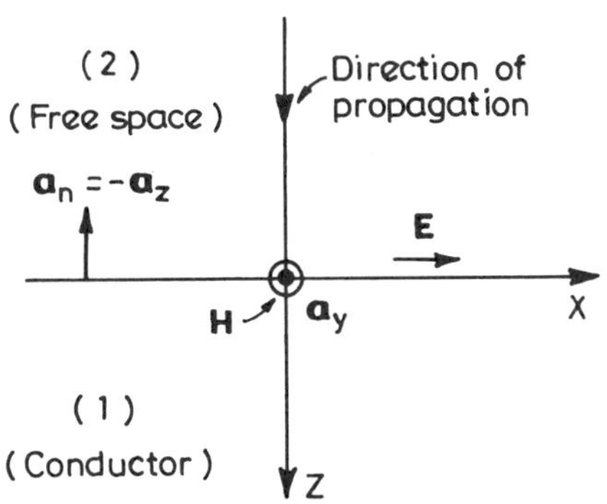

Fig. 4.8.2 The conductor-free-space interface showing the directions of the field intensities, wave propagation and $\mathbf{a}_n$ (= - $\mathbf{a}_z$), a unit vector directed from the region 1 (conductor) to region 2 (free space).

The average power lost through an element of area dS, say, at the conducting boundary (z = 0) may be found using (4.8.6) as equal to $-\mathcal{P}_{av} \cdot \mathbf{a}_n$ dS which, on putting $\mathbf{a}_n = -\mathbf{a}_z$, and with the help of (4.8.42), becomes equal to (1/2) *Re* (E_0) (H_0^*) dS, so that dividing the latter by the element of area dS, one may now write the average power lost in the conductor per unit area, P_{LA}, say, as:

$$P_{LA} = \frac{1}{2} Re\, (E_0)(H_0{}^*)\ . \qquad (4.8.43)$$

Using (4.5.41), one may then express (4.8.43) as:

$$P_{LA} = \frac{1}{2} Re\, (\eta H_0)(H_0{}^*)\ , \qquad (4.8.44)$$

where η is the intrinsic impedance of the conducting medium which has been shown in Example 4.5.5 to be equal to the surface impedance Z_S for a good conductor. The real part of Z_S (= η) being the surface resistance R_S, one obtains from (4.8.44) :

$$P_{LA} = \frac{1}{2} R_S\, (H_0)(H_0{}^*) = \frac{1}{2} R_S\, \mathbf{H} \cdot \mathbf{H}^*\ . \qquad (4.8.45)$$

Further, one may choose to express (4.8.45) in terms of the surface current density $\mathbf{J}_S$ by invoking the boundary condition (4.7.8) which has to be read in the present context, with $\mathbf{H}_2 = \mathbf{H} = H\mathbf{a}_y$, $\mathbf{H}_1 = 0$ (with time dependence understood) and $\mathbf{a}_n = -\,\mathbf{a}_z$ (see Table 4.7.1 with reference to Example 4.7.1), as:

$$-\mathbf{a}_z \times H\,\mathbf{a}_y = H\mathbf{a}_x = \mathbf{J}_s$$

which yields

$$\mathbf{H} \cdot \mathbf{H}^* = H\,\mathbf{a}_x \cdot H^*\,\mathbf{a}_x = \mathbf{J}_S \cdot \mathbf{J}_S{}^*$$

the substituton of which in (4.8.45) gives:

$$P_{LA} = (1/2)\, R_S\, \mathbf{J}_S \cdot \mathbf{J}_S{}^*\ ,$$

the required expression for power lost per unit area at a conducting boundary in terms of the surface resistance and current density.

<u>Example 4.8.5</u> Show that, for an infinitesimal oscillating dipole (antenna), while the power is radiative at a large distance from the antenna, it is reactive in the close vicinity of the antenna.

The electric and magnetic field intensities at a distance r from an infinitesimal oscillating dipole are, in general, given by (4.6.30) and (4.6.31), respectively (see Example 4.6.1). As a special case, for large distance from the dipole, these field expressions may be put together in the form of (4.6.32). The latter may be used to find $\mathcal{P}_{complex}$ as follows:

$$\mathcal{P}_{complex} = \mathbf{E} \times \mathbf{H}^* = (E_\theta H_\phi{}^* - E_\phi H_\theta{}^*)\mathbf{a}_r = E_\theta H_\phi{}^*\, \mathbf{a}_r\ , \qquad (4.8.46)$$

since, in this case, as can be seen either from (4.6.30) and (4.6.31), or from (4.6.32), $E_\phi = H_\theta = 0$. Substituting (4.6.32) in (4.8.46) one then obtains:

$$\mathcal{P}_{complex} (= E_\theta H_\phi^* \mathbf{a}_r) = \eta \left(\frac{I\, dl\, \beta \sin\theta}{4\pi r} \right)^2 \mathbf{a}_r\ , \qquad (4.8.47)$$

where the symbols have been explained in Example 4.6.1. Thus, $\mathcal{P}_{complex}$ is purely real at large distances from the dipole, and one obtains from (4.8.40) and (4.8.41):

$$\mathcal{P}_{av} = \frac{1}{2}\eta\left(\frac{I\,dl\,\beta\sin\theta}{4\pi r}\right)^2 \mathbf{a}_r \quad \text{and} \quad \mathcal{P}_{reactive} = 0 . \qquad (4.8.48)$$

In the close vicinity of the dipole, one may consider, in the field expressions (4.6.30) and (4.6.31), the terms varying inversely as higher index of power of r as predominantly important, enabling one to write:

$$E_r = \frac{-jI\,dl}{2\pi}\frac{\eta}{\beta}\frac{1}{r^3}\cos\theta\exp(-j\beta r), \quad E_\theta = \frac{-jI\,dl}{4\pi}\frac{\eta}{\beta}\frac{1}{r^3}\sin\theta\exp(-j\beta r),$$

$$E_\phi = 0; \quad H_r = 0, \quad H_\theta = 0, \quad H_\phi = \frac{I\,dl\sin\theta}{4\pi r^2}\exp(-j\beta r). \qquad (4.8.49)$$

It is of interest to compare the fields given by (4.8.49) with those due to a short electrostatic dipole and to an infinitesimal steady current element. For this purpose, let us put $\beta = \omega(\mu_0\,\varepsilon_0)^{1/2}$, $\eta = (\mu_0/\varepsilon_0)^{1/2}$ (see section *4.5.1*), $I = dq/dt = j\omega q$ (for an understood time dependence of the form $\exp j\omega t$), where q is the oscillating charge of the dipole, to express the electric field intensity given by (4.8.49) in the following vectorial form:

$$\mathbf{E} = \frac{p}{4\pi\varepsilon_0 r^3}(2\cos\theta\,\mathbf{a}_r + \sin\theta\,\mathbf{a}_\theta)\exp j(\omega t - \beta r) , \qquad (4.8.50)$$

where $p = q\,dl$ is the dipole moment. The amplitude of (4.8.50) is found to be identical with (2.5.13), the expression for the electrostatic field intensity due to a short electrostatic dipole. The amplitude of magnetic field intensity

$$\mathbf{H} = \frac{I dl\sin\theta}{4\pi r^2}\mathbf{a}_\phi \exp j(\omega t - \beta r) \qquad (4.8.51)$$

written from (4.8.49), is found to be identical with Biot-Savart's law (3.2.1), for a steady-current element, which may be read, in the present context, interpreting **dl** as dl $\mathbf{a}_z$, **r** as r $\mathbf{a}_r$, and hence, $\mathbf{dl}\times\mathbf{r}$ as $(dl)(r)\,\mathbf{a}_z\times\mathbf{a}_r = (dl)(r)\sin\theta\,\mathbf{a}_\phi$. Further, for such fields close to the oscillating dipole, one may, with the help of (4.8.49), see that $\mathcal{P}_{complex}$ is purely reactive, and that

$$\mathcal{P}_{av} = \frac{1}{2}\,Re\,(\mathbf{E}\times\mathbf{H}^*) = 0$$

and

$$\mathcal{P}_{reactive} = \frac{1}{2}\,Im\,(\mathbf{E}\times\mathbf{H}^*) = \frac{1}{2}\frac{\eta}{\beta}\left(\frac{I dl}{4\pi}\right)^2\frac{1}{r^5}(-\sin\theta\,\mathbf{a}_r + 2\sin\theta\cos\theta\,\mathbf{a}_\theta) . \qquad (4.8.52)$$

It is clear from (4.8.48) that there exists a non-zero value of $\mathcal{P}_{av}.\mathbf{a}_r$ (since $\mathcal{P}_{av} = \mathcal{P}_{av}\,\mathbf{a}_r \neq 0$) in the radial direction (directed from the dipole to the point where the power density is found), corresponding to a radiated power in that direction, at large distances from the oscillating dipole. Also, it is clear from (4.8.52) that, in view of $\mathcal{P}_{av} = 0$, there is no outward flow of power in the close vicinity of such a dipole and that, in view of $\mathcal{P}_{reactive} \neq 0$, there is a storage of reactive energy near the dipole, such energy oscillating back and forth between the dipole and the region immediately surrounding it.

In describing the time-harmonic phenomena here (in section 4.8.1), and developing relevant concepts like complex, average and reactive Poynting vector, reference has been made to the general statement of Poynting theorem (4.8.5)-(4.8.7). Now that for the time-harmonic fields complex Poynting vector appears to be more relevant, one may re-formulate the Poynting theorem using the complex Poynting vector, in (4.8.1). In view of this then (4.8.2) is modified as:

$$\nabla.(\mathbf{E}\times\mathbf{H}^*) = -\mu\mathbf{H}^*.\frac{\partial\mathbf{H}}{\partial t} - \varepsilon\mathbf{E}.\frac{\partial\mathbf{E}^*}{\partial t} - \mathbf{E}.\mathbf{J}_c^* . \qquad (4.8.53)$$

Now, in the phasor notation, assuming $\mathbf{E}$ and $\mathbf{H}$ to vary as exp ($j\omega t$) and hence $\mathbf{E}^*$ and $\mathbf{H}^*$ as exp ($-j\omega t$), one may take $\partial E/\partial t$, $\partial E^*/\partial t = (\pm)\, j\omega E^{(*)}$ and $\partial H/\partial t$, $\partial H^*/\partial t = (\pm)\, j\omega H^{(*)}$ in (4.8.53) thereby obtaining

$$\nabla.(\mathbf{E}\times\mathbf{H}^*) = -j\omega\mu\, \mathbf{H}.\mathbf{H}^* + j\omega\varepsilon\, \mathbf{E}.\mathbf{E}^* - \mathbf{E}.\, \mathbf{J}_c^* . \qquad (4.8.54)$$

Then following the same method as done in deriving (4.8.6) from (4.8.4), that is, by first taking the volume integral of (4.8.54) and then applying Gauss's theorem, one may easily obtain:

$$-\oint_S (\mathbf{E}\times\mathbf{H}^*).\mathbf{a}_n\, dS = j\omega\int_\tau (\mu\mathbf{H}.\mathbf{H}^* - \varepsilon\mathbf{E}.\mathbf{E}^*)\, d\tau + \oint_\tau \mathbf{E}.\mathbf{J}_c^*\, d\tau . \qquad (4.8.55)$$

Separating (4.8.55) into the real and imaginary parts and taking their half values, one may then, respectively, write:

$$-\oint_S \frac{1}{2} Re(\mathbf{E}\times\mathbf{H}^*).\mathbf{a}_n\, dS = -\oint_S \mathcal{P}_{av}.\mathbf{a}_n\, dS = \oint_\tau \frac{1}{2}\mathbf{E}.\mathbf{J}_c^*\, d\tau = \oint_\tau \frac{1}{2}\sigma E^2\, d\tau \qquad (4.8.56)$$

and

$$-\oint \frac{1}{2} Im(\mathbf{E}\times\mathbf{H}^*).\mathbf{a}_n\, dS = -\oint \mathcal{P}_{reactive}.\mathbf{a}_n\, dS = 2\omega\oint_\tau \left(\frac{1}{2}\frac{\mu\mathbf{H}.\mathbf{H}^*}{2} - \frac{1}{2}\frac{\varepsilon\mathbf{E}.\mathbf{E}^*}{2}\right) d\tau , \qquad (4.8.57)$$

where we have used Ohm's law to write $\mathbf{E}\,.\,\mathbf{J}_c = \mathbf{E}\,.\,\sigma\mathbf{E}^* = \sigma E^2$.

Physically, the left hand side of (4.8.56) represents the average power entering the closed volume, while its right hand side represents the average ohmic loss of power in the volume (the factor of 1/2 appearing out of averaging of the power, for instance, (see (4.8.25)). We notice that, unlike in the general formulation (4.8.6), the stored energy term is absent in (4.8.55) as is expected in the time-periodic case in which, on the average, there would be no increase in the energy storage. The imaginary part of (4.8.55), namely (4.8.57), which balances the reactive power, states that the reactive power flowing into the volume enclosure is equal to 2ω times the difference between the average energies stored in electric and magnetic field, the quantities $(1/2)[(1/2)\mu\mathbf{H}.\mathbf{H}^*]$ and $(1/2)[(1/2)\varepsilon\mathbf{E}.\mathbf{E}^*]$ being the time-independent quantities representing the average energy densities in electric and magnetic fields (the factor of 1/2 again appearing out of averaging of each of these quantities).

A similar procedure, in the presence of a power generator inside the closed volume, and for time-periodic fields, would make the general formulation (4.8.7) read as follows (c.f. (4.8.55)):

$$-\oint_\tau \mathbf{E}.\mathbf{J}^*_{\text{ext}}d\tau = \oint_S (\mathbf{E}\times\mathbf{H}^*).\mathbf{a}_n\,dS + j\omega\oint_\tau(\mu\,\mathbf{H}.\mathbf{H}^* - \varepsilon\,\mathbf{E}.\mathbf{E}^*)d\tau + \oint_\tau \mathbf{E}.\mathbf{J}^*_c d\tau. \tag{4.8.58}$$

As explained following (4.8.7), the left hand side of (4.8.58) would essentially be a positive quantity if $\mathbf{J}_{\text{ext}}$ is interpreted as the generator current density which would have a direction opposite to **E**. Now (4.8.58), just as (4.8.55), may be separated into the real and imaginary parts showing the details of power balance. An interesting result follows by taking half the real part of (4.8.58) and using the relation $\mathbf{E}.\mathbf{J}_c{}^* = \sigma E^2$, as in (4.8.55), as:

$$\frac{1}{2}Re\oint_\tau(\mathbf{E}.\mathbf{J}_{\text{ext}})d\tau = \frac{1}{2}Re\oint_S(\mathbf{E}\times\mathbf{H}^*).\mathbf{a}_n\,dS + \oint\frac{1}{2}\sigma E^2\,d\tau\,. \tag{4.8.59}$$

The left hand side of (4.8.59) represents the time-averaged generated power, while the first and the second terms of its right hand side represent the power leaving the volume and the average ohmic loses within the volume. When applied to a radiating system (for instance, see Example 4.8.5), in free space (where ohmic losses are absent so that one may take as zero the second term of the right hand side of (4.8.59), the latter simply states that the power fed into the radiating system (represented by its left hand side) is radiated out into space (represented by its right hand side).

Thus it becomes clear that the complex Poynting vector and its surface integration over the geometry of interest would give one a picture of power balance in a practical problem. It also obviously turns out that, for time-periodic electromagnetic waves, one may find as in Example 4.8.2, the power propagating down a structure supporting these waves by taking half the real part of the integration of the complex Poynting vector over the structure cross section.

4.9 Summary

The time variation of charge density at a point may be related to the spatial variation of current density by what is known as the continuity equation which follows from the principle of conservation of charges. The continuity equation is valid for both the conduction and convection currents. In an interesting example the distribution of charges inside a conductor is described.

The time-varying electric field intensity is associated with a displacement current and a consequent magnetic field intensity. Similarly, the time-varying magnetic field intensity is associated with an electric field intensity governed by Faraday's law. Thus, in time-varying phenomena, the electric and magnetic fields are coupled. The coupling of these fields are described by two of the four Maxwell's equations, the remaining two of them being essentially the electrostatic and magnetostatic Gauss's law or Poisson's equations which remain unchanged in form in time-varying situations. Maxwell's equations may be expressed in either integral forms, involving the line, surface and volume integrals, or differential forms, involving the divergence and curl operations. Maxwell's equations in the differential form may be decoupled to form wave equations in electric and magnetic fields. The wave equations may be solved to study the propagation of electromagnetic waves through an unbounded media. For instance, it has been shown that a uniform, plane electromagnetic wave is supported by a homogeneous, isotropic medium. For both free-space and conducting regions, the electric and magnetic fields are found to be orthogonal to each other and each perpendicular to the direction of propagation. The concerned wave is called the transverse electromagnetic (TEM) wave. The study gives expressions for the phase velocity of propagation and intrinsic impedance of the medium. The wave is found to be attenuated in a conducting medium and the study of such phenomenon gives the concept of the skin effect. The skin depth as well as the surface resistance of a semi-infinite, planar conductor are found to be dependent on the conductivity of the medium and the operating frequency. Further, the surface impedance is found to be equal to the intrinsic impedance, for a good conductor. The study for a poor conductor like see-water reveals that one should go for relatively lower operating frequencies from the standpoint of reduced attenuation for sub-marine communication; the size of the radiating element would also be less in such a medium than in free-space.

Just as one develops, in electrostatics, the electric field intensity from a scalar potential, so also in time-varying phenomena can one develop the electric field intensity from a scalar, and, in addition, a vector potential. Subject to the Lorentz condition, it is also possible to develop the electric field intensity only from the vector potential. The magnetic field intensity can also be developed from the vector potential. The vector and scalar potentials both can be found from the solutions of the wave equations in these quantities. These solutions which can be obtained in terms of volume charge and current-density distributions give the concepts of retarded scalar and vector potentials. The applications of these concepts are found, for instance, in the characterisation of a radiating element for which the current distribution is known.

The propagation of electromagnetic waves through a bounded medium or through a physical structure supporting an electromagnetic wave may be studied by solving Maxwell's equations, and more precisely, wave equations derived therefrom, subject to a set of boundary conditions at points of discontinuity in the medium or structure. These boundary conditions can be developed from the integral form of Maxwell's equations. The general boundary conditions developed may further be studied as special cases for time-independent and time-dependent situations at the interfaces between specific media such as dielectric-dielectric and conductor-dielectric interfaces. The boundary conditions can be used to find the field distribution in one region in terms of a known field distribution in another medium adjacent to it. Alternatively, the surface-charge and current distributions can be found with the help of these conditions at the interface between two medium if the field distributions in these media are specified.

Finally, it is necessary from a practical point of view to interpret field quantities for the estimate of power propagating through a medium or down a physical structure. Thus the concept of a power density vector or Poynting vector comes into play, the flux of such vector estimating the power. The power flow into a closed volume due to power sources outside the volume as well as the power flowing out of a closed volume containing power sources may be stated by a power-balance theorem called the Poynting theorem. The theorem brings into its purview both the time-independent and time-dependent (periodic) situations. However, for the latter situations, it is more relevant to introduce the complex Poynting vector and re-formulate the Poynting theorem in terms of it to get an insight into the time-averaged power flow problems. Power propagating down a practical structure supporting an electromagnetic wave is obtained by integrating the complex Poynting vector over the structure cross section and then taking half of its real part.

Problems

4.1 Show that about half a day is the relaxation time in mica ($\sigma = 10^{-15}$ mho/m, $\varepsilon_r = 6$).

4.2 What relation instead of (4.5.16) (as found in Example 4.5.2) would be obtained had the waves propagated in the negative, instead of positive, z direction.

Answer: $E_x/H_y = -E_y/H_x = -\eta$.

4.3 Calculate the attenuation in dB's of a uniform plane electromagnetic wave at 1 MHz which propagates a kilometer through dry earth, the loss-tangent of which at the given frequency is 0.06. Take the relative permittivity of the medium as 3.

Answer: 9.55 dB.

4.4 The inner and the outer conductors of a coaxial cable are made of a highly conducting metal of conductivity σ. The region between the conductors is made of a lossy dielectric of loss tangent, $\tan\delta$, at the operating frequency, f. If a is the radius of the central conductor and b is the inner radius of the outer conductor, show that R, the 'longitudinal' resistance per unit length, and G, the 'transverse' conductance per unit length, are given by:

$$R = (\pi f \mu_0/\sigma)^{1/2}\,[(2\pi a)^{-1} + (2\pi b)^{-1}] \text{ and } G = \omega C \tan\delta,$$

where C, the capacitance per unit length of the cable, is given by (2.7.13) (G being the transverse (or radial) leakage current through the dielectric between the conductors per unit potential difference between them).

4.5 Assuming the rf dependence as $\exp j(\omega t - \beta z)$ and using relevant Maxwell's equations in the cylindrical system of coordinates (r, θ, z), obtain the following expressions for the radial and the azimuthal components of electric and magnetic field components in terms of their axial field components:

$$E_r = [1/(\beta^2 - \omega^2\mu\varepsilon)][j\beta(\partial E_z/\partial r) + (j\omega\mu/r)(\partial H_z/\partial\theta)]\,,$$

$$E_\theta = -[1/(\beta^2 - \omega^2\mu\varepsilon)][j\omega\mu(\partial H_z/\partial r) - (j\beta/r)(\partial E_z/\partial\theta)]\,,$$

$$H_r = -[1/(\beta^2 - \omega^2\mu\varepsilon)][(j\omega\mu/r)(\partial E_z/\partial\theta) - j\beta(\partial H_z/\partial r)]\,,$$

and

$$H_\theta = [1/(\beta^2 - \omega^2\mu\varepsilon)][j\omega\varepsilon(\partial E_z/\partial r) + (j\beta/r)(\partial H_z/\partial\theta)]\,.$$

4.6 The electric field intensity at a point on the surface of a conductor placed in free space is:

$$2\mathbf{a}_x - 3\mathbf{a}_y + 6\mathbf{a}_z \text{ V/m}\,.$$

Show that the surface charge density on the point is $7\varepsilon_0$ coulombs/m^2.

Hint: Apply the boundary condition $\mathbf{D}.\mathbf{a}_n = \rho_s$, at the conducting surface, where $\mathbf{D}$ (= D $\mathbf{a}_n$) is normal to the surface.

4.7 Show that a peak surface current density of 1 amp/m is developed at a conducting plate, at z = 0, which bounds a dielectric medium, of relative permittivity 4, the electric field intensity in the latter medium being given as:

$$\mathbf{E} = 60\,\pi \sin\beta z \sin \omega t\, \mathbf{a}_x \,.$$

Hint: Apply one of the Maxwell's equations to obtain an expression for **H** in the dielectric, and hence the value of the latter at the boundary z=0. Then, applying the relevant boundary condition involving the surface current density, the desired result follows.

4.8 In a free-space region, the electrostatic potential V is given by the function: V = 300 (5 - r)/r. If a point r = 300 cm, $\theta = 30^0$, $\phi = 120^0$ lies on a conducting surface placed in the region, find the equation of the surface.

Answer: r = 3.

Hint: At a conducting surface, V = constant.

4.9 For a plane-polarised wave incident at the interface between two dielectric media of permittivities ε_1 and ε_2, respectively, show i) that there will be no reflection for *parallel* polarisation (for which the electric field intensity vector is parallel to the plane of incidence), if the angle of incidence is equal to $\tan^{-1} (\varepsilon_2/\varepsilon_1)^{1/2}$, called the polarising angle (also called the Brewster's angle); and ii) that there will be *total* internal reflection if the angle of incidence is greater than the critical angle $\sin^{-1} (\varepsilon_2/\varepsilon_1)^{1/2}$, for polarisation of both *parallel* and *perpendicular* (for which the electric field intensity vector is normal to the plane of incidence) types, the medium of permittivity ε_1 containing the incident and reflected waves while the medium of permittivity ε_2 containing the refracted wave.

4.10 A current flows through the cross section of a composite structure consisting of two solid rectangular conducting blocks forming a common interface between them. Show that the current densities perpendicular to the cross section of the two conductors (or parallel to their interface) are distributed in the conductors in proportion to their conductivity values.

4.11 A current flows from a conducting medium of conductivity σ_1 to another conducting medium of conductivity σ_2 being obliquely incident at the interface between the two conducting media. If the angles of the current density vectors in these two media, at the interface as made with the normal to the latter, are θ_1 and θ_2, respectively, show that i)tan θ_1/(tan θ_2) = σ_1/σ_2 and ii) $\rho_S = (\tau_2 - \tau_1)\, J_n$, where ρ_S is the surface charge density at the interface; τ_1 and τ_2 are the relaxation time constants in the two media, respectively; and J_n is the normal component of current density at the interface in either of the media, being continuous at the interface.

4.12 Using the result i) of problem 4.11, appreciate that the electric field intensity at the surface of a good conductor of conductivity σ_1 , which is exposed to a dielectric of conductivity σ_2 ($<< \sigma_1$), is normal to the interface.

4.13 Apply the boundary conditions for time-independent fields to show that a static magnetic field in free-space is normal to the boundary of a high-permeability medium.

4.14 For a uniform plane electromagnetic wave propagating through free space, establish that i) the instantaneous Poynting vector is equal to i) the phase velocity times the instantaneous energy density, while ii) the latter is equal to the instantaneous pressure of the wave acting in the direction of propagation.

4.15 A uniform plane electromagnetic wave of rms electric field intensity $6\sqrt{\pi}$ V/m is totally reflected at normal incidence from a conducting planar surface. Calculate the average pressure exerted on the plate.

Hint: The desired pressure is equal to the pressure of the wave totally absorbed by the plate plus and equal amount of pressure corresponding to the wave being re-emitted from the surface in the direction of the reflected wave.

4.16 Power is transmitted along z through a rectangular hollow pipe (called a waveguide). Considering the side walls of the pipe to be located at x = 0, x = a, y = 0 and y = b, respectively, (so that the cross-sectional area the waves propagate through is axb), find a) the power transmitted and b) the power loss per unit length in the side walls of the pipe, in terms of the resistivity of the wall ρ_r, it being given that the fields inside the pipe are:

$$\mathbf{H} = (j\beta a/\pi)\,(H_0)\,(\sin(\pi x/a))\,\mathbf{a}_x + H_0\,(\cos(\pi x/a))\,\mathbf{a}_z \exp j(\omega t - \beta z)$$

and

$$\mathbf{E} = (-j\omega\mu_0\, a/\pi)\,(H_0)\,(\sin(\pi x/a))\,\mathbf{a}_y \exp j(\omega t - \beta z)$$

Answer: (a) $(\omega\mu_0\beta\, a^3 b/(4\pi^2))\, H_0^2$

(b) $(\omega\mu_0\, \rho_r/2)^{1/2}\, (b+(a/2)\,(1+(\beta a/\pi)^2)\, H_0^2$.

4.17 A uniform plane electromagnetic wave propagating in free space in the positive z direction suffers a reflection at normal incidence from a plane conducting surface at z = 0. Find **E** and **H**, the resultants of the incident and reflected electric field intensities, in terms of E_0, the peak value of the amplitude of the electric field intensity, supposedly in the positive x direction, and β, the axial propagation constant, it being assumed that the field quantities, for the waves propagating in $\pm$ z direction, vary as $\exp j(\omega t \mp \beta z)$.

Answer: $\mathbf{E} = (-2jE_0 \sin\beta z)\,(\exp j\omega t)\,\mathbf{a}_x$ and $\mathbf{H} = (2E_0\,(\mu_0/\varepsilon_0)^{-1/2} \cos\beta z)\,(\exp j\omega t)\,\mathbf{a}_y$.

4.18 (a) Locate the nodes and antinodes in the standing-wave pattern exhibited by the field quantities obtained in problem (4.17). (b) Also, show that no power is transported by such a standing wave.

Answer: (a) Nodes for **E** and antinodes for **H** : $z = 0, -\lambda/2, -\lambda, -(3/2)\lambda, \ldots$

Antinodes for **E** and nodes for **H** : $z = -\lambda/4, -3\lambda/4, -5\lambda/4, \ldots,$

(b) Poynting vector $\mathbf{E}\times\mathbf{H}^*$ is purely imaginary having no real part and hence there would be no power flow.

4.19 What is a) the incident power density and b) the power absorbed per unit area in a sheet of brass ($\sigma = 1.5\times10^7$ mho/m) on which a uniform plane wave is incident normally with a peak electric field intensity of 1V/cm at 10 GHz?

Answer: (a) 13.26 W/m^2, (b) 7.21 mW/m^2 .

References

[1] L. Solymar, *Lectures on Electromagnetic Theory: A Short course for Engineers* (Oxford Press, London, 1976).

[2] R. Plonsey and R.E. Collin, *Principles and Applications of Electromagnetic Fields* (Mc Graw-Hill, New York, 1961).

[3] W.H. Hayt, Jr., *Engineering Electromagnetics* (Mc Graw-Hill, New York, 1989).

[4] M. Faraday, *Experimental Researches in Electricity,* (3 volumes, 1839, 1944, 1855: Taylor and Francis, vols. 1 and 3; Richard and John Edward Taylor, vol. 2; Dover, New York, 1965).

[5] G.E. Owen, *Electromagnetic Theory* (Allyn and Bacon, Boston, 1963).

[6] D. Halliday and R. Resnick, *Physics, Part II,* (John Wiley, New York, 1966).

[7] J.C. Maxwell, *A Treatise on Electricity and Magnetism* (Clarendon, Oxford, 1873).

[8] J.C. Maxwell, "A dynamical theory of the electromagnetic field," *Phil. Trans. Roy. Soc.* **155** (1865) 459-512.

[9] J. Hendry, *James Clerk Maxwell and the Theory of the Electromagnetic Field* (Adam Hilger, Bristol, 1986).

[10] S. Ramo, J.R. Whinnary, and T.V. Duzer, *Fields and Waves in Communication Electronics* (John Wiley, New York, 1965).

[11] M.A. Plonus, *Applied Electromagnetics* (Mc Graw-Hill, New York, 1978).

[12] J.A. Stratton, *Electromagnetic Theory* (Mc Graw-Hill, New York, 1941).

[13] J.H. Poynting, "On the transfer of energy in the electromagnetic field," *Phil. Trans. Roy. Soc.* **175** (1884) 343.

[14] E.C. Jordon and K.G. Balmain, *Electromagnetic Waves and Radiating Systems* (Printice-Hall, Englewood Cliffs, 1968).

PART TWO

APPLICATION OF ELECTROMAGNETIC THEORY TO BEAM-WAVE ELECTRONICS PROBLEMS

CHAPTER 5

PROPAGATION THROUGH A SLOW-WAVE STRUCTURE

5.0 Introduction

In this chapter which markes the beginning of part *two* of the volume let us study the behaviour of a slow-wave structure(SWS) by applying the time-varying field concepts developed in chapter *4* of part *one*. A SWS is an electromagnetic structure through which electromagnetic waves can propagate with a phase velocity $v_p < c$, where c $(= (\mu_0 \varepsilon_0)^{-1/2})$ is the velocity of light in free space (see section *4.5.1*). This makes an SWS different from a conventional waveguiding structure used in the microwave or millimeter-wave range like a rectangular or circular hollow metal-pipe waveguide for which $v_p > c$. Such an SWS finds applications in devices where it is desired that an electron beam should interact in velocity synchronism with an electromagnetic wave resulting in an exchange of energy either from the wave to the beam, as in an electron beam accelerator, or from the beam to the wave, as in a microwave tube, like the traveling-wave tube (TWT) (see chapter *8*).

The phase velocity of an SWS, which is to be synchronous, say, with a typical 2 kV linear beam of electrons (which has an axial beam velocity $\approx 2.65 \times 10^7$ m/s) would have to be about 1/11 of c (see Example 5.1.1). Such an order of phase-velocity is realizable in an SWS by providing some sort of axial periodicity in the structure. Such structures are usually of three categories: i) helix or helix-derived structure; ii) periodically-obstacled waveguide/cavity chain; and iii) serpentine line. The common structures belonging to these categories have been briefly outlined in appendix *a5.0.1*.

A metallic wire or tape wound in the form of a helix belongs to the first of the above three categories of the SWS [1]. Although it is the simplest, yet the helix exhibits the maximum potential for the control of its dispersion and hence it yields the widest bandwidth of all the structures. With proper care in the design of the helix as well as with the advent of recent technological innovations, other performance characteristics of the helix like the power, frequency, etc. can also be considerably improved. As such the helix is extensively used as an SWS of a TWT.

The 'untaxed' bandwidth potential of the helix together with the scope for improving its other performance characteristics like the power, frequency, efficiency, etc. makes the helix an increasingly important SWS (see appendix *a5.0.1* and also appendix *a8.2.3*). This motivates the author to take up in this chapter the analysis of a helical structure in some details. It can be appreciated that the helix presents complex boundary conditions to electromagnetic waves, and as such it is necessary to physically model it before taking up its actual analysis. In this practical boundary-value problem chosen one thus gets enough opportunities to apply and exemplify the various concepts of time-varying fields of electromagnetics developed in the preceding chapter.

Here, two analytical approaches have been made to the problem of a helical SWS. These are the field and the equivalent circuit analyses. Both these techniques, as we would see, yield one and the same dispersion relation, a relation which tells one how the phase velocity of the electromagnetic wave supported by the structure varies with frequency, or how ω (the angular frequency of the wave) is related to β (the phase propagation constant of the wave). The dispersion relation can be used to find the pertinent design parameters of the structure.

It is also envisaged here to find two more important parameters of the structure, namely, the characteristic and the interaction impedances. The first of these is important from the consideration of matching the helix with the system of coupling of rf power in and out of the helix. The second one is a measure of how much axial electric field is available (for the interaction with an electron beam) for a given power propagating down the structure. The quantity (as we would see later in chapter *8*) has a relevance to the gain of the device that utilizes the structure as an SWS.

Appendix a5.0.1 A brief overview on common slow-wave structures

The helix continues to be an extensively used SWS since its inception in a TWT by Kompfner[1]. A bifilar helix, also called a folded helix structure, made of two contra-wound helices of equal but reversed pitches, and a ring -and -bar structure are the two examples of a structure derived from the basic helix (Fig. a5.0.1(a)). The helix as well as the helix-derived structures belongs to the first of the three categories of the SWSs mentioned in the text [2], [3].

A helix-derived structure can have a larger transverse dimension while interacting with a higher voltage electron beam than a simple helix, without giving rise to a backward mode that can cause oscillations in a TWT. Also, such a structure would be better than a simple helix from thermal considerations since the former would be more massive and have more thermal contacts with the supporting structure and the envelope than the latter. The interaction impedance (a figure of merit defined in section 5.5) of a helix-derived structure would also be higher, corresponding to a higher gain per axial length of the device, than a simple helix. However, the performance in which the helix-derived structure, and for that matter any other SWSs cannot match a simple helix is the bandwidth which can be widened by suitably tailoring the design of the helix supports and that of the envelope of the structure.

A conventional waveguide, say, a circular-cylindrical type, which behaves, as such, as a high-pass *filter may be obstacled by circular washers placed at regular axial intervals, thereby making the structure exhibit band-pass filter characteristics. Such obstacles slow down a wave propagating through the waveguide very much like they would slow down water flowing through a pipe. Smaller the size of the holes of the washers, more would be the velocity reduction and narrower the frequency band of such a structure which comes under the second category of the SWSs mentioned in the text [4] - [8].*

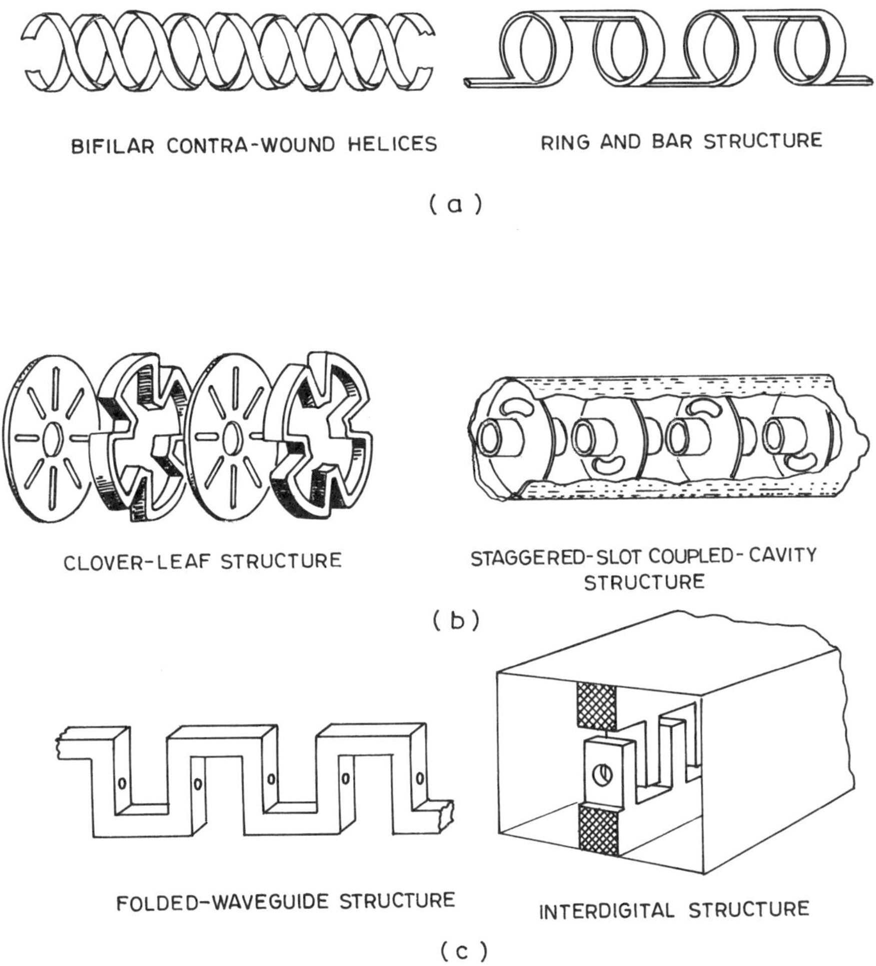

Fig. a5.0.1 Two examples of each of the three categories of slow-wave structures: helix or helix-derived [2], [3] **(a)**, periodically obstacled waveguide/cavity chain [6]-[8] **(b)** and serpentine line [8], [9] **(c)**.

By reducing the hole size to a minimum, and just allowing an electron beam to pass through the holes, one can use such a structure as the SWS of a single-frequency linear-beam accelerator. Under the same category, one may name a cavity-chain structure which consists of a chain of resonator cavities each coupled to its adjacent next such that the cavity-to-cavity phase delay is of the order of $\pi/2$, in one of the two types of such a structure, while it is of the order of $3\pi/2$, in the other. The cavity coupling in these two types are thus capacitive and inductive, respectively. The typical examples of the cavity chain structure (Fig. 5.0.1(b)) are the centipede and the clover-leaf structures belonging to the first of these types and the staggered-slot coupled cavity structure to the second. More details about the cavity-chain structures including the explanations for how a desired type of coupling is implemented inthem arre available in the literature [5], [7], [8]. There, one can see how, in a centipede structure, the S-shaped metal loops are arranged at regular angular intervals on each and successive cavity walls; how, in a clover-leaf structure, the radial components of fields are coupled by radial slots from one four-lobed cavity to its adjacent next, which is rotated by an angle of $\pi/4$; how, in a staggered-slot coupled-cavity structure, kidney-shaped, azimuthal slots provide the inductive coupling between successive cavities, and so on. It may be mentioned that, for the velocity synchronism in an amplifying device, the beam voltage ought to be somewhat more in the first than in the second type of the cavity-chain structure. Consequently, more output power is obtainable, say, from a clover-leaf TWT (the first-type) than from a coupled-cavity TWT (the second type).

The examples of the serpentine line structure (Fig. a5.0.1(c)), under the third category of the SWSs mentioned in the text, are the folded waveguide and the inter-digital structure [8], [9]. Each of these structures is used as an SWS of a backward-wave oscillator (BWO). These structures also find applications as SWS's in their planar version. For instance, both a meander-line, which is a folded planar line, and a planar inter-digital line have been used in crossed-field amplifier (CFA) tubes [5].

Out of the various structures under the three categories discussed above, the helix has the widest bandwidth. With proper care in the helix loading design, for instance, by anisotropically loading the helix (say, by providing metallic, radially-inward vanes with an overall metal envelope)[10] or/and by inhomogeneously loading the helix (say, by bundling the helix by tapered-geometry supports like those of a half-moon shaped cross section)[11], one may control the dispersion of the helix (that is the variation of its phase velocity with frequency). Such a controlled-dispersion helix is unrivaled as an SWS as far as bandwidths are concerned, and can be employed in an ultra-band (multi-octave) electronic warfare (EW) TWT. The limitations of the helix with respect to its transverse dimension for an effective interaction with a high-voltage electron beam have been overcome by providing a resonant attenuation using a meander-line attenuator on the helix-supporting rods between the helix turns. This permits now the helix to be used in high beam-voltage, and hence high-power, TWTs. The output power of a helix TWT is also limited by thermal considerations. However, the present-day technological innovations such as the brazed-helix technology, the block-supported helix, the fabrication of the helix as a part of the tube envelope, the use of high-thermal conductivity supports together with an effective cooling mechanism, etc. have increased the power capability as well as the life and reliability of a helix TWT [4], [12], [13]. The efficiency degradation of a TWT, due to the de-synchronization caused by the reduction of the electron beam velocity as the electrons deliver their kinetic energy to electromagnetic waves, may be compensated for by tapering the pitch of the helix for velocity re-synchronization [14]. Thus the design of a helix has a scope for efficiency improvement. Also, the operating frequency of a helix can be increased to make it suitable for employment in a millimeter-wave TWT by using II-A diamond or plasma-sprayed beryllia supports for the helix in pressure-fitted assemblies [5] (see also appendix a8.2.3).

5.1 Modelling of a Practical Helical Structure

A practical helical slow-wave structure typically consists of a metal wire or tape wound in the form of a helix (Fig. 5.1.1), supported by discrete longitudinal dielectric support rods/bars, the whole enclosed in a dielectric/metal envelope (Fig. 5.1.2). Such a structure poses a rather complex electromagnetic boundary-value problem. However, the problem becomes much simplified if the actual structure is replaced by a simple physical model that could be treated as an easier boundary-value problem.

The actual helix is replaced by a cylindrical sheath in a physical model called the *sheath-helix* model [15] which is described in section *5.1.1*. The discrete dielectric supports are smoothed out into a continuous dielectric tube in a model described in section *5.1.2*.

Though many useful properties of a helical SWS could be predicted with the help of the sheath-helix model described in this section, the space-harmonic effects caused by the axial periodicity of the structure remain unexplained by the model. Such effects can be studied in a model called the *tape-helix* model which has been considered later in this chapter (see section *5.6*).

5.1.1 Sheath-Helix Model

In this model (Fig. 5.1.1(b)) the actual helix of a finite thickness, whether it is made of a wire or tape, is replaced by a circular cylindrical sheath which has [4], [15]

i) an infinitesimal thickness;

ii) a radius equal to the mean radius of the actual helix of a finite thickness;

iii) an anisotropic conductivity such that the sheath conductivities are infinite and zero in directions parallel and perpendicular to the helix winding direction, respectively.

It is obvious that the sheath-helix model would be closer to the actual situation if the helix is more tightly wound. Quantitatively, one may put this condition of the validity of the model as that there are large number of turns per guide wavelength of the electromagnetic wave supported by the helix. Denoting this wavelength as λ_g the *guide* wavelength, remembering that the helix is a slow-waveguiding structure, one may write the validity condition for the sheath-model, more quantitatively, as:

$$\lambda_g / p >> 1 , \tag{5.1.1}$$

where p is the pitch of the helix, being equal to the turn-to turn axial distance. The pitch of the helix, p, is related to the sheath-helix radius, a (that is, the mean radius of the actual helix of a finite thickness), by the helix pitch angle ψ as:

$$\tan\psi = p / (2\pi a) . \tag{5.1.2}$$

λ_g is related to β, the axial phase propagation constant or to v_p, the phase velocity, and to the frequency of the wave, as (see appendix *a 4.5.1*):

$$\lambda_g = 2\pi / \beta = 2\pi / (\omega/v_p) = v_p / f \ . \tag{5.1.3}$$

One may have the following approximate relation for v_p by identifying it as the axial component of the velocity, c, with which the wave is supposed to advance in the winding direction (Fig. 5.1.1(b)):

$$v_p \cong c \sin\psi \ . \tag{5.1.4}$$

A more rigorous proof of (5.1.4) may be seen in Example 5.3.1.

Now, for relatively small values of pitch angle ψ, one may take $\sin\psi \approx \tan\psi$ and with this approximation combine (5.1.1)-(5.1.4) and write the following three alternative forms of the condition for the validity of the sheath-helix model (including (5.1.1), re-written):

$$\lambda_g/p >> 1 \qquad \text{(a)}$$

$$v_p/c >> p/\lambda \qquad \text{(b)}$$

and

$$\lambda >> 2\pi a \ , \qquad \text{(c)} \qquad (5.1.5)$$

where λ (= c/f) is the free-space wavelength.

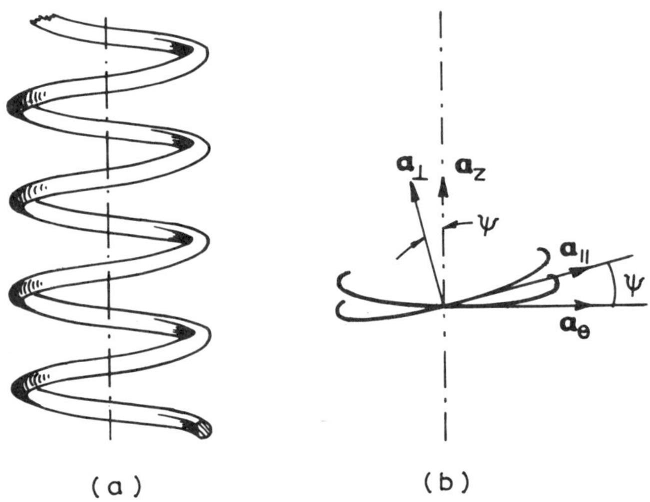

Fig. 5.1.1. Actual helix **(a)** and its sheath-helix model showing the unit-vectors in the axial and azimuthal directions as well as in directions parallel and perpendicular to the winding direction of the helix **(b)**.

Example 5.1.1 What should be the phase velocity v_p of the wave to be supported by an SWS which is to be synchronous with a 2kV electron beam?

Let the electrons of the beam starting from a zero dc beam velocity be accelerated by a potential gradient such that they all acquire a dc velocity v_0. The kinetic energy $(1/2)mv_0^2$ thus gained by a beam electron may be equated to the fall in its potential energy, $|e|V_0$, where V_0 is the accelerating potential; and e and m are the electronic charge and mass, respectively. Thus one has

$$\frac{1}{2}mv_0^2 = |e|\, V_0 ,$$

whence the beam velocity v_0 is obtained as:

$$v_0 = (2|\eta|V_0)^{1/2} , \tag{5.1.6}$$

where η (= e/m) is the charge-to-mass ratio of an electron. For synchronism, $v_p = v_0$, which may then be calculated from (5.1.6) by putting $V_0 = 2000$V, $|\eta| = |e|/m = (1.6\text{x}10^{-19}$ Coul.)/(9.1x10^{-31} kg), as: v_p (= v_0) ≈ 2.65x10^7 m/s, thus giving v_p/c = (2.65x10^7 m/s)/(3x10^8 m/s) ≈ 1/11.3 .

Example 5.1.2 A helix has a pitch of 0.8mm and a mean radius equal to 1.2 mm. Calculate the helix pitch angle ψ and compare $\sin\psi$ with $\tan\psi$. Also find the phase velocity relative to the velocity of light in free space and justify further that for such a helix one may use the sheath-helix abstraction at 6GHz.

With the help of (5.1.2), one may calculate

$$\psi = \tan^{-1}(p/2\pi a) = \tan^{-1}[(0.8 \times 10^{-3})/((2\pi \times 1.2\times10^{-3})] = 6° .$$

At this value of ψ, the values of sin 6° and tan 6° are each approximately equal to 0.11. Thus in this case one has

$$\sin\psi \approx \tan\psi . \tag{5.1.7}$$

One may then combine (5.1.2), (5.1.4) and (5.1.7) to obtain the following expression for the phase velocity reduction factor

$$v_p/c \cong p/2\pi a . \tag{5.1.8}$$

This factor which, according to (5.1.4), is: $v_p/c = \sin\psi \cong \sin 6° = 0.11$, may be alternatively calculated from (5.1.8) to yield the same value. Further, in this problem, for the given parameters: $p = 0.8 \times 10^{-3}$ m, $a = 1.2 \times 10^{-3}$ m, $f = 6 \times 10^9$ Hz, one has the following values by calculation:

$$\lambda\,(= c/f) = (3 \times 10^8 \text{ m/s})/(6 \times 10^9 \text{ Hz}) \approx 50 \times 10^{-3} \text{ m},$$

$$2\pi a = 2 \times 3.14 \times 1.2 \times 10^{-3} \approx 7.54 \times 10^{-3} \text{ m} ,$$

$$p/\lambda = (0.8 \times 10^{-3} \text{ m}) / (50 \times 10^{-3} \text{ m}) \approx 0.016 , \text{ and}$$

$$v_p/c \;(= \sin\psi \approx \tan\psi = p/2\pi a) \approx 0.11 .$$

Thus one may see that either of (5.1.5(b)) and (5.1.5(c)) are fairly true here, λ and v_p /c being each approximately 7 times as great as $2\pi a$ and p/λ, respectively. Therefore, it should be reasonable to approximate the present helix by the sheath-helix model.

5.1.2 A Model for Discrete Helix Supports

In a practical device the helix is held in position with the help of a dielectric-support system. The simplest configuration is a helix closely fitting in a dielectric, say, a glass tube. The proximity of a dielectric medium around the helix affects the phase velocity as well as the interaction impedance of the structure (the latter being defined in section *5.5*). The helix may also be bundled by means of a number of (usually three) identical longitudinal dielectric rods/bars arranged symmetrically around the helix at a regular angular interval. The simplest support system consists of dielectric wedge bars which may be considered as a dielectric tube with portions radially cut out from it. Considerable simplification is achieved in the analysis of such a structure if one physically models the structure as the one in which the discrete dielectric support bars are azimuthally smoothed out into a single continuous dielectric tube of an effective relative permittivity ε_r', say, which may be found, rather heuristically, from the relative volume occupied by the supports in the structure [16] (Fig. 5.1.2). This may be done for a helix supported by dielectric wedge bars in an envelope as follows.

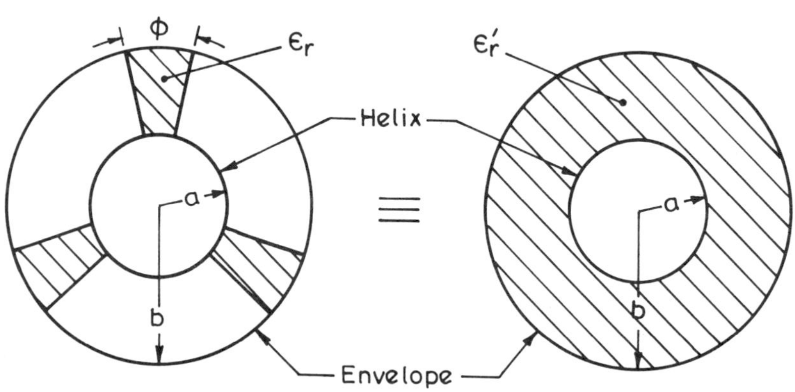

Fig. 5.1.2. Cross section of a helix supported by dielectric wedge bars in an envelope **(a)** and its equivalent dielectric model **(b)**.

Let $\hat{A}$ be the area between the helix and the envelope and $\hat{A}_s$ be the area occupied by the supports alone in the structure cross section (Fig. 5.1.2). Then the area of cross section of the free-space region between the helix and the envelope is $\hat{A} - \hat{A}_s$. Therefore, one may find the effective relative permittivity ε'_r of the equivalent dielectric tube from the following relation:

$$\varepsilon'_r = \frac{\sum_i \varepsilon_{ri} \hat{A}_i}{\sum_i \hat{A}_i} , \tag{5.1.9}$$

where i extends over the various regions of the structure cross section, between the helix and the envelope. In the present case (Fig. 5.1.2), (5.1.9) takes the form

$$\varepsilon'_r = \frac{(\varepsilon_r)(\hat{A}_s) + (1)(\hat{A} - \hat{A}_s)}{(\hat{A}_s) + (\hat{A} - \hat{A}_s)} = 1 + (\varepsilon_r - 1)\frac{\hat{A}_s}{\hat{A}} , \tag{5.1.10}$$

where ε_r is the relative permittivity of the support bars. In the case of wedge-shaped support bars, one may replace $\hat{A}_s/\hat{A}$ in (5.1.10) by $\phi N/2\pi$, where N is the number of bars and ϕ is the wedge angle of each bar, enabling one to write

$$\varepsilon'_r = 1 + (\phi N/2\pi)(\varepsilon_r - 1) . \tag{5.1.11}$$

A rigorous analysis [17], [18] considering angular harmonic effects caused by the positioning of the helix-support bars at a regular angular interval establishes that the method of smoothing out the effects of discrete bars into *a single* continuous dielectric tube of an effective relative permittivity given by (5.1.10), though it is accurate for dielectric wedge support bars, is not exactly so for other commonly used rectangular or circular rods/bars or supports of tapered geometry cross section used for wide band applications. For the support cross section deviating from a simple wedge geometry, it would be more appropriate to azimuthally smooth out the supports into *a number* of dielectric tubes of appropriate effective relative permittivities instead of *a single* dielectric tube (see section *5.8*).

5.2 Field Expressions and Sheath-Helix Boundary Conditions

Due to anisotropic conductivity of the sheath helix (and the consequent skew boundary conditions, see section *5.2.2*), one expects both transverse electric (TE)($H_z \neq 0$) and transverse magnetic (TM)($E_z \neq 0$) modes to be present in the sheath helix. Then one may write the following wave equation in the cylindrical system of coordinates for the axial electric field component E_z, as relevant to the cylindrical sheath, with the help of (4.5.1), through (4.5.2), under the source-free condition $\mathbf{J_c} = \rho = 0$:

$$\nabla^2 E_z - \mu\varepsilon \frac{\partial^2 E_z}{\partial t^2} = 0 \tag{5.2.1}$$

Similarly, the source-free ($\mathbf{J_c} = 0$) wave equation for the axial magnetic field component H_z may be written for the sheath helix, with the help of (4.5.3), through (4.5.2), as:

$$\nabla^2 H_z - \mu\varepsilon \frac{\partial^2 H_z}{\partial t^2} = 0 \quad . \tag{5.2.2}$$

Spelling out the Laplacian ∇^2 in the cylindrical system of coordinates with the help of (2.6.4(b)), one may then write from (5.2.1) and (5.2.2):

$$\left(\frac{\partial^2}{\partial r^2} + \frac{1}{r}\frac{\partial}{\partial r} + \frac{1}{r^2}\frac{\partial^2}{\partial \theta^2} + \frac{\partial^2}{\partial z^2} - \mu\varepsilon\frac{\partial^2}{\partial t^2}\right)(E_z, H_z) = 0 \ . \tag{5.2.3}$$

Taking the fields to vary as exp $j(\omega t - \beta z)$, where β is the axial phase propagation constant, one may replace $\partial/\partial t$ by $j\omega$ and $\partial/\partial z$ by $-j\beta$ in (5.2.3), and write, for the non-azimuthally varying mode, $\partial/\partial\theta = 0$:

$$\left(\frac{\partial^2}{\partial r^2} + \frac{1}{r}\frac{\partial}{\partial r}\right)(E_z, H_z) + \Gamma^2(E_z, H_z) = 0. \tag{5.2.4}$$

where

$$\Gamma^2 = \omega^2 \mu\varepsilon - \beta^2 \ . \tag{5.2.5}$$

5.2.1 Field Expressions

Equation (5.2.4) may be identified as the zeroth-order Bessel equation (see appendix *a5.2.1*). For anticipated slow-waves supported by the helix (which can be made to interact with an electron beam in a practical device like TWT (see chapter *8*)), the phase velocity would be much less than that of a uniform, plane electromagnetic wave propagating through an unbounded, homogeneous isotropic, source-free medium (see section *4.5.1*). In other words, one would have:

$$\omega/\beta << 1/(\mu\varepsilon)^{1/2} \ . \tag{5.2.6}$$

(5.2.6) would make Γ^2, given by (5.2.5), essentially a negative quantity and hence Γ an imaginary quantity $j\gamma$, say, where

$$\gamma = (\beta^2 - \omega^2 \mu\varepsilon)^{1/2} \ . \tag{5.2.7}$$

The solution of (5.2.4) then may be written as:

$$E_z = P\, J_0\{\Gamma r\} + Q\, Y_0\{\Gamma r\} = P\, J_0\{j\gamma r\} + Q\, Y_0\{j\gamma r\} \ , \tag{5.2.8}$$

where P and Q are arbitrary constants. J_0 and Y_0 are the zeroth-order Bessel functions of the first and second kinds, respectively (appendix *a5.2.1*). Choosing to express the ordinary Bessel functions of complex arguments occurring in (5.2.8), in terms of the modified Bessel functions, one may express (5.2.8) as (appendix *a5.2.1*):

$$E_z = A\, I_0\{\gamma r\} + B\, K_0\{\gamma r\} \ , \tag{5.2.9}$$

where I_0 and K_0 are the zeroth-order modified Bessel functions of the first and second kinds, respectively; A and B are the field constants related to P and Q as (appendix *a5.2.1*):

$$A = P + jQ \quad \text{and} \quad B = -2\,Q/\pi \ . \tag{5.2.10}$$

Similarly, the solution of (5.2.4) for the magnetic field intensity H_z may be expressed in terms of the corresponding field constants C and D as:

$$H_z = C\, I_0\{\gamma r\} + D\, K_0\{\gamma r\} \ . \tag{5.2.11}$$

In (5.2.9) and (5.2.11), γ occurs in the arguments of the modified Bessel functions as a factor of the radial coordinate r, and hence γ determines how the field quantities vary with r. Therefore γ, which is given by (5.2.7), is called the radial propagation constant. It may be mentioned that the factor $\exp j(\omega t - \beta z)$, corresponding to a wave propagating in the positive z-direction with a phase velocity ω/β, is understood in the field expressions (5.2.9) and (5.2.11) (see the discussion following (5.2.3)).

Once the axial electric and magnetic field intensities E_z and H_z have been expressed in terms of the field constants A, B, C and D, it is now quite easy to obtain, with the help of Maxwell's equations (4.4.1)-(4.4.4), the following expressions for the azimuthal and the radial field components (for the details of deduction, see appendix *a5.2.2*):

$$E_\theta = -(j\,\omega\mu/\gamma)\,[C\, I_1\{\gamma r\} - D\, K_1\{\gamma r\}] \tag{5.2.12}$$

$$H_\theta = (j\,\omega\varepsilon/\gamma)\,[A\, I_1\{\gamma r\} - B\, K_1\{\gamma r\}] \tag{5.2.13}$$

$$E_r = (j\,\beta/\gamma)\,[A\, I_1\{\gamma r\} - B\, K_1\{\gamma r\}] \tag{5.2.14}$$

and

$$H_r = (j\,\beta/\gamma)\,[C\, I_1\{\gamma r\} - D\, K_1\{\gamma r\}] \ , \tag{5.2.15}$$

where I_1 and K_1 are the first-order modified Bessel functions of the first and second kinds, respectively. In field expressions (5.2.9), (5.2.11) - (5.2.15), the dependence of the form $\exp j(\omega t - \beta z)$ is understood.

5.2.2 Sheath-Helix Boundary Conditions

Remembering the anisotropic-conductivity of the sheath helix (section *5.1.1*), one may proceed to write the boundary conditions at the inner as well as outer surface of the helical sheath. For this purpose one has to recall the boundary conditions at the interface between a dielectric and a metal (see Table 4.7.1). In this case, the dielectric inside the helix has to be taken as a free space ($\varepsilon = \varepsilon_0$), while that outside the helix as a free-space or a dielectric, according as the helix considered is immersed in a free space or is supported by a dielectric. Besides, one has the sheath helix itself as an intermediate conducting medium.

Let us represent the electric and magnetic intensity vectors each in terms of their components in three mutually perpendicular directions as follows. The first of these components is tangential to the surface of the helix-sheath and parallel to the winding directions; the second is also tangential to the sheath surface but perpendicular to the winding direction; while the third is perpendicular to sheath surface, being radially outward. Denoting the three unit vectors in these mutually perpendicular directions as $\mathbf{a}_{\parallel}$, $\mathbf{a}_{\perp}$ and $\mathbf{a}_r$, respectively, one may express the electric and magnetic field intensity vectors $\mathbf{E}$ and $\mathbf{H}$, in terms of their components, as (see Fig. 5.1.1(b)):

$$\mathbf{E} = E_{\parallel}\mathbf{a}_{\parallel} + E_{\perp}\mathbf{a}_{\perp} + E_r\mathbf{a}_r$$

and

$$\mathbf{H} = H_{\parallel}\mathbf{a}_{\parallel} + H_{\perp}\mathbf{a}_{\perp} + H_r\mathbf{a}_r \quad , \tag{5.2.16}$$

where the subscripts '$\parallel$' and '$\perp$' are used to denote the components parallel and perpendicular to the winding direction, respectively.

Let us now apply the boundary condition (4.7.10) to the inner surface of the helix. In (4.7.10) one has then to take, for this surface, $\mathbf{a}_n = \mathbf{a}_r$, $\mathbf{E}_1 = \mathbf{E}_1$, $\mathbf{E}_2 = \mathbf{E}_h$, where the subscripts 1 and h refer to the regions inside the helix winding and in the interior of the sheath-helix, respectively. Then one can write, using the component-wise representation of field quantities as (5.2.16), the following boundary condition from (4.7.10):

$$\mathbf{a}_r \times [(E_{\parallel h} - E_{\parallel 1})\,\mathbf{a}_{\parallel} + (E_{\perp h} - E_{\perp 1})\,\mathbf{a}_{\perp} + (E_{rh} - E_{r1})\,\mathbf{a}_r] = 0 \tag{5.2.17}$$

whence, using the vector relations: $\mathbf{a}_r \times \mathbf{a}_{\parallel} = \mathbf{a}_{\perp}$, $\mathbf{a}_{\perp} \times \mathbf{a}_r = \mathbf{a}_{\parallel}$, etc.(Fig. 5.1.1), one gets

$$(E_{\parallel h} - E_{\parallel 1})\,\mathbf{a}_{\perp} - (E_{\perp h} - E_{\perp 1})\,\mathbf{a}_{\parallel} = 0 \ ,$$

which yields

$$E_{\parallel h} - E_{\parallel 1} = 0 \qquad \text{(a)}$$

and

$$E_{\perp h} - E_{\perp 1} = 0 \ . \qquad \text{(b)} \tag{5.2.18}$$

Similarly, for the interface taken at the outer surface of the helix sheath, one has $\mathbf{a}_n = \mathbf{a}_r$, $\mathbf{E}_1 = \mathbf{E}_h$, $\mathbf{E}_2 = \mathbf{E}_2$, where the subscript 2 refers to the region outside the helix winding. This gives a relation similar to (5.2.17) as follows:

$$\mathbf{a}_r \times [(E_{\|2} - E_{\|h})\,\mathbf{a}_{\|} + (E_{\perp 2} - E_{\perp h})\,\mathbf{a}_{\perp} + (E_{r2} - E_{rh})\,\mathbf{a}_r] = 0 \tag{5.2.19}$$

whence, applying the relations: $\mathbf{a}_r \times \mathbf{a}_{\|} = \mathbf{a}_{\perp}$, etc., one has

$$(E_{\|2} - E_{\|h})\,\mathbf{a}_{\perp} - (E_{\perp 2} - E_{\perp h})\,\mathbf{a}_{\|} = 0 \tag{5.2.20}$$

which gives

$$E_{\|2} - E_{\|h} = 0 \qquad \text{(a)}$$

and

$$E_{\perp 2} - E_{\perp h} = 0\,. \qquad \text{(b)} \tag{5.2.21}$$

Similarly, if one now starts from (4.7.8) instead of (4.7.10), one obtains the following boundary conditions, corresponding to (5.2.17) and (5.2.19), at the inner and outer surfaces of the sheath, respectively:

$$\mathbf{a}_r \times [(H_{\|h} - H_{\|1})\,\mathbf{a}_{\|} + (H_{\perp h} - H_{\perp 1})\,\mathbf{a}_{\perp} + (H_{rh} - H_{r1})\,\mathbf{a}_r] = \mathbf{J}_{si}$$

and

$$\mathbf{a}_r \times [(H_{\|2} - H_{\|h})\,\mathbf{a}_{\|} + (H_{\perp 2} - H_{\perp h})\,\mathbf{a}_{\perp} + (H_{r2} - H_{rh})\,\mathbf{a}_r] = \mathbf{J}_{so}\,,$$

which when added yield:

$$\mathbf{a}_r \times [(H_{\|2} - H_{\|1})\mathbf{a}_{\|} + (H_{\perp 2} - H_{\perp 1})\mathbf{a}_{\perp} + (H_{r2} - H_{r1})\mathbf{a}_r] = \mathbf{J}_{si} + \mathbf{J}_{so} = \mathbf{J}_s\,, \tag{5.2.22}$$

where $\mathbf{J}_{si}$ and $\mathbf{J}_{so}$ are the surface current densities at the inner and the outer surfaces of the helix-sheath, respectively, and $\mathbf{J}_s$ (= $\mathbf{J}_{si} + \mathbf{J}_{so}$) is the total surface current density at the two surfaces of the sheath taken together. Now, one may choose to express $\mathbf{J}_s$ in terms of the three components, as in (5.2.16). This would enable one to write from (5.2.22), invoking the relations $\mathbf{a}_r \times \mathbf{a}_{\|} = \mathbf{a}_{\perp}$, etc., the following boundary condition for the discontinuity of the magnetic field intensities at the sheath:

$$(H_{\|2} - H_{\|1})\,\mathbf{a}_{\perp} - (H_{\perp 2} - H_{\perp 1})\,\mathbf{a}_{\|} = J_{s,\|}\,\mathbf{a}_{\|} + J_{s,\perp}\,\mathbf{a}_{\perp}$$

which would then give the component-wise relations:

$$H_{\|2} - H_{\|1} = J_{s,\perp} \qquad \text{(a)}$$

and

$$H_{\perp 1} - H_{\perp 2} = J_{s,\|}\,. \qquad \text{(b)} \tag{5.2.23}$$

At this juncture, let us recall the anisotropic conductivity of the helix sheath model (section *5.1.1*). In this model, it is assumed that the sheath has an infinite conductivity in the winding direction, which would make $E_{\|,h} = 0$ (see Table 4.7.1, for the boundary condition at the metal-dielectric interface). The model also considers no conductivity of the sheath in a direction perpendicular to the winding direction; this would make $J_{s,\perp} = 0$. Then, putting $E_{\|,h} = J_{s,\perp} = 0$, we have from (5.2.18(a)), (5.2.21(a)) and (5.2.23(a)):

$$E_{\|1} = 0 \qquad \text{(a)}$$

$$E_{\|2} = 0 \qquad \text{(b)}$$

and

$$H_{\|1} = H_{\|2} \, . \qquad \text{(c)} \qquad (5.2.24)$$

The boundary conditions (5.2.24), expressed in terms of the parallel components of field intensities, may also be expressed in terms of the axial and azimuthal components of field intensities. Let us take, for instance, $E_{\|}$ which may be put as

$$E_{\|} = \mathbf{E} \cdot \mathbf{a}_{\|} \, . \qquad (5.2.25)$$

Now, if we substitute in (5.2.25)

$$\mathbf{E} = E_r \mathbf{a}_r + E_\theta \mathbf{a}_\theta + E_z \mathbf{a}_z \, , \qquad (5.2.26)$$

the expression in terms of the components in the usual cylindrical system of coordinates, we have

$$E_{\|} = E_r \mathbf{a}_r \cdot \mathbf{a}_{\|} + E_\theta \mathbf{a}_\theta \cdot \mathbf{a}_{\|} + E_z \mathbf{a}_z \cdot \mathbf{a}_{\|} \, . \qquad (5.2.27)$$

Now, in view of the relations: $\mathbf{a}_r \cdot \mathbf{a}_{\|} = 0$, $\mathbf{a}_\theta \cdot \mathbf{a}_{\|} = \cos\psi$ and $\mathbf{a}_z \cdot \mathbf{a}_{\|} = \sin\psi$ (Fig. 5.1.1(b)), the parallel component of $\mathbf{E}$ given by (5.2.27) takes the following form:

$$E_{\|} = E_\theta \cos\psi + E_z \sin\psi \, ,$$

a form that can be used for $H_{\|}$ as well, which then can be used to express (5.2.24(a), (b), and (c)), the boundary conditions at the helix sheath r = a, as:

$$E_{\theta 1} \cos\psi + E_{z1} \sin\psi = 0 \, , \qquad (5.2.28)$$

$$E_{\theta 2} \cos\psi + E_{z2} \sin\psi = 0 \qquad (5.2.29)$$

and

$$H_{\theta 1} \cos\psi + H_{z1} \sin\psi = H_{\theta 2} \cos\psi + H_{z2} \sin\psi \, , \qquad (5.2.30)$$

respectively. Besides, the above three boundary conditions, another boundary condition is usually required to be considered. The fourth condition is obtained without making any reference to the parallel component. For this purpose, the boundary condition (4.7.10) may be recalled to write

$$\mathbf{a}_r \times [(E_{rh} - E_{r1}) \mathbf{a}_r + (E_{\theta h} - E_{\theta 1}) \mathbf{a}_\theta + (E_{zh} - E_{z1}) \mathbf{a}_z] = 0 \, , \qquad (5.2.31)$$

at the inner surface of the sheath; and

$$\mathbf{a}_r \times [(E_{r2} - E_{rh}) \mathbf{a}_r + (E_{\theta 2} - E_{\theta h}) \mathbf{a}_\theta + (E_{z2} - E_{zh}) \mathbf{a}_z] = 0 \, , \qquad (5.2.32)$$

at the outer surface of the sheath. (5.2.31) and (5.2.32) may be further simplified, invoking the relations $\mathbf{a}_r \times \mathbf{a}_r = 0$, $\mathbf{a}_r \times \mathbf{a}_\theta = \mathbf{a}_z$, etc., as:

$$(E_{\theta h} - E_{\theta 1})\,\mathbf{a}_z \;-\; (E_{zh} - E_{z1})\,\mathbf{a}_\theta \;= 0$$

and

$$(E_{\theta 2} - E_{\theta h})\,\mathbf{a}_z \;-\; (E_{z2} - E_{zh})\,\mathbf{a}_\theta \;= 0\;,$$

which when added would give:

$$(E_{\theta 2} - E_{\theta 1})\,\mathbf{a}_z \;+\; (E_{z1} - E_{z2})\,\mathbf{a}_\theta \;= 0\;. \tag{5.2.33}$$

It can be appreciated from (5.2.33) that $(E_{\theta 2} - E_{\theta 1})$ and $(E_{z1} - E_{z2})$ are each equal to zero giving the following boundary conditions at $r = a$:

$$E_{z1} \;=\; E_{z2} \tag{5.2.34}$$

and

$$E_{\theta 1} \;=\; E_{\theta 2}\;. \tag{5.2.35}$$

Thus, one may take either (5.2.34) or (5.2.35) as the fourth boundary condition at the helix sheath ($r = a$), the other three conditions being (5.2.28)-(5.2.30). It may also be easily seen that, out of (5.2.34) and (5.2.35), one condition is redundant since it can be deduced with the help of the other condition, and using the conditions (5.2.28) and (5.2.29). The set of these four boundary conditions at the helix sheath ($r = a$), namely, (5.2.28)-(5.2.30) and either (5.2.34) or (5.2.35) are used, as can be seen in section *5.3*, in predicting the behaviour of a helical SWS by using an analytical method known as the field analysis. In an alternative method called the equivalent circuit analysis, taken up in section *5.4*, one would, however, require to use another set of four boundary conditions at the sheath. The two of these conditions have already been obtained as (5.2.34) and (5.2.35), while the remaining two are derived from the basic relation (4.7.8) using the same approach as followed in obtaining (5.2.22), however, now choosing to express the field components in the usual cylindrical system of coordinates r, θ, z, instead of (5.2.16). (It may be noted that, in the manner in which the fields are represented as (5.2.16), the r, θ and z components have been replaced by the $\|$, $\perp$ and r components, respectively). Thus, replacing the $\|$, $\perp$ and r components by the r, θ and z components, respectively, in (5.2.22), one obtains the following condition:

$$\mathbf{a}_r \times [(H_{r2} - H_{r1})\,\mathbf{a}_r + (H_{\theta 2} - H_{\theta 1})\,\mathbf{a}_\theta + (H_{z2} - H_{z1})\,\mathbf{a}_z] \;=\; \mathbf{J}_s\;,$$

which simplifies to:

$$(H_{\theta 2} - H_{\theta 1})\,\mathbf{a}_z \;+\; (H_{z1} - H_{z2})\,\mathbf{a}_\theta \;=\; J_{sz}\,\mathbf{a}_z \;+\; J_{s\theta}\,\mathbf{a}_\theta$$

yielding the following boundary conditions in terms of surface-current densities at $r = a$:

$$H_{\theta 2} \;-\; H_{\theta 1} \;=\; J_{sz} \tag{5.2.36}$$

and $$H_{z1} - H_{z2} = J_{s\theta} \; . \tag{5.2.37}$$

In this section, two sets of boundary conditions valid at the cylindrical sheath of an anisotropic conductivity, which replaces an actual helix, have been identified. These are i) (5.2.28) - (5.2.30) and either (5.2.34) or (5.2.35) and ii) (5.2.34) - (5.2.37). They have been used in the field and the equivalent circuit analyses of the helical SWS, considered in sections *5.3* and *5.4* , respectively, to follow.

Appendix a5.2.1 Modified Bessel functions

A linear second-order differential equation in y as a function of x of the following form:

(a5.2.1) $$x^2 \frac{d^2 y}{dx^2} + x\frac{dy}{dx} + (x^2 - n^2)y = 0$$

is called the Bessel's differential equation. An alternative form of (a5.2.1) is obtained by dividing (a5.2.1) by x^2:

(a5.2.2) $$\left(\frac{d^2}{dx^2} + \frac{1}{x}\frac{d}{dx}\right)y + \left(1 - \frac{n^2}{x^2}\right)y = 0 .$$

If we define

(a5.2.3) $$x = \Gamma r ,$$

and substitute (a5.2.3) in (a5.2.2), we obtain

(a5.2.4) $$\left(\frac{d^2}{dr^2} + \frac{1}{r}\frac{d}{dr}\right)y + \left(\Gamma^2 - \frac{n^2}{r^2}\right)y = 0 .$$

The solution of (a5.2.1) is known in the literature [19] as follows:

(a5.2.5) $$y = P J_n\{x\} + Q Y_n\{x\} ,$$

where P and Q are arbitrary constants. $J_n\{x\}$ *and* $Y_n[x\}$, *called the* n^{th} *-order ordinary Bessel functions of argument x, of the first and second kinds, respectively, are:*

$$J_n\{x\} = \sum_{s=0}^{\infty} \frac{(-1)^s (x/2)^{n+2s}}{s!(n+s)!} ,$$

and

$$Y_n\{x\} = \frac{2}{\pi}(\gamma' + \ln\frac{x}{2})J_n\{x\} - \frac{1}{\pi}\sum_{s=0}^{n-1}\frac{(n-s-1)!}{s!}\left(\frac{2}{x}\right)^{n-2s}$$

$$-\frac{1}{\pi}\sum_{s=1}^{\infty}\frac{(-1)^s (x/2)^{n+2s}}{s!(n+s)!}\left(1+\frac{1}{2}+\frac{1}{3}+\cdots+\frac{1}{s}+1+\frac{1}{2}+\frac{1}{3}+\cdots\right)$$

where γ' is Euler's constant defined by [19]

$$\gamma' = \underset{n\to\infty}{Lt}\left(1+\frac{1}{2}+\frac{1}{3}+\ldots+\frac{1}{n}-\log n\right) = 0.5772157 \ .$$

(We have used here the symbol γ' for the Euler's constant instead of the usual γ since, the symbol γ has been used for the radial propagation constant).

If we choose to express (a5.2.1) in the form: (a5.2.4), then the solution of the latter may be written directly from (a5.2.5),on substituting (a5.2.3) in it, as follows:

(a5.2.6) $$y = P J_n\{\Gamma r\} + Q Y_n\{\Gamma r\} \ .$$

A special case of interest is the one in which Γ is imaginary:

(a5.2.7) $$\Gamma = j\gamma, \ \text{say},$$

where γ is real. Then the solution y may be read on substituting (a5.2.7) into (a5.2.6) as:

(a5.2.8) $$y = P J_n\{j\gamma r\} + Q Y_n\{j\gamma r\} \ .$$

Thus the arguments of J_n and Y_n in (a5.2.8) are found to be imaginary. Let us now define the following two functions $I_n\{x\}$ and $K_n\{x\}$, called th n^{th} -order 'modified' Bessel functions of the first and second kinds, respectively:

(a5.2.9) $$I_n\{x\} = j^{-n} J_n\{j x\} \ \text{and}$$

(a5.2.10) $$K_n\{x\} = (\pi/2)\, j^{n+1}\left(J_n\{j x\} + j Y_n\{j x\}\right) ,$$

such that these functions are real when x is real, a linear combination of $J_n\{j x\}$ and $Y_n\{j x\}$ being chosen in the definition of $K_n\{x\}$ in order to make the latter a decaying exponential function for larger values of x. Interpreting x as γr, one may express $J_n\{j \gamma r\}$ and $Y_n\{j \gamma r\}$ in terms of the modified Bessel functions with the help of (a5.2.9) and (5.2.10), and thus write (a5.2.8) finally as

(a5.2.11) $$y = A\, I_n\{\gamma r\} + B\, K_n\{\gamma r\}, \quad \text{where}$$

(a5.2.12) $$A = (P + j Q)\, j^n \ \text{and}\ B = (2Q/\pi)\, j^{-(n+2)} \ .$$

Interpreting E_z or H_z as y, one may compare (5.2.4) with (a5.2.4) and identify the former as the Bessel equation of order n = 0, with the complex argument $\Gamma = j\gamma$, where γ is given by (5.2.7). The solution (a5.2.11), with the field constants given by (a5.2.12), may be seen to be identical, for n = 0, with (5.2.9) with the field constants given by (5.2.10). It may be mentioned here that the solution of (5.2.4) gives only its r-dependent part (5.2.9). This would make one interpret the partial derivatives occurring in (5.2.4) as the complete derivatives while making a comparison of (5.2.4) with the Bessel equation (a5.2.4) of order n = 0.

Appendix a5.2.2 Azimuthal electric and magnetic field intensities

With the help of (a3.2.25) and (a3.2.26) one may write (4.4.3(b)) and (4.4.4(b)) (with $\boldsymbol{J}_C = 0$), component-wise, as follows:

(a5.2.13) $$\frac{1}{r}\frac{\partial E_z}{\partial \theta} - \frac{\partial E_\theta}{\partial z} = -\mu\frac{\partial H_r}{\partial t},$$

(the r-component)

(a5.2.14) $$\frac{1}{r}\frac{\partial H_z}{\partial \theta} - \frac{\partial H_\theta}{\partial z} = \varepsilon\frac{\partial E_r}{\partial t},$$

(a5.2.15) $$\frac{\partial H_r}{\partial z} - \frac{\partial H_z}{\partial r} = \varepsilon\frac{\partial E_\theta}{\partial t} \quad \text{and}$$

(the θ-component)

(a5.2.16) $$\frac{\partial E_r}{\partial z} - \frac{\partial E_z}{\partial r} = -\mu\frac{\partial H_\theta}{\partial t}.$$

Taking $\partial/\partial t = j\omega$, $\partial/\partial z = -j\beta$ and $\partial/\partial\theta = 0$ (see the discussion following (5.2.4)), and using (5.2.9) and (5.2.11), one may express (a5.2.13) - (a5.2.16) as follows:

(a5.2.17) $$j\beta E_\theta = -j\omega\mu H_r .$$

(a5.2.18) $$j\beta H_\theta = j\omega\varepsilon E_r .$$

(a5.2.19) $$-j\beta H_r - [C I'_0\{\gamma r\} + D K'_0\{\gamma r\}]\gamma = j\omega\varepsilon E_\theta \quad \text{and}$$

(a5.2.20) $$-j\beta E_r - [A I'_0\{\gamma r\} + B K'_0\{\gamma r\}]\gamma = -j\omega\mu H_\theta ,$$

where the prime means the derivative with respect to argument.

Eliminating H_r from (a5.2.17) and (a5.2.19) and E_r from (a5.2.18) and (a5.2.20), one obtains

(a5.2.21) $$E_\theta = (-j\omega\mu/\gamma)[C I'_0\{\gamma r\} + D K'_0\{\gamma r\}] \quad \text{and}$$

(a5.2.22) $$H_\theta = (j\omega\varepsilon/\gamma)[A I'_0\{\gamma r\} + B K'_0\{\gamma r\}] .$$

Using the relations: $I'_0\{x\} = I_1\{x\}$ and $K'_0\{x\} = -K_1\{x\}$, (a5.2.21) and (a5.2.22) pass on to (5.2.12) and (5.2.13), respectively. (5.2.14) and (5.2.15), the expressions for the radial components of electric (E_r) and magnetic (H_r) field intensities obtained on substituting (a5.2.22) and (a5.2.21) in (a5.2.18) and (a5.2.17), respectively.

5.3 Field Analysis and Dispersion Relation in the Sheath-Helix Model

Propagation characteristics of a helical SWS depend upon the dimensions, namely, the pitch, diameter and thickness of the helix and upon the physical property, namely, the resistivity of the material of the helix. The materials such as tungsten, tungsten-rhenium alloy, molybdenum etc. usually make the helix. (The resistivity of tungsten is 5.6×10^{-6} ohm cm, to mention typically). The dimensions, geometries and materials of the supports and the envelope also influence the behaviour of the structure. For a helix-support material, one should choose a dielectric of good thermal conductivity. Some such materials in use are beryllia, alumina, isotropic boron nitride(IBN), anisotropic pyrolytic boron nitride (APBN), diamond etc. (The relative permittivity and the thermal conductivity of alumina are 9.1 and 0.3 W/cm/°C, respectively, to mention typically). The envelope may be made of a glass/ceramic or a metal.

Let us first take up in section *5.3.1* a simple case of a helix in free space — an isolated or unloaded helix, without supports and the envelope, and find by field analysis [15], [20] a dispersion relation — a relation which would tell one how the phase velocity of electromagnetic waves varies with frequency. The dispersion relation of the structure is modified by the *loading* of the helix as caused by the dielectric supports and the envelope. This modification, as can be seen in section *5.3.2*, manifests itself in the form of a factor called the dielectric loading factor (DLF) which would appear in the dispersion relation.

5.3.1 Helix in Free Space

The electric and magnetic fields inside (r < a) and outside(r > a) the helical sheath may be re-written here as (see (5.2.9), (5.2.11) - (5.2.13)):

$$E_{zp} = A_p I_0\{\gamma_p r\} + B_p K_0\{\gamma_p r\} \qquad \text{(a)}$$

$$H_{zp} = C_p I_0\{\gamma_p r\} + D_p K_0\{\gamma_p r\} \qquad \text{(b)}$$

$$E_{\theta p} = (-j\omega\mu_p/\gamma_p)[C_p I_1\{\gamma_p r\} - D_p K_1\{\gamma_p r\}] \qquad \text{(c)}$$

and

$$H_{\theta p} = (j\omega\varepsilon_p/\gamma_p)[A_p I_1\{\gamma_p r\} - B_p K_1\{\gamma_p r\}] \, , \qquad \text{(d)} \qquad (5.3.1)$$

where the subscript p refers to the region of the structure. For a helix in free space under consideration, p = 1 and 2 refer to the regions inside and outside the helix, respectively. Here, both for p = 1 and 2, $\varepsilon_p = \varepsilon_0$, $\mu_p = \mu_0$, $\gamma_p = \gamma = (\beta^2 - \omega^2\mu_0\varepsilon_0)^{1/2}$ (see (5.2.7)), the regions referred to being each a free space. In field expressions (5.3.1), the dependence $\exp j(\omega t - \beta z)$ is understood.

Out of the eight field constants (A_p, B_p, C_p, and D_p, with p = 1 and 2), four constants have zero values as follows. It follows from (5.3.1a) that the value of E_{z1} (inside the helix) blows up to infinity at r = 0 (at the axis of the helix) since $K_0\{x\} \to \infty$ as $x \to 0$, unless one would put $B_1 = 0$. Also, as can be again seen from (5.3.1(a)), the finiteness of E_{z2} (outside the helix) at $r = \infty$ demands that one should put $A_2 = 0$ since $I_0\{x\} \to \infty$ as $x \to \infty$. By making similar arguments one may see from (5.3.1(b)) that $D_1 = 0$ and $C_2 = 0$. The remaining field constants, for the present case of a helix in free space, namely, A_1, C_1, B_2 and D_2 have non-zero values. Therefore, the fields inside (1) and outside (2) the helix may be written from (5.3.1) in terms of these non-zero field constants as follows.

$$E_{z1} = A_1 I_0\{\gamma r\} \quad \text{and} \quad E_{z2} = B_2 K_0\{\gamma r\} \ ,$$

$$H_{z1} = C_1 I_0\{\gamma r\} \quad \text{and} \quad H_{z2} = D_2 K_0\{\gamma r\} \ ,$$

$$E_{\theta 1} = (-j\omega\mu_0/\gamma) C_1 I_1\{\gamma r\} \quad \text{and} \quad E_{\theta 2} = (j\omega\mu_0/\gamma) D_2 K_1\{\gamma r\} \ ,$$

and

$$H_{\theta 1} = (j\omega\varepsilon_0/\gamma) A_1 I_1\{\gamma r\} \quad \text{and} \quad H_{\theta 2} = (-j\omega\varepsilon_0/\gamma) B_2 K_1\{\gamma r\} \ . \tag{5.3.2}$$

If one now substitutes (5.3.2) into the boundary conditions (5.2.29), (5.2.30), (5.2.28) and (5.2.34), one obtains the following simultaneous equations in A_1, C_1, B_2 and D_2, respectively:

$$a_{11} A_1 + a_{12} C_1 + a_{13} B_2 + a_{14} D_2 = 0 \ ,$$

$$a_{21} A_1 + a_{22} C_1 + a_{23} B_2 + a_{24} D_2 = 0 \ ,$$

$$a_{31} A_1 + a_{32} C_1 + a_{33} B_2 + a_{34} D_2 = 0 \ ,$$

and

$$a_{41} A_1 + a_{42} C_1 + a_{43} B_2 + a_{44} D_2 = 0 \ , \tag{5.3.3}$$

where

$$a_{11} = 0, \quad a_{12} = 0, \quad a_{13} = K_0\{\gamma a\}\sin\psi, \quad a_{14} = (j\omega\mu_0/\gamma) K_1\{\gamma a\}\cos\psi \ ;$$

$$a_{21} = (j\omega\varepsilon_0/\gamma) I_1\{\gamma a\}\cos\psi, \quad a_{22} = I_0\{\gamma a\}\sin\psi, \quad a_{23} = (j\omega\varepsilon_0/\gamma) K_1\{\gamma a\}\cos\psi,$$

$$a_{24} = -K_0\{\gamma a\}\sin\psi; \quad a_{31} = I_0\{\gamma a\}\sin\psi, \quad a_{32} = (-j\omega\mu_0/\gamma) I_1\{\gamma a\}\cos\psi,$$

$$a_{33} = 0, \quad a_{34} = 0 \ ; \quad a_{41} = I_0\{\gamma a\}, \quad a_{42} = 0, \quad a_{43} = -K_0\{\gamma a\}, \quad a_{44} = 0 \ . \tag{5.3.4}$$

The condition that the set of equations (5.3.3) shall have a non-trivial solution is that the (4 x 4) determinant formed by the coefficient of the field constants (A_1, C_1, B_2 and D_2) should equal zero, that is,

$$|a_{ij}|_{(i,\, j\, =\, 1,\, 2,\, 3,\, 4)} = 0 \ . \tag{5.3.5}$$

We notice that some of the elements of (5.3.5), as given by (5.3.4), are equal to zero. This would enable us to simplify (5.3.5) as:

$$a_{13}a_{24}a_{41}a_{32} + a_{14}a_{21}a_{32}a_{43} + a_{14}a_{22}a_{31}a_{43} + a_{14}a_{23}a_{41}a_{32} = 0 \ . \tag{5.3.6}$$

Substituting (5.3.4) in (5.3.6), and carrying out further simplification, one can easily show:

$$-\sin^2\psi\, I_0\{\gamma a\}K_0\{\gamma a\}(j\omega\mu_0/\gamma)[I_0\{\gamma a\}K_1\{\gamma a\} - K_0\{\gamma a\}I_1\{\gamma a\}]$$

$$= \cos^2\psi\, I_1\{\gamma a\}K_1\{\gamma a\}(j\omega\mu_0/\gamma)^2(j\omega\varepsilon_0/\gamma)\,[I_0\{\gamma a\}K_1\{\gamma a\} - K_0\{\gamma a\}I_1\{\gamma a\}]$$

which, since $I_0\{\gamma a\}K_1\{\gamma a\} \neq K_0\{\gamma a\}I_1\{\gamma a\}$, takes the following form after a little re-arrangement:

$$\frac{k_0 \cot\psi}{\gamma} = \left[\frac{I_0\{\gamma a\}\, K_0\{\gamma a\}}{I_1\{\gamma a\}\, K_1\{\gamma a\}}\right]^{1/2} . \tag{5.3.7}$$

Here we have put $\omega(\mu_0\varepsilon_0)^{1/2} = k_0$, the free-space propagation constant. Since the radial propagation constant γ is related to the axial propagation constant β by (5.2.7), one may see that (5.3.7) gives one the ω-β relationship, or a relation wherefrom to find how the phase velocity v_p (= ω/β) varies with frequency ω. (5.3.7) is, therefore, called the dispersion relation of the structure, in this case, a helix in free-space.

The left hand side of (5.3.7) may be written with the help of (5.2.7) as:

$$\frac{k_0 \cot\psi}{\gamma} = \frac{k_0 \cot\psi}{(\beta^2 - \omega^2\mu_0\varepsilon_0)^{1/2}} = \frac{k_0 \cot\psi}{(\beta^2 - k_0^2)^{1/2}} . \tag{5.3.8}$$

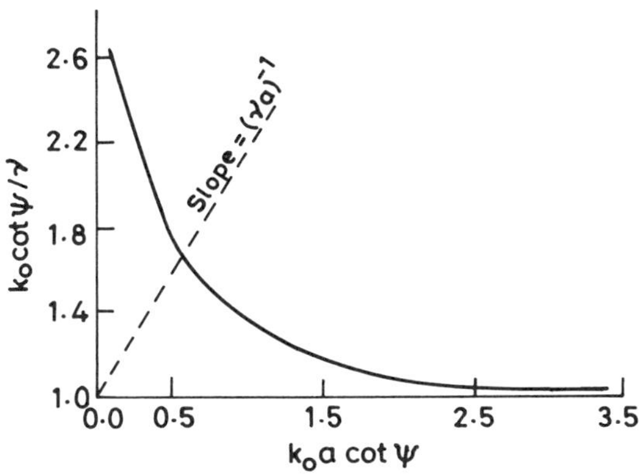

Fig. 5.3.1 The dispersion curve for a helix in free space.

Under slow-wave assumption, one has $v_p \ll c$ giving β (= ω/v_p) $\gg k_0$ (= ω/c). This would make (5.3.8) read as:

$$\frac{k_0 \cot\psi}{\gamma} \cong \frac{k_0 \cot\psi}{\beta} = \frac{(\omega / c)\cot\psi}{\omega / v_p} \cong \frac{v_p}{c}\cot\psi . \tag{5.3.9}$$

The method of generating a dispersion curve: $k \cot\psi/\gamma$, a normalized phase velocity (see (5.3.9)), versus $k_0 a \cot\psi$ (= $\omega a \cot\psi/c$), a normalized frequency, could be as follows [20]. First, let us take an arbitrary value of γa and evaluate the right hand side of (5.3.7) and hence the value of $k_0\cot\psi/\gamma$. Then let us find the product of this value of $k_0\cot\psi/\gamma$ with the value of γa taken, to obtain the value $(k_0\cot\psi/\gamma)(\gamma a) = k_0 a \cot\psi$.

Thus we get a set of values of $k_0 a \cot\psi$ and $k_0 \cot\psi/\gamma$, corresponding to a given value of γa, which may be located as a point on the dispersion curve (Fig. 5.3.1). Similarly, another set of values of $k_0 a \cot\psi$ and $k_0 \cot\psi/\gamma$ may be generated, from another arbitrarily chosen value of γa; these values may be then located as another point on the dispersion curve. In this way a number of points may be generated, corresponding to a number of values of γa, which may then be connected to obtain the desired dispersion curve of the helix (see Fig. 5.3.1). We may notice that the slope of the line joining the origin of the plot with any of these points is equal to the reciprocal of the value of γa which the point corresponds to, and also that, the higher the value of γa, the higher would be the value of the normalized frequency $k_0 a \cot\psi$.

<u>*Example 5.3.1*</u> Verify the validity of the approximate relation $v_p \cong c \sin\psi$ (5.1.6), using the dispersion relation (5.3.9), taking relatively high values of γa.

For values of $\gamma a >> 10$, it may be seen with the help of a mathematical table of modified Bessel functions [21] that the quantity $[I_0\{\gamma a\} K_0\{\gamma a\}/(I_1\{\gamma a\} K_1\{\gamma a\}] \rightarrow 1$, enabling one to write from (5.3.7):

$$k_0 \cot\psi / \gamma = 1 \ . \tag{5.3.10}$$

Substituting $\gamma = (\beta^2 - k_0^2)^{1/2}$ (see (5.2.7)) in (5.3.10), squaring it, and then re-arranging the terms, one obtains

$$\beta^2 = k_0^2 (1 + \cot^2\psi) = k_0^2 \operatorname{cosec}^2\psi \ ,$$

whence we get

$$k_0/\beta = (\omega/c)/(\omega/v_p) = v_p/c = \sin\psi \ ,$$

which is seen identical with (5.1.6).

5.3.2. Dielectric-Loaded Helix in a Metal Envelope

Let us now take the case of a practical helix supported by a number of identical dielectric wedge bars in a metal envelope as a typical example of a loaded helix for field analysis (Fig. 5.1.2). The structure may be considered to be equivalent to a helix surrounded by a continuous dielectric tube of an effective permittivity (ε_r') given by (5.1.11)(see section *5.1.2*).

In this case the fields continue to be given by (5.3.1), p = 1 referring to the region inside the helix, while p = 2 to the region outside the helix but extending upto r = b (instead of ∞), where b is the radius of the metal envelope (Fig. 5.1.2(b)). Here, $\varepsilon_p = \varepsilon_0$ and $\varepsilon_0\varepsilon_r'$ for p = 1 and 2, respectively; and $\mu_p = \mu_0$ for both p = 1 and 2. Then, as can be seen from (5.2.7), $\gamma_1 = (\beta^2 - \omega^2\mu_0\varepsilon_0)^{1/2}$ and $\gamma_2 = (\beta^2 - \omega^2\mu_0\varepsilon_0\varepsilon_r')^{1/2}$, though, under the slow-wave assumption, one may continue to take $\gamma_1 \approx \gamma_2 (\approx \beta) = \gamma = (\beta^2 - \omega^2\mu_0\varepsilon_0)^{1/2} = (\beta^2 - k_0^2)^{1/2}$ (as argued while arriving at the approximate relation (5.3.9)). Also, as explained following (5.3.1), one has in this case the field constants $B_1 = D_1 = 0$. However, the field constants A_2 and C_2 do not continue to be equal to zero as the relevant reason (section *5.3.1*) to make these constants zero does no longer hold good now, since the fields are shielded in this case at $r = b$, unlike in the case of a helix in free space where they extend upto infinity. Thus A_2 and C_2 would now add to the list of non-zero constants A_1, C_1, B_2 and D_2 to make their number equal to six in this problem.

The method of deriving the dispersion relation would then be same as that in section *5.3.1* for an unloaded helix. The fields inside the helix, which continue to be given by (5.3.2), and those outside the helix may be written as:

$$E_{z1} = A_1 I_0\{\gamma r\} \text{ and } E_{z2} = A_2 I_0\{\gamma r\} + B_2 K_0\{\gamma r\} ;$$

$$H_{z1} = C_1 I_0\{\gamma r\} \text{ and } H_{z2} = C_2 I_0\{\gamma r\} + D_2 K_0\{\gamma r\} ;$$

$$E_{\theta 1} = (-j\omega\mu_0/\gamma)\, C_1 I_1\{\gamma r\} \text{ and } E_{\theta 2} = (-j\omega\mu_0/\gamma)[C_2 I_1\{\gamma r\} - D_2 K_1\{\gamma r\}] ;$$

and $$H_{\theta 1} = (j\omega\varepsilon_0/\gamma) A_1 I_0\{\gamma r\} \text{ and } H_{\theta 2} = (j\omega\varepsilon_0\varepsilon'_r/\gamma)[A_2 I_1\{\gamma r\} - B_2 K_1\{\gamma r\}] . \quad (5.3.11)$$

In (5.3.11), the dependence $\exp j(\omega t - \beta z)$ is understood. The six field quantities given by (5.3.11) may be substituted in the six boundary conditions: four at the sheath helix, $r = a$: (5.2.28), (5.2.29),(5.2.30) and (5.2.34) and two at the metal envelope, $r = b$: $E_{z2} = 0$ and $E_{\theta 2} = 0$, to obtain six simultaneous equations, similar to (5.3.3), now in A_1, C_1, A_2, B_2, C_2, D_2 (instead of in A_1, C_1, B_2 and D_2). For a non-trivial solution, a determinantal dispersion relation, similar to (5.3.5), is obtained as follows:

$$|a_{ij}|_{(i, j = 1, 2, 3, 4, 5, 6)} = 0 , \quad (5.3.12)$$

where

$$a_{11} = I_0\{\gamma a\}\sin\psi,\ a_{12} = -(j\omega\mu_0/\gamma)\, I_1\{\gamma a\}\cos\psi,\ a_{13} = a_{14} = a_{15} = a_{16} = 0 ;$$

$$a_{21} = a_{22} = 0,\ a_{23} = I_0\{\gamma a\}\sin\psi,\ a_{24} = K_0\{\gamma a\}\sin\psi,\ a_{25} = -(j\omega\mu_0/\gamma)\, I_1\{\gamma a\}\cos\psi ,$$

$$a_{26} = (j\omega\mu_0/\gamma)\, K_1\{\gamma a\}\cos\psi ;\quad a_{31} = (j\omega\varepsilon_0/\gamma)\, I_1\{\gamma a\}\cos\psi,\ a_{32} = I_0\{\gamma a\}\sin\psi ,$$

$$a_{33} = -(j\omega\varepsilon_0\varepsilon'_r/\gamma) I_1\{\gamma a\}\cos\psi,\ a_{34} = (j\omega\varepsilon_0\varepsilon'_r/\gamma)\, K_1\{\gamma a\}\cos\psi,\ a_{35} = -I_0\{\gamma a\}\sin\psi,$$

$$a_{36} = -K_0\{\gamma a\}\sin\psi ;\quad a_{41} = I_0\{\gamma a\},\ a_{42} = 0,\ a_{43} = -I_0\{\gamma a\},\ a_{44} = -K_0\{\gamma a\},$$

$$a_{45} = a_{46} = 0 ;\ a_{51} = a_{52} = 0,\ a_{53} = I_0\{\gamma b\},\ a_{54} = K_0\{\gamma b\},\ a_{55} = a_{56} = 0 ;$$

$$a_{61} = a_{62} = a_{63} = a_{64} = 0 ,\ a_{65} = I_1\{\gamma b\},\ a_{66} = -K_1\{\gamma b\} . \quad (5.3.13)$$

At this juncture let us recall how the dispersion relation for an unloaded helix is derived in a convenient form similar to (5.3.7) which was obtained by the algebraic simplification of (5.3.5) which involved a 4x4 determinant. In the present case of a loaded helix one may proceed in a similar way to simplify (5.3.12) read through (5.3.13), which now involves a 6 x 6 determinant. The details of algebra, left as an exercise to one who is interested, with a hint that one should also now make use of the recurrence relation:

$$I_0\{x\}K_1\{x\} + K_0\{x\}I_1\{x\} = 1/x \ ,$$

where, $x = \gamma a$, would yield the following convenient form of the dispersion relation for a loaded helix:

$$\frac{k_0 \cot\psi}{\gamma} = \left[\frac{I_0\{\gamma a\}\, K_0\{\gamma a\}}{I_1\{\gamma a\}\, K_1\{\gamma a\}}\right]^{1/2} D_{lf} \ , \tag{5.3.14}$$

where D_{lf} is a function of the structure parameters, called the dielectric loading factor, also abbreviated as DLF, given by:

$$D_{lf} = \left[\left(1 - \frac{I_0\{\gamma a\}\, K_0\{\gamma b\}}{K_0\{\gamma a\}\, I_0\{\gamma b\}}\right)^{-1}\left(1 - \frac{I_1\{\gamma a\}\, K_1\{\gamma b\}}{K_1\{\gamma a\}\, I_1\{\gamma b\}}\right) \mathrm{x}\right.$$

$$\left. \mathrm{x} \left\{1 + (\varepsilon_r' - 1)\,\gamma a\, I_0\{\gamma a\} K_1\{\gamma a\}\left(1 + \frac{I_1\{\gamma a\}\, K_0\{\gamma b\}}{K_1\{\gamma a\}\, I_0\{\gamma b\}}\right)\right\}\right]^{-1/2} . \tag{5.3.15}$$

5.4 Equivalent Circuit Analysis and Dispersion Relation in the Sheath-Helix Model

Now let us turn towards an alternative analytical approach to deriving the dispersion relation of a helical SWS [22]-[24]. In this approach, the actual SWS is treated as a transmission line characterized by a set of four distributed *line parameters* (see appendix *a5.4.1*). The propagation characteristics of the SWS can then be expressed in terms of these line parameters. The problem then resolves down to expressing these line parameters in terms of the structure parameters.

For the sake of simplicity, however, the losses in the structure will be ignored as done in the preceding section, too. In this case the axial phase propagation is given by the simple relation (see appendix *a5.4.1* where the relation is derived as (a5.4.16)):

$$\beta = \omega\,(L_e C_e)^{1/2}, \tag{5.4.1}$$

where C_e is the shunt capacitance per unit length and L_e is the series inductance per unit length of the equivalent line. One may see from the analysis to be developed here that C_e and L_e are obtainable from the axial and the azimuthal components of electric field intensity at the sheath helix radius $r = a$, expressed in terms of the axial and the azimuthal components of sheath helix current, respectively. It will be seen that, in the equivalent circuit analysis, one has at a time to deal with only half the total number of boundary conditions — one half while finding C_e and the remaining half while finding L_e. Thus, in this approach, unlike in the field analysis, it is not required to simplify a complex dispersion relation which involves an $n \times n$ determinant, where n is the number of boundary conditions, which is also equal to the number of non-zero field constants, relevant to the problem. Thus the equivalent circuit analysis makes the study of loaded structures somewhat simple which, if the field analysis were used, would be quite involved for complex structures for which the value of n would be high.

Let us develop here the equivalent circuit analysis of the loaded helix studied earlier by the field analysis in section *5.3.2*. It will be of interest to see whether one obtains or not, using the equivalent circuit analysis, the same dispersion relation as (5.3.14) and the same dielectric loading factor as (5.3.15).

5.4.1 Equivalent Shunt Capacitance per Unit Length

The expression for C_e may be found in the following steps [22]-[24].

1) An expression for the axial electric field intensity at the sheath helix, r = a, is found in terms of the axial sheath helix current. For this purpose, we proceed as follows.

a) The field expressions (5.3.11) are substituted into the boundary conditions (5.2.34) and (5.2.36) both at r = a, and also into the boundary condition: $E_{z2} = 0$ at r = b, to obtain the following simultaneous equations in A_1, A_2 and B_2:

$$A_1 I_0\{\gamma a\} = A_2 I_0\{\gamma a\} + B_2 K_0\{\gamma a\} \tag{5.4.2}$$

$$\frac{j\omega\varepsilon_0\varepsilon_r'}{\gamma}[A_2 I_1\{\gamma a\} - B_2 K_1\{\gamma a\}] - (\frac{j\omega\varepsilon_0}{\gamma}) A_1 I_1\{\gamma a\} = J_{sz} = \frac{I_{za}}{2\pi a} \tag{5.4.3}$$

and

$$A_2 I_0\{\gamma b\} + B_2 K_0\{\gamma b\} = 0 . \tag{5.4.4}$$

It may be noted that in (5.4.3), J_{sz}, the axial surface current density or the axial current per unit width of the sheath helix has been put as

$$J_{sz} = I_{za}/(2\pi a) ,$$

where I_{za} is the total sheath helix current and '$2\pi a$' is the *width* of the sheath helix, treating the problem as *planar*, as has been done in Example 4.5.6, for a situation in which the skin depth is small compared to the structure dimension, here, the sheath helix radius.

b) Now, one may solve the simultaneous equations (5.4.2), (5.4.3) and (5.4.4) to obtain an expression for A_1 in terms of I_{za}, say, in the following form:

$$A_1 = M I_{za} , \tag{5.4.5}$$

where M would come out as a function of structure parameters.

c) One may next use (5.3.11) to obtain

$$E_{za} = (E_{z1})_{r=a} = A_1 I_0\{\gamma a\} , \tag{5.4.6}$$

the axial electric field intensity at the sheath helix, r = a.

d) (5.4.5) is substituted in (5.4.6) to obtain an expression for E_{za} in terms of I_{za},

$$E_{za} = M I_0\{\gamma a\} I_{za} ,$$

which, interpreting $M I_0(\gamma a)$ as P_C, say, may be put as

$$E_{za} = P_C I_{za} . \tag{5.4.7}$$

2) E_{za}, the axial electric field intensity at the sheath helix, r = a, is expressed in terms of the circuit potential V. For this, we proceed as follows.

The axial component of the electric field intensity E_{za} is related to the scalar and vector potential with the help of (4.6.8) as follows:

$$E_{za} = -(\nabla V)_z - \partial A_z/\partial t . \tag{5.4.8}$$

Noting from (2.5.12(b)) that in the cylindrical system of coordinates $(\nabla V)_z = \partial V/\partial z$, (5.4.8) may be written as:

$$E_{za} = -\frac{\partial V}{\partial z} - \frac{\partial A_z}{\partial t}$$

which, assuming that the rf quantities vary as exp $j(\omega t - \beta z)$, takes the following form:

$$E_{za} = j\beta V - j\omega A_z . \tag{5.4.9}$$

Further, the vector potential **A** is related to the scalar potential V through (4.6.13) which may be written as:

$$\nabla . \mathbf{A} + \mu\varepsilon\, \partial V/\partial t = 0 , \tag{5.4.10}$$

which, on expansion in the cylindrical system of coordinates with the help of (2.3.5(b)), becomes:

$$\frac{1}{r}(rA_r) + \frac{1}{r}\frac{\partial A_\theta}{\partial \theta} + \frac{\partial A_z}{\partial z} + \mu\varepsilon\frac{\partial V}{\partial t} = 0 . \tag{5.4.11}$$

Now, since there exists no sheath-helix current in the radial direction, one has $A_r = 0$ (see (4.6.20)). This, together with the consideration of non-azimuthally-varying mode ($\partial/\partial\theta = 0$), makes (5.4.11) read as

$$\frac{\partial A_z}{\partial z} + \mu\varepsilon\frac{\partial V}{\partial t} = 0$$

which, again taking the variation of rf quantities as exp $j(\omega t - \beta z)$ becomes

$$-j\beta A_z + j\omega\mu\varepsilon V = 0$$

whence $$A_z = (\omega\mu\varepsilon/\beta)\,V\ . \tag{5.4.12}$$

Substituting (5.4.12) into (5.4.9) one then gets

$$E_{za} = j(\beta - \omega^2\mu\varepsilon/\beta)\,V = j[(\beta^2 - \omega^2\mu\varepsilon)/\beta]\,V$$

which, with the help of (5.2.7), becomes

$$E_{za} = j\,(\gamma^2/\,\beta)\,V\ . \tag{5.4.13}$$

This is the required expression for the axial electric field intensity in terms of the circuit potential.

3) The circuit potential is then related to the axial component of sheath helix current by (a5.4.4), (taking $G_e = 0$ for a lossless structure) which is one of the telegraphist's equations (see appendix *a5.4.1*):

$$\frac{\partial I_{za}}{\partial z} + C_e\frac{\partial V}{\partial t} = 0$$

which taking the dependence of rf quantities as exp $j(\omega t - \beta z)$ may be read as

$$-j\,\beta\,I_{za} + j\,\omega C_e\,V = 0$$

whence one has

$$C_e = \frac{I_{za}}{V}\frac{\beta}{\omega}\ . \tag{5.4.14}$$

Substituting I_{za} from (5.4.7) and V from (5.4.13) into (5.4.14), one obtains the expression for C_e as:

$$C_e = j\frac{\gamma^2}{\omega P_C}\ . \tag{5.4.15}$$

Now, working out the above steps 1(b)-(d) with the help of a simple algebra, one may first easily obtain an expression for P_C, and then substitute it in (5.4.15) to get the following expression for C_e, the shunt capacitance per unit length of the loaded helix:

$$C_e = C_{e0}\,\alpha_C\ , \tag{5.4.16}$$

where

$$C_{e0} = 2\pi\varepsilon_0/[I_0\{\gamma a\}K_0\{\gamma a\}]\ , \tag{5.4.17}$$

and α_C is a dimension less parameter given by

$$\alpha_C = \left(1 - \frac{I_0\{\gamma a\}\,K_0\{\gamma b\}}{K_0\{\gamma a\}\,I_0\{\gamma b\}}\right)^{-1}\left[1+(\varepsilon_r' - 1)\,\gamma a\,I_0\{\gamma a\}\,K_1\{\gamma a\}\left(1+\frac{I_1\{\gamma a\}\,K_0\{\gamma b\}}{K_1\{\gamma a\}\,I_0\{\gamma b\}}\right)\right]\ . \tag{5.4.18}$$

It is of interest to note that, in the case of $\varepsilon_r = 1$, and hence $\varepsilon'_r = 1$ (see (5.1.10)), and $b \to \infty$, the problem would refer to a helix in free-space or an unloaded helix. In this case, since $I_0\{\gamma b\} \to \infty$ and $K_0\{\gamma b\} \to 0$ (as $b \to \infty$) and $\varepsilon'_r - 1 = 0$, it follows, from (5.4.18), that $\alpha_C = 1$, and, from (5.4.16), that $C_e = C_{e0}$. Thus C_{e0}, given by (5.4.17), is the expression for the capacitance per unit length of a helix in free space. It may be mentioned that one would have obtained the same expression as (5.4.17), if one had analyzed a helix in free space using the same approach as followed here for a loaded helix, for the shunt capacitance per unit length. α_C may be called as the capacitive loading factor which when multiplied by C_{e0} would give one the value of C_e, the capacitance per unit length of a loaded helix.

5.4.2 Equivalent Series Inductance per Unit Length

Now let us proceed to find the expression for L_e, in the following steps [22]-[24].

1) An expression for the azimuthal electric field intensity at the sheath helix, $r = a$, is found in terms of the azimuthal sheath-helix current. For this purpose, we now proceed as follows.

a) The field expressions (5.3.11) are substituted into the boundary conditions (5.2.35) and (5.2.37), both at $r = c$, and also into the boundary condition: $E_{\theta 2} = 0$ at $r = b$, to obtain the following simultaneous equations in C_1, C_2 and D_2:

$$C_1 I_1\{\gamma a\} = C_2 I_1\{\gamma a\} - D_2 K_1\{\gamma a\} \tag{5.4.19}$$

$$C_1 I_0\{\gamma a\} - C_2 I_0\{\gamma a\} - D_2 K_0\{\gamma a\} = J_{s\theta} = I_{\theta a}/(2\pi a) \tag{5.4.20}$$

and

$$C_2 I_1\{\gamma b\} - D_2 K_1\{\gamma b\} = 0 . \tag{5.4.21}$$

In (5.4.20), $J_{s\theta}$ has been put equal to $I_{\theta a}/(2\pi a)$ using the same reasoning as given following (5.4.4).

b) Now, one may solve the simultaneous equations (5.4.19), (5.4.20) and (5.4.21) for C_1 in terms of $I_{\theta a}$, say, to obtain an expression in the following form:

$$C_1 = S I_{\theta a} , \tag{5.4.22}$$

where S would be a function of structure parameters.

c) One may now use (5.3.11) to obtain

$$E_{\theta a} = (E_{\theta 1})_{r=a} = -(j\,\omega\mu_0/\gamma)\, C_1 I_1\{\gamma a\} , \tag{5.4.23}$$

the azimuthal electric field intensity at the sheath helix, $r = a$.

d) (5.4.22) is then substituted in (5.4.23) to obtain

$$E_{\theta a} = -(j\,\omega\mu_0/\gamma)\, S I_1\{\gamma a\} I_{\theta a} ,$$

which on putting $-(j\omega\mu_0/\gamma)S\, I_1\{\gamma a\} = Q_L$, say, yields

$$E_{\theta a} = Q_L\, I_{\theta a}\ . \tag{5.4.24}$$

2) We then recall the expression for the axial electric field intensity in terms of the circuit potential:

$$E_{za} = (j\,\gamma^2/\beta)\, V\ , \tag{5.4.25}$$

which is (5.4.13) re-written.

3) The next step is to relate the axial component of sheath-helix current with the circuit potential by one of the telegraphist's equations (a5.4.3), taking $R_e = 0$ for a lossless case, (see appendix *a5.4.1*), as follows:

$$\frac{\partial V}{\partial z} + L_e \frac{\partial I_{za}}{\partial t} = 0$$

which, in view of the dependence of rf quantities as $\exp j(\omega t - \beta z)$, may be expressed as

$$-j\,\beta\, V + j\,\omega L_e\, I_{za} = 0$$

whence one has

$$L_e = \frac{V}{I_{za}}\frac{\beta}{\omega}\ . \tag{5.4.26}$$

3) Now, let us relate $E_{\theta a}$ with E_{za} with the help of the boundary condition (5.2.28) as:

$$E_{\theta a}\cos\psi + E_{za}\sin\psi = 0\ . \tag{5.4.27}$$

Further, no sheath-helix current perpendicular to the helix winding direction, in view of the zero conductivity of the sheath in this direction (section *5.1.1*), enables one to write (Fig. 5.1.1(b)):

$$I_{\theta a}\sin\psi = I_{za}\cos\psi\ . \tag{5.4.28}$$

Substituting V / I_{za}, obtained by combining (5.4.24), (5.4.25), (5.4.27) and (5.4.28), into (5.4.26), the following expression for L_e is obtained

$$L_e = \frac{j\beta^2 Q_L}{\gamma^2\omega\tan^2\psi}\ . \tag{5.4.29}$$

Now, working out the above steps 1(b)-(d) one can easily find the expression for Q_L and substitute it in (5.4.29) to find the following expression for L_e, the series inductance per unit length of the loaded helix:

$$L_e = L_{e0}\,\alpha_L\ , \tag{5.4.30}$$

where

$$L_{e0} = \frac{\mu_0}{2\pi}\left(\frac{\beta}{\gamma}\right)^2 \cot^2\psi\, I_1\{\gamma a\}K_1\{\gamma a\}\ , \tag{5.4.31}$$

and α_L is a dimensionless parameter given by

$$\alpha_L = 1 - \frac{I_1\{\gamma a\} K_1\{\gamma b\}}{K_1\{\gamma a\} I_1\{\gamma b\}} \quad . \tag{5.4.32}$$

We notice from (5.4.32) that, as $b \to \infty$, $\alpha_L \to 1$ and $L_e \to L_{e0}$, the expression that one would have obtained from first principles for the equivalent series inductance per unit length of an unloaded helix or a helix in free space, following the same approach as used here for the loaded helix. Thus α_L may be called the inductive loading factor which when multiplied by L_{e0} would give one the value of L_e, the inductance per unit length of a loaded helix.

5.4.3 Dispersion Relation from Equivalent Line Parameters

If we substitute (5.4.16) and (5.4.30) in (5.4.1), and make use of (5.4.17) and (5.4.31), and re-arrange terms, we may show:

$$\frac{k_0 \cot\psi}{\gamma} = \left(\frac{I_0\{\gamma a\} K_0\{\gamma a\}}{I_1\{\gamma a\} K_1\{\gamma a\}}\right)^{1/2} (\alpha_L \, \alpha_C)^{-1/2}. \tag{5.4.33}$$

Now, we may choose to express (5.4.33) as

$$\frac{k_0 \cot\psi}{\gamma} = \left(\frac{I_0\{\gamma a\} K_0\{\gamma a\}}{I_1\{\gamma a\} K_1\{\gamma a\}}\right)^{1/2} D_{lf} \quad , \tag{5.4.34}$$

by putting

$$D_{lf} = (\alpha_L \alpha_C)^{-1/2} \quad . \tag{5.4.35}$$

It is of interest to see that (5.4.34) and (5.4.35), through (5.4.18) and (5.4.32), deduced here by equivalent circuit analysis are exactly identical with the dispersion relation (5.3.14) and (5.3.15), respectively, which were deduced in section *5.3* by the field-analytical approach.

Example 5.4.1 Calculate the pitch of a helix of a helical SWS of DLF = 0.8 for a TWT which employs a 2 kV electron beam and operates at 6GHz. Assume that the gain of the device is optimized at a value $\gamma a = 1.5$. The values of the relevant modified Bessel functions are given as: $I_0\{1.5\} = 1.6467$, $I_1\{1.5\} = 0.9817$, $K_0\{1.5\} = 0.2138$ and $K_1\{1.5\} = 0.2774$.

Under the slow-wave assumption ($\beta >> k$), one has from (5.2.7):

$$\gamma a = \beta a = (\omega/v_p)a = (2\pi f/v_p)a = 1.5 \quad . \tag{5.4.36}$$

For a synchronous 2 kV electron beam, one has $v_p = 2.65\times10^7$ m/s(see Example 5.1.1). Substituting this value of v_p and $f = 6$ GHz $= 6\times10^9$ Hz in (5.4.37), one may calculate a as:

$$a = \frac{1.5 v_p}{2\pi f} = \frac{1.5\times 2.65\times 10^7}{2\times 3.14\times 6\times 10^9}\ \text{m} = 1.05\,\text{mm} .$$

Making use of (5.3.9), as well as the relation $\cot\psi = 2\pi a/p$, and putting $D_{lf} = 0.8$, in (5.3.14), one has

$$\frac{2.65\times10^7}{3\times10^8}\times\frac{2\pi a}{p} = \left(\frac{I_0\{1.5\}K_0\{1.5\}}{I_1\{1.5\}K_1\{1.5\}}\right)^{1/2}\times 0.8 ,$$

whence substituting a = 1.05 mm (already calculated) and the values of the modified Bessel functions given, one calculates

$$p = \frac{2.65\times10^7}{3\times10^8}\times\frac{2\times3.14\times1.05}{0.8}\times\left(\frac{0.9817\times0.2774}{1.6467\times0.2138}\right)^{1/2}\ \text{m} = 0.64\ \text{mm}.$$

Appendix a5.4.1 Telegraphist's equations of a transmission line

A transmission line may be represented by a set of equivalent circuit parameters distributed uniformly over its length. An infinitesimal length dz of the line is considered and shown as a four-terminal network in Fig. a5.4.1

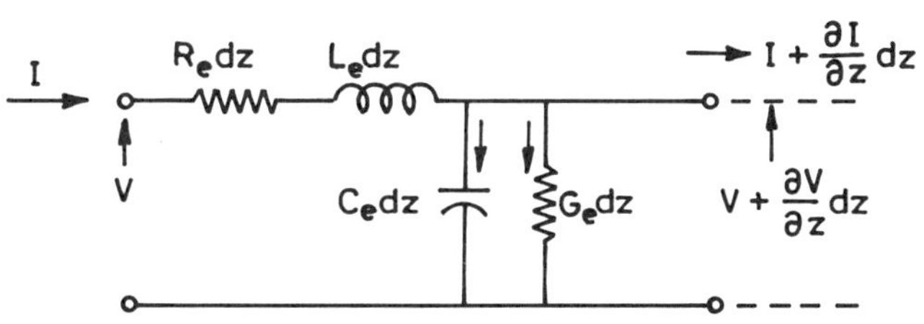

Fig. a5.4.1 The equivalent circuit of a transmission line of an infinitesimal length dz .

The losses in the conductors of the line made of materials of finite resistivities are taken into account by a series resistance $R_e\,dz$, for the infinitesimal length considered, where R_e is the equivalent series resistance per unit length of the line. The magnetic flux linked up with the circuit due to currents flowing through the conductors of the line may be considered by providing a series inductance $L_e\,dz$, where L_e is the equivalent series inductance per unit length of the line. The proximity of the conductors gives rise to a shunt capacitance $C_e\,dz$, where C_e is the equivalent shunt capacitance per unit length of the line. The dielectric, if present, between the conductors of the line may not be perfect and as such would contribute to a shunt conductance $G_e\,dz$ of the line, where G_e is the equivalent shunt conductance per unit length of the line. These line parameters, namely, R_e, L_e, C_e and G_e define the propagation characteristics of the line, in terms of the propagation constant, referring to the circuit-

voltage and -current waves set up on the line. Two basic coupled equations in circuit voltage V and circuit current I are obtained by applying Kirchoff's voltage and current laws to the equivalent network for the infinitesimal length line (Fig. a5.4.1) to obtain, respectively, the following two equations:

$$V-(V+\frac{\partial V}{\partial z}dz) = I R_e dz + L_e dz\frac{\partial I}{\partial t} \quad \text{and} \tag{a5.4.1}$$

$$I = (I+\frac{\partial I}{\partial z}dz)+C_e dz\frac{\partial V}{\partial t}+G_e dz V . \tag{a5.4.2}$$

It may be mentioned that the third and the last terms of the right hand side of (a5.4.2) represent the currents through the shunt-capacitive and the shunt-conducting paths of the line, respectively. Upon simplification, (a5.4.1) and (a5.4.2) may be written as:

$$\frac{\partial V}{\partial z} = -(R_e I + L_e\frac{\partial I}{\partial t}) \quad \text{and} \tag{a5.4.3}$$

$$\frac{\partial I}{\partial z} = -(G_e V + C_e\frac{\partial V}{\partial t}) . \tag{a5.4.4}$$

Taking the rf dependence of the form $\exp j(\omega t - \beta z)$, *one may write (a5.4.3) and (a5.4.4) as:*

$$\frac{\partial V}{\partial z} = -(R_e + j\omega L_e) I \quad \text{and} \tag{a5.4.5}$$

$$\frac{\partial I}{\partial z} = -(G_e + j\omega C_e) V . \tag{a5.4.6}$$

The set of equations: either (a5.4.3) and (a5.4.4) or (a5.4.5) and (a5.4.6) are known as the telegraphist's equations. These equations are coupled in V and I. They may be decoupled to obtain a wave equation, say, for the circuit voltage V, by differentiating (a5.4.5) and then substituting (a5.4.6) there into obtain:

$$\frac{\partial^2 V}{\partial z^2} = \gamma'^2 V , \tag{a5.4.7}$$

where

$$\gamma' = [(R_e + j\omega L_e)(G_e + j\omega C_e)]^{1/2} . \tag{a5.4.8}$$

The solution of (a5.4.7) for V may be put as:

$$V = V^{+}\exp(-\gamma' z) + V^{-}\exp(\gamma' z) , \tag{a5.4.9}$$

in which the time dependence $\exp(j\omega t)$ *is understood. It may be seen from (a5.4.8) that* γ' *is complex which may represented in terms of its real*(α) *and imaginary*(β) *parts as:*

$$\gamma' = \alpha + j\beta . \tag{a5.4.10}$$

Making use of (a5.4.10) and remembering the time-dependent part $\exp(j\omega t)$, *(a5.4.9) may be written as:*

(a5.4.11) $$V = V^{+}\exp - \alpha z \ \exp j(\omega t - \beta z) + V^{-} \exp \alpha z \ \exp j(\omega t + \beta z) .$$

Thus the first and the second terms of (a5.4.11) (or (a5.4.9)) represent the forward and the backward waves, respectively, V^{+} and V^{-} being their respective voltage amplitudes which are attenuated by the factors $\exp -\alpha z$ and $\exp \alpha z$ for waves traveling in the +z and -z directions, respectively (see discussions following (4.5.34)). α is called the attenuation constant and β the phase propagation constant of the line. These are respectively given, with the help of (a5.4.8) and (a5.4.10), (separating the real and imaginary parts of (a5.4.8)), by:

(a5.4.12) $$\alpha = [(R_e^2 + \omega^2 L_e^2)(G_e^2 + \omega^2 C_e^2)]^{1/4} \cos\frac{\theta_1 + \theta_2}{2} \quad \text{and}$$

(a5.4.13) $$\beta = [(R_e^2 + \omega^2 L_e^2)(G_e^2 + \omega^2 C_e^2)]^{1/4} \sin\frac{\theta_1 + \theta_2}{2}, \quad \text{where}$$

(a5.4.14) $$\theta_1 = \tan^{-1}(\omega L_e/R_e) \text{ and } \theta_2 = \tan^{-1}(\omega C_e/G_e) .$$

It is easy to see that, for a lossless line for which $R_e = G_e = 0$, one would have, from (a5.4.14), $\theta_1 = \pi/2$ and also $\theta_2 = \pi/2$. These values of θ_1 and θ_2 would make the cosine and sine terms of (a5.4.12) and (a5.4.13) equal to zero and unity, respectively. Consequently, for a lossless line ($R_e = G_e = 0$), one has from (a5.4.12) and (a5.4.13):

(a5.4.15) $$\alpha = 0 \quad \text{and}$$

(a5.4.16) $$\beta = \omega (L_e C_e)^{1/2} .$$

One may see that (a5.4.16) is identical with (5.4.1).

It is also instructive to know how the circuit current I varies with z. This may be done by differentiating (a5.4.9) and substituting it in (a5.4.5) to obtain

(a5.4.17) $$I = \gamma' \frac{[V^{+} \exp(-\gamma' z) - V^{-} \exp(\gamma' z)]}{(R_e + j\omega L_e)} .$$

Substituting (a5.4.8) into (a5.4.17) one then gets

(a5.4.18) $$I = \frac{V^{+} \exp(-\gamma' z) - V^{-} \exp(\gamma' z)}{Z_0}, \quad \text{where}$$

(a5.4.19) $$Z_0 = \left(\frac{R_e + j\omega L_e}{G_e + j\omega C_e}\right)^{1/2}$$

is called the characteristic impedance of the line.

Let a line of length l be terminated in a load impedance at $z = 0$ and be fed by a source at $z = -l$, which may be called the sending or source or input end. Then the impedance at the sending end or the input impedance Z_{in} is obtained by substituting $z = -l$ in each of (a5.4.9) and (a5.4.18) and then by performing division as follows:

(a5.4.20) $$Z_{in} = (\frac{V}{I})_{z=-l} = Z_0 \frac{V^+ \exp(\gamma' l) + V^- \exp(-\gamma' l)}{V^+ \exp(\gamma' l) - V^- \exp(-\gamma' l)} \quad .$$

It is also useful to know the input impedance of the line in terms of the load impedance. For this purpose, let the load voltage V_L and the load current I_L be written with the help of (a5.4.9) and (a5.4.18) as:

(a5.4.21) $$V_L = (V)_{z=0} = V^+ + V^- \text{ and } I_L = (V^+ - V^-)/Z_0 \quad .$$

From (a5.4.21), one may then write

(a5.4.22) $$V^+ = (V_L + I_L Z_0)/2 \text{ and } V^- = (V_L - I_L Z_0)/2 \quad .$$

Now, substituting (a5.4.22) in (a5.4.20) one then has

(a5.4.23) $$Z_{in} = Z_0 \frac{V_L \cosh \gamma' l + I_L Z_0 \sinh \gamma' l}{V_L \sinh \gamma' l + I_L Z_0 \cosh \gamma' l} \quad .$$

Dividing the numerator and the denominator of (a5.4.23) by $I_L \cosh\gamma' l$ and putting $V_L/I_L = Z_L$, the load impedance, one may write

(a5.4.24) $$Z_{in} = Z_0 \frac{Z_L + Z_0 \tanh \gamma' l}{Z_0 + Z_L \tanh \gamma' l} \quad ,$$

which is the expression for the input impedance in terms of the load impedance.

In the case the load impedance $Z_L = Z_0$, one has $Z_{in} = Z_0$. This enables one to define the characteristic impedance as such an impedance that, when the line is terminated in it, the input impedance becomes equal to the terminating impedance itself. Alternatively, putting $l \to \infty$ in (a5.4.24), one has $Z_{in} = Z_0$. Thus the input impedance of an infinitely long line would be equal to the characteristic impedance itself. Therefore, from the sending or source end of a line, which is terminated in its characteristic impedance, the line would appear to be infinitely long.

The characteristic impedance of a line may be related to the terminating or load impedance through the reflection coefficient of the line Γ_L which is given by

(a5.4.25) $$\Gamma_L = V^- / V^+ \quad ,$$

noting from (a5.4.21) that, at the load end $z = 0$, one has $V = V^+ + V^-$, the first term of which is the voltage amplitude of the forward or incident wave and the second term the voltage amplitude of the backward or reflected wave. Then making use of (a5.4.22), one has from (a5.4.25)

(a5.4.26) $$\Gamma_L = \frac{V_L - I_L Z_0}{V_L + I_L Z_0} \quad .$$

Dividing (a5.4.26) by I_L and noting that $V_L/I_L = Z_L$, one has then

(a5.4.27) $$\Gamma_L = \frac{Z_L - Z_0}{Z_L + Z_0} \quad .$$

The reflection coefficient Γ_L of a line terminated in the characteristic impedance ($Z_L = Z_0$) is thus zero. Thus to avoid reflection it is necessary that the line be terminated in its characteristic impedance. The characteristic impedance of a transmission line in terms of the line parameters is given by (a5.4.19) whence, for a lossless line ($R_e = G_e = 0$), one has

(a5.4.28) $$Z_0 = (L_e/C_e)^{1/2}.$$

5.5 Helix Impedances

The systems which are used for coupling of rf power in and out of a helical SWS — the input and output couplers — should have their impedances matched with the characteristic impedance of an SWS, treated as a transmission line, for maximum transfer of power as well as for the sake of avoiding reflection at the coupler ends which would otherwise cause undesirable oscillations in a TWT that employs such an SWS. The characteristic impedance also appears in the circuit equation describing the beam-wave interaction in a TWT and hence also in the expression for the gain of such a device (see chapter *8*). The characteristic impedance Z_0 of a lossless SWS treated as a transmission line is given in term of the distributed line parameters by:

$$Z_0 = (L_e/C_e)^{1/2}, \tag{5.5.1}$$

which is (a5.4.28) rewritten.

From the consideration of the gain of a TWT(section *8.2.4*), however, a more realistic impedance parameter than the characteristic impedance is the interaction impedance K of the SWS, which may be defined as:

$$K = \frac{|V_z|^2}{2P}, \tag{5.5.2}$$

where P is the power propagating down the structure. V_z is the longitudinal voltage which may be found, as follows [25], by taking the negative line integral of the axial electric field intensity $E_z = E_z\{0\}\sin\beta z$ (see (2.5.11)), between the limits $z = 0$, the reference point where the field intensity is zero and $z = \lambda_g/4$, the point where the intensity is maximum, where $E_z\{0\}$ is the peak value of the intensity, and λ_g ($= 2\pi/\beta$) is the wavelength of the wave supported by the SWS. Thus

$$V_z = -\int_{z=\lambda_g/4}^{z=0} E_z\,dz = -\int_{z=\lambda_g/4}^{z=0} E_z\{0\}\sin\beta z\,dz = \frac{E_z\{0\}}{\beta}. \tag{5.5.3}$$

Substituting (5.5.3) in (5.5.2), one gets

$$K = \frac{E_z^2\{0\}}{2\beta^2 P}. \tag{5.5.4}$$

In a more appropriate expression, however, one ought to replace $E_z^2\{0\}$ in (5.5.4) by $E_{z,av}^2$, the square of the axial electric field intensity averaged over the cross section of the electron beam [15] which is supposedly placed in the structure used in a practical device (see problem 5.12).

Power propagating down the structure, P, as required for evaluating K given by (5.5.4), may be formed by integrating the axial (z-) component of the average Poynting vector, given by (4.8.40), over the structure cross section, as follows:

$$P = (1/2)\,Re\int_{\text{(structure cross section)}} (\mathbf{E}\times\mathbf{H}^*)_z.\mathbf{a}_n ds \ . \qquad (5.5.5)$$

For a cylindrical SWS, one may take ds = 2πr dr, the element of area, being the area of an annular ring of radius r and infinitesimally radial thickness dr ; and $\mathbf{a}_n = \mathbf{a}_z$. This enables one to write from (5.5.5):

$$P = (1/2)\,Re\int_{\text{(structure cross section)}} (\mathbf{E}\times\mathbf{H}^*)_z 2\pi r\, dr \ . \qquad (5.5.6)$$

It is also of interest to relate the amount of dispersion of the structure, $dv_p/d\omega$, with the interaction impedance K. For this purpose, let us put P, the energy flow per second, as $P = W v_g$ and substitute it in (5.5.4) to obtain

$$K = \frac{E_z^2\{0\}}{2\beta^2 W v_g} \ , \qquad (5.5.7)$$

where v_g (= dω/dβ) is the group velocity of the wave (see appendix *a5.5.1*), and W is the energy stored per unit axial length of the structure. Substituting v_g expressed in terms of v_p from (a5.5.5) in (5.5.7), one obtains

$$K = \frac{E_z^2\{0\}}{2\beta^2 W v_p}\left(1-\beta\frac{dv_p}{d\omega}\right), \qquad (5.5.8)$$

a relation which shows the dependence of K on the amount of dispersion, $dv_p/d\omega$.

5.5.1 Characteristic Impedance of a Loaded Helix

Substituting (5.4.16) and (5.4.30) in (5.5.1), one obtains for the loaded helix (Fig. 5.1.2):

$$Z_0 = \left(\frac{L_0}{C_0}\right)\!\left(\frac{\alpha_L}{\alpha_C}\right)^{1/2} \ , \qquad (5.5.9)$$

where the quantities in the right hand side are given by the relations (5.4.17), (5.4.18), (5.4.31) and (5.4.32).

5.5.2 Interaction Impedance of a Loaded Helix

For the loaded helix (Fig. 5.1.2) one may write P with the help of (5.5.6) as:

$$P = \frac{1}{2} Re\int_0^a (E_{r1}H_{\theta 1}^* - E_{\theta 1}H_{r1}^*)\, 2\pi r\, dr + \frac{1}{2} Re\int_a^b (E_{r2}H_{\theta 2}^* - E_{\theta 2}H_{r2}^*)\, 2\pi r\, dr \,. \quad (5.5.10)$$

The azimuthal components of fields as required to be substituted in (5.5.10) are obtainable from (5.3.11). Besides, the radial field components to be substituted therein may be had from (5.2.14) and (5.2.15) as:

$$E_{r1} = \frac{j\beta}{\gamma}[A_1 I_1\{\gamma r\}] \ ,$$

$$H_{r1} = \frac{j\beta}{\gamma}[C_1 I_1\{\gamma r\}] \ ,$$

$$E_{r2} = \frac{j\beta}{\gamma}[A_2 I_1\{\gamma r\} - B_2 K_1\{\gamma r\}], \quad \text{and}$$

$$H_{r2} = \frac{j\beta}{\gamma}[C_2 I_1\{\gamma r\} - D_2 K_1\{\gamma r\}] \ . \quad (5.5.11)$$

The six field constants which appear in the field expressions (5.3.11) and (5.5.11) to be substituted in (5.5.10) may thus be identified as: C_1, A_2, B_2, C_2, D_2 and A_1. Out of these, the first five constants each may be linearly related to A_1, with the help of any five out of the six boundary conditions mentioned following (5.3.11). A_1, however, may be identified as $E_z\{0\}$ as follows:

$$E_z\{0\} = (E_{z1})_{r=0} = A_1 I_0\{0\} = A_1 \ . \qquad \text{(since } I_0\{0\} = 1)$$

Thus each field component occurring in (5.5.10) becomes linearly related to A_1, that is, to $E_z\{0\}$ as follows:

$$E_{r1} = \frac{j\beta}{\gamma} E_z\{0\} I_1\{\gamma r\},$$

$$H_{r1} = \frac{\beta}{\omega\mu_0}\left(\frac{I_0\{\gamma a\}}{I_1\{\gamma a\}}\tan\psi\right) E_z\{0\} I_1\{\gamma r\} \ ,$$

$$E_{\theta 1} = -\left(\frac{I_0\{\gamma a\}}{I_1\{\gamma a\}}\tan\psi\right) E_z\{0\} I_1\{\gamma r\} \ ,$$

$$H_{\theta 1} = \frac{j\omega\varepsilon_0}{\gamma} E_z\{0\} I_1\{\gamma r\} \quad ,$$

$$E_{r2} = -\frac{j\beta}{\gamma}E_z\{0\}\left[\frac{\xi_{00}}{1-\xi_{00}}I_1\{\gamma r\} + \frac{1}{1-\xi_{00}}\frac{I_0\{\gamma a\}}{K_0\{\gamma a\}}K_1\{\gamma r\}\right],$$

$$H_{r2} = -\frac{\beta}{\omega\mu_0}E_z\{0\}\left[\frac{\xi_{11}}{1-\xi_{11}}\left(\frac{I_0\{\gamma a\}}{I_1\{\gamma a\}}\tan\psi\right)I_1\{\gamma r\} - \frac{1}{1-\xi_{11}}\left(\frac{I_0\{\gamma a\}}{K_1\{\gamma a\}}\tan\psi\right)K_1\{\gamma r\}\right],$$

$$E_{\theta 2} = E_z\{0\}\left[\frac{\xi_{11}}{1-\xi_{11}}\left(\frac{I_0\{\gamma a\}}{I_1\{\gamma a\}}\tan\psi\right)I_1\{\gamma r\} - \frac{1}{1-\xi_{11}}\left(\frac{I_0\{\gamma a\}}{K_1\{\gamma a\}}\tan\psi\right)K_1\{\gamma r\}\right],$$

and
$$H_{\theta 2} = -\frac{j\omega\varepsilon_0\varepsilon'_r}{\gamma}E_z\{0\}\left[\frac{\xi_{00}}{1-\xi_{00}}I_1\{\gamma r\} + \frac{1}{1-\xi_{00}}\frac{I_0\{\gamma a\}}{K_0\{\gamma a\}}K_1\{\gamma r\}\right], \tag{5.5.12}$$

where

$$\xi_{00} = \frac{I_0\{\gamma a\}K_0\{\gamma b\}}{K_0\{\gamma a\}I_0\{\gamma b\}}$$

and
$$\xi_{11} = \frac{I_1\{\gamma a\}K_1\{\gamma b\}}{K_1\{\gamma a\}I_1\{\gamma b\}} .$$

In view of (5.5.12) it is clear from (5.5.10) that P would be proportional to $E_z^2\{0\}$. Thus it can be seen from (5.5.4) that K becomes independent of $E_z\{0\}$, since $E_z^2\{0\}$ which occurs both in the numerator and the denominator of the right hand side of (5.5.4) would cancel out.

Substituting (5.5.12) in (5.5.10), and evaluating the integral, one may find P and substitute the latter in (5.5.4) to obtain the following expression for K:

$$K = \left(\frac{\beta}{k}\right)\left(\frac{\gamma}{\beta}\right)^4\frac{(\mu_0/\varepsilon_0)^{1/2}}{(\gamma a)^2\pi(M_1G_1 + M_2G_2 + M_3G_3 + M_4G_4)}, \tag{5.5.13}$$

where M_1, M_2, M_3 and M_4 are functions of structure parameters and G_1, G_2, G_3 and G_4 are definite integrals given by:

$$M_1 = 1 + \frac{1}{D_{lf}^2}\frac{I_0\{\gamma a\}K_1\{\gamma a\}}{K_0\{\gamma a\}I_1\{\gamma a\}},$$

$$M_2 = \left[\left(\frac{\xi_{00}}{1-\xi_{00}}\right)^2 + \frac{1}{D_{lf}^2}\frac{I_0\{\gamma a\}K_1\{\gamma a\}}{K_0\{\gamma a\}I_1\{\gamma a\}}\left(\frac{\xi_{11}}{1-\xi_{11}}\right)^2\right]\varepsilon'_r,$$

$$M_3 = \left(\frac{I_0\{\gamma a\}}{K_0\{\gamma a\}}\right)^2\left(\frac{1}{1-\xi_{00}}\right)^2\varepsilon'_r + \frac{1}{D_{lf}^2}\frac{I_0\{\gamma a\}I_1\{\gamma a\}}{K_0\{\gamma a\}K_1\{\gamma a\}}\left(\frac{1}{1-\xi_{11}}\right)^2,$$

and $$M_4 = \frac{2I_0\{\gamma a\}}{K_0\{\gamma a\}}\left[\frac{\xi_{00}}{(1-\xi_{00})^2}\varepsilon'_r - \frac{1}{D_{lf}^2}\frac{\xi_{11}}{(1-\xi_{11})^2}\right] \; ;$$

$$G_1 = \frac{2}{a^2}\int_0^a I_1^2\{\gamma r\}\, r\, dr = I_0\{\gamma a\} I_1\{\gamma a\}\left[\left(\frac{I_1\{\gamma a\}}{I_0\{\gamma a\}} - \frac{I_0\{\gamma a\}}{I_1\{\gamma a\}}\right) + \frac{2}{\gamma a}\right] ,$$

$$G_2 = \frac{2}{a^2}\int_a^b I_1^2\{\gamma r\}\, r\, dr = \left(\frac{\gamma b}{\gamma a}\right)^2 I_0\{\gamma b\} I_1\{\gamma b\}\left[\left(\frac{I_1\{\gamma b\}}{I_0\{\gamma b\}} - \frac{I_0\{\gamma b\}}{I_1\{\gamma b\}}\right) + \frac{2}{\gamma b}\right]$$

$$- I_0\{\gamma a\} I_1\{\gamma a\}\left[\left(\frac{I_1\{\gamma a\}}{I_0\{\gamma a\}} - \frac{I_0\{\gamma a\}}{I_1\{\gamma a\}}\right) + \frac{2}{\gamma a}\right] ,$$

$$G_3 = \frac{2}{a^2}\int_a^b K_1^2\{\gamma r\}\, r\, dr = K_0\{\gamma a\} K_1\{\gamma a\}\left[\left(\frac{K_0\{\gamma a\}}{K_1\{\gamma a\}} - \frac{K_1\{\gamma a\}}{K_0\{\gamma a\}}\right) + \frac{2}{\gamma a}\right]$$

$$- \left(\frac{\gamma b}{\gamma a}\right)^2 K_0\{\gamma b\} K_1\{\gamma b\}\left[\left(\frac{K_0\{\gamma b\}}{K_1\{\gamma b\}} - \frac{K_1\{\gamma b\}}{K_0\{\gamma b\}}\right) + \frac{2}{\gamma b}\right] ,$$

$$G_4 = \frac{2}{a^2}\int_a^b I_1\{\gamma r\} K_1\{\gamma r\}\, r\, dr = \left(\frac{\gamma b}{\gamma a}\right)^2 (I_0\{\gamma b\} K_0\{\gamma b\} + I_1\{\gamma b\} K_1\{\gamma b\})$$

$$+ \frac{\gamma b}{(\gamma a)^2}(I_0\{\gamma b\} K_1\{\gamma b\} - K_0\{\gamma b\} I_1\{\gamma b\}) - [I_0\{\gamma a\} K_0\{\gamma a\} + I_1\{\gamma a\} K_1\{\gamma a\}$$

$$+ \frac{1}{\gamma a}(I_0\{\gamma a\} K_1\{\gamma a\} - I_1\{\gamma a\} K_0\{\gamma a\})] .$$

Appendix a5.5.1 *Relation between the phase velocity, group velocity and dispersion*

The phase velocity of a wave is the velocity with which a plane of constant phase progresses. For a time-harmonic wave travelling in the positive z-direction, the phase velocity has been found by setting the phase of the wave $\omega t - \beta z$ *= a constant (see section 4.5.1 and also appendix a4.5.1). This upon differentiation gives the phase velocity* v_p *as*

(a5.5.1) $$v_p \left(= \frac{dz}{dt}\right) = \frac{\omega}{\beta} .$$

The phase velocity, however, has a meaning as long as it has a reference to an 'infinite' monochromatic wave train. The information or intelligence can not be conveyed by such a single infinite wave train; this has to be done by a "group" of such wave trains at different frequencies which 'Fourier'-combine to form a finite wave train or a wave packet, say, in the form of a modulated wave, the modulation envelope containing the information. In passage through a "dispersive" medium or a propagating structure like the SWS under discussion, the shape of the wave packet would, in general, change since different frequency components would travel with different phase velocities in such a medium/structure (for instance, see Fig. 5.3.1). However, the distortion in the wave shape would be small and the wave packet would retain its identity, thus retaining the intelligence being conveyed, if the dispersion is low and the frequency spread is small compared to the center of the group. One may then attempt to conceive the velocity of such a wave packet or group of waves, called the group velocity. Since the energy of the wave carried in the wave packet, the group velocity is also the velocity with which the energy is transported.

The group velocity may be found as the velocity of the "group amplitude" of a wave packet. This may be illustrated with reference to a generic example of two 'Fourier' components of a group, one at $\omega - \Delta\omega$ *and the other at* $\omega + \Delta\omega$ *(separated in the frequency range by* 2ω *Let, at these frequencies, the corresponding phase propagation constants be* $\beta - \Delta\beta$ *and* $\beta + \Delta\beta$, *respectively. For the sake of simplicity, let the amplitudes of these components be the same, say,* E_0, *the latter being chosen as the magnitude of the electric field intensity of the wave. Thus the combination of the field quantities, E, may be written as:*

$$E = E_0 \exp j[(\omega - \Delta\omega)t - (\beta - \Delta\beta)z] + E_0 \exp j[(\omega + \Delta\omega)t - (\beta + \Delta\beta)z]$$

$$= E_0 \exp j(\omega t - \beta z)[\exp -j(\Delta\omega t - \Delta\beta z) + \exp j(\Delta\omega t - \Delta\beta z)] ,$$

which may be written, invoking the relation $e^{\pm j\psi} = \cos\psi \pm j \sin\psi$ *with* $\psi = \Delta\omega t - \Delta\beta z$, *as*

$$E = 2E_0 \cos(\Delta\omega t - \Delta\beta z) \exp j(\omega t - \beta z) . \tag{a5.5.2}$$

Clearly, the amplitude of the field function exhibits the "wave like" *variation with z and t. The plane of constant amplitude of the combination or group travels in the positive z-direction with the group velocity* v_g *which may be found by setting the argument of the cosine function of the amplitude part in (a5.5.2):*

$$\Delta\omega t - \Delta\beta z = \text{a constant} ,$$

which, upon differentiation, gives

$$\frac{dz}{dt} = \frac{\Delta\omega}{\Delta\beta} ,$$

which for small values of $\Delta\omega$ *in the limit would give the following expression for* v_g *:*

$$v_g \left(= \frac{dz}{dt}\right) = \frac{d\omega}{d\beta} . \tag{a5.5.3}$$

Now, re-writing (a5.5.1) as

$$\omega = v_p \beta ,$$

and differentiating it with respect to β, *one gets*

$$\frac{d\omega}{d\beta} = v_p + \beta \frac{dv_p}{d\beta} = v_p + \beta \frac{dv_p}{d\omega}\frac{d\omega}{d\beta} . \tag{a5.5.4}$$

Substituting $v_g = d\omega/d\beta$ *in (a5.5.4), and re-arranging terms, one then obtains*

$$v_g = \frac{v_p}{1 - \beta \dfrac{dv_p}{d\omega}} . \tag{a5.5.5}$$

5.6 Analysis in the Tape-Helix Model

The limitation of the sheath-helix model (section *5.1.1*) that it does not reveal in the analysis (sections *5.3 - 5.5*) the space harmonics arising from the axial periodicity of a helical structure is removed by using a physical model of the actual helix called the tape-helix model [4], [26].

In this model the helix is considered as a perfectly conducting tape of finite width and of infinitesimal thickness. The problem has been studied for two cases [26] — for a narrow-tape case, for which the tape width δ, say, is smaller than the helix pitch p as well as for a narrow-gap case, for which the axial gap between two consecutive turns δ', say, is smaller than helix pitch p (Fig. 5.6.1). In the narrow-tape case (δ/p small) the current is assumed to be on the tape surface and be only in the direction of the tape (winding direction), and the electric field intensity at the surface of the tape, at r = a, parallel to the winding direction and along center-line of the tape is set equal to zero. In the narrow-gap case (δ'/p small), the electric field intensity in the gap between consecutive turns at the surface flushing with the tape, r = a, is assumed to consist only of a component perpendicular to the tape edges, and the surface current density at this surface (r = a) in a direction perpendicular to the winding direction is set equal to zero . In the present section, however, we have restricted ourselves to the narrow-tape case and carried out the analysis to obtain the dispersion relation . We have initially considered the case of a helix in free-space (section *5.6.2*), and, subsequently, the loaded helix has also been dealt with (section *5.6.3*).

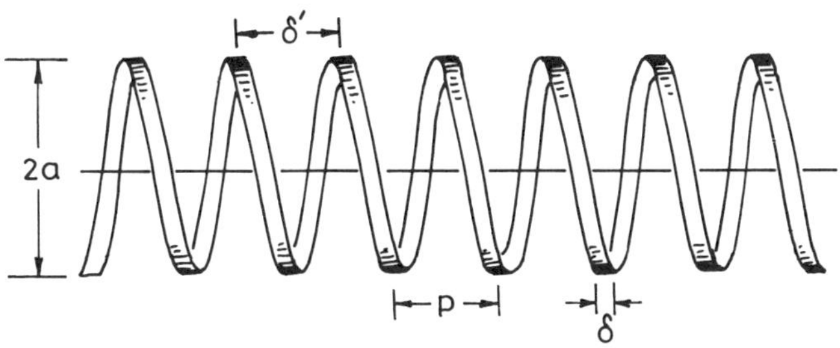

Fig. 5.6.1 Tape helix model showing the helix pitch p, diameter 2a, tape width δ, and the gap between helix turns, δ' .

5.6.1 Expressions for Fields Comprised of Space-Harmonics

For a non-azimuthally varying mode ($\partial/\partial\theta \neq 0$) the solution of wave equation (5.2.3) for a helix in free-space is modified from that given by (5.3.2) (see appendix *a5.2.2*) as follows (see problems 5.13 and 5.14):

$$E_{z1} = [A_{1m}\, I_m\{\gamma r\}]\exp(-j\beta z)\ \exp(jm\phi) \tag{5.6.1}$$

$$E_{\theta 1} = \left[\frac{-n\beta}{\gamma^2 r}A_{1m}I_m\{\gamma r\} - \frac{j\omega\mu_0}{\gamma}C_{1m}\, I'_n\{\gamma r\}\right]\exp(-j\beta z)\ \exp(jm\theta) \tag{5.6.2}$$

$$H_{z1} = [C_{1m}\, I_m\{\gamma r\}]\exp(-j\beta z)\ \exp(jm\theta) \tag{5.6.3}$$

$$H_{\theta 1} = \left[\frac{j\omega\varepsilon_0}{\gamma}A_{1m}I'_m\{\gamma r\} - \frac{n\beta}{\gamma^2 r}C_{1m}\, I_m\{\gamma r\}\right]\exp(-j\beta z)\ \exp(jm\theta) \tag{5.6.4}$$

$$E_{z2} = [B_{2m}\, K_m\{\gamma r\}]\exp(-j\beta z)\ \exp(jm\theta) \tag{5.6.5}$$

$$E_{\theta 2} = \left[\frac{-n\beta}{\gamma^2 r}B_{2m}K_m\{\gamma r\} - \frac{j\omega\mu_0}{\gamma}D_{2m}\, K'_m\{\gamma r\}\right]\exp(-j\beta z)\ \exp(jm\theta) \tag{5.6.6}$$

$$H_{z2} = [D_{2m}\, K_m\{\gamma r\}]\exp(-j\beta z)\ \exp(jm\phi) \tag{5.6.7}$$

and
$$H_{\theta 2} = \left[\frac{j\omega\varepsilon_0}{\gamma}B_{2m}K'_m\{\gamma r\} - \frac{n\beta}{\gamma^2 r}D_{2m}\, K_m\{\gamma r\}\right]\exp(-j\beta z)\ \exp(jm\phi)\ . \tag{5.6.8}$$

Here, the subscripts 1 and 2 refer to field quantities in the regions inside and outside the helix, respectively, both considered to be the free-space regions, the region outside extending upto infinity. I_m and K_m are the m*th*-order modified Bessel's functions of the first and second kinds, respectively, m being an integer. It may be seen that for m = 0 these expressions (5.6.1) - (5.6.8) pass on to those given by (5.3.2) (remembering $I'_0\{\gamma r\} = I_1\{\gamma r\}$ and $K'_0\{\gamma r\} = -K_1\{\gamma r\}$ (see appendix *a5.2.2*) and interpreting $A_{10} = A_1$, etc.).

In the tape-helix model care is taken to include the space-periodicity of the structure in terms of the helix coinciding with itself, if i) either it is translated through an axial period p (axial translation) ii) or it is translated through an arbitrary axial distance $z' < p$ and then rotated through an angle $2\pi z'/p$ (or, alternatively, it is rotated through an arbitrary angle $\theta' < 2\pi$ and then translated through an axial distance $p\theta'/2\pi$) (differential skew transformation).

The invariance to the axial translation through a period p suggests the following functional dependence f{z} of rf field quantities:

$$f\{z\} = \exp(-j\,\beta_n z) = \exp(-j\,\beta_0 z)\exp(-j\,2\pi\, n\, z/p)\ , \tag{5.6.9}$$

according to Floquet's theorem [4], which states that for a given mode of propagation at a given steady-state frequency the fields at one cross section differ from those one period (= p) away only by a complex constant, where β_n is the n*th* space harmonic axial propagation constant, n being an integer, and is given by:

$$\beta_n = \beta_0 + 2\pi n z/p\ . \tag{5.6.10}$$

This can be checked from (5.6.9) to see that

$$\begin{aligned} f\{z+p\} &= \exp[-j\,\beta_n(z+p)] = \exp[-j\,\beta_0(z+p)]\exp[-j\,2\pi n\,(z+p)/p] \\ &= [\exp(-j\,\beta_0 z)\exp(-j\,2\pi nz/p)]\,[\exp -j\,\beta_0 p]\,[\exp(-j\,2\pi n)] \\ &= \exp(-j\,\beta_0 p)\,f\{z\}\ . \end{aligned} \tag{5.6.11}$$

Further, choosing to express (5.6.9) in terms of a periodic function

$$f_p\{z\} = \exp(-j\,2\pi\,n\,z\,/\,p)\ ,$$

one may write

$$f\{z\} = f_p\{z\}\exp(-j\,\beta_0\,z)$$

giving, as can be seen with the help of (5.6.11),

$$f_p\{z+p\} = f_p\{z\},$$

as expected of a periodic function.

Combining the z-dependence of the form (5.6.9) with the angular dependence as $\exp(jm\theta)$ (see (5.6.1) - (5.6.8)), one may express rf field intensities, for instance, the axial component of electric field intensity inside the helix, with the help of (5.6.1), considering fields to be comprised of harmonic components, as follows:

$$\begin{aligned} E_{z1}\{r,\theta,z\} &= \sum_{m=-\infty}^{\infty}\sum_{n=-\infty}^{\infty} A_{1m,n}\, I_m\{\gamma_n r\}\exp(-j\beta_n z)\exp(jm\theta) \\ &= \exp(-j\beta_0 z)\sum_{m=-\infty}^{\infty}\sum_{n=-\infty}^{\infty} A_{1m,n}\, I_m\{\gamma_n r\}\exp(-j2\pi nz/p)\exp(jm\theta)\ , \end{aligned} \tag{5.6.12}$$

where $\gamma_n\ (=(\beta_n^2-k_0^2)^{1/2})$ is the *nth* space harmonic radial propagation constant, β_n being substituted for by (5.6.10). Also, the periodicity of the helix with respect to an axial translation $z' < p$ followed by a rotation $2\pi z'/p$ (skew transformation) demands that E_{z1} as given by (5.6.12) should satisfy the following invariance:

$$E_{z1}\{r,\phi+2\pi z'/p,z+z'\} = \exp(-j\beta_0 z')\,E_{z1}\{r,\phi,z\}$$

which on substituting (5.6.12) may be re-stated as:

$$E_{z1}\{r,\phi+2\pi z'/p,z+z'\} = \exp(-j\,\beta_0 z')\sum_{m=-\infty}^{\infty}\sum_{n=-\infty}^{\infty} A_{1m,n}\, I_m\{\gamma_n r\}\exp(-j2\pi nz/p)\exp(jm\theta)\ . \tag{5.6.13}$$

Now, with the help of (5.6.12) one can see

$$E_{z1}\{r, \phi+2\pi z'/p, z+z'\} = \exp(-j\beta_0 z')\,[\exp(-j\,\beta_0 z) \sum_{m=-\infty}^{\infty} \sum_{n=-\infty}^{\infty} A_{1m,n}\, I_m\{\gamma_n r\}\times$$

$$\times \exp(-j\,2\pi nz/p)\exp(\,jm\theta)]\,[\exp j2\pi(m-n)\,z'/p]$$

which satisfies the invariance to the skew transformation, given by (5.6.13), only if one puts

$$m = n\ . \tag{5.6.14}$$

Substitution of (5.6.14) into (5.6.12) then yields

$$E_{z1} = \exp(-j\beta_0 z) \sum_{n=-\infty}^{\infty} A_{1n}\, I_n\{\gamma_n r\} \exp(-j\,2\pi nz/p)\exp(\,jn\phi)]\ . \tag{5.6.15}$$

One may further express (5.6.15) as

$$E_{z1} = \sum_{n=-\infty}^{\infty} E_{z1,n}\ . \tag{5.6.16}$$

By comparing (5.6.16) with (5.6.15) $E_{z1,n}$ can be easily identified. The expression for $E_{z1,n}$ thus obtained would exactly be the same as the one obtained from (5.6.1), however, interpreting m, β and γ occurring therein as n, β_n and γ_n, respectively. Thus one gets:

$$E_{z1,n} = [A_{1n}\, I_n\{\gamma_n r\}]\exp(-j\,\beta_n z)\exp(\,jn\theta) \quad \text{(a)}$$

$$E_{\theta 1,n} = \left[\frac{-n\beta_n}{\gamma_n^2 r} A_{1n} I_n\{\gamma_n r\} - \frac{j\omega\mu_0}{\gamma_n} C_{1n}\, I'_n\{\gamma_n r\}\right]\exp(-j\beta_n z)\ \exp(jm\theta) \quad \text{(b)}$$

$$H_{z1,n} = [C_{1n}\, I_n\{\gamma_n r\}]\exp(-j\beta_n z)\exp(\,jn\theta) \quad \text{(c)}$$

$$H_{\theta 1,n} = \left[\frac{j\omega\varepsilon_0}{\gamma_n} A_{1n}\, I'_n\{\gamma_n r\} - \frac{n\beta_n}{\gamma_n^2 r} C_{1n}\, I_n\{\gamma_n r\}\right]\exp(-j\beta_n z)\exp(j\,n\theta) \quad \text{(d)}$$

$$E_{z2,n} = [B_{2n}\, K_n\{\gamma_n r\}]\exp(-j\,\beta_n z)\exp(\,jn\theta) \quad \text{(e)}$$

$$E_{\theta 2,n} = \left[\frac{-n\beta_n}{\gamma_n^2 r} B_{2n} K_n\{\gamma_n r\} - \frac{j\omega\mu_0}{\gamma_n} D_{2n} K'_n\{\gamma_n r\}\right]\exp(-j\beta_n z)\exp(jn\theta) \quad \text{(f)}$$

$$H_{z2,n} = [D_{2n}\, K_n\{\gamma_n r\}]\exp(-j\,\beta_n z)\exp(\,jn\theta) \quad \text{(g)}$$

and
$$H_{\theta 2,n} = \left[\frac{j\omega\varepsilon_0}{\gamma_n} B_{2n}\, K'_n\{\gamma_n r\} - \frac{n\beta_n}{\gamma_n^2 r} D_{2n}\, K_n\{\gamma_n r\}\right]\exp(-j\beta_n z)\exp(j\,n\theta)\,. \quad \text{(h)}$$

$$(5.6.17)$$

Now that the expressions for the individual space-harmonic components are identified as (5.6.17) one may then use the latter to obtain the expressions for the desired field intensities in the summation form as in (5.6.16).

5.6.2 Dispersion Relation for a Helix in Free Space in the Tape-Model

It is of interest to find the component of the electric field intensity inside the helix parallel to the winding direction, $E_{1,\|}$, as:

$$E_{1,\|} = \sum_{n=-\infty}^{\infty} E_{1,\| n} = \sum_{n=-\infty}^{\infty} E_{\phi 1,n} \cos\psi + E_{z1,n} \sin\psi ,$$

which, with the help of (5.6.17a) and (5.6.17(b)) through (5.6.10), may be found at the tape surface, r = a, as:

$$E_{1,\|}\{a\} = \sum_{n=-\infty}^{\infty}\left[\left(\frac{-n\beta_n}{\gamma_n^2 a} I_n\{\gamma_n a\}\cos\psi + I_n\{\gamma_n a\}\sin\psi\right)A_{1n} - \left(\frac{j\omega\mu_0}{\gamma_n} I'_n\{\gamma_n a\}\cos\psi\right)C_{1n}\right]\times$$

$$\times \exp(-j\beta_0 z)\exp -jn((2\pi z/p)-\theta) \;. \tag{5.6.18}$$

The constants A_{1n} and C_{1n} occurring in (5.6.18) may be found in terms of the axial and the azimuthal components of surface current densities J_{zn} and $J_{\theta n}$, respectively. In the narrow-tape analysis considered here the tape-surface current is assumed to exist only parallel to the winding direction, as mentioned before, giving

$$J_{zn} = J_{\| n} \sin\psi$$

and

$$J_{\theta n} = J_{\| n} \cos\psi \;. \tag{5.6.19}$$

One may now form four simultaneous equations in four field constants A_{1n}, C_{1n}, B_{2n} and D_{2n} by substituting field expressions (5.6.17) into the following four boundary conditions (see (5.2.34) - (5.2.37)):

$$E_{z1,n}\{a\} = E_{z2,n}\{a\}, \qquad E_{\theta 1,n}\{a\} = E_{\theta 2,n}\{a\} ,$$

$$H_{\theta 2,n}\{a\} - H_{\theta 1,n}\{a\} = J_{zn} \text{ and } H_{z1,n}\{a\} - H_{z2,n}\{a\} = J_{\theta n} \;. \tag{5.6.20}$$

The simultaneous equations so obtained may be solved for the desired field constants as:

$$A_{1n} = \frac{\gamma_n K_n\{\gamma_n a\}}{j\omega\varepsilon_0}\,\frac{\hat{J}_{zn} - (n\beta_n/\gamma_n^2 a)\hat{J}_{\theta n}}{I_n\{\gamma_n a\}K'_n\{\gamma_n a\} - K_n\{\gamma_n a\}I'_n\{\gamma_n a\}}$$

and
$$C_{1n} = \frac{K'_n\{\gamma_n a\} I_n\{\gamma_n a\}\hat{J}_{\theta n}}{I_n\{\gamma_n a\}K'_n\{\gamma_n a\} - K_n\{\gamma_n a\}I'_n\{\gamma_n a\}}, \qquad (5.6.21)$$

where $\hat{J}_{zn}$ and $\hat{J}_{\theta n}$ are the amplitudes of the surface current densities J_{zn} and $J_{\theta n}$, respectively, which must have the same rf dependence as the field components, that is, as in (5.6.17). In fact, one may see that the surface current density is the discontinuity of the magnetic field intensities at the surface (see (5.6.20)). Thus one may express J_{zn} and $J_{\theta n}$ as:

$$J_{zn} = \hat{J}_{zn} \exp(-j\beta_n z)\exp(jn\theta) = \hat{J}_{zn}\exp(-j\beta_0 z)\exp(-j2\pi nz/p)\exp(jn\theta)$$

and
$$J_{\theta n} = \hat{J}_{\theta n} \exp(-j\beta_n z)\exp(jn\theta) = \hat{J}_{\theta n}\exp(-j\beta_0 z)\exp(-j2\pi nz/p)\exp(jn\theta),$$

remembering expression (5.6.10) for β_n. Consequently one may also write (see (5.6.19))

$$J_{\parallel n} = \hat{J}_{\parallel n} \exp(-j\beta_n z)\exp(jn\theta) = \hat{J}_{\parallel n}\exp(-j\beta_0 z)\exp(-j2\pi nz/p)\exp(jn\theta). \qquad (5.6.22)$$

Now, one may choose to express the constants A_{1n} and C_{1n} in terms of $\hat{J}_{\parallel n}$ from (5.6.21) with the help of (5.6.19), and using the relation $I_n\{\gamma_n a\}K'_n\{\gamma_n a\} - K_n\{\gamma_n a\}I'_n\{\gamma_n a\} = 1/\gamma_n a$, as:

$$A_{1n} = -\frac{\gamma_n^2 a K_n\{\gamma_n a\}}{j\omega\varepsilon_0}[\hat{J}_{\parallel n}\sin\psi - (n\beta_n/\gamma_n^2 a)\hat{J}_{\parallel n}\cos\psi]$$

and
$$C_{1n} = -\gamma_n a K'_n\{\gamma_n a\}\hat{J}_{\parallel n}\cos\psi. \qquad (5.6.23)$$

Then plugging (5.6.23) into (5.6.18) and after a little algebra one obtains

$$E_{1,\parallel}\{a\} = \sum_{n=-\infty}^{\infty}\frac{j\sin^2\psi\,\hat{J}_{\parallel n}}{\omega\varepsilon_0 a}\Big[\Big(\frac{n\beta_n a}{\gamma_n a}\cot\psi - \gamma_n a\Big)^2 I_n\{\gamma_n a\}K_n\{\gamma_n a\}$$
$$+ k_0^2 a^2\cot^2\psi\, I'_n\{\gamma_n a\}K'_n\{\gamma_n a\}\Big]\exp(-j\beta_0 z)\exp -jn((2\pi z/p)-\theta). \qquad (5.6.24)$$

Considerable simplification is achieved in finding $\hat{J}_{\parallel n}$ occurring in (5.6.24) if it is assumed that the amplitude of the tape-surface current density $J_{\parallel}$, which has been assumed to be comprised of only the parallel component, is constant at a value J, say, over the tape width δ, though its phase does vary in the direction of the winding of the tape according to the phase factor $\exp(j\beta_0 z)$, where z corresponds to a point moving along the centre-line of the tape [4]. The value of z, corresponding to a point on the centre line of the tape characterised by an azimuthal angle θ, is given $z = p\theta/2\pi$. Thus over a period of the structure, that is, the helix pitch p, one may write:

$$J_{\|} = J\exp(-j\beta_0 z)$$

$$= J\exp(-j\beta_0\, p\theta/2\pi)\,,\quad p\theta/2\pi - \delta/2 < z < p\theta/2\pi + \delta/2 \qquad \text{(a)}$$

and $$J_{\|} = 0\,,\qquad p\theta/2\pi + \delta/2 < z < (p\theta/2\pi - \delta/2) + p\ . \qquad \text{(b)} \qquad (5.6.25)$$

In (5.6.25(a)) and (5.6.25(b)), the ranges of z cover the tape width, δ, and the gap width, $p - \delta$, respectively. These two ranges put together cover a single axial period p. $J_{\|}$ assumed as (5.6.25) may be used to find $\hat{J}_{\|n}$ occurring in (5.6.24). For this purpose, $J_{\|}$ is first written as the summation of its components $J_{\|n}$ as:

$$J_{\|} = \sum_{n=-\infty}^{\infty} J_{\|n}$$

which with the help of (5.6.22) may be written as

$$J_{\|} = \exp(-j\beta_0 z) \sum_{n=-\infty}^{\infty} \hat{J}_{\|n} \exp - jn(2\pi z/p - \theta)\ . \qquad (5.6.26)$$

$\hat{J}_{\|n}$ is then formed using (5.6.25) and (5.2.26) by the usual method of finding Fourier components[4] as (see appendix *a5.6.1* for the deduction):

$$\hat{J}_{\|n} = (J)\left(\frac{\sin(\beta_n\delta/2)}{\beta_n\delta/2}\right)\left(\frac{\delta}{p}\right)\ . \qquad (5.6.27)$$

In the narrow-tape analysis, followed here to find out the dispersion relation, what one has to do is to substitute (5.6.27) into (5.6.24) and evaluate the electric field intensity at the tape surface (r = a) parallel to the tape (winding direction), $E_{1,\|}\{a\}$, at a point on the centre-line of the tape , $z = p\theta/2\pi$, and finally put this equal to zero (see the discussion preceding the beginning of section *5.6.1*). The procedure then gives in a straightforward manner:

$$E_{1,\|}\{a\}\ (z = p\theta/2\pi) = \sum_{n=-\infty}^{\infty} \frac{j\sin^2\psi}{\omega\varepsilon_0 a}(J)\left(\frac{\delta}{p}\right)\left(\frac{\sin(\beta_n\delta/2)}{\beta_n\delta/2}\right)\left[\left(\frac{n\beta_n}{\gamma_n}\cot\psi - \gamma_n a\right)^2 \times\right.$$

$$\left.\times\ I_n\{\gamma_n a\}K_n\{\gamma_n a\} + k_0^2a^2\cot^2\psi\ I_n'\{\gamma_n a\}K_n'\{\gamma_n a\}\right]\exp(-j\beta_0 z)\exp - jn((2\pi z/p) - \theta) = 0$$

which may be simplified to:

$$\frac{j\sin^2\psi}{\omega\varepsilon_0 a}(J)\left(\frac{\delta}{p}\right)\exp(-j\beta_0 p\theta/2\pi)\sum_{n=-\infty}^{\infty}\left[\left(\frac{n\beta_n}{\gamma_n}\cot\psi - \gamma_n a\right)^2 I_n\{\gamma_n a\}K_n\{\gamma_n a\}\right.$$

$$\left. + k_0^2a^2\cot^2\psi\ I_n'\{\gamma_n a\}K_n'\{\gamma_n a\}\right]\left(\frac{\sin(\beta_n\delta/2)}{\beta_n\delta/2}\right) = 0\ . \qquad (5.6.28)$$

Since the quantity outside the summation (5.6.28) can not be zero, one must put the quantity under the summation as zero and express it using the relation $\gamma_n = (\beta_n^2 - k_0^2)^{1/2}$ (as defined following (5.6.12)) as follows:

$$\sum_{n=-\infty}^{\infty} [(\frac{n\beta_n}{(\beta_n^2 - k_0^2)^{1/2}} \cot\psi - (\beta_n^2 - k_0^2)^{1/2} a)^2 \, I_n\{(\beta_n^2 - k_0^2)^{1/2} a\} K_n\{(\beta_n^2 - k_0^2)^{1/2} a\}$$

$$+ k_0^2 a^2 \cot^2\psi \, I'_n\{(\beta_n^2 - k_0^2)^{1/2} a\} K'_n\{(\beta_n^2 - k_0^2)^{1/2} a\}] \left(\frac{\sin(\beta_n \delta/2)}{\beta_n \delta/2}\right) = 0 \,. \quad (5.6.29)$$

Thus in (5.6.29) one finds the dispersion relation in the tape-model under the narrow-tape approximation considered here.

Further, by expanding the square in (5.6.29) and using the relation (5.6.10), which with the help of (5.1.2) may also be put as

$$\beta_n a = \beta_0 a + n \cot\psi \,, \quad (5.6.30)$$

one may obtain by a simple algebra the following alternative form of the dispersion relation:

$$\sum_{n=-\infty}^{\infty} [\left(\beta_0^2 a^2 - k_0^2 a^2 + \frac{k_0^2 a^2 n^2 \cot^2\psi}{(\beta_o a + n \cot\psi)^2 - k_0^2 a^2}\right) I_n\{(\beta_n^2 - k_0^2)^{1/2} a\} K_n\{(\beta_n^2 - k_0^2)^{1/2} a\}$$

$$+ k_0^2 a^2 \cot^2\psi \, I'_n\{(\beta_n^2 - k_0^2)^{1/2} a\} K'_n\{(\beta_n^2 - k_0^2)^{1/2} a\}] \left(\frac{\sin(\beta_n \delta/2)}{\beta_n \delta/2}\right) = 0 \,. \quad (5.6.31)$$

It may be mentioned that (5.6.31) is identical with the narrow-tape dispersion relation (12) of [26] where it is also mentioned that under the narrow-gap approximation the relation (5.6.31) is modified only in that the *nth* term is multiplied by a factor

$$1 / (\beta_n^2 - k_0^2) a^2 \, I_n\{(\beta_n^2 - k_0^2)^{1/2} a\} K_n\{(\beta_n^2 - k_0^2)^{1/2} a\} I'_n\{(\beta_n^2 - k_0^2)^{1/2} a\} K'_n\{(\beta_n^2 - k_0^2)^{1/2} a\},$$

and δ' (gap width) replaces δ (tape width).

<u>*Example 5.6.1*</u> Verify that the zero*th* term (n = 0) in the summation form of tape-helix dispersion relation yields the dispersion relation of the sheath-helix model.

Collecting the terms corresponding to n = 0 in either (5.6.29) or (5.6.31) one obtains since $\sin(\beta_0\delta/2)/(\beta_0\delta/2) \neq 0$:

$$(\beta_0^2 a^2 - k_0^2 a^2) \, I_0\{\gamma_0 a\} K_0\{\gamma_0 a\} + k_0^2 a^2 \cot^2\psi \, I'_0\{\gamma_0 a\} K'_0\{\gamma_0 a\} = 0 \,. \quad (5.6.32)$$

where $\gamma_0 = (\beta_0^2 - k_0^2)^{1/2}$, as define following (5.6.12). Using the relation $I'_0\{\gamma_0 a\} = I_1\{\gamma_0 a\}$ and $K'_0\{\gamma_0 a\} = -K_1\{\gamma_0 a\}$, one may then express (5.6.32) as

$$\frac{k_0 \cot\psi}{\gamma_0} = \left(\frac{I_0\{\gamma_0 a\}K_0\{\gamma_0 a\}}{I_1\{\gamma_0 a\}K_1\{\gamma_0 a\}}\right)^{1/2} . \tag{5.6.33}$$

It may be seen by interpreting γ_0 ($= (\beta_0^2 - k_0^2)^{1/2}$) as γ ($= (\beta^2 - k_0^2)^{1/2}$) that (5.6.33) is identical with (5.3.7) — the dispersion relation obtained by the sheath-helix model.

The dispersion characteristics may be represented in the form of a (ω-β) diagram for the desired space-harmonic component $\beta = \beta_g$, say. However, for this purpose it will be useful to express (5.6.29) separating the *gth* term from the remaining terms in the summation. Here, g can take the positive as well as the negative integral and zero values. Thus the explicit form of the dispersion relation which would be suitable to obtain the (ω-β_g) plot would be:

$$\left(\frac{g\beta_g}{(\beta_g^2 - k_0^2)^{1/2}}\cot\psi - (\beta_g^2 - k_0^2)^{1/2} a\right)^2 I_g\{(\beta_g^2 - k_0^2)^{1/2} a\} K_g\{(\beta_g^2 - k_0^2)^{1/2} a\}$$

$$+\ k_0^2 a^2 \cot^2\psi\ I_g'\{(\beta_g^2 - k_0^2)^{1/2} a\}\ K_g'\{(\beta_g^2 - k_0^2)^{1/2} a\} \left(\frac{\sin(\beta_g \delta/2)}{\beta_g \delta/2}\right)$$

$$+ \sum_{h(\neq 0)=-\infty}^{\infty} \left[\left(\frac{(g+h)\beta_{g+h}}{(\beta_{g+h}^2 - k_0^2)^{1/2}}\cot\psi - (\beta_{g+h}^2 - k_0^2)^{1/2} a\right)^2 I_{g+h}\{(\beta_{g+h}^2 - k_0^2)^{1/2} a\} \times\right.$$

$$\left. K_{g+h}\{(\beta_{g+h}^2 - k_0^2)^{1/2} a\} + k_0^2 a^2 \cot^2\psi\ I_{g+h}'\{(\beta_{g+h}^2 - k_0^2)^{1/2} a\} K_{g+h}'\{(\beta_{g+h}^2 - k_0^2)^{1/2} a\}\right] \times$$

$$\times \left(\frac{\sin(\beta_{g+h}\delta/2)}{\beta_{g+h}\delta/2}\right) = 0 , \tag{5.6.34}$$

where β_{g+h} is given, with the help of (5.6.10), by:

$$\beta_{g+h} = \beta_g + 2\pi h/p , \tag{5.6.35}$$

h ($\neq$ 0) being an integer that can take only the positive or negative values. For instance, one may get, with the help of (5.6.35), $\beta_{-2} = \beta_0 - 4\pi/p$, $\beta_{-1} = \beta_0 - 2\pi/p$, $\beta_1 = \beta_0 + 2\pi/p$, $\beta_2 = \beta_0 + 4\pi/p$, etc. in terms of β_0 ; $\beta_{-2} = \beta_1 - 6\pi/p$, $\beta_{-1} = \beta_1 - 4\pi/p$, $\beta_0 = \beta_1 - 2\pi/p$, $\beta_2 = \beta_1 + 2\pi/p$, etc. in terms of β_1 ; $\beta_{-2} = \beta_{-1} - 2\pi/p$, $\beta_0 = \beta_{-1} + 2\pi/p$, $\beta_1 = \beta_{-1} + 4\pi/p$, $\beta_2 = \beta_{-1} + 6\pi/p$, etc. in terms of β_{-1} ; and so on.

An interesting feature expected to be observed in the dispersion characteristics plotted with the help of the dispersion relation deduced in this section is the appearance of regions where the solutions are forbidden (section 5.7). In general, those solution of the dispersion relation are permitted for which the arguments of the modified Bessel functions $\gamma_n a$ (= $(\beta_n^2 - k_0^2)^{1/2}$ a) are real corresponding to $k_0 < \pm\beta_n$ and, consequently, the boundaries separating the permitted regions from the forbidden ones corresponding to $k_0 > \pm\beta_n$ are given by $k_0 = \pm\beta_n$ which with the help of (5.6.30) may be put as

$$k_0a/\cot\psi \;=\; \pm(\beta_0a/\cot\psi + n)\,. \tag{5.6.36}$$

These regions are depicted in Fig. 5.6.2 in which the hatched region shows the forbidden region, and the boundaries between the allowed (unhatched) and forbidden regions are shown as straight lines given by (5.6.36), of positive and negative unity slopes (±1), for different values of n (= 0, 1, -1, 2, -2, etc.). The dispersion plot using the tape-model is presented in section *5.7*.

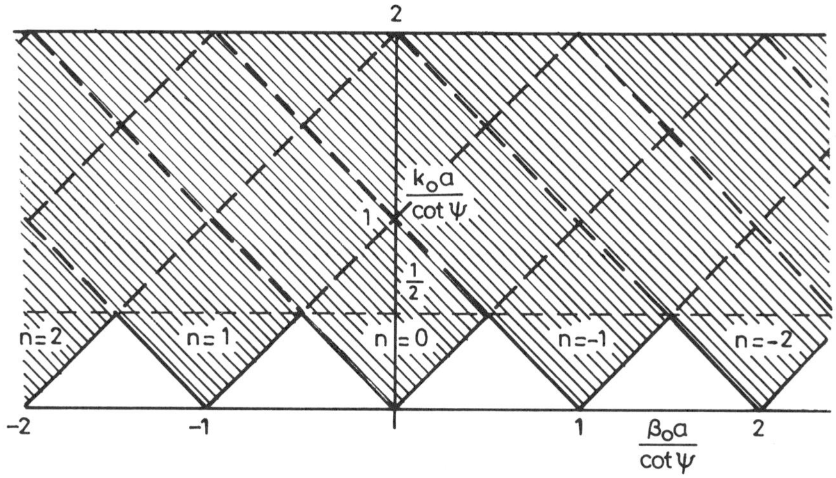

Fig. 5.6.2 Forbidden and allowed regions in the ω-β plot shown as hatched and unhatched, respectively.

5.6.3 Dispersion Relation for a Loaded Helix in the Tape-Model

Let us now take up for analysis in the tape-helix model a loaded helix that has been studied by the sheath-helix model in section *5.3.2* as well as by the equivalent circuit analysis in section *5.4*. But instead of directly putting the loaded helix to the tape-model analysis which becomes quite tedious, what we intend to do is to marry the results of the tape-model analysis of a helix in free-space with those of the sheath-helix model analysis of a loaded helix. The approach has been tested to be valid and found as extremely useful even for a helix loaded in a complex manner [27].

One may choose to write the dispersion relation of a helix in free space in the tape model, (5.6.29), as

$$\sum_{n=-\infty}^{\infty} (M_n + N_n)\left(\frac{\sin(\beta_n\delta/2)}{\beta_n\delta/2}\right) = 0 , \tag{5.6.37}$$

where

$$M_n = \left(\frac{n\beta_n}{\gamma_n}\cot\psi - \gamma_n a\right)^2 I_n\{\gamma_n a\}K_n\{\gamma_n a\} \tag{5.6.38}$$

and

$$N_n = k_0^2 a^2 \cot^2\psi\ I'_n\{\gamma_n a\}K'_n\{\gamma_n a\}, \tag{5.6.39}$$

remembering the relation $\gamma_n = (\beta_n^2 - k_0^2)^{1/2}$ as defined following (5.6.12).

Squaring both sides of (5.3.7), more precisely of (5.6.33) in which γ and β have been interpreted as γ_0 and β_0, respectively; re-arranging terms; and making use of the relation $I'_0\{\gamma a\}= I_1\{\gamma a\}$ and $K'_0\{\gamma a\}= - K_1\{\gamma a\}$, the dispersion relation for a helix in free space in the sheath-helix model may be written as

$$\gamma_0^2 a^2 I_0\{\gamma a\}K_0\{\gamma a\} + k_0^2 a^2 \cot^2\psi\ I'_0\{\gamma a\}K'_0\{\gamma a\} = 0 . \tag{5.6.40}$$

Further in view of (5.6.38) and (5.6.39), one may express (5.6.40) as

$$M_0 + N_0 = 0 , \tag{5.6.41}$$

(helix in free space; sheath-helix model)

where M_0 and N_0 can be had from (5.6.38) and (5.6.39), respectively, taking n = 0, as

$$M_0 = \gamma_0^2 a^2 I_0\{\gamma_0 a\}K_0\{\gamma_0 a\} \tag{5.6.42}$$

and

$$N_0 = k_0^2 a^2 \cot^2\psi\ I'_0\{\gamma_0 a\}K'_0\{\gamma_0 a\} . \tag{5.6.43}$$

For a loaded helix, the dispersion relation in the sheath-helix model is given by (5.4.34). The expression for α_C and α_L which enter (5.4.34) through (5.4.35) are given by (5.4.18) and (5.4.32), respectively. Then using the relation $I'_0\{\gamma a\}= I_1\{\gamma a\}$ and $K'_0\{\gamma a\} = - K_1\{\gamma a\}$, interpreting γ as γ_0 (see Example 5.6.1), and by a simple algebra, one may express (5.4.34) as:

$$M_0\, \alpha_C^{-1} + N_0\, \alpha_L = 0 . \tag{5.6.44}$$

It may be noted that that the expression for α_C given by (5.4.18) involves $I_0\{\gamma a\}$and $K_0\{\gamma a\}$ and that the expression for α_L given by (5.4.32) involves $I_1\{\gamma a\}$ and $K_1\{\gamma a\} = - K'_0\{\gamma a\}$. Thus the expressions for α_C and α_L involve the zero*th* order modified Bessel functions and their derivatives. So choosing to add the subscript 0 to represent the order of Bessel functions and their derivatives, one may re-write (5.4.18) and (5.4.32) interpreting γ as γ_0 therein as follows:

$$\alpha_{C_0} = \left(1 - \frac{I_0\{\gamma_0 a\}K_0\{\gamma_0 b\}}{K_0\{\gamma_0 a\}I_0\{\gamma_0 b\}}\right)^{-1} \times$$

$$\times\left[1 - (\varepsilon_r' - 1)\gamma_0 a\, I_0\{\gamma_0 a\}\, K_0'\{\gamma_0 a\}\left(1 - \frac{I_0'\{\gamma_0 a\}K_0\{\gamma_0 b\}}{K_0'\{\gamma_0 a\}I_0\{\gamma_0 b\}}\right)\right] \qquad (5.6.45)$$

and
$$\alpha_{L_0} = 1 - \frac{I_0'\{\gamma_0 a\}\, K_0'\{\gamma_0 b\}}{K_0'\{\gamma_0 a\}\, I_0'\{\gamma_0 b\}} \,. \qquad (5.6.46)$$

In terms of the new nomenclature α_{Co} and α_{Lo}, one may then express (5.6.44) as

$$M_0\, \alpha_{Co}^{-1} + N_0\, \alpha_{Lo} = 0 \,. \qquad (5.6.47)$$

(loaded helix; sheath-helix model)

It may be noted that, as discussed following (5.4.18) and (5.4.32), respectively, the values of α_{Co} and α_{Lo} for a helix in free space are each equal to unity. This enables one to put (5.6.37) in a form similar to (5.6.47), as:

$$\sum_{n=-\infty}^{\infty} (M_n\, \alpha_{Co}^{-1}{}_{\text{free-space}} + N_n\, \alpha_{Lo\ \text{free-space}})\left(\frac{\sin(\beta_n\delta/2)}{\beta_n\delta/2}\right) = 0 \,. \qquad (5.6.48)$$

(helix in free space; tape-helix model)

Let us then make an intuitive proposition of combining the dispersion relation of a loaded helix in the sheath-helix model given by (5.6.47) with that of a helix in free space in the tape-helix model, given by (5.6.48), to obtain the following expression for the dispersion relation of a loaded helix in the tape-helix model:

$$\sum_{n=-\infty}^{\infty} (M_n\, \alpha_{Cn}^{-1} + N_n\, \alpha_{Ln})\left(\frac{\sin(\beta_n\delta/2)}{\beta_n\delta/2}\right) = 0 \,, \qquad (5.6.49)$$

(loaded helix; tape-helix model)

where M_n and N_n are given by (5.6.38) and (5.6.39), respectively, and the expression for α_{Cn} and α_{Ln} are obtained by replacing 0 by n in (5.6.45) and (5.6.46) as:

$$\alpha_{C_n} = \left(1 - \frac{I_n\{\gamma_n a\}K_n\{\gamma_n b\}}{K_n\{\gamma_n a\}I_n\{\gamma_n b\}}\right)^{-1} \times$$

$$\times\left[1 - (\varepsilon_r' - 1)\gamma_n a\, I_n\{\gamma_n a\}\, K_n'\{\gamma_n a\}\left(1 - \frac{I_n'\{\gamma_n a\}\, K_n\{\gamma_n b\}}{K_n'\{\gamma_n a\}\, I_n\{\gamma_n b\}}\right)\right] \qquad (5.6.50)$$

and
$$\alpha_{L_n} = 1 - \frac{I_n'\{\gamma_n a\}\, K_n'\{\gamma_n b\}}{K_n'\{\gamma_n a\}\, I_n'\{\gamma_n b\}} \,. \qquad (5.6.51)$$

It may be noted that the loaded helix has not been analysed here in the tape-helix model from first principles, as the helix in free space was done in section *5.6.2*, though such a loaded helix was analysed from first principles in the sheath-helix model in section *5.3.2* (by field analysis) and in section *5.4* (by equivalent-circuit analysis). But, interestingly, had the loaded helix been analysed in the tape-helix model from first principles, which would become tedious though, one would obtain exactly the same dispersion relation as (5.6.49) which has to be read with the help of (5.6.50) and (5.6.51)[27]. The simple method of tape-helix model suggested here is rather general and is applicable to the analysis of a helix loaded in a complex manner for which the results in the simpler sheath-helix model are known apriori. The dispersion characteristics of a helical structure obtained in the tape-model have been discussed in section *5.7*.

Appendix a5.6.1 *Amplitude of the space-harmonic tape current density*

Let us multiply (5.6.26) by $\exp(j\beta_0 z)\exp jm(2\pi z/p - \theta)$ *and integrate it over an axial period p, extending between* $z = p\theta/2\pi - \delta/2$*, referring to one edge of a turn, and* $z = (p\theta/2\pi - \delta/2) + p$*, referring to the corresponding edge of the next turn. The axial period p may be seen to be comprised of the tape width extending between* $z = p\theta/2\pi - \delta/2$ *and* $z = p\theta/2\pi + \delta/2$ *and the gap width extending between* $z = p\theta/2\pi + \delta/2$ *and* $(p\theta/2\pi - \delta/2) + p$ *(see (5.6.25)). Thus we get*

$$\text{(a5.6.1)}\quad \int_{z=p\theta/2\pi-\delta/2}^{p\theta/2\pi+\delta/2} J_{\|}\exp j\beta_0 z\,\exp jm(2\pi z/p-\theta)\,dz + \int_{z=p\theta/2\pi+\delta/2}^{(p\theta/2\pi-\delta/2)+p} J_{\|}\exp j\beta_0 z\,\exp jm(2\pi z/p-\theta)dz$$

$$= \int_{z=p\theta/2\pi-\delta/2}^{(p\theta/2\pi-\delta/2)+p}\sum_{n=-\infty}^{\infty}\hat{J}_{\|n}\exp j(m-n)(2\pi z/p-\theta)\,dz\,.$$

In the left hand side of (a5.6.1), $J_{\|}$ *has to be taken as* $J\exp(-j\beta_0\, p\theta/2\pi)$*, according to (5.6.25(a)), and 0, according to (5.6.25(b)). In the right hand side of (a5.6.1), the integration is carried out isolating the term corresponding to* $n = m$ *from those corresponding to* $n \neq m$*. Thus (a5.6.1) simplifies to:*

$$\text{(a5.6.2)}\quad J\exp[-j(\beta_0\,p/2\pi+m)\theta]\int_{z=p\theta/2\pi-\delta/2}^{p\theta/2\pi+\delta/2}\exp j(\beta_0+2\pi m/p)z\,dz$$

$$= \int_{z=p\theta/2\pi-\delta/2\ (n=m)}^{(p\theta/2\pi-\delta/2)+p}\hat{J}_{\|m}\,dz + \int_{z=p\theta/2\pi-\delta/2}^{(p\theta/2\pi-\delta/2)+p}\sum_{n\neq m=-\infty}^{\infty}\hat{J}_{\|n}\exp[-j(m-n)\theta]\exp j(m-n)(2\pi z/p)dz.$$

Using the relation $\beta_m = \beta_0 + 2\pi m/p$ *(see (5.6.10)), and hence also* $\beta_m\,p/2\pi = \beta_0\,p/2\pi + m$*, one may express (a5.6.2) as*

$$(a5.6.3) \quad J \exp(-j\beta_m p\theta/2\pi) \int_{z=p\theta/2\pi-\delta/2}^{p\theta/2\pi+\delta/2} \exp(j\beta_m z)\, dz$$

$$= \hat{J}_{\|m} \int_{z=p\theta/2\pi-\delta/2}^{(p\theta/2\pi-\delta/2)+p} dz + \sum_{n\neq m=-\infty}^{\infty} \hat{J}_{\|n} \int_{z=p\theta/2\pi-\delta/2}^{(p\theta/2\pi-\delta/2)+p} \exp[-j(m-n)\theta] \exp j(m-n)\, z/p\, dz\,.$$

Carrying out the straightforward integrations (a5.6.3) becomes

$$J[\exp(-j\beta_m p\theta/2\pi)\exp(j\beta_m p\theta/2\pi)]\left(\frac{\exp(j\beta_m\delta/2)-\exp(-j\beta_m\delta/2)}{j\beta_m}\right)$$

$$= \hat{J}_{\|m}\, p + \sum_{n\neq m=-\infty}^{\infty} \hat{J}_{\|n} \frac{\exp -j(m-n)\theta}{j(m-n)2\pi/p} [\exp\{j(m-n)(2\pi/p)(p\theta/2\pi-\delta/2)\}][\exp j(m-n)2\pi - 1]$$

$$= \hat{J}_{\|m}\, p \qquad [\because \exp j(m-n)2\pi - 1 = 0 \text{ for } n \neq m]$$

which further simplifies to:

$$(a5.6.4) \qquad (J/\beta_m)\ 2j \sin(\beta_m \delta/2) = \hat{J}_{\|m}\, p \ [\because \exp(j\beta_m\delta/2) - \exp(-j\beta_m\delta/2) = 2j \sin \beta_m \delta/2]\,.$$

By re-arrangement of terms, one may then express (a5.6.4) as

$$\hat{J}_{\|m} = (J)\left(\frac{\sin(\beta_m\delta/2)}{\beta_m\delta/2}\right)\left(\frac{\delta}{p}\right)$$

which for the nth harmonic may be read by replacing m by n as:

$$\hat{J}_{\|n} = (J)\left(\frac{\sin(\beta_n\delta/2)}{\beta_n\delta/2}\right)\left(\frac{\delta}{p}\right)\,,$$

which is identical with (5.6.27).

5.7 Structure Characteristics

We have already discussed (section *5.3.1*) the method of plotting the dispersion characteristics of an unloaded helix using the dispersion relation obtained in the sheath-helix method. In this section let us, following the same method, study the dispersion characteristics of a loaded helix using the dispersion relation obtained first in the sheath-helix model (sections *5.3* and *5.4*) and then in the tape-helix model (section *5.7*). We also intend to examine the impedance expressions (section *5.5*) to study the effects of helix loading.

The relevant helix-support parameters may be combined in a single parameter, namely, ε_r' — the effective relative permittivity of the continuous dielectric tube into which the discrete supports can be azimuthally smoothed out . The approach is accurate for wedge supports for which ε_r' is given by (5.1.11), in terms of the relative permittivity of the supports ε_r, the number of supports N, and the wedge angle ϕ. For supports deviating from wedge geometry, the method is, however, approximate (see, for instance, problem 5.6), and a reasonable approach would be to azimuthally smooth-out the supports into a number of dielectric tubes of appropriate relative permittivity values (section *5.8*). Besides the support parameters, which are accounted for by ε_r', another important parameter relevant to helix loading is the proximity of the envelope which may be defined in terms of the ratio of the envelope-to-mean helix radii, b/a.

The effects of helix loading both by the helix dielectric supports and the overall metal envelope are taken into account by the dielectric loading factor D_{lf} which is given in terms of the structure parameters either by (5.3.15) or by (5.4.35). The value of D_{lf} which is equal to unity for an unloaded helix becomes less than unity as the helix is loaded, and more the helix loading the less would be the value of D_{lf}. The variation of D_{lf} with the dimensionless frequency γa ($\cong \beta a = \omega a/v_p$, under the slow wave assumption ; see (5.3.9)) (Fig. 5.7.1) shows that the value of D_{lf} increases with frequency. In fact, it can be shown, by considering a free space gap between the helix and the beginning of the dielectric to account for the finite radial thickness of the helix (section *5.8*), that at high operating frequencies the value of D_{lf} tends to become unity — the value corresponding to a helix in free space. In other words, this would mean that the helix fails to "see" the effect of helix loading at large operating frequencies. It can also be seen that the value of D_{lf} decreases with the increase of ε_r' (Fig. 5.7.1(a)), say, corresponding to an increase in the value of ε_r as well as with the increase in the value of b/a (Fig. 5.7.1(b)).

The dispersion characteristics of a loaded helix may be plotted taking Y as the ordinate and X as the abscissa, as has been done (Fig. 5.3.1) in the case of a helix in free-space, where

$$Y = k_0 \cot\psi / \gamma, \qquad \text{(a)}$$

and

$$X = \gamma a Y = k_0 a \cot\psi, \qquad \text{(b)} \qquad (5.7.1)$$

Y being obtainable from the dispersion relation (5.3.14) or (5.4.34) in the sheath-helix model. With the help of (5.7.1) one may express

$$\beta = ((\gamma a)^2/a^2 + \omega^2/c^2)^{1/2} = ((X/Y)^2/a^2 + \omega^2/c^2)^{1/2} \qquad \text{(a)}$$

and

$$\omega = cp\,\gamma a Y / 2\pi a^2 = cpX / 2\pi a^2 . \qquad \text{(b)} \qquad (5.7.2)$$

We may also write

$$v_p = \omega / \beta \qquad \text{(a)}$$

and

$$f = \omega / 2\pi , \qquad \text{(b)} \qquad (5.7.3)$$

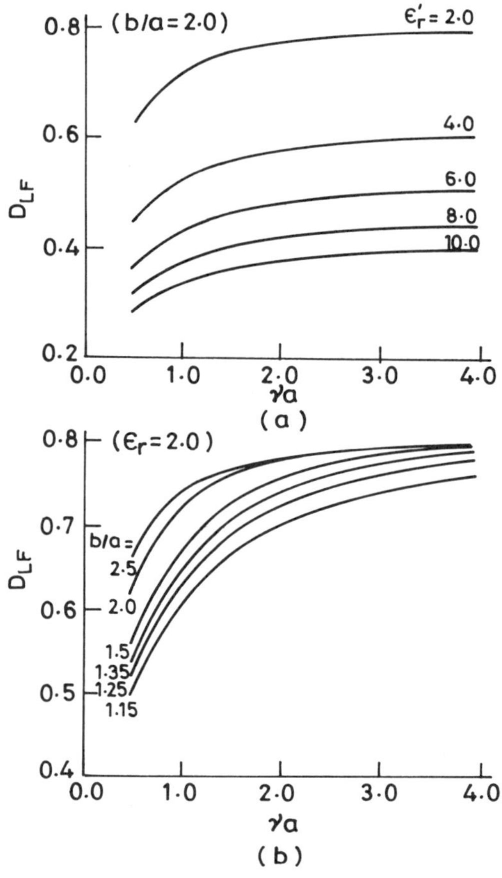

Fig 5.7.1 Variation of the dielectric loading factor with normalised frequency γa of a dielectric-supported helix in a metal envelope, taking the effective relative permittivity ε_r' of the support system (a) and b/a measuring the proximity of the envelope (b) as the parameters .

where ω and β are obtainable from (5.7.2). Thus the frequency value which a particular value of γa corresponds to can be ascertained with the help of (5.7.1)-(5.7.3) (problem 5.8). Further, instead of plotting the normalized quantities Y against X (Fig 5.7.2), one may choose to plot (ω-β) - diagram, with the help of (5.7.2), or, alternatively, v_p -versus - f characteristics (for instance, Fig. 5.7.8) with the help of (5.7.3), provided the helix dimensions, namely, p and a, and hence $\psi = \cot^{-1}(2\pi a/p)$ are specified.

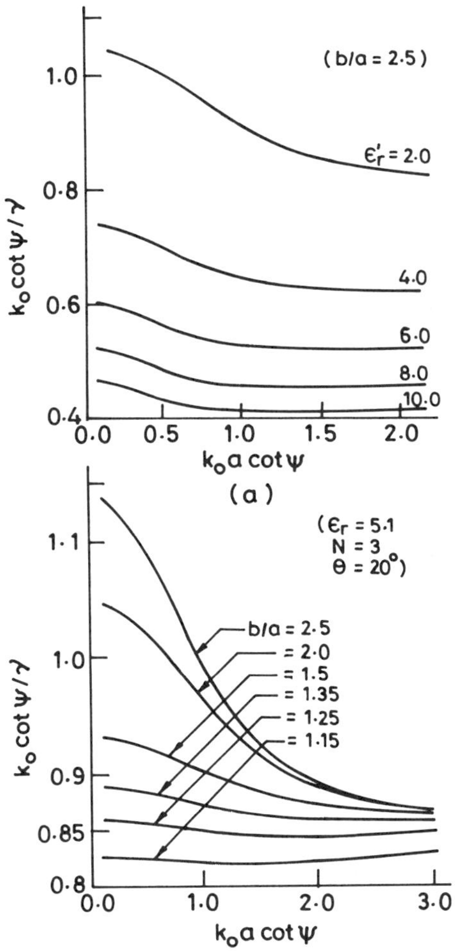

Fig. 5.7.2 Dispersion characteristics of a dielectric-supported helix in a metal envelope, taking the effective relative permittivity ε_r' of the support system **(a)** and b/a measuring the proximity of the envelope **(b)** as the parameters.

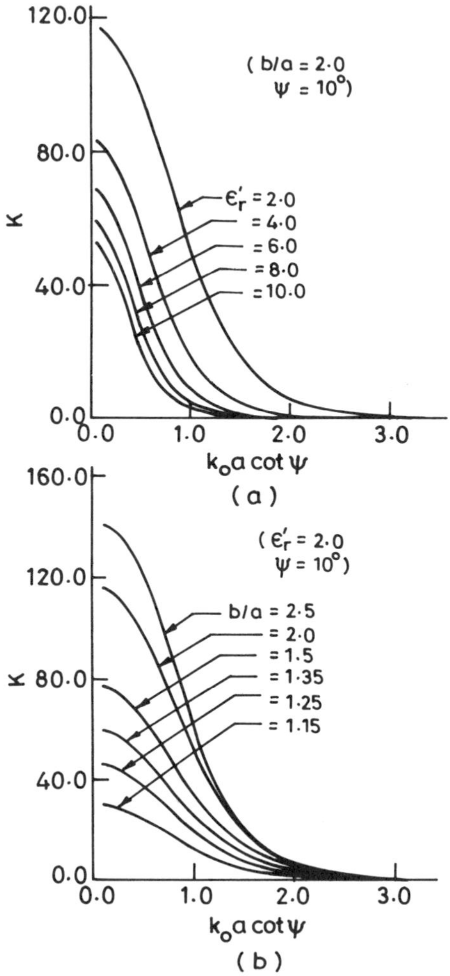

Fig. 5.7.3 Variation of the interaction impedance K with the normalised frequency $k_0a \cot\psi$, taking the effective relative permittivity ε_r' of the support system **(a)** and b/a measuring the proximity of the envelope **(b)** as the parameters.

It is found that the phase velocity decreases as the value of ε_r' is increased (Fig. 5.7.2(a)), say, by increasing the value of ε_r (problem 5.10(a)), and also as the value of b/a decreased, that is, the metal envelope brought closer to the helix (Fig. 5.7.2(b)) (problem 5.10(b)). The shape of the dispersion is, however, seen to be more sensitive to the proximity of the envelope than to the value of ε_r' (hence to the value of ε_r, ϕ and N, for wedge supports). One must also monitor the value of K which like the phase velocity decreases as ε_r' is increased (Fig. 5.7.3(a)), and as b/a is decreased (Fig. 5.7.3(b)). It may be mentioned that for the estimate of device gain the interaction impedance K is a more realistic parameter than the characteristic impedance Z_0. For practical structures, the value of K is found to be typically of the order of $Z_0/2$ (problem 5.11).

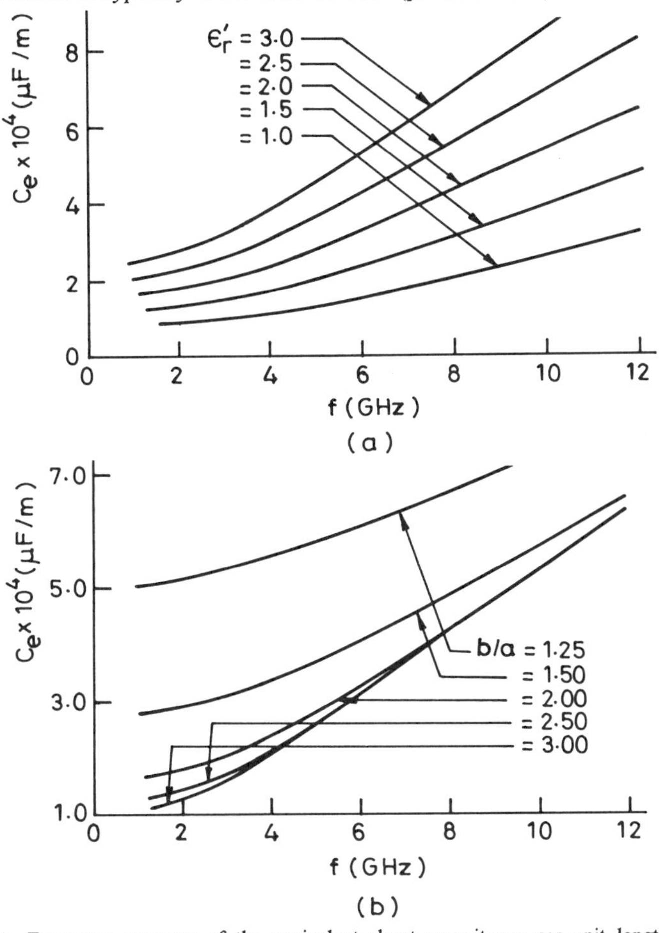

Fig. 5.7.4 Frequency response of the equivalent shunt capacitance per unit length of a dielectric-supported helix in a metal envelope, taking the effective relative permittivity ε_r' of the support system (for a = 1.2 mm, b = 2.0 and $\cot\psi$ = 10.0) **(a)** and b/a, measuring the proximity of the envelope (for a = 1.2 mm, ε_r' = 2.0 and $\cot\psi$ = 10.0) **(b)** as the parameters.

It is of interest to study the nature of variation of the equivalent line parameters of the structure, considered as lossless for the sake of simplicity, namely, C_e and L_e . (The methods of taking losses into account have been discussed in section *5.8*). The dependence of C_e on ε_r' (hence on ε_r, N etc.) is explicitly clear though that of L_e is not so from their respective expressions. Interestingly, L_e also depends on ε_r' at a given frequency f since, say, from material to material the value of f which a particular value of γa corresponds to (problem 5.8) varies as can be seen from the dependence of ω on Y (see(5.7.2(b))) which, in turn, does depend on ε_r' (that is, on ε_r, etc.) .

It may be seen that C_e increases but L_e decreases with the increase of f (Figs 5.7.4 and 5.7.5) as well as with the increase of ε_r' (Figs. 5.7.4(a) and 5.7.5(a)) and with the decrease of b/a (Figs. 5.7.4(b) and 5.7.5(b)), that is, with the increase of helix loading. Thus $Z_o = L_e/C_e$ (see (5.5.1)) also decreases with the increase of f as well as with the increase of helix loading. Further, the increase of C_e will more than compensate for the decrease in L_e so as to reduce v_p (= ω/β) = $1/(L_eC_e)^{1/2}$, (see (5.4.1)), as the helix loading is increased.

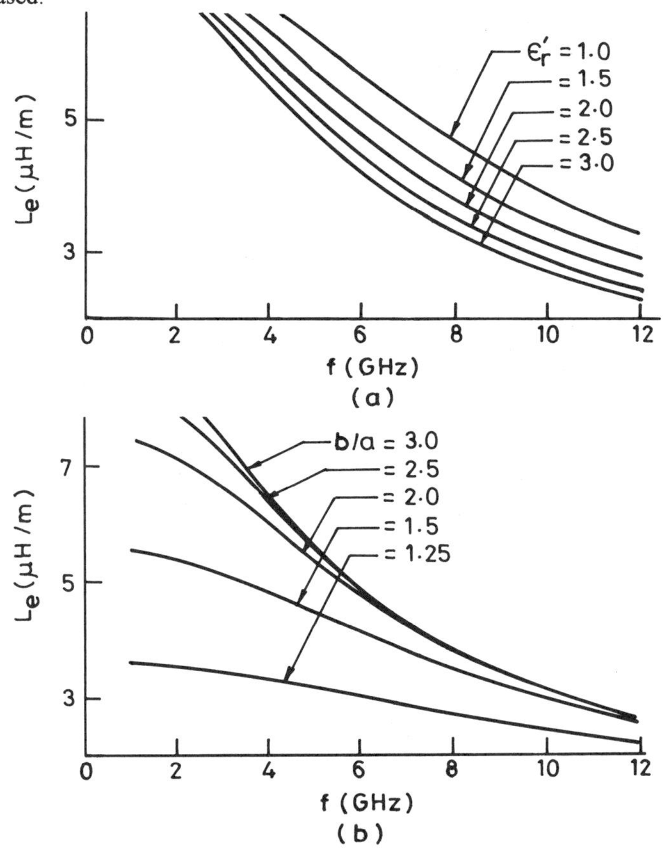

Fig. 5.7.5 Frequency response of the equivalent series inductance per unit length of a dielectric-supported helix in a metal envelope taking ε_r' **(a)** and b/a **(b)** as the parameters, for situations identical with those given in Fig. 5.7.4.

The space-harmonics in the tape model cause forbidden zones to appear in the (ω-β) diagram plotted for a helix in free space (Fig 5.7.6). It is found that a particular forbidden zone (shown hatched) is assigned to a particular negative space-harmonic number.

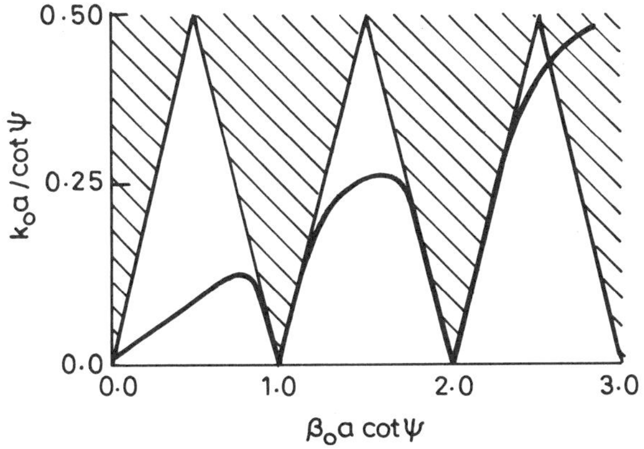

Fig. 5.7.6 Normalised ω-β diagram of a helix in free space in the tape-helix model.

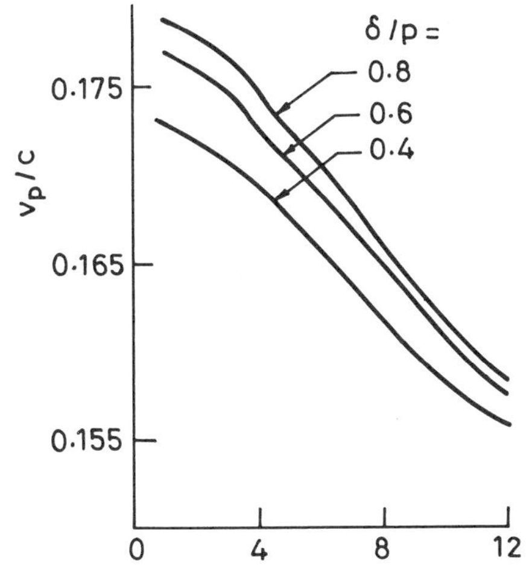

Fig. 5.7.7 Normalized phase velocity (v_p/c) - versus - frequency (f) characteristics of a helix supported by dielectric wedge bars in a metal envelope, taking the ratio of the tape-width to the helix pitch (δ/p) as the parameter (for $\phi = 20^0$, $\psi = 10^0$, N = 3, a = 1.0 mm, b/a = 2, $\varepsilon_r = 5.1$ (APBN)), in the tape-helix model.

In the typical plot shown (Fig. 5.7.6), the space harmonic numbers n = ±1 and ±2 have been included. This plot, therefore, shows two forbidden zones which owe to n = -1, and -2, respectively. Further, the analysis in the tape-helix model reveals that the tape width is an important parameter since the increase in its value causes the phase velocity to increase, as can be seen from the v_p/c - versus - f plot for a typical loaded helix (Fig. 5.7.7).

Finally, it is of interest to note that the phase velocity calculated by the tape model is somewhat less than that calculated by the sheath-helix model as can be seen for a typical loaded structure (Fig. 5.7.8). It is found that with the increase of the helix pitch (p) the difference between the phase velocities calculated by the sheath and the tape models increases. The comparison of the result for a loaded helix (Fig. 5.7.8) with that for a helix in free space (not shown here) shows that the difference between the two models becomes more pronounced for a loaded helix than for an unloaded one.

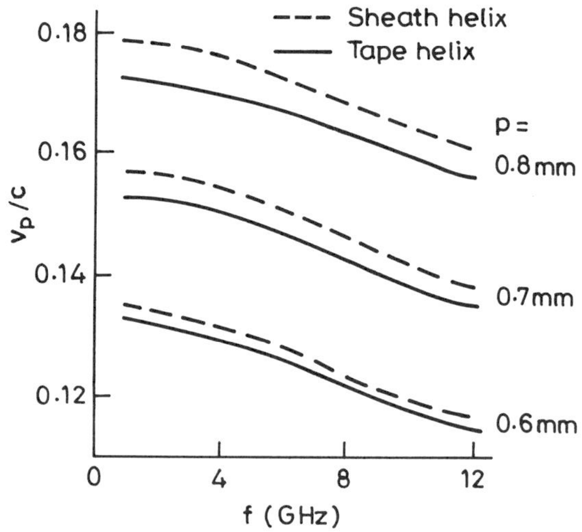

Fig. 5.7.8 The comparison between the dispersion characteristics of a helix supported by a dielectric wedge bars in a metal envelope obtained by the tape-helix (solid line) and sheath-helix (broken line) models, taking the helix pitch (p) as the parameter (for a = 0.75 mm, b = 1.6 mm, δ = 0.4 mm, ε_r = 5.1 (APBN), ϕ = 20 ^{0}, N = 3).

5.8 Refinement of the Model for Practical Relevance

Considering the various points of practical relevance one may further include in the physical models of the actual helical SWS the following aspects:

i) finite radial thickness of the helix wire/tape,
ii) inhomogeneous loading of the helix,
iii) anisotropic loading of the helix,
iv) finite resistivity of the metallic parts (like the helix and the envelope),
v) resistive attenuator coating over the helix (say, on the dielectric helix-support rods), etc.

In this section, these considerations have been reviewed in brief making reference to some available literature. In the sheath-helix or tape-helix model, the sheath helix is considered to be of infinitesimal thickness. But in practice the helix, whether it is made of a wire or a tape, does have a finite thickness and this plays an important role in controlling the propagation characteristics of the structure. If the discrete supports for the helix are replaced by a continuous dielectric tube (section *5.1.2*), then the effect of the finite wire thickness may be taken into account by considering a free-space gap between the physical model of the helix and the dielectric tube [20], [28]. This gap may be taken as half the thickness of the actual helix. The effect of the helix thickness is found to be more pronounced at relatively high frequencies. At such frequencies, rf fields become more concentrated near the helix and, as a result, the helix fails to see the dielectric. The theory and the experiment on helix dispersion more closely match when this effect is considered by introducing the additional free-space gap region into the model. In view of the addition of this region one, however, need to consider four additional boundary conditions for the analysis of the structure. These conditions arise from the continuity of the axial and the azimuthal electric and magnetic field intensities, at the boundary between the free space (of half the helix thickness) and the beginning of the dielectric region [20], [28].

The helix may be considered as homogeneously loaded if it either closely fits into a dielectric tube or is supported by a number of dielectric *wedge* bars arranged at a regular interval around the helix. In the latter case, the discrete bars may be azimuthally smoothed into *a single* continuous dielectric tube [17], [18]. However, if the helix is supported by dielectric bars/rods which have a geometry other than that of a *wedge*, it is not proper to smooth out the supports into *a single* dielectric tube. Instead, for such supports, which may be circular cylindrical or rectangular, it would be more appropriate to azimuthally smoothout the discrete supports into *a number of* dielectric tubes of appropriate effective permittivity values, and increase the number of such tubes till converging results are obtained. The same approach is applicable to tapered-geometry supports for the helix such as supports of half-moon shaped or double-curved cross sections which are used to inhomogeneously load the helix to control the helix dispersion, for ultra-band TWTs [29]-[31].

Further, the model of the structure may be improvised to take into consideration the anisotropic loading of the helix, a technique which is also used to dispersion-shape a helix for the purpose of broadbanding a TWT [10], [16], [24], [32]-[34]. Such type of loading may be brought about by providing an anisotropically conducting metal envelope for the helix, for instance, an envelope provided with metal vanes or segments projecting radially inward. There are also instances where vane/segment loading has been simulated by metal-coating the dielectric rods/bars which are used to support the helix [35], [36]. A simple model for such an envelope is an anisotropically conducting circular-cylindrical sheath which conducts only in the axial direction, and positioned at the tips of the vanes. Such a cylinder while it would shield the axial electric field is transparent to the azimuthal electric field [16], [33], [34]. This model is reasonably valid for those structures in which there are a large number of thin vanes present. One has, however, to go for a rigorous field analysis considering angular harmonics generated, if one desires to include the effects of finite number and angular extent of vanes [32]; alternatively, one may try out an empirical model in which the location of the axially conducting cylinder replacing the vane tips is modified by an extent which is a function of the number and angular extent of vanes in an otherwise simple model for a large number of thin vanes [24].

The analysis of the helical SWS may be refined by considering the losses due to finite resistivities of the metallic parts of the structure, say, the helix and the envelope [37]-[39]. The method of estimating the attenuation of waves caused by losses in the metallic parts is similar to one usually followed in evaluating wall losses of a conventional metallic waveguide. The approach would be to treat the radial propagation as purely real, and unchanged by losses, while writing the first-order field quantities. With the help of these expressions, then the power loss per unit axial length as well as the power propagating down the structure are found. These power quantities are then interpreted to find the attenuation constant as well as the phase propagation constant in the presence of resistive losses [39].

In the presence of an attenuator coating on the supporting rods for the helix, which is used to suppress oscillations in a TWT (chapter *8*), the radial propagation constant would become a complex quantity. Moreover, one should in such a problem consider the appropriate boundary condition related to the discontinuity of magnetic field intensities at the coating surface attributable to a finite coating-resistivity. The approach leads to a dispersion relation with modified Bessel functions of a complex argument. The solution of such a dispersion relation, which becomes an interesting numerical problem, gives one the attenuation caused by the coating [40], [41].

5.9 Summary

The slow-wave structure (SWS) is a periodic structure which supports an electromagnetic wave which is slower than light in free space. Such a structure finds an important application in a microwave/millimeter-wave electron beam device called the traveling-wave tube which has been taken up for study in chapter *8*. The SWS may be categorized into three groups: helix and helix-derived, physically obstacled waveguide/cavity-chain and serpentine line structures. The helix is certainly superior to its counterparts with respect to its untaxed bandwidth. The present study is concentrated on the helical SWS, though a brief outline of other structures with relevant references of the literature is given in appendix *a5.0.1*. With proper loading, the dispersion of a helix may be controlled rendering the structure suitable for ultra-band TWTs. With the advent of recent technological innovations, other performance characteristics like the power, frequency, etc. of the helix can also be considerably enhanced making the structure increasingly important. Further, the helix poses itself as an excellent boundary-value problem. In this problem one finds an ample opportunity to exemplify the various concepts of time-varying fields such as Maxwell's equations, electromagnetic boundary conditions, Poynting vector for power flow, and so on.

For the sake of simplicity in the analysis, the actual helical structure has been replaced here by its physical model. In this model, the actual helix is replaced by an infinitesimally thin cylindrical sheath of an anisotropic conductivity; and the discrete helix supports made of a dielectric are azimuthally smoothed out into a single continuous dielectric tube of an effective permittivity. The need for replacing the discrete supports by a number of such dielectric tubes instead of a single tube, in the case of the geometry of the supports deviating from that of a wedge, has also been discussed. Also, some other aspects which have not been brought within the purview of the present analysis have been touched upon giving suitable references to the available literature. These are the effects of the finite thickness of the helix, the finite conductivity of the metallic parts, the inhomogeneous and the anisotropic loading of the helix for ultra-band TWTs and the effects of resistive attenuator coating and the space-periodicity of the structure, the latter calling for the tape model in place of the sheath model.

The sheath-helix approximation is seen to be valid for many practical situations in which the circumference of the helix would be less than the free-space wavelength. Analytical studies on the structure have been carried out by two approaches — the field and the equivalent circuit analyses. It is of interest to note that both these analyses yield one and the same dispersion relation.

The analysis has been developed here from first principles. The wave equation is set up in the cylindrical system of coordinates. It is then solved under slow-wave assumptions subject to the boundary conditions to be satisfied at the sheath-helix as well as at the overall envelope of the structure. In the field analysis of the structure, the dispersion relation is obtained in the form of a determinant set equal to zero. The number of elements of the determinant increases as one goes from an unloaded to a loaded helix, and, in fact, for more and more complex structures for which more and more boundary conditions would have to be involved. It often becomes quite tedious to simplify a dispersion relation thus obtained to a workable form if the number of elements of the determinant is high. In the equivalent circuit analysis, on the other hand, no such determinant need to be handled by a lengthy algebra. Here, at a time, one deals with only half the boundary conditions to find a line parameter of relevance. The line parameters thus found are substituted in the transmission line equation to obtain the desired dispersion relation. The equivalent circuit analysis also yields another important parameter of the SWS, namely, the characteristic impedance which is important from the standpoint of matching of the SWS with an external circuit which may be used for coupling of power, either in or out of the SWS.

The dispersion relation of a loaded helix is found to differ from that of a helix in free space by a factor called the dielectric loading factor (DLF). The dispersion relation may be used to design a helix to be used in a device. One may plot a normalized dispersion curve and optimize the structure dimensions for the desired shape of the curve. Such control of dispersion has, however, to be monitored to see that there is no deterioration in the value of the interaction impedance, a parameter which is more realistic than the characteristic impedance from the point of view of device performance (chapter *8*). The interaction impedance may be found by integrating the average complex Poynting vector over the cross section of the structure. It has been found that the control of dispersion though it is needed for broadbanding a device has to be examined from the standpoint of the interaction impedance of the structure and hence the gain of the device.

It may be mentioned that initially, for the sake of simplicity, the dispersion relation of a helical SWS was derived in a sheath-helix model which ignores the space-harmonic effects. A rigorous tape-helix model subsequently developed takes into account these aspects. The (ω-β) plot shows the existence of forbidden zones — a particular zone being caused by the presence of a particular negative space-harmonic component. Also, it is found that the phase velocities predicted by the sheath- and tape- helix models differ, more so for larger helix pitches.

The sheath-helix model results, however, continue to be important since it has been shown that such results for a loaded helix could be combined with those for a simple helix in free-space analysed in the tape-model to obtain the results for a loaded helix in the tape-model.

The analytical methods, though it is developed here with particular reference to a simple structure of a loaded helix, is rather general, and may be applied to even complex structures and also could be improvised to incorporate many other factors of practical relevance such as the finite radial thickness of the helix wire/tape, the inhomogeneous as well as the anisotropic helix loading for wide-band applications, the finite resistivity of the metallic parts, the resistive attenuator coating, and so on.

Problems

5.1 Which out of two helices (a) of the same diameter but of different pitches (b) of identical pitch and diameter, the first of which is unloaded while the second is loaded, would support slower waves?

Answer: (a) Lower-pitch helix (b) Loaded helix.

5.2 Two helices one loaded with DLF = D_{lf} and the other unloaded have the same diameter. If they are to support waves of the same phase velocity, what should be the relation between their pitches?

Answer: The pitch of the loaded helix has to be greater than that of the unloaded one by a factor of $1/D_{lf}$.

5.3 Calculate the value of the pitch of the helix of the structure of Example 5.4.1 using the approximate relation (5.1.8) and incorporating the answer to problem 5.2. Compare this value with that found by the more rigorous method of Example 5.4.1, using (5.1.8).

Answer: 0.73 mm. The value is found to be greater than that found in Example 5.4.1.

5.4 The gain of a helix-TWT peaks at a value of $\gamma a = 1.6$ (see the discussion preceding section *8.2.4.A*). Taking the velocity reduction factor of the helix as $v_p/c = 1/10$, show that the helix radius, under slow-wave assumption, is approximately (1/40)*th* of the wavelength in free-space.

Hint: Take $\gamma a \approx \beta a = (\omega/v_p)a = 2\pi fa/\{(v_p/c)c\} = 2\pi a/\{(v_p/c)\lambda_{\text{free-space}}\}$.

5.5 A helix of mean radius 1.0 mm and pitch 0.6 mm is loaded with a DLF of 0.75 such that it supports a wave of normalized radial propagation constant γa=1.6. Find the axial propagation constant β and the corresponding $\omega(=2\pi f)$. $[I_0\{1.6\} = 1.750, K_0\{1.6\} = 0.188, I_1\{1.6\} = 1.085, K_1\{1.6\} = 0.241]$

Answer: $\beta(\approx\gamma) = 1.6$ rad/mm, $\omega = 2\pi \times 6.15\times10^9$ rad/sec (f = 6.15 GHz).

5.6 A helix is bundled by N identical dielectric rods/bars each of relative permittivity ε_r arranged symmetrically around the helix. If, for the sake of simplicity of analysis, these supports are azimuthally smoothed out into an equivalent continuous dielectric tube of an effective relative permittivity ε'_r, show that

(a) $\varepsilon'_r = 1 + (\varepsilon_r-1)Nd/\{4(d+2a)\}$,

for circular-cylindrical support rods each of diameter d;

and (b) $\varepsilon'_r = 1+(\varepsilon_r-1)N[(a+y)^2(\psi_1-0.5\sin2\psi_1)-a^2(\psi_0-0.5\sin2\psi_0)+XY]/[2Ya+Y^2]\,\pi$,

where $\psi_0 = \sin^{-1}(0.5X/a)$ and $\psi_1 = \sin^{-1}(0.5X/(a+Y))$, for rectangular support bars each sized to closely fit into the helix-envelope assembly, where Y is the dimension of the bar in the radial direction, being the helix-envelope clearance, and X is the dimension transverse to the former.

5.7 The helix of an SWS is supported by dielectric bars in a glass envelope. The relative permittivity of the dielectric tube into which the discrete helix supports are supposedly smoothed out, azimuthally, is ε'_r and that of the glass tube is ε_{rg}. Show that the DLF (D_{lf}) of the structure is given by:

$$D_{lf} = \left[\frac{P_1S_1 + P_2S_2}{Q_1S_1 + Q_2S_2}\right]^{1/2},$$

where

$$P_1 = \frac{1-\xi_{ab}}{\gamma a I_1\{\gamma a\}K_0\{\gamma a\}}, \qquad P_2 = \frac{\varepsilon'_r(1+\chi_{ab})}{\gamma a I_1\{\gamma a\}K_0\{\gamma a\}};$$

$$Q_1 = 1+\xi_{ab}(\varepsilon'_r - 1) + \chi_{aa}\varepsilon'_r, \qquad Q_2 = \varepsilon'_r[1+\varepsilon'_r\chi_{aa}(1-\varsigma_{ab}) + \chi_{ab}];$$

$$S_1 = \varepsilon_{rg}\chi_{bb}[\varsigma_{bc}(1+\chi_{cb}) + \varepsilon_{rg}(1-\varsigma_{bc})] \quad \text{and} \quad S_2 = \varepsilon_{rg}(1+\chi_{bc}) + \chi_{cc}(1-\xi_{bc}),$$

where

$$\xi_{ij} = \frac{I_0\{\gamma i\}K_0\{\gamma j\}}{K_0\{\gamma i\}I_0\{\gamma j\}}, \qquad \varsigma_{ij} = \frac{I_1\{\gamma i\}K_1\{\gamma j\}}{K_1\{\gamma i\}I_1\{\gamma j\}}$$

and

$$\chi_{ij} = \frac{I_0\{\gamma i\}K_1\{\gamma j\}}{K_0\{\gamma i\}I_1\{\gamma j\}}, \qquad (i, j = a, b, c),$$

where a is the mean helix radius, and b and c are the inner and outer radii of the glass tube, respectively.

5.8 A structure consisting of a helix supported by wedge bars in a metal envelope has the following specifications: a = 1.0 mm, $\psi = 10°$, b/a = 1.5, N = 3, $\phi = 20°$, $\varepsilon_r = 5.1$ (APBN). Give a plot to interpret γa in terms of f.

<u>Answer</u>:

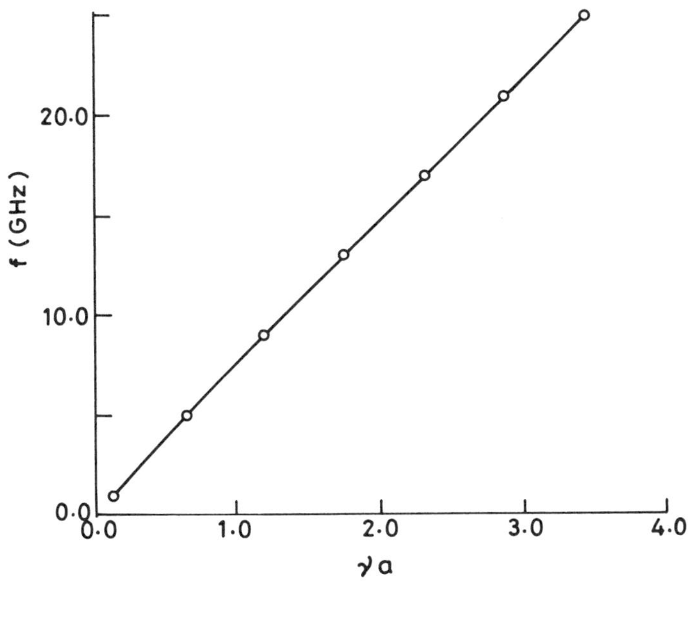

5.9 (a) A structure has the same parameters as in problem 5.8 except that now b/a is changed to 2.0 (from 1.5). Calculate D_{lf} and K at f = 10 GHz . (b) A structure has the same parameters as in problem 5.8 except that now ϕ is changed to 30° (from 20°). Calculate D_{lf} and K at f = 10 GHz. (c) A structure has the same parameters as in problem 5.8 except that now the material is changed to alumina (ε_r = 9.0) from APBN (ε_r = 5.1). Calculate D_{lf} and K. (d) A structure has the same parameters as in problem 5.8. Calculate D_{lf} and K at f = 5, 9 and 13 GHz.

Answer: (a) 0.80 and 48.4 ohms (b) 0.72 and 25.0 ohms (c) 0.68 and 20.3 ohms (d) 0.65, 0.75 and 0.80, and 66.6, 38.6, and 17.2 ohms.

5.10 (a) For a helix supported by wedge bars in a metal envelope, specified by ϕ = 20°, N = 3, b/a = 2.5, obtain the dispersion plot of $k_0 \cot\psi/\gamma$ - versus - $k_0 a \cot\psi$, taking ε_r as the parameter varying from 4.5 to 9.5, at the interval of 1.0. (b) In the specifications of (a), take ε_r = 6.65 for beryllia as the helix-support material and plot the dispersion characteristics as in (a), however, now taking (instead of ε_r) b/a as the parameter varying from 1.25 to 2.0 at the interval of 0.25.

Answer:

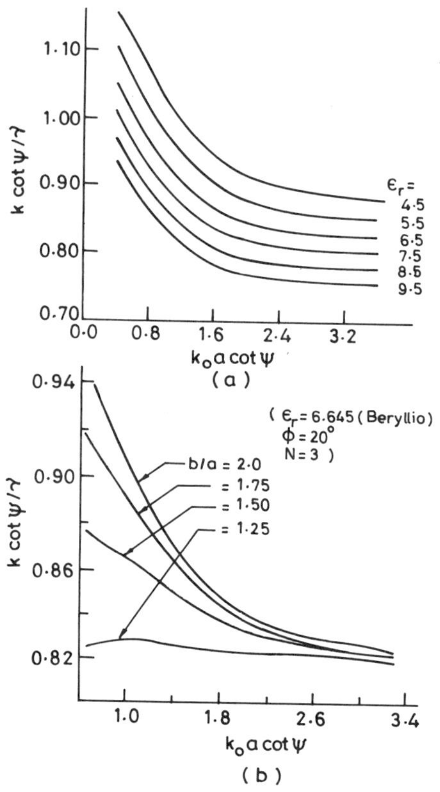

5.11 Find the values of the characteristic impedance Z_0 and the interaction impedance K and appreciate that $Z_0 \sim K/2$ with reference to a wedge-bar supported helix in a metal envelope, taking $\gamma a = 1.6$, cot $\psi = 8$, $b/a = 1.25$, $\phi = 20°$, N = 3, and $\varepsilon_r = 6.65$ (beryllia).

Answer: $Z_0 \approx 31$ ohms, K ≈ 62 ohms.

5.12 If $E_z^2(0)$ in the definition of K is replaced by $E^2_{z,av}$, as discussed following (5.5.4), in what way would the expression for K be modified from (5.5.4) for a 'solid' circular-cylindrical electron beam of radius r_B placed at the helix axis.

Answer A factor $I_0^2(\gamma r_B) - I_1^2(\gamma r_B)$ would have to be added.

5.13 Show by solving wave equation (5.2.3) by the method of separation of variables that , for a non-azimuthally-varying mode ($\partial/\partial\theta \neq 0$), the azimuthal rf fields vary as $\exp(\pm j\, m\theta)$, where m is an integer.

5.14 Derive, for a non-azimuthally-varying mode ($\partial/\partial\theta \neq 0$), the field quantities inside and outside the helix by solving wave equation (5.2.3) and hence appreciate how the field expressions are modified from those given by (5.3.2).

Answer: Expressions (5.6.1) - (5.6.8). These expressions, for m = 0, pass on to those given by (5.3.2).

5.15 Solve the wave equation in the cylindrical coordinate system (r, θ, z) as applied to a hollow metal pipe cylinder (waveguide) and obtain an expression for the axial component of magnetic field intensity and for the azimuthal and the radial components of electric and magnetic field intensities in a waveguide mode called the TE-mode in which the electric field intensity inside the pipe parallel to its axis is zero ($E_z = 0$), assuming that the structure is a 'fast' waveguide supporting an electromagnetic wave whose phase velocity is greater than c ($\omega(\mu_0\varepsilon_0)^{1/2} > \beta$) and taking the rf fields vary as $\exp j(\omega t - \beta z - m\theta)$, where m is an integer.

Hint: Write the wave equation (5.2.3) for H_z for the given rf dependence of fields as

$$[(d^2/dr^2) + (1/r)(d/dr)]H_z + (k_t^2 - m^2/r^2)H_z = 0$$

which has a solution (see appendix *a5.2.1*)

$$H_z = C\, J_m(k_t r) + D\, Y_m(k_t r)$$

which, because D = 0 in order to make fields finite at the axis (r = 0)(since at r = 0, $Y_m(\gamma r) \to -\infty$), may be expressed invoking the given rf dependence as follows:

$$H_z = C\, J_m(k_t r) \exp j(\omega t - \beta z - m\theta),$$

where C and D are field constants. $K_t\ (= (k_0^2 - \beta^2)^{1/2})$ is a radial propagation constant which is a real quantity for a fast-wave mode ($k_0 > \beta$, given), $k_0 = \omega(\mu_0\varepsilon_0)^{1/2}$ being the free-space propagation constant. From the expression for H_z, then obtain the expressions for E_θ, E_r, H_θ and H_r using Maxwell's equation (see problem 4.5) remembering that in the present case (TE-mode)$E_z = 0$.

Answer: $E_\theta = (j\omega\mu_0/k_t)\, C\, J_m'(k_t r) \exp j(\omega t - \beta z - m\theta)$,

$E_r = -(\omega\mu_0/k_t)\, C\, (m/k_t r)\, J_m(k_t r) \exp j(\omega t - \beta z - m\theta)$,

$H_\theta = (\beta/\mu_0\omega)E_r$

and $H_r = (-\beta/\mu_0\omega)\, E_\theta$, with H_z as in the above hint.

References

[1] R. Kompfner, "Travelling-wave tube as amplifier at microwave frequencies," *Proc. IRE* **35** (1947) 124-127.

[2] *Hughes TWT and TWTA Handbook* (A bulletin for business circulation), Electron Dynamics Division, Hughes Aircraft Co., California.

[3] M. Chodorow and E.L. Chu, "Cross-wound twin helices for travelling-wave tubes,"., *J. Appl. Phys* **26** (1955) 33-43.

[4] D.A. Watkins, *Topics in Electromagnetic Theory* (John Wiley, New York, 1958).

[5] A.S. Gilmour, Jr., *Microwave Tubes* (Artech House, Washington, 1986).

[6] M. Chodorow and R.A. Craig, "Some new circuits for high power traveling-waves tubes", *Proc. IRE* **45** (1957) 1106-1118.

[7] R.M. Bevense, *Electromagnetic slow-wave systems* (John Wiley, New York, 1958).

[8] J.F. Gittins, *Power Travelling -Wave Tubes* (English Univ. Press, London, 1964).

[9] F.E. Terman, *Electronic and Radio Engineering* (McGraw-Hill, New York, 1955).

[10] J.L. Putz and M.J. Cascone, "Effective use of dispersion shaping in broad-band helix TWT circuits", *IEDM Tech. Dig.* ,1979, pp. 422-424.

[11] E.F. Belohoubek, "Helix-support structures for ultra-band traveling-wave tubes", *RCA Review* **26**, (1965) 106-117.

[12] P. Gosset and E.D. Maloney, "Brazed-helix TWT's for TV-broadcasting and high power telecommunication satellites" , *National Telecomm. Conf.*, Washington DC, 1979.

[13] Y. Hiramatsu, "Study of block-supported helix circuits for high-power TWT's", *IEDM Tech. Dig.*, 1979, pp. 416-417.

[14] H.G. Kosmhal, "Modern multi-stage depressed collectors — a review" , *Proc. IEEE* **70** (1982) 1325-1334.

[15] J.R. Pierce, *Traveling-Wave Tubes* (Van Nostrand, Princeton, 1950).

[16] S.F. Paik, "Design formulas for helix dispersion shaping" , *IEEE Trans. Electron Dev.* **ED-16**. (1969) 1010-1014.

[17] L.N. Loshakov and E.B. Ol'derogge, "Propagation of slow electromagnetic waves along a helix with dielectric supports" , *Radio Engg. and Electron Phys.* **13** (1968) 45-51.

[18] B.N. Basu, R.K. Jha, A.K. Sinha, and L. Kishore, "Electromagnetic wave propagation through an azimuthally perturbed helix" , *J. Appl. Phys.* **58** (1985) 3625-3627.

[19] L.A. Pipes and L.R. Harvill, *Applied Mathematics for Engineers and Physicists (Mc Graw-Hill Kagakusha, Tokyo, 1970).*

[20] D.T. Swift-Hook, "Dispersion curves for a helix in a glass tube" , *Proc. IEE* ***105b*** (1958) 747-755.

[21] E. Jhanke, F. Emde, and F. Lösch, *Tables of Higher Functions* (McGraw-Hill, New York, 1960).

[22] J.E. Rowe, *Nonlinear Electron-Wave Interaction Phenomena*, (Academic, New York, 1965).

[23] B.N. Basu, "Equivalent circuit analysis of a dielectric-supported helix in a metal shell", *Int. J. Electronics* **47** (1979) 311-314.

[24] L. Kumar, R. S. Raju, S. N. Joshi, and B. N. Basu, "Modelling of a vane-loaded helical slow-wave structure for broadband traveling-wave tubes," *IEEE Trans. Electron Dev.* **ED-9** (1989)1991-1999.

[25] I. Lebedev, *Microwave Engineering* (Mir, Moscow, 1973).

[26] S. Sensiper, "Electromagnetic wave propagation on helical structures," *Proc. IRE* **43** (1955) 149-161.

[27] A. K. Sinha, R. Verma, R.K. Gupta, L. Kumar, S. N. Joshi, P. K. Jain, and B. N. Basu, "Simplified tape model of arbitrarily-loaded helical slow-wave structures for travelling-wave tubes," *IEE Proc.-H* **139** (1992) 347-350.

[28] P. K. Jain, K. V. R. Murty, S. N. Joshi, and B. N. Basu, "Effects of finite thickness of the helix wire on the characteristics of the helical slow-wave structure of a traveling-wave tube," *IEEE Trans. Electron Dev.* **ED-34** (1987) 1209-1213.

[29] B.N. Basu and A.K Sinha, "Dispersion-shaping using an inhomogeneous dielectric support for the helix in a travelling-wave tube" , *Int. J. Electron.* **50** (1981) 235-238.

[30] P. K. Jain and B.N. Basu, "The inhomogeneous loading effect of practical dielectric supports for the helical slow-wave structure of a TWT" , *IEEE Trans. Electron Dev.* **ED-34** (1987) 2643-2648.

[31] S. Kapoor, R.S. Raju, R.K. Gupta, S. N. Joshi, and B. N. Basu , "Analysis of an inhomogeneously loaded helical slow-wave structure for broadband TWT's" , *IEEE Trans. Electron Dev.* **ED-36** (1989) 2000-2004.

[32] N.P. Kravchenko, L.N. Loshakov, and Yu.N. Pchel'nikov, "Computation of dispersion characteristics of a spiral placed in a screen with longitudinal ribs", *Radio Engg. Electron. Phys.* **21** (1976) 33-39.

[33] A.K. Sinha and B.N. Basu, "Dispersion-shaping in a helix slow-wave structure using metal fins", *J. Instn. Electron. Telecomm. Engrs.* **26** (1980) 318-320.

[34] B.N. Basu, B.B. Pal, V.N. Singh, and N.C. Vaidya, "Optimum design of a potentially dispersion-free helical slow-wave circuit of a broad-band TWT", *IEEE Trans. Microwave Theory & Tech.* **32** (1984) 461-463.

[35] P. Gallupi and M. D. Salvatore, "Evaluation of three techniques of controlling phase velocity dispersion in helix TWT's", *Proc. Int. Conf. Microwave Tubes and Systems: Problems and Prospects (Conf. Pub. no.241)* , London, Oct. 1984, pp.488-490.

[36] P. Gallupi and G. Lamesa, "A new technique for ultra-band high-power TWT's, *Military Microwave Conf. Proc.* **MM-80** (1980) 501-505.

[37] J. H. Bryant and E.J. White, "Attenuation and power handling capacity of helical radio-frequency lines", *IRE Trans. Microwave Theory & Tech.* **PG MTT-1** (1953) 33-38.

[38] A. S. Gilmour, Jr., M.R. Gillet, and J.T. Chen, "Theoretical and experimental TWT helix loss determination", *IEEE Trans. Electron Dev.* **ED-26** (1979) 1581-1588.

[39] P.K. Jain and B.N. Basu, "The effect of conductivity losses on propagation through the helical slow-wave structure of a traveling-wave tube", *IEEE Trans. Electron Dev.* **ED-35** (1988) 549-558.

[40] S. E. Webber, "Calculation of wave propagation on a helix in the attenuator region", *IRE Trans. Electron Dev.* **ED-1** (1954) 35-39.

[41] P. K. Jain and B. N. Basu, "Theory of the attenuator-coated helical slow-wave structure of a travelling-wave tube", *IEEE Trans. Electron Dev.* **ED-35** (1988) 1750-1757.

CHAPTER 6

FORMATION OF AN ELECTRON BEAM

6.0 Introduction

The flow of electrons forming an electron beam can be made to interact with electromagnetic waves in a practical electron beam device (see chapter *8*) provided the beam so formed has well-defined parameters for this purpose. In this chapter of part two of the present volume we apply some of the fundamental concepts of electrostatics developed in chapter *2* of part one to find methods of forming an electron beam of given parameters.

First, the electron beam to be formed has to have a given geometry, say, be linear and solid cylindrical of circular cross section or hollow cylindrical of tubular cross section. Thus the cross-sectional dimension, for instance, the *beam radius* in the case of a solid circular-cross-section linear beam has to be specified as one of the beam parameters of relevance. Further, the beam electrons, all supposedly of the same velocity, need to be accelerated to this velocity by an electrostatic field which is established by applying an accelerating potential on a relevant electrode in the vicinity of source of electrons (to be described later in this chapter). This gives the *beam potential* as the second relevant beam parameter. The third important beam parameter is the *beam current* constituted by the flow of electrons. The present chapter is devoted to describing the principle of formation of a solid, circular-cross-section electron beam when these beam parameters, namely, the beam voltage, current and radius are known. We consider here two types of electron flow — the simple rectilinear parallel-flow of electrons derived from a *flat* cathode (see sections *6.1* and *6.2*) and the convergent conical flow derived from a *curved* spherical-cup cathode (see section *6.3*). An overview on the cathode — the emitter of electrons — is given in appendix *a6.0.1*. It is obvious that the operating cathode *emission current density* would also emerge out as a relevant parameter in the study of formation of an electron beam.

Considerable simplification to the problem is achieved if the electron beam is treated as a charge fluid and the electrons over its cross section are considered to be uniformly distributed, all having the same unperturbed (dc) velocity in the longitudinal direction. Further, it is assumed that the beam cross sections at two different planes at the same instant of time as well as those at two different time instants at the same plane are indistinguishable from each other. Therefore, such an electron beam can now be treated under the purview of electrostatics and as such one may easily apply the concepts like Poisson's and Laplace's equations developed in section *2.6* to this problem.

Rectilinear flow of electrons may be established ideally between two infinite flat planes, the electrons emitting from and leaving one of them, the cathode, and reaching the other, the anode. From practical considerations however one would not obviously welcome *infinite* flat cathode and anode, and therefore look forward to establishing the desired beam of finite cross-sectional area using finite-dimension electrodes. This would lead to a design concept due to Pierce [1] involving a system of electrodes named as the Pierce electron *gun* — so called as it *throws* an electron beam into an interaction region where the beam could be made to interact with an electromagnetic wave for a gainful purpose (see chapter *8*). The electrons beyond the anode if allowed to pass through the latter would, however, diverge out due to space-charge and other external fields, unless a system called the focussing structure (see chapter *7*), different from the electron gun system, is separately provided to keep the beam confined. The concepts of Pierce gun using a flat cathode are developed in sections *6.1* and *6.2*. In section *6.3* these concepts are extended to a gun which uses a spherical-cup cathode.

Appendix a6.0.1 *Cathodes: an overview*

Cathode — the emitter of electrons — is described as the heart *of a microwave tube. The type of the microwave tube in which the beam-wave interaction has been demonstrated in this treatise (chapter 8) uses a cathode in which the mechanism of electron emission is thermionic. It may be mentioned that there are some crossed-field tubes like magnetrons in which the secondary emission also adds to the mechanism of electron emission. In such tubes* secondary *electrons are emitted from the cathode when the latter is bombarded by* primary *electrons.*

The thermionic cathodes may be of two types — directly heated and indirectly heated. A directly heated cathode is made in the form of a filament of an emitting material like tungsten and tantalum. The filament is heated to a high temperature by passing electric current through it. The material therefore should be chosen to have a high melting point. Tungsten has a melting point of 3370 °C. Tantalum has a lower melting point — 2850 °C. The work function, which is the energy required to liberate an electron from the surface of the material, is lower, namely, 4.1eV for tantalum than for tungsten, namely, 4.6eV. Thus more emission is expected from the surface of tantalum, say, ~10A/cm^2*, than from that of tungsten, say, ~ 0.5A/cm*^{2}*, at around 2200 °C, though the former is more prone to being poisoned by the residual gases present in the device which is essentially a vacuum tube (chapter 8). Tantalum is used as an electron emitter in X-ray tubes and high voltage transmitting and diode rectifier tubes. In some devices like magnetrons for a microwave oven, thorium is mixed with tungsten to obtain a composite material — thoriated tungsten which has a relatively low value of work function, namely, 2.6 eV — lower than those of both tungsten (4.6 eV) and thorium (3.5 eV). The reduction in work function may be attributed to the formation of a dipole layer reducing the electrostatic forces on the tungsten surface. In order to reduce the evaporation of thorium and hence increase the cathode life, thoriated tungsten is subject to carbonization(heat-treated in a hydrocarbon) to form tungsten carbide to which thorium adheres more than it does to pure tungsten. One can obtain an emission density ~3A/cm*^{2} *at around 2200 °C from a thoriated tungsten cathode.*

Majority of today's microwave tubes, however, use indirectly-heated cathodes. In such version of cathodes, the emitting surface is formed on a thin metal cylinder which is indirectly *heated by an insulated heater in the form of a filament. This allows all the parts of the cathode surface to attain the same potential. The design of the heater-cathode assembly of the device should also provide for* *expansion and contraction without over-stressing the heater wire or the insulator between the heater and the cathode. The heater material should be so chosen that its recrystalisation temperature is well above its operating temperature. Typically, tungsten/tungsten-rhenium has been used as a heater material. The heater is often potted with alumina into a cavity on the bottom of the cathode to ensure high temperature cycling stability, stability with time, high rigidity to withstand rigors of environment, less warm-up time, etc. In an extensively used type of indirectly heated cathode known as oxide-coated cathode, a mixture of barium, strontium and calcium carbonates (typically in the percentage proportion by weight of 57.3:42.2:0.5) is coated on the emitting portion of the metal cylinder usually of nickel which is doped with a small fraction of zinc, tungsten/zirconium or magnesium which acts as an activator. The carbonates are reduced to oxides by the process of activation which involves heating the cathode by passing large current through the heater filament and drawing current to the anode (the electrode at a positive potential in the vicinity of the cathode). The combination of barium oxide and free barium (work function ~1.8 eV), which is obtained by partial reduction, lowers the work function of the cathode to provide space-charge limited electron emission around quite a low temperature ~600 °C. The increased emission could also be attributed to the available free electrons in the oxide semiconductors. The emission density in an oxide-coated cathode at 650 - 700 °C could be hundreds of mA/cm^2 under dc or cw operation and tens of A/cm^2 under pulsed operation. Thus by the choice of a suitable duty cycle, that is, given a time to recover, such a cathode can yield high emission at a relatively low temperature. However, such a cathode is likely to be easily poisoned by residual gases in the device, more so if operated at a low temperature. On the other hand the operation at a high temperature may cause the evaporation of the active cathode material, namely, barium from the cathode surface.*

The search for cathodes providing a higher emission current density, a higher pulse-length operation, a longer life, a lower heater power requirement, a reduced susceptibility to damage by residual gases in the tube, a potential for reactivation if exposed to leakage, etc. led to the development of a class of cathodes known as dispenser cathodes — so named because such a cathode takes care in continually replenishing/dispensing the active barium from the interior of the cathode which eventually evaporates from the emitting surface.

In its simplest form, known as the L-cathode, the dispenser cathode consists of barium-strontium carbonates in a cavity behind a porous tungsten plug. There are many variations from this simple form of cathode, namely, pressed cathode, A-, B-, M-, MM (mixed-metal)- types, CPC (coated particle cathode), etc.

The pressed cathode is made by blending and mixing base powder, namely, tungsten or tungsten-iridium, emissive mix, namely, the mixture of barium, calcium and strontium, either oxides or carbonates, and an activator, typically, zirconium hydride. The mixture is pressed (140 -1400x10^6 N/cm^2) into a molybdenum sleeve, or alternatively, into pellets and then crimped into a molybdenum cylinder; sintered in hydrogen and then lathe-finished.

The A-type cathode is first formed from a porous tungsten and then impregnated with barium aluminate. The B-type cathode is a modified version of the A-type in which calcium oxide is added to the impregnants to reduce the barium sublimation rate as well as to enhance emission properties of the cathode. The proportions of the mix impregnated into porous tungsten are, typically, 5 BaO : 3 CaO : 2 Al_2O_3 (5:3:2), 4:1:1, 3:1:1, etc. It may be mentioned that the aluminate stabilizes calcium oxide which is otherwise hygyroscopic. The chemical reaction showing barium as a product may be put as [2],[3]:

$$W + 3\,Ba_3Al_2O_6 + 6\,CaO = 3\,Ba_2CaAl_2O_6 + Ca_3WO_6 + 3\,Ba\,.$$

A typical flow-chart for making such a cathode consists in (i) taking a base powder of typically 2-14 μm size high-purity tungsten / tungsten-iridium, (ii) pressing it isostatically in a sealed rubber mould (~ 140x10^6 N/cm^2) so as to increase the density of particles; (iii) sintering it (~ 2500°C for 30 minutes) to increase the density further; plastic -or copper-infiltering it, in order to make it easily *machinable; (iv) shaping it in a machine as desired; (v) removing the infiltered plastic/copper; (vi) melting the impregnants or emissive mix into the porous tungsten; finally, (vii) cleaning it, say, by grit-blasting with alumina. Some of the control areas are emitter temperature, pore size, pore per unit area, uniformity of pores, etc. One may obtain several amps/cm^2 at 1100°C or higher with a B-type cathode.*

The M-type cathode came into being while aiming at reducing the cathode operating temperature and obtaining an emission current density comparable with or greater than that of the B-type cathode, and a longer life. In an M-type cathode, which is essentially a B-type cathode with a thin metal film coated on its emitting surface, the porous tungsten impregnated with barium-calcium aluminate is coated (~ 2000 - 10,000 AU) with a thin layer of osmium-iridium or osmium-ruthenium, by the process of sputtering or vapor depositing or ion-plating, and then hydrogen-fired for several minutes to sinter the coating onto the tungsten substrate. Surprisingly, although the film material has a higher work function than tungsten, yet its effect is to reduce the overall work function and yield the same current density as that of the B-type but with a cathode operating temperature reduced by ~ 90°C. The coating thickness should be optimum so that neither it should be so thin as to degrade by diffusion into the tungsten matrix, thereby converting the cathode into a B-type, nor it should be so thick as to impede the supply of barium to the emitting surface. The emission delay ~ 4 min. has been reported as a disadvantage of the M-type if used as a fast warm-up cathode.

In an attempt to remove the demerit of the M-type cathode in respect of cathode degradation, another cathode, namely, the mixed metal matrix (MM) cathode has come into existence. In such a cathode, the enhancing metal is put into the tungsten matrix itself. The other innovations in cathode technology include the coated particle cathode (CPC) which is made of specially coated particles bounded to a nickel surface. Recently, the deposition of tungsten-osmium alloy on the surface of porous tungsten matrix has been tried out. In another version of dispenser cathodes, pores have been provided on a thin foil of tungsten by laser drilling or ion-etching. Thus the present-day cathode technology has been making continuous progress aiming at high emission density, low operating temperature, less susceptibility to damage and surface degradation due to residual gases and diffusion of film coating; ruggedness; reduction of the electrical breakdown, RF losses, grid emission caused by sublimed materials; less warm-up time, etc.

6.1 Rectilinear Flow between Parallel Planes

In most of the practical microwave electron beam devices (microwave tubes), and of course for the particular type of beam-wave interaction studied later in this book (chapter *8*), the electron beam is realized under what is known as space-charge limited condition. The operation under this condition can be understood with reference to a simple diode consisting of two electrodes — a cathode and an anode, the latter being raised to a higher potential with respect to the former. In a space-charge limited diode, the cathode temperature is raised to a relatively high value and the anode potential to such a value that the number of emitted electrons exceeds the number of electrons reaching the anode. Under this condition the anode current does not increase with cathode temperature thus reaching a temperature saturation; the anode current could however be increased by increasing the anode potential under this condition. (Under another condition known as the temperature-limited condition, all the emitted electrons reach the anode, and as such by increasing the anode potential one can not increase the anode current thus reaching the voltage saturation; under this condition the anode current could be increased by increasing the cathode temperature).

A basic law known as Child-Langmuir's law [4] tells one for a simple planar geometry how under the space-charge limited condition the anode current depends on the anode potential as well as on the separation between the cathode and the anode. This law can be deduced by solving Poisson's equation (2.6.2) subject to the relevant boundary conditions of the problem to be described later in this section.

But first let us consider how the potential is distributed in the region between the cathode and the anode in the absence of electrons in the region. The problem is to solve Laplace's equation (2.6.5) for potential V in the one-dimensional case to obtain

$$V = A_1 z + B_1 \qquad (6.1.1)$$

which is in fact equation (2.6.9) re-written, where z is the distance measured from the cathode towards the anode. Putting $V = 0$ at the cathode, $z = 0$, and $V = V_0$ at the anode, $z = d$, (that is, by considering V_0 as the anode potential with respect to the cathode and, d as the anode-to-cathode separation), one obtains from (6.1.1): $B_1 = 0$ and $A_1 = V_0/d$. Substituting the values of these constants in (6.1.1) one obtains

$$V = (V_0/d)\, z \ . \qquad (6.1.2)$$

The linear variation of V given by (6.1.2) obtained for the case of absence of space charge is shown by the broken line of constant slope in Fig. 6.1.1.

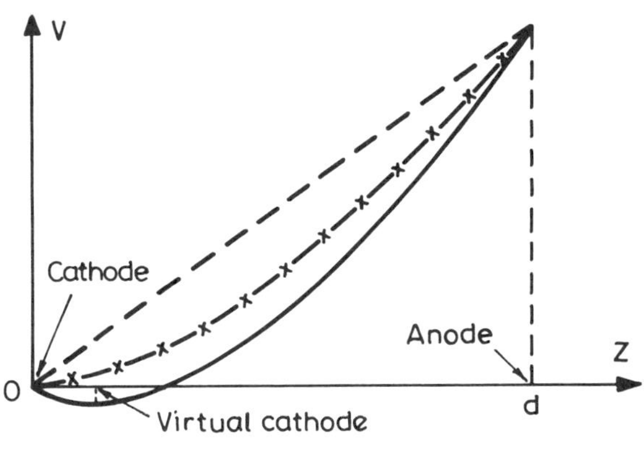

Fig. 6.1.1 Potential distribution in the cathode-anode region. The broken line shows the linear variation of potential in the absence of space charge. The line with crosses gives the variation of potential in the presence of space charge considering a zero velocity of emission at the cathode showing a zero slope at the cathode. The solid line gives the potential variation in the presence of space charge considering a finite velocity of emission at the cathode showing a voltage minimum and zero slope at the *virtual cathode*.

Under the space-charge limited condition, the number of electrons in flight between the cathode and the anode is such that the effect of the *negative* space-charge field due to them at the cathode is neutralized by that of the electrostatic field due to the *positive* potential applied on the anode. The electrons in excess over this number are repelled back into the cathode. This would correspond to a zero slope in the potential variation, and hence to a zero electric field intensity, at the cathode. In other words, this would mean that the electric flux lines would terminate on the electrons rather than thread into the cathode. The distribution of potential in this case is shown as the line with crosses in Fig. 6.1.1. That at equilibrium the slope of this potential distribution at the cathode would be zero may be appreciated as follows [5]. If the slope of potential variation is positive, more electrons would leave the vicinity of cathode which would increase the negative space charge in the region and hence depress the potential distribution curve towards the zero slope at the cathode. On the other hand, if the slope overshoots to a negative value, the emitted electrons would be forced back to the cathode which would reduce the negative space charge in the region and consequently lift the potential distribution curve to have a zero slope at the cathode.

Let us now attempt to solve Poisson's equation under the potential distribution at equilibrium corresponding to a zero slope at the cathode, as described above. This gives the boundary condition at the cathode as

$$dV/dz\Big|_{z=0} = 0 \tag{6.1.3}$$

in addition to the zero reference potential defined at the cathode as

$$V\Big|_{z=0} = 0 \ . \tag{6.1.4}$$

Taking the velocity of emission at the cathode (z = 0), that is, the initial velocity of electrons as zero, one may relate the potential V, at a distance z from the cathode, with the velocity of electrons v at z by balancing the kinetic energy gained by an electron when it *falls* through a potential V, as follows:

$$\tfrac{1}{2} mv^2 = |e|\, V \tag{6.1.5}$$

whence one has

$$v = (2|\eta|\, V)^{1/2} \ , \tag{6.1.6}$$

where m is the electronic mass; e and η (= e/m) are the charge and the specific charge of an electron both carrying their negative sign. (6.1.6) may be used in conjunction with the following relation between the current density J and volume charge density ρ contributed by the negative charges of electrons (for the deduction of the relation, see appendix *a6.1.1*):

$$J = \rho v \tag{6.1.7}$$

to read ρ as

$$\rho = \frac{J}{(2|\eta|)^{1/2}} V^{-1/2} . \qquad (6.1.8)$$

Substituting (6.1.8) into Poisson's equation (2.6.2) one then obtains:

$$\nabla^2 V = \frac{|J|}{(2|\eta|)^{1/2} \varepsilon_0} V^{-1/2} \qquad (6.1.9)$$

in writing which we have taken care to remember that J as contributed by the flow of electrons carries a negative sign. Further, here, the problem is considered as one-dimensional, J and v both being directed along z. Considering V to vary only along z, the one-dimensional form of (6.1.9) would be (see(2.6.4(a))):

$$\frac{d^2 V}{d z^2} = \frac{|J|}{(2|\eta|)^{1/2} \varepsilon_0} V^{-1/2} . \qquad (6.1.10)$$

(Here, we have replaced the partial derivative as the complete derivative since V is now a function of only one variable, z). Multiplying both sides of (6.1.10) by 2(dv/dz) and integrating we have

$$\left(\frac{dV}{dz}\right)^2 = \frac{4|J|}{(2|\eta|)^{1/2} \varepsilon_0} V^{1/2} , \qquad (6.1.11)$$

where the integration constant becomes zero in view of the boundary conditions (6.1.3) and (6.1.4). Taking the square root of (6.1.11),

$$\frac{dV}{dz} = \left(\frac{4|J|}{(2|\eta|)^{1/2} \varepsilon_0}\right)^{1/2} V^{1/4} . \qquad (6.1.12)$$

(Since V increases along z (Fig. 6.1.1), only the positive value of the square root is admissible here). Expressing (6.1.12) as

$$\frac{dV}{V^{1/4}} = \left(\frac{4|J|}{(2|\eta|)^{1/2} \varepsilon_0}\right)^{1/2} dz ,$$

and integrating it we have

$$V^{3/4} = \frac{3}{4}\left(\frac{4|J|}{(2|\eta|)^{1/2} \varepsilon_0}\right)^{1/2} z \qquad (6.1.13)$$

where the integration constant becomes zero in view of the boundary condition (6.1.4). The alternative forms of (6.1.13) are obtained by a simple re-arrangement of terms as follows:

$$|J| = \left(\frac{4}{9}\right)(2|\eta|)^{1/2}\varepsilon_0 \frac{V^{3/2}}{z^2} \tag{6.1.14}$$

and

$$V = \left(\frac{9|J|}{4(2|\eta|)^{1/2}\varepsilon_0}\right)^{2/3} z^{4/3} . \tag{6.1.15}$$

Thus we have deduced Child-Langmuir law in the form of (6.1.14) or (6.1.15). This is also known as 3/2-power law because of the index of power of V in the expression for J given by (6.1.14).

Example 6.1.1 Numerically appreciate how the anode current per unit area in a diode with its anode raised to a potential 1kV with respect to the cathode would change if the cathode -to-anode separation is reduced from 15cm to 1mm.

From (6.1.14), putting $J = J_0$, $V = V_0$ and $z = d$ one may write

$$|J_0| = \left(\frac{4}{9}\right)(2|\eta|)^{1/2}\varepsilon_0 \frac{V_0^{3/2}}{d^2} , \tag{6.1.16}$$

where $| J_0 |$ is the anode current per unit area, V_0 is the anode potential with respect to the cathode, d is the cathode -to- anode distance. By a simple calculation from (6.1.16) one obtains $| J_0 | = 3.3\ \mu A/mm^2$ at $V_0 = 1kV$, taking d = 15cm. For the same anode potential $V_0 = 1kV$, this quantity, $| J_0 |$, would increase to $73.8\ mA/mm^2$, if the separation is reduced to d = 1mm. (The value of the constant, the quantity independent of V and z in (6.1.14) (or, of V_0 and d in (6.1.16)), comes out to be $\approx 2.33 \times 10^{-6}$ amp/(volt)$^{3/2}$ (on substituting the values of $|\eta| = |e|/m$ and ε_0).

It is clear from (6.1.16) that the anode current which could be drawn from a diode of a given geometry (that is, for given anode area and anode-to-cathode separation) divided by 3/2-power of anode potential is a constant. If an electron beam could be taken beyond the anode, say, through a perforation in the anode, then the anode voltage and the anode current would be interpreted as the beam voltage and beam current, respectively, and the quantity beam current divided by 3/2-power of beam voltage is referred to as the beam perveance and considered as an important parameter in the study of an electron gun.

In the foregoing analysis leading to the deduction of Child-Langmuir's law, given by (6.1.14) or (6.1.15), it has been assumed that the initial velocity of electrons and the electric field intensity both are zero at the cathode. In fact, however, the electrons at the cathode have a velocity distribution from zero to infinity. A reasonable understanding of the phenomenon could be developed by assuming that all the electrons emit from the cathode with the same initial velocity. By utilizing the kinetic energy of the electrons, the emitted electrons can overcome the space-charge barrier near the cathode to reach a plane close to the cathode where the electrons would have now a zero velocity and the potential contour would reach a minimum as indicated by the solid-line curve in Fig. 6.1.1. Obviously the slope of the potential variation and hence the electric field intensity would also be zero at this plane. Thus the location of the plane of zero electron velocity and zero electric field intensity would now shift from the actual cathode to a plane close to the cathode called the virtual cathode. The potential minimum $V_{min.}$ at the virtual cathode (measured below the zero reference at the actual cathode) may be found by the energy balance equation (6.1.5) in which v has to be interpreted now as the initial velocity of electrons and V as $V_{min.}$. The emitted electrons which all come to rest at the virtual cathode may proceed either towards the cathode or towards the anode. It can be shown that the virtual cathode approaches closer to the actual cathode if the anode-to-cathode spacing is reduced, the anode potential is increased, the potential minimum at the virtual cathode ($V_{min.}$) or the initial velocity of electrons at the actual cathode is less, and a greater fraction of emitted electrons proceed towards the anode from the virtual cathode (see (6.1.20) in Example 6.1.2).

Example 6.1.2 Considering that the electrons could move from the virtual cathode either towards the actual cathode or towards the anode, apply the concepts leading to Child-Langmuir's law separately to the region between the virtual cathode and the anode as well as to the region between the virtual cathode and the actual cathode, and hence derive an expression for the distance of the virtual cathode from the actual cathode (see the solid line curve of Fig. 6.1.1).

All the emitted electrons come to a stop at the virtual cathode and let a fraction α of them proceed toward the anode from the virtual cathode, while the remaining fraction $(1-\alpha)$ proceed towards the actual cathode.

In the region between the virtual cathode and the anode, the volume charge density ρ is given by the following expression, instead of (6.1.8):

$$\rho = \frac{\alpha J_e}{(2|\eta|)^{1/2}} V'^{-1/2} , \tag{6.1.17}$$

where J_e is the emitted current density and V' is the potential in the region considered, measured, however, with respect to the bottom of the potential well, that is, with respect to the virtual cathode (Fig. 6.1.1). Therefore, using (6.1.17) and the same method as followed in obtaining (6.1.13) and hence also (6.1.15), one obtains at the anode ($V' = V_0 + V_{min}$)

$$\alpha J_e = \left(\frac{4}{9}\right)(2|\eta|)^{1/2} \varepsilon_0 \frac{(V_0 + V_{min})^{3/2}}{(d-\Delta)^2}, \tag{6.1.18}$$

where Δ is the distance of the virtual cathode from the actual cathode, so that $(d - \Delta)$ measures the distance of the anode from the virtual cathode, d being the actual cathode-to-anode distance. Further, here the anode is at $V_0 + V_{min}$ with respect to the trough of the potential (since it is at V_0 with respect to the cathode while the latter itself is at V_{min} with respect to the potential trough).

In the region between the virtual cathode and the actual cathode, there exist two components of current — J_e, as constituted by the flow of electrons from the actual cathode to the virtual cathode, and $(1 - \alpha)J_e$, as constituted by the returning electrons from the virtual cathode to the actual cathode. As far as the volume charge density, ρ, in this region is concerned these components would however, add irrespective of the direction of flow of electrons. Therefore ρ would be given by (see (6.1.7) and (6.1.8) and (6.1.17)):

$$\rho = \frac{J_e + (1-\alpha)J_e}{(2|\eta|)^{1/2}} V'^{-1/2} \quad ,$$

which then yields an expression similar to (6.1.18), but now at the cathode, as:

$$J_e + (1 - \alpha)\, J_e = \left(\frac{4}{9}\right)(2|\eta|)^{1/2} \varepsilon_0 \frac{V_{min}^{3/2}}{\Delta^2} \quad . \tag{6.1.19}$$

Combining (6.1.18) and (6.1.19) and by a little re-arrangement of terms one then obtains

$$\Delta = \frac{d}{1 + (\frac{V_0}{V_{min}} + 1)^{3/4} (\frac{2}{\alpha} - 1)^{1/2}} \quad , \tag{6.1.20}$$

the desired expression giving the position of the virtual cathode with respect to the actual cathode.

Appendix a6.1.1 *Relation between beam parameters*

Let us assume that all the electrons move with the same velocity v along z, and constitute a convection current density J, also in the same direction. Let us now consider a cylindrical volume such that the length of the cylinder is along the direction of electron velocities and is of numerical value v, and that the two faces of the cylinder, normal to the length, are identical and each of unit area. The volume(length × area) of the cylinder then equals v. Clearly then in unit time the electrons lying on one of its faces would cross the distance v to reach the other face. Therefore, all the electrons in the cylindrical volume thus considered would in the period of unit time leave the cylinder crossing one of its faces of unit area. Such number, which is equal to the number density (number per unit volume) n, say, times the volume of the cylinder v then represents the number of electrons crossing unit area per unit time. Multiplying this by the charge e each electron carries we then have the current density J, the current per unit area, as follows:

$$J = e\, n\, v$$

which on putting $ne = \rho$, the volume charge density, gives

$$J = \rho v \ ,$$

which is the relation (6.1.7) used in the text.

6.2 Parallel-Flow Pierce Gun

Let us take up here a method due to Pierce [1] of forming a rectangular strip beam using finite-dimension electrodes and considering the emitting surface to be flat. The problem is the simplest and most fundamental of its kind in demonstrating, in general, Pierce's method of realizing an electron gun. Further the results of this problem, as we may see later, would have applications in developing concepts related to forming a cylindrical beam of circular cross section as well. The method is to simulate conditions such that the electrons at the edge of the beam do not *feel* that the region outside the beam is devoid of charges. So, the beam electrons including those at its edge behave as if they are the components of a parallel flow between infinite, parallel cathode and anode each of planar geometry. To be more specific the conditions to be simulated must ensure [5]

i) that both inside and at the edge of the beam, the potential varies according to Child Langmuir's law (6.1.15); and

ii) that there is no electrostatic force transverse to the rectilinear flow on the beam-edge electrons.

The above conditions are realized in this method by providing electrodes, usually two — one at the cathode potential while the other at the anode potential, such that these electrodes coincide with the equipotential surfaces at these two potentials. These equipotentials, inside the beam, are each parallel planes transverse to the rectilinear electronic motion (according to equation (6.1.15)). However, the equipotentials outside the beam are to be found by interpreting the solution of Laplace's equation (2.6.5), subject to the two conditions (i) and (ii) stated above.

The electrode outside the beam coinciding with the equipotential at the cathode potential (taken as the zero reference potential) is called the beam-forming (or focusing) electrode (BFE)(Fig. 6.2.1). The BFE, though it is at the cathode potential and it starts from the cathode is kept thermally insulated from the latter and is essentially a non-emitting electrode. Also, the material separating the BFE from the cathode is chosen to be a good electrical insulator as well so that one could if necessary provide a potential difference between these electrodes. The anode in an electron gun system is not supposed to obstruct the flow of electrons. As such in the electron flow region it could be at best in the form of a grid which, in the present problem, would be parallel to the cathode. The electrode outside the beam coinciding with the equipotential at the anode potential is not physically separated from the anode inside the beam of electrons. In fact, in many applications the anode structure in the form of a grid inside the beam is done away with so that the electrode at the anode potential outside the beam , to be called as the anode now, would have a hole or aperture to allow the electron beam to pass through it. (The effect of introducing an anode aperture has been considered in section *6.3*).

Let us now formulate the problem in the rectangular system of coordinates (x, y, z) in which the origin is chosen on the cathode and at the beam edge (Fig. 6.2.1). The flow of electrons is considered to be along z, and for the sake of simplicity the dimension of the cathode in the x-direction, and hence that of the cross section of the beam derived therefrom as well, along the same direction are considered to be large compared to their dimensions measured along y. This would essentially make zero the potential variation along x and hence enable one to write Laplace's equation (2.6.5), through (2.6.4(a)), in the beam-free region $(y > 0)$, in the following two-dimensional form:

$$\frac{\partial^2 V}{\partial z^2} + \frac{\partial^2 V}{\partial y^2} = 0 \ . \tag{6.2.1}$$

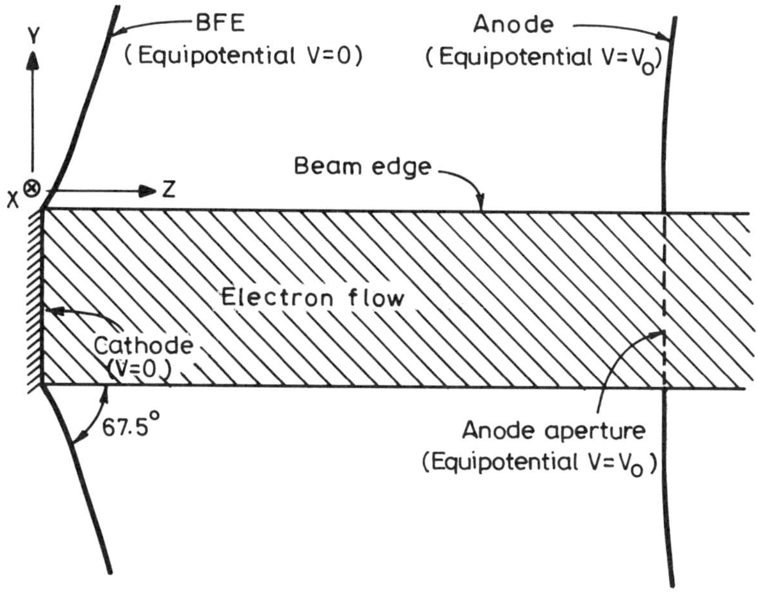

Fig. 6.2.1 Cross section of Pierce-gun for a rectangular strip beam showing the BFE and the anode as well as the system of coordinates in which the problem is formulated.

Equation (6.2.1) may be solved subject to the two already mentioned *boundary* conditions which may be mathematically stated as:

$$V\{z\}\Big|_{y=0} = \left(\frac{9|J|}{4(2|\eta|)^{1/2}\varepsilon_0}\right)^{2/3} z^{4/3} , \tag{6.2.2}$$

which indicates that the potential along the beam edge is given by Child-Langmuir's law (6.1.15); and

$$\frac{\partial V}{\partial y}\Big|_{y=0} = 0 , \tag{6.2.3}$$

which means that there is no electrostatic force $-\partial V/\partial y$, along y at the beam edge $(y = 0)$ which would be responsible for deviating the beam electrons from the rectilinear flow.

Now, since it is known (see appendix *a6.2.1*) that the real part of a complex, analytical function $f\{z + j\,y\} = u_r + j\,u_i$, say, would satisfy the two-dimensional Laplace's equation

$$\frac{\partial^2 u_r}{\partial z^2} + \frac{\partial^2 u_r}{\partial y^2} = 0 \quad ,$$

one may choose the following function V{z, y} as the one which would obey Laplace's equation (6.2.1) and, at the same time, satisfy the conditions (6.2.2) and (6.2.3), both at the beam edge:

$$\begin{aligned} V\{z,y\} &= Re\ f\{z + j\,y\} \\ &= Re\left(\frac{9\,|J|}{4\,(2|\eta|)^{1/2}\varepsilon_0}\right)^{2/3} (z + j\,y)^{4/3} \ . \end{aligned} \qquad (6.2.4)$$

Thus the potential at any point outside the beam is given by (6.2.4) which may also be put as

$$\begin{aligned} V\{z,y\} &= Re\left(\frac{9\,|J|}{4\,(2|\eta|)^{1/2}\varepsilon_0}\right)^{\frac{2}{3}} \left[(z^2 + y^2)^{\frac{1}{2}}\left(\cos\theta + j\,\sin\theta\right)\right]^{\frac{4}{3}} \\ &= Re\left(\frac{9\,|J|}{4\,(2|\eta|)^{1/2}\varepsilon_0}\right)^{\frac{2}{3}} (z^2 + y^2)^{\frac{2}{3}}\left(\cos\frac{4\theta}{3} + j\,\sin\frac{4\theta}{3}\right) \\ &= \left(\frac{9\,|J|}{4\,(2|\eta|)^{1/2}\varepsilon_0}\right)^{\frac{2}{3}} (z^2 + y^2)^{\frac{2}{3}} \cos\frac{4\theta}{3} \ , \quad (y > 0), \end{aligned} \qquad (6.2.5)$$

where $\theta = \tan^{-1} y/z$.

Thus, in principle, knowing the potential values at various points (z, y) with the help of (6.2.5) one should be able to find the equipotential at the desired potential value. For example, one may find, and that is also the simplest thing to do, the equipotential at zero potential outside the beam which would also give the shape of the BFE. Thus putting equal to zero the left hand side of (6.2.5) one has $\cos(4\theta/3) = 0$ giving

$$\theta = \tan^{-1} y/z = 3\pi/8 = 67.5^\circ \qquad (6.2.6)$$

suggesting that the cross section of the BFE would essentially be a straight line that makes an angle of 67.5° with the beam edge (Fig. 6.2.1). The complete shape of the BFE in three dimensions could now be generated by moving the line thus obtained along $\pm$ x directions. It may be seen that the result (6.2.5) is independent of beam-cross-sectional dimensions, and so an identical shape could be predicted for the BFE in the vicinity of the remaining edge of the beam strip (Fig. 6.2.1). Thus, by this method one could easily obtain the shape of the complete BFE *hat*. However, finding the shape of the anode hat would not be so much simple since the solution of (6.2.5) for a non-zero value in its left hand side corresponding to a non-zero anode potential would not be as straightforward as the solution (6.2.6) for a BFE. A simple approach could be to find the potential values with the help of (6.2.5) in the vicinity of the anode and then generate the locus of points having the same value of potential as that of the anode.

We have considered here a rather simple problem of a strip beam of one of its cross-sectional dimensions taken as very much larger than the other. A problem of more practical relevance would be to form a cylindrical beam of circular cross section. An approximate solution to this problem for points close to the beam edge would continue to be given by the solution of the rectangular-strip beam case, since such close-by points would fail to *see* the curvature of the circular cross section beam. For instance, the BFE would continue meet the beam edge in this case, too, at an angle of 67.5° given by (6.2.6). The complete shape of an electrode now could be generated by rotating the section of the electrode about the z-axis. Thus near the beam edge, the BFE would be a section of cone of a half angle of 67.5°. At this stage, however, we do not intend to take up this problem at so much rigor. Later in section *6.3*, we have considered in details the problem of forming a circular-cross section beam starting from a convergent conical beam which in turn has been derived from a curved, instead of a flat, cathode.

Appendix a6.2.1 *Properties of a complex analytic function*

The potential function is essentially an analytic function which has the property of having a derivative that is independent of the direction of approach to the point [6]. In order to arrive at some of its properties let us take a complex function of $z + jy = Z$, *say:*

$$W = f\{Z\} = f\{z + jy\} = u_r + j\,u_i, \quad \text{say.} \tag{a6.2.1}$$

If the function (a6.2.1) is analytic, then the derivative

$$\frac{dW}{dZ} = \mathop{Lt}_{\Delta Z \to 0} \frac{\Delta W}{\Delta Z}$$

is the same no matter how the limit ΔZ $(= \Delta z + j\Delta y) \to 0$ *is arrived at —whether by first putting* $\Delta z = 0$ *and then letting* $\Delta y \to 0$ *or by first putting* $\Delta y = 0$ *and then letting* $\Delta z \to 0$. *Thus the condition for the function to be analytic may be put as:*

$$\left(\frac{dW}{dZ} =\right) \mathop{Lt}_{\substack{\Delta z = 0 \\ \Delta y \to 0}} \frac{\Delta W}{j\,\Delta y} = \mathop{Lt}_{\substack{\Delta y = 0 \\ \Delta z \to 0}} \frac{\Delta W}{\Delta z}$$

which then becomes

(a6.2.2) $$\frac{1}{j}\frac{\partial W}{\partial y} = \frac{\partial W}{\partial z} \quad .$$

Substituting (a6.2.1) into (a6.2.2), we get

$$-j\frac{\partial u_r}{\partial y} + \frac{\partial u_i}{\partial y} = \frac{\partial u_r}{\partial z} + j\frac{\partial u_i}{\partial z}$$

which, on equating the real and imaginary parts, yields

(a6.2.3) $$\frac{\partial u_i}{\partial y} = \frac{\partial u_r}{\partial z} \qquad \textit{and}$$

(a6.2.4) $$-\frac{\partial u_r}{\partial y} = \frac{\partial u_i}{\partial z} ,$$

called the Cauchy-Riemann conditions. Taking the derivatives of (a6.2.3) and (a6.2.4) one can further show

$$\frac{\partial^2 u_r}{\partial z^2} = \frac{\partial^2 u_i}{\partial z\, \partial y} , \quad \frac{\partial^2 u_r}{\partial y^2} = -\frac{\partial^2 u_i}{\partial y\, \partial z} \quad ; \textit{ and}$$

$$\frac{\partial^2 u_i}{\partial z^2} = -\frac{\partial^2 u_r}{\partial z\, \partial y} , \quad \frac{\partial^2 u_i}{\partial y^2} = \frac{\partial^2 u_r}{\partial y\, \partial z} ,$$

whence invoking the property of partial differentiation:

$$\frac{\partial^2 u_{r,i}}{\partial z\, \partial y} = \frac{\partial^2 u_{r,i}}{\partial y\, \partial z} ,$$

we obtain

(a6.2.5) $$\frac{\partial^2 u_r}{\partial z^2} + \frac{\partial^2 u_r}{\partial y^2} = 0$$

as well as

(a6.2.6) $$\frac{\partial^2 u_i}{\partial z^2} + \frac{\partial^2 u_i}{\partial y^2} = 0 ,$$

to show that both the real (u_r) and the imaginary (u_i) parts of the complex analytic function $W = f\{z + j\, y\}$ separately satisfy Laplace's equation in two dimensions (z, y).

6.3 Convergent-Flow Pierce Gun

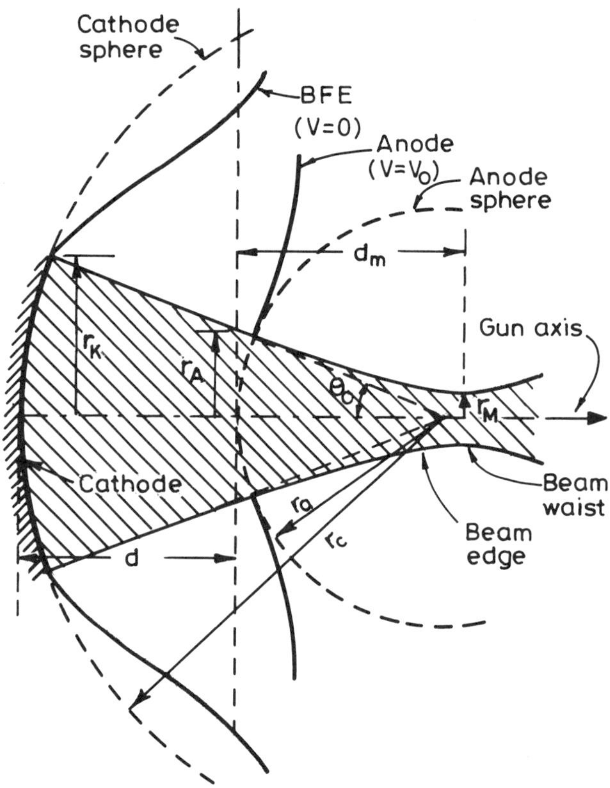

Fig. 6.3.1 Cross-sectional geometry of the convergent Pierce-gun showing the spherical-cup cathode and the anode aperture/grid inside the beam; the BFE and the anode, outside the beam; the beam edge and the beam waist (or throat); and the concentric cathode and anode spheres, also indicating the relevant parameters: cathode and anode radii of curvature (r_c, r_a, respectively), cathode-disc and anode-aperture radii (r_K, r_A, respectively), cathode-to-anode distance ($d=r_c-r_a$), beam-waist radius (r_M) and beam-waist distance from the anode (d_m).

In a parallel-flow electron gun derived from a flat cathode (section *6.2*) the beam cross section has the same area as the cathode. Therefore, more beam current from such a gun would call for a larger cathode and hence a larger beam-cross-sectional area as well. But, unfortunately, from the consideration of the design of the device in which such a beam could be employed, there would be a restriction in the increase of the beam cross-sectional area particularly at higher values of operating frequency. At such frequencies, it is known (see chapter *8*), the *transverse* cross-sectional dimensions of the electromagnetic structure and hence those of the electron beam interacting with the electromagnetic waves supported by such a structure would have to be reduced. An alternative method to combat this problem would be to extract more beam current from a large, *curved* cathode and then *converge* the beam to the desired cross section of reduced area. Further, if a larger beam current is not required, such a method would give one an option of lightly loading the cathode for its better life.

Let us now take the cathode in the form of a spherical cup, which emits electrons radially and perpendicular to its surface (Fig. 6.3.1). But let us take up before that a more basic problem in which we have two concentric spheres of which the entire *outer sphere* emits electrons radially as the cathode and that the entire *inner sphere* receives them radially as the anode. The problem which is obviously to be studied now in the spherical system of coordinates enjoys the *one-dimensional* similarity with the case of the rectilinear flow of electrons between large parallel cathode and anode planes (section *6.1*). There, we may recall, we had solved the one-dimensional Poisson's equation in the rectangular system to obtain what is known as Child-Langmuir's law in the form of (6.1.14) or (6.1.15). Here, in the case of radial flow between concentric cathode and anode spheres, we have to proceed to solve Poisson's equation (2.6.2) which may be written with the help of (2.6.4(c)) taking now potential to vary only with the radial coordinate r ($\partial V/\partial\theta = \partial V/\partial\varphi = 0$) as follows:

$$\frac{d^2V}{dr^2} + \frac{2}{r}\frac{dV}{dr} = -\frac{\rho}{\varepsilon_0} \tag{6.3.1}$$

(in which care has been taken to replace the partial derivative by the complete derivative, V being a function of only r).

6.3.1 Langmuir and Blodgett's Solution

Let us try a series solution of (6.3.1). An important clue to this problem would be that, whatever be the solution proposed, it should as a special case pass on to the solution for a large flat and parallel anode-cathode assembly given by (6.1.15) which may be re-written as:

$$V = \left(\frac{9I_0}{4(2|\eta|)^{1/2}\varepsilon_0 A_c}\right)^{2/3} z^{4/3} , \tag{6.3.2}$$

replacing $|J|$ by I_0/A_c, where I_0 is the beam current interpreted as a positive current and A_c is the cathode area, noting that, in the special case of a parallel flow between large flat planes, $|J|$ does not change with z, and has the same value as it has, namely, I_0/A_c, at the cathode (z = 0). Therefore, taking (6.3.2) as the guideline one may suggest a series solution for (6.3.1) as:

$$V = \left(\frac{9I_0}{4(2|\eta|)^{1/2}\varepsilon_0 A_c}\right)^{2/3} r_c^{4/3} G\{u\}^{4/3}, \qquad (6.3.3)$$

where r_c is the radius of curvature of the cathode which is related to the cathode area A_c as

$$A_c = 4\pi\, r_c^2 \; ; \qquad (6.3.4)$$

and G{u} is a dimensionless series in

$$u = \ln(r/r_c) \qquad (6.3.5)$$

which, in the Tailor's form, may be put as

$$G\{u\} = G\{0\} + uG'\{0\} + \frac{u^2}{2!}G''\{0\} + \frac{u^3}{3!}G'''\{0\} + \frac{u^4}{4!}G''''\{0\} + \ldots, \qquad (6.3.6)$$

where G', G", G"', ... etc. are, respectively, the first and higher derivative of G with respect to u. Obviously, here care has been taken while writing (6.3.3) that the latter is dimensionally correct which can be easily appreciated by comparing (6.3.3) with (6.3.2). Substituting (6.3.4) into (6.3.3), an alternative form of the suggested potential function (6.3.3) is obtained as follows:

$$V = \left(\frac{9I_0}{16\pi(2|\eta|)^{1/2}\varepsilon_0}\right)^{2/3} G^{4/3}\{u\}, \qquad (6.3.7)$$

which, further, introducing a constant

$$K' = \frac{I_0}{4\pi\varepsilon_0(2|\eta|)^{1/2}} \qquad (6.3.8)$$

may be expressed as

$$V = \left(\frac{9}{4}K'\right)^{2/3} G^{4/3}\{u\} \; . \qquad (6.3.9)$$

It is now necessary to see whether (6.3.7), as a special case, would pass on to (6.3.2) and, if so, under what conditions. For this purpose, let us take the radii of curvature of the concentric cathode and anode spheres to be much larger than the cathode-to-anode distance so that the system could approach the planar cathode-anode system(section *6.1*). Then one may put

$$z/r_c \ll 1 \tag{6.3.10}$$

where z is the distance of a point in the cathode-anode region measured from, and perpendicular to, the cathode. z may be expressed as

$$z = r_c - r \tag{6.3.11}$$

where r is the radial coordinate of the point considered in the region. Using (6.3.10) and (6.3.11), therefore, one may write for u from (6.3.5) as:

$$u = \ln(r/r_c) = \ln(r_c - z)/r_c = \ln(1 - z/r_c) \cong -z/r_c\ , \tag{6.3.12}$$

ignoring higher powers of z/r_c in the expansion of the logarithmic function. With the help of (6.3.12) one may then write (6.3.6) as

$$G\{u\} \cong G\{0\} - (z/r_c)\, G'\{0\}\ , \tag{6.3.13}$$

ignoring higher powers of z/r_c in view of (6.3.10). Further at the cathode, $r = r_c$, one has, as can be seen from (6.3.5), $u = \ln 1 = 0$. So, choosing the cathode potential $V = 0$ and putting $u = 0$ at the cathode, we have from (6.3.3), at the cathode:

$$G\{0\} = 0\ . \tag{6.3.14}$$

Substituting (6.3.14) in (6.3.13) one then has

$$G\{u\} \cong (-z/r_c)\, G'\{0\}$$

which may be now substituted in (6.3.3) to give

$$V = \left(\frac{9 I_0}{4(2|\eta|)^{1/2} \varepsilon_0 A_c}\right)^{2/3} G'\{0\}^{4/3}\, z^{4/3}$$

whence we get

$$V = \left(\frac{9 I_0}{4(2|\eta|)^{1/2} \varepsilon_0 A_c}\right)^{2/3} z^{4/3}\ , \tag{6.3.15}$$

provided we choose to put $G'\{0\}^{4/3} = 1$, that is,

$$G'\{0\} = \pm 1\ . \tag{6.3.16}$$

Thus we see that (6.3.15) is identical with (6.3.2) under the approximation (6.3.10). So we have a justification to support the use of the proposed potential function (6.3.7), which involves G{u} which in turn is given by (6.3.6) in which one has to take G'{0} as given by

(6.3.16). It remains, therefore, to find the other higher-order derivatives G"{0}, G'''{0}, etc. of the series expansion, before (6.3.7) could be used as the solution for Poisson's equation (6.3.1) being sought for.

For this purpose, first Poisson's equation (6.3.1) is written with the help of (6.1.8) as

$$\frac{d^2V}{dr^2} + \frac{2}{r}\frac{dV}{dr} = \frac{|J|\,V^{-1/2}}{(2|\eta|)^{1/2}\,\varepsilon_0} \quad . \tag{6.3.17}$$

Writing current density $|J|$ at a point of radial coordinate r between the cathode and the anode($r_a < r < r_c$) as

$$|J| = \frac{I_0}{4\pi r^2} \quad , \tag{6.3.18}$$

where I_0 is the beam current constituted by the electron flow from the cathode to the anode sphere, $4\pi r^2$ being the area of the sphere at radius r, and substituting (6.3.18) in (6.3.17), we have

$$\frac{d^2V}{dr^2} + \frac{2}{r}\frac{dV}{dr} = \frac{I_0}{4\pi(2|\eta|)^{1/2}\,\varepsilon_0}\,\frac{V^{-1/2}}{r^2}$$

$$= \frac{K'}{r^2}V^{-1/2} \quad . \tag{6.3.19}$$

It may be seen that (6.3.19) involves the first and the second derivatives of V which may be obtained by differentiating (6.3.9) as follows:

$$\frac{dV}{dr} = \left(\frac{4}{3}\right)\left(\frac{9}{4}K'\right)^{\frac{2}{3}} G^{\frac{1}{3}}\frac{dG}{dr} \quad ,$$

$$\frac{d^2V}{dr^2} = \left(\frac{4}{3}\right)\left(\frac{9}{4}K'\right)^{\frac{2}{3}}\left[\frac{1}{3}G^{-2/3}\left(\frac{dG}{dr}\right)^2 + G^{1/3}\frac{d^2G}{dr^2}\right] . \tag{6.3.20}$$

Substituting (6.3.20) into (6.3.19), we obtain

$$r^2\left(\frac{dG}{dr}\right)^2 + 3r^2 G\frac{d^2G}{dr^2} + 6rG\frac{dG}{dr} - 1 = 0 \quad . \tag{6.3.21}$$

Further, in view of :

$$u = \ln \frac{r}{r_c} \qquad \text{(see (6.3.5))},$$

$$\frac{du}{dr} = \left(\frac{1}{r/r_c}\right)\left(\frac{1}{r_c}\right) = \frac{1}{r},$$

$$\frac{dG}{dr} = \left(\frac{dG}{du}\right)\left(\frac{du}{dr}\right) = \frac{G'\{u\}}{r},$$

and
$$\frac{d^2G}{dr^2} = \frac{1}{r}G''\{u\}\frac{du}{dr} - \frac{1}{r^2}G'\{u\} = \frac{G''\{u\} - G'\{u\}}{r^2},$$

(6.3.21) may be expressed as

$$G'\{u\}^2 + 3\,G\{u\}\,G''\{u\} + 3\,G\{u\}\,G'\{u\} - 1 = 0\,. \tag{6.3.22}$$

Now, at the cathode ($r = r_c$, $u = 0$), one has from (6.3.14) $G\{u\} = G\{0\} = 0$ which when substituted into (6.3.22) immediately yields

$$G'\{u\} = \pm 1\,, \tag{6.3.23}$$

a result which agrees with (6.3.16).

Out of the two solutions (6.3.23) or (6.3.16) the one with the minus sign pertains to the convergent beam in which the cathode is outside the anode($r_c > r_a$). This is so because, here, V decreases with r (say, from $V = V_0$ at $r = r_a$ to $V = 0$ at $r = r_c > r_a$) and hence G which is related to V as (6.3.9) also decreases with r and hence with u. (We notice that u (= ln r/r_c, equation (6.3.5)) decreases with r, for instance, from $u = 0$ at $r = r_c$ to a negative value, ln r_a/r_c, at $r = r_a$, where $r_a < r_c$). Similarly, it can be appreciated that the solution (6.3.23) or (6.3.16) with a plus sign would refer to the divergent beam case in which the anode and the cathode spheres would have their positions interchanged giving $r_a > r_c$. Thus (6.3.23) may be elaborated as:

$$G'\{0\} = -1, \quad \text{for a convergent beam} \qquad \text{(a)}$$

and
$$G'\{0\} = +1, \quad \text{for a divergent beam}\,. \qquad \text{(b)} \qquad (6.2.24)$$

In order to determine $G''\{0\}$ what we have to do is to differentiate (6.3.22) as follows

$$2\,G'\{u\}\,G''\{u\} + 3\,G\{u\}\,G'''\{u\} + 3\,G''\{u\}\,G'\{u\} + 3\,G\{u\}\,G''\{u\} + 3\,G'\{u\}^2 = 0 \tag{6.3.25}$$

and put $u = 0$, (6.3.14) and (6.3.24) in (6.3.25), to obtain

$$G''\{0\} = +\frac{3}{5}, \quad \text{for a convergent beam} \qquad \text{(a)}$$

and
$$G''\{0\} = -\frac{3}{5}, \quad \text{for a divergent beam} \qquad \text{(b)} \qquad (6.3.26)$$

Similarly, $G'''\{0\}$ may be determined, by differentiating (6.3.25) and making use of (6.3.14), (6.3.24) and (6.3.26), to obtain

$$G'''\{0\} = -\frac{18}{40}, \quad \text{for a convergent beam} \qquad (a)$$

$$G'''\{0\} = +\frac{18}{40}, \quad \text{for a divergent beam}; \qquad (b) \qquad (6.3.27)$$

and the same procedure could be extended to yield the values of higher order derivatives

$$G''''\{0\} = +\frac{189}{550}, \quad \text{for a convergent beam} \qquad (a)$$

$$G''''\{0\} = -\frac{189}{550}, \quad \text{for a divergent beam}, \qquad (b) \qquad (6.3.28)$$

and so on.

Therefore, the function $G\{u\}$ may now be spelt out from (6.3.6) by making use of (6.3.14), (6.3.24), (6.3.26), (6.3.27), (6.3.28), etc., as:

$$G\{u\} = -u + \frac{3}{10}u^2 - \frac{3}{40}u^3 + \frac{63}{4400}u^4 - \ldots, \quad \text{for a convergent beam} \quad (a)$$

$$G\{u\} = u - \frac{3}{10}u^2 + \frac{3}{40}u^3 - \frac{63}{4400}u^4 + \ldots, \quad \text{for a divergent beam}. \quad (b) \qquad (6.3.29)$$

Finally, expressing u as (6.3.5) the series function (6.3.29) may be written as:

for a convergent beam,

$$G\{u\} = G\{\ln\frac{r}{r_c}\} = -\ln\frac{r}{r_c} + \frac{3}{10}\ln^2\frac{r}{r_c} - \frac{3}{40}\ln^3\frac{r}{r_c} + \frac{63}{4400}\ln^4\frac{r}{r_c} - \ldots$$

$$= G\{\ln\frac{r}{r_c}\} = \ln\frac{r_c}{r} + \frac{3}{10}\ln^2\frac{r_c}{r} + \frac{3}{40}\ln^3\frac{r_c}{r} + \frac{63}{4400}\ln^4\frac{r_c}{r} + \ldots \qquad (6.3.30)$$

and

for a divergent beam,

$$G\{u\} = G\{\ln\frac{r}{r_c}\} = \ln\frac{r}{r_c} - \frac{3}{10}\ln^2\frac{r}{r_c} + \frac{3}{40}\ln^3\frac{r}{r_c} - \frac{63}{4400}\ln^4\frac{r}{r_c} + \ldots$$

$$= G\{\ln\frac{r}{r_c}\} = -\ln\frac{r_c}{r} - \frac{3}{10}\ln^2\frac{r_c}{r} - \frac{3}{40}\ln^3\frac{r_c}{r} - \frac{63}{4400}\ln^4\frac{r_c}{r} - \ldots \qquad (6.3.31)$$

Thus in this sub-section we have deduced the solution of the one-dimensional Poisson's equation, for the problem of electron flow between the concentric spherical cathode and anode, in the form (6.3.7) or (6.3.9), in terms of a series function G which in turn for a convergent beam (for a cathode outside an anode) is given by (6.3.30), and for a divergent beam (for a cathode inside an anode) by (6.3.31). The solution (6.3.7) or (6.3.9) when read through (6.3.30) or (6.3.31) is known as Langmuir and Blodgett's solution [7].

6.3.2 Beam Current from a Spherical-Cup Cathode

For the complete spherical geometry of the cathode-anode system, that is, considering the entire outersphere as the electron emitter (see Fig. 6.3.1), one may write from (6.3.7), putting $V = V_0$ at $r = r_a$ (at the anode):

$$V_0 = \left(\frac{9I_0}{16\pi(2|\eta|)^{1/2}\varepsilon_0}\right)^{2/3} G_0^{4/3}\{u\} , \qquad (6.3.32)$$

where $G_0\{u\}$ is the value of $G\{u\}$ corresponding to $r = r_a$ in (6.3.30), that is,

$$G_0\{u\} = \ln\frac{r_c}{r_a} + \frac{3}{10}\ln^2\frac{r_c}{r_a} + \frac{3}{40}\ln^3\frac{r_c}{r_a} + \frac{63}{4400}\ln^4\frac{r_c}{r_a} + \ldots . \qquad (6.3.33)$$

One may re-arrange (6.3.32) as

$$I_0 = \frac{(4/9)(4\pi\varepsilon_0(2|\eta|)^{1/2})V_0^{3/2}}{G_0^2\{u\}} , \qquad (6.3.34)$$

the expression for beam current for the complete cathode sphere emitting electrons. Now if the cathode is in the form of a spherical cup being part of the complete cathode sphere of radius r_c, then the beam current would obviously be reduced from its value given by (6.3.34) by a factor which is equal to the ratio of the area of the spherical-cup emitting portion, $2\pi r_c^2(1-\cos\theta_0)$, to the area of the cathode sphere, $4\pi r_c^2$, giving

$$I_0 = \left[\frac{(4/9)(4\pi\varepsilon_0(2|\eta|)^{1/2})V_0^{3/2}}{G_0^2\{u\}}\right]\left[\frac{2\pi r_c^2(1-\cos\theta_0)}{4\pi r_c^2}\right] , \qquad (6.3.35)$$

where θ_0 is the half cone angle subtended by the spherical cup at the common center of curvature of the cathode and the anode spheres, and hence also is the half beam-cone angle of the convergent conical beam intended to be derived from the spherical-cup cathode. Therefore, from (6.3.35) one may write for the beam perveance:

$$\frac{I_0}{V_0^{3/2}} = [2\pi\varepsilon_0(2|\eta|)^{1/2}]\left[\left(\frac{4}{9}\right)\left(\frac{1-\cos\theta_0}{G_0^2\{u\}}\right)\right], \tag{6.3.36}$$

an alternative form of which in terms of a dimensionless perveance k^2, may be put as:

$$k^2 = \left(\frac{4}{9}\right)\left(\frac{1-\cos\theta_0}{G_0^2\{u\}}\right) \tag{6.3.37}$$

where

$$k = \left(\frac{\text{perv}}{2\pi\varepsilon_0(2|\eta|)^{1/2}}\right)^{1/2}, \tag{6.3.38}$$

expressed in terms of $I_0/V_0^{3/2}$, the beam perveance abbreviated as *perv*. It will be seen in section *6.3.3* that k has the significance of being a normalisation factor for a distance over which the electron beam would spread usually in a field-free region beyond the anode aperture.

Example 6.3.1 A spherical-cup cathode of radius of curvature 29.2mm forms a convergent conical beam of half cone angle 7° in a Pierce-gun set-up in which the anode is separated by a distance of 12.9mm from the cathode. Show that an anode current ~20mA is obtainable from such a gun for an anode potential ~2 kV.

Here, r_c = 29.2 mm and the inter-electrode spacing d = $r_c - r_a$ = 12.9 mm. From this, r_a = (29.2 - 12.9) mm = 16.3 mm, and so r_c/r_a = 1.79, which gives $\ln r_c/r_a = 2.3 \log_{10} 1.79 = 2.3 \times 0.255 = 0.5865$. Using this value, $G_0\{u\}$ can be calculated from (6.3.33). From this we get $G_0^2\{u\} \approx 0.5$ which may be plugged into (6.3.36), along with V_0 = 2 kV = 2000 V and $\theta_0 = 7°$ giving $\cos\theta_0 = \cos 7° = 0.9925$, to get

$$I_0 = [2\pi\varepsilon_0(2|\eta|)^{1/2}][(4/9)(1-\cos\theta_0)/G_0^2][V_0^{3/2}]$$

$$= 14.648\text{x}10^{-6}\text{x}\frac{(2000)^{3/2}}{0.5}\text{x}0.0075\,\text{amp} \approx 20\text{ mA}.$$

6.3.3 Beam Spread under Space-Charge Field

An electron beam would spread due to coulomb repulsion between electrons in a field-free region. Let us quantitatively estimate to what extent such a beam would spread if left over a given distance. This will be helpful in studying the behaviour of the beam beyond the anode aperture of a Pierce gun, and particularly for a convergent conical-flow beam, in predicting the parameters related to the beam waist (Fig. 6.3.1).

For this purpose, let us consider a cylindrical beam of radius r, say. The space-charge electric field intensity E_S experienced by the electrons situated at the beam edge is obtainable from (2.2.2) or (2.3.7) as:

$$E_s = \frac{\rho r}{2\varepsilon_0} . \tag{6.3.39}$$

The force equation for such a beam edge electron may be written as:

$$\frac{d^2 r}{dt^2} = \eta E_s$$

which with the help of (6.3.39) becomes

$$\frac{d^2 r}{dt^2} = \eta \left(\frac{\rho r}{2\varepsilon_0} \right) . \tag{6.3.40}$$

In order to know the beam trajectory one has to express the left hand side of (6.3.40) in terms of a derivative with respect to z, a distance along the gun axis, instead of one with respect to t as:

$$\frac{d^2 r}{dt^2} = \frac{d}{dt}\left(\frac{dr}{dt} \right) = \frac{d}{dt}\left[\left(\frac{dr}{dz} \right)\left(\frac{dz}{dt} \right) \right] . \tag{6.3.41}$$

Putting v = dz/dt, the axial velocity of electrons, and treating it as a constant in the field-free region where the beam is placed, one may write (6.3.41) as

$$\begin{aligned} \frac{d^2 r}{dt^2} &= \frac{d}{dt}\left(v \frac{dr}{dz} \right) = v \frac{d}{dt}\left(\frac{dr}{dz} \right) \\ &= v \left(\frac{d}{dz}\left(\frac{dr}{dz} \right) \right)\left(\frac{dz}{dt} \right) = (v)\left(\frac{d^2 r}{dz^2} \right)(v) \\ &= v^2 \frac{d^2 r}{dz^2} . \end{aligned} \tag{6.3.42}$$

Making use of (6.3.42) one may then express (6.3.40) as:

$$v^2 \frac{d^2 r}{d z^2} = \left(\frac{\eta \rho}{2\varepsilon_0}\right) r \quad . \tag{6.3.43}$$

Further, using (6.1.7), which is the relation between J and ρ, and remembering that η and J each carry a negative sign, one may write from (6.3.43)

$$\frac{d^2 r}{d z^2} = \left(\frac{|\eta||J|}{2 v^3 \varepsilon_0}\right) r \quad . \tag{6.3.44}$$

In the present study we are interested in knowing the beam spread in the region beyond the anode aperture assuming the beam to be in the field-free region. In this region the beam electrons will maintain the same velocity as the one to which they have been accelerated by the anode potential V_0. The velocity v of such electrons will, therefore, be given by (6.1.6) as:

$$v = (2|\eta|V_0)^{1/2} . \tag{6.3.45}$$

Further, writing

$$|J| = \frac{I_0}{(\pi r)^2} \quad , \tag{6.3.46}$$

and making use of (6.3.45) and (6.3.46), one may express (6.3.44) as:

$$\frac{d^2 r}{d z^2} = \frac{k^2}{2 r} \quad , \tag{6.3.47}$$

where k is given by (6.3.38).

Considerable simplification in the prediction of the beam trajectory is achieved if the following normalised quantities are defined:

$$R = r / r_0$$

and

$$Z = k\,(z / r_0) \quad , \tag{6.3.48}$$

where r_0 is a reference value of r. One may further easily show from (6.3.48):

$$r = r_0 R \qquad \text{(a)}$$

$$\frac{dr}{dz} = k \frac{dR}{dZ} \qquad \text{(b)}$$

$$\frac{d^2 r}{d z^2} = \left(\frac{k^2}{r_0}\right) \frac{d^2 R}{d Z^2} \quad . \qquad \text{(c)} \tag{6.3.49}$$

Substituting (6.3.49(c)) and (6.3.49(a)) in the left and the right hand sides of (6.3.47), respectively, one obtains the following equation for beam trajectory in terms of the normalised quantities:

$$\frac{d^2R}{dZ^2} = \frac{1}{2R} \,. \tag{6.3.50}$$

In order to solve (6.3.50) let us first multiply both the left and the right hand sides of (6.3.50) by $2\,dR/dZ$ to obtain:

$$\left(2\frac{dR}{dZ}\right)\left(\frac{d^2R}{dZ^2}\right) = \left(2\frac{dR}{dZ}\right)\left(\frac{1}{2R}\right)$$

which may be written as

$$\frac{d}{dZ}\left(\frac{dR}{dZ}\right)^2 = \left(\frac{dR}{dZ}\right)\left(\frac{1}{R}\right)$$

which upon integration gives

$$\int d\left(\frac{dR}{dZ}\right)^2 = \int\frac{dR}{R} + A$$

whence we get

$$\left(\frac{dR}{dZ}\right)^2 = \ln R + A \quad , \tag{6.3.51}$$

where A is an integration constant which may be set equal to zero if the reference is so chosen that $dR/dZ = 0$, and hence $dr/dz = 0$ (see 6.3.49(b)), at $\ln R = 0$, that is, at $R = 1$ and hence at $r = r_0$ (see (6.3.49(a))). So, taking $dr/dz = 0$ at the reference $r = r_0$, one has $A = 0$ which makes (6.3.51) read as

$$\left(\frac{dR}{dZ}\right)^2 = \ln R$$

giving the following two possible solutions

$$\frac{dR}{dZ} = \pm\,(\ln R)^{1/2} \,. \tag{6.3.52}$$

6.3.3.A Throw of the gun

In the present problem of a conical-flow beam it is expected that the radial electron beam, though it maintains its converging nature, with a half cone angle θ_0 at the beam edge upto the anode (see Fig. 6.3.1), would diverge out due to space-charge forces beyond the aperture of the anode. Thus the beam instead of converging to a common center of the cathode and anode spheres would pass through a minimum beam radius r_M, say, where clearly the slope of the beam trajectory would be zero, that is, $dr/dz = 0$. Thus if we choose $r_M = r_0$, the radius reference, then in view of the discussion following (6.3.51), we may apply (6.3.52) to describe the beam trajectory in the vicinity of the beam waist (or throat) where the beam radius has the minimum value r_M (see Fig. 6.3.1). Clearly, the plus and the minus signs in the right hand side of (6.3.52) refer to the positive and negative slopes of beam trajectory (corresponding to the positive and negative values of dR/dZ, and hence those of dr/dz (see 6.3.49(b)), respectively), and hence, respectively, to the regions to the right and to the left of the beam waist of radius, r_M. Then it should be possible now to find d_m, the distance of the beam waist from the anode, with the help of (6.3.52), with its minus sign, which may be written as:

$$dZ = \frac{-dR}{(\ln R)^{1/2}} \quad . \tag{6.3.53}$$

(6.3.53) may be integrated between the limits $Z = 0$ (corresponding to $z = 0$), $R = 1$ (corresponding to $r = r_0 = r_M$), at the beam waist, and $Z = -k\, d_m/r_M$ (corresponding to $z = -d_m$), $R = r/r_0 = r_A/r_M$ (corresponding to $r = r_A$) at the anode aperture, to get:

$$\int_0^{-k(d_m/r_M)} dZ = \int_1^{r_A/r_M} \frac{-dR}{(\ln R)^{1/2}} \; , \tag{6.3.54}$$

where d_m is the distance of the beam waist from the anode along the gun axis, and r_A is the radius of the anode aperture. Evaluating the integral in the left hand side of (6.3.54) in a straightforward manner, we then have

$$d_m = \left(\frac{r_M}{k}\right) \int_1^{r_A/r_M} \frac{dR}{(\ln R)^{1/2}} \quad . \tag{6.3.55}$$

Carrying out the definite integral as shown in *appendix a6.3.1*, (6.3.55) may now be written as:

$$d_m = \left(\frac{2r_M}{k}\right) \left[\ln^{1/2}\frac{r_A}{r_M} + \frac{\ln^{3/2}\frac{r_A}{r_M}}{3(1!)} + \frac{\ln^{5/2}\frac{r_A}{r_M}}{5(2!)} + \frac{\ln^{7/2}\frac{r_A}{r_M}}{7(3!)} + \cdots \right] . \tag{6.3.56}$$

The location of the beam waist as given by the value of d_m referred to as the *throw* of the gun is an important gun parameter. It may be mentioned that the throw of the gun is also relevant from the standpoint of the operation of a focussing structure to be considered in the following chapter.

Example 6.3.2 Calculate the throw of a convergent Pierce gun that has an anode aperture of 2 mm radius, forming a 10 kV, 1 A, 1 mm (radius) beam. Also appreciate at what distance beyond the waist, the beam would have its radius doubled due to beam spread.

The value of beam perveance $I_0/V_0^{3/2}$ may be calculated as 10^{-6} amp/(volt)$^{3/2}$ (that is, 1 micro perv). Substituting this into (6.3.38), k may be calculated as

$$k = [\text{perv}/\{(2\pi\varepsilon_0)(2|\eta|)^{1/2}\}]^{1/2} = (\text{perv})^{1/2}/(5.72 \times 10^{-3}) = (10^{-6})^{1/2}/(5.72 \times 10^{-3})$$

$$= 1/5.72 \ .$$

Substituting the above value of k together with the given values $r_M = 1\text{mm}$, $r_A = 2\text{mm}$ into (6.3.56), we may calculate the throw d_m as

$$d_m = [2 \times 1 \times 5.72]\,[\ln^{1/2} 2 + (1/3)(\ln^{1/2} 2)^3 + (1/10)(\ln^{1/2} 2)^5 + ...]$$

$$\cong [2 \times 1 \times 5.72]\,[0.83 + 0.19 + 0.04] = 12.13 \text{ mm.}$$

Now, the beam trajectory has a symmetry about the beam waist on its two sides according to (6.3.52). Thus the beam trajectory beyond the beam waist may be considered as the mirror image of the one between the beam-waist and the anode. In this particular problem the distance between the beam-waist and the anode where the beam would have a radius ($= r_A = 2\text{mm}$) equal to double the beam-waist radius ($= r_M = 1\text{mm}$) has already been calculated above as $d_m = 12.13$ mm. In view of the symmetry, therefore, beyond the beam-waist the beam will spread and double its radius from its value at the beam-waist at exactly the same distance, namely, 12.13 mm away from the beam-waist on the opposite side of the anode.

6.3.3.B *Relation between the anode aperture and beam-waist radii*

The beam transit angle at the anode is measured with respect to a direction parallel to the gun axis, and it is equal to the half-cone angle θ_0 itself, unless it is modified due to lens effects at the anode aperture in the presence of dissimilar fields on the two sides of the aperture (section *6.3.3.C*). Taking this beam transit angle (θ_0), then one may write the slope of the beam trajectory dr/dz at the anode as (Fig. 6.3.1)

$$\left(\frac{dr}{dz}\right)_A = -\tan\theta_0 \ , \qquad (6.3.57)$$

where the subscript A denotes that the derivative has been taken at the anode. With the help of (6.3.49(b)), one may express (6.3.57) as:

$$k\left(\frac{dR}{dZ}\right)_A = -\tan\theta_0 \quad . \qquad (6.3.58)$$

Now one may substitute (6.3.52), with its negative sign for the region between the anode and the beam waist(see the discussion preceding (6.3.53)), into (6.3.58) to get

$$k\,(\log R_A)^{1/2} = \tan\theta_0 \ , \qquad (6.3.59)$$

where R_A is the value of R at the anode given by

$$R_A = \frac{r_A}{r_0} = \frac{r_A}{r_M} \ , \qquad (6.3.60)$$

r_M being chosen as the reference radius, r_0. Squaring (6.3.59) and taking its exponent and making use of (6.3.60) we then obtain

$$\frac{r_A}{r_M} = \exp\left(\frac{\tan^2\theta_0}{k^2}\right) \ , \qquad (6.3.61)$$

an expression for the ratio of the anode aperture-to-beam waist radii, in terms of the beam transit angle, here taken as θ_0, the half beam cone angle.

6.3.3C Lens effect at the anode aperture

The anode aperture presents itself as an *electrostatic lens* [8] to the beam of electrons which are accelerated by the anode potential (V_0). The *second* focal length f_2 of such a lens is given by (see appendix *a6.3.2*):

$$f_2 = \frac{4V_0}{E_1 - E_2} \ , \qquad (6.3.62)$$

where E_1 and E_2 are the electric field intensities along the gun axis in the vicinity of the anode aperture in the regions to the left and to the right of the anode, respectively, both directed from the left to the right of the lens. The lens would be converging or diverging, according as sign of f is positive or negative. Clearly, if E_1 is negative, for the anode at a positive potential with respect to the cathode, and $E_2 = 0$, that is, if the beam is launched into a field-free region, then the lens would have a diverging effect, and as a result the path of the electron would suffer a deviation at the anode aperture, say, by an angle of δ. This would modify the transit angle from θ_0 to θ_0', say, which may be expressed as (see Fig. 6.3.2(a)):

$$\theta_0' = \theta_0 - \delta \ . \qquad (6.3.63)$$

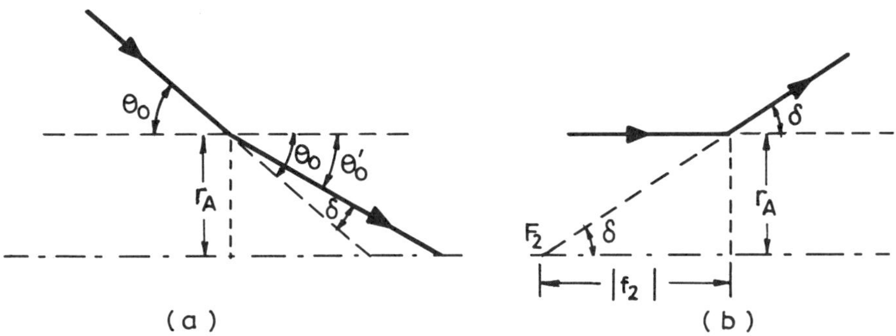

Fig. 6.3.2. Electron path showing the modified beam transit angle, θ_0', caused by the deviation of the electron path, δ, at the anode-aperture lens **(a)** and the "paraxial-ray" diagram to indicate the second focus of the lens, F_2, and relate the magnitude of the second focal length, $|f_2|$, with the deviation, δ, and the anode-aperture radius, r_A**(b)** .

The deviation δ may be estimated with reference to a paraxial-ray diagram of the lens in which a ray parallel to the axis would appear to diverge out from the *second* focus of the lens, giving an approximate relation for small deviation δ, as (see Fig. 6.3.2(b))

$$\delta = \frac{r_A}{|f_2|}, \tag{6.3.64}$$

where f_2 is the second focal length, given by (6.3.62) which, however simplifies, for a field-free region beyond the anode aperture($E_2 = 0$) as:

$$f_2 = \frac{4 V_0}{E_1}. \tag{6.3.65}$$

E_1 may be expressed as a negative potential gradient(see section *2.5*) as:

$$E_1 = -\left(\frac{\partial V}{\partial z}\right)_0, \tag{6.3.66}$$

where the positive z is taken from the left to the right along the axis of the gun, and the subscript '0' denotes the derivative at the gun anode. In the present case, the value of the radial coordinate of a point on the gun axis increases from the right to the left (Fig. 6.3.1) which would, therefore, give an alternative form of (6.3.66) as

$$E_1 = \left(\frac{\partial V}{\partial r}\right)_0 = \left(\frac{dV}{dr}\right)_0, \tag{6.3.67}$$

in which the partial derivatives has been replaced by the complete derivative, V being a function of only r. Before finding the derivative, V is first expressed as follows

$$\frac{V}{V_0} = \frac{G^{4/3}}{G_0^{4/3}} , \tag{6.3.68}$$

which is obtained from (6.3.7) and (6.3.32) by division. Then by differentiating (6.3.68) we get

$$\frac{dV}{dr} = \left(\frac{4}{3}\right)\left(\frac{V_0}{G_0^{4/3}}\right)G^{1/3}\left(\frac{dG}{dr}\right)$$

which, in view of the relations

$$\frac{d}{dr}\left(G^2\right) = 2G\,\frac{dG}{dr}$$

and

$$\frac{d}{dr}\left(G^2\right) = \left(\frac{dG^2}{d(r_c/r)}\right)\left(\frac{d(r_c/r)}{dr}\right)$$

$$= -\left(\frac{dG^2}{d(r_c/r)}\right)\left(\frac{r_c}{r^2}\right)$$

becomes

$$\frac{dV}{dr} = -\left(\frac{2}{3}\right)\left(\frac{V_0}{G_0^{4/3}}\right)\left(\frac{1}{G^{2/3}}\right)\left(\frac{r_c}{r^2}\right)\left(\frac{dG^2}{d(r_c/r)}\right) ,$$

evaluated at the anode ($r = r_a$) as:

$$\left(\frac{dV}{dr}\right)_0 = -\left(\frac{2}{3}\right)\left(\frac{V_0}{G_0^2}\right)\left(\frac{r_c}{r_a^2}\right)\left(\frac{dG^2}{d(r_c/r)}\right)_0$$

which may be substituted into (6.3.65) through (6.3.67) to obtain the following expression for f_2:

$$f_2 = \frac{-6G_0^2}{\left(\frac{r_c}{r_a^2}\right)\left(\frac{dG^2}{d(r_c/r)}\right)_0} . \tag{6.3.69}$$

If the function G^2 is calculated using (6.3.30) for various values of $r_c/r > 1$, it is found that the value of G^2 increases with r_c/r, rendering a positive value to $[dG^2/d(r_c/r)]_0$ at the anode. This would make f_2 have a negative value, as can be seen from (6.3.69), and hence the anode aperture behave essentially as a diverging lens. The result may be used to estimate the modified transit angle θ_0' which may be expressed with the help of (6.3.63) and (6.3.64) as:

$$\theta_0' = \theta_0 - \delta = \theta_0 - \frac{r_A}{|f_2|} = \theta_0\left(1 - \frac{r_A}{|f_2|\theta_0}\right) . \qquad (6.3.70)$$

Further, in view of the relation (see Fig. 6.3.1):

$$\theta_0 \cong \frac{r_A}{r_a} \quad (= r_K / r_c) \ , \qquad (6.3.71)$$

where r_K is the cathode-disc radius, (6.3.70) becomes

$$\theta_0' = \theta_0\left(1 - \frac{r_a}{|f_2|}\right) . \qquad (6.3.71)$$

Substituting (6.3.69) into (6.3.71), we then have

$$\theta_0' = \theta_0\left[1 - \left(\frac{1}{6}\right)\left(\frac{r_c}{r_a}\right)\left(\frac{1}{G_0^2}\right)\left|\frac{dG^2}{d(r_c/r)}\right|_0\right] , \qquad (6.3.72)$$

remembering though that, in the present case of a diverging lens (see the discussion following (6.3.69)), $[dG^2/d(r_c/r)]_0$ would have a positive value.

In view of the modification of the beam transit angle from θ_0 to θ'_0 caused by the lens effect of the anode aperture, the relation between the anode aperture and beam waist radii, (6.3.61), would also modify as:

$$\frac{r_A}{r_M} = \exp\left(\frac{\tan^2\theta_0'}{k^2}\right), \qquad (6.3.73)$$

where θ'_0 is given by (6.3.72).

6.3.3.D Expression for beam convergence

The beam convergence, r_K/r_M, may be defined as the ratio of the cathode aperture radius r_K to the beam-waist radius r_M (the latter being the beam radius that has to be maintained in the interaction structure with the help of a separate focussing structure (chapter *7*)). (The parameter,area convergence, is similarly used to represent the ratio of the beam cross-sectional area at the cathode to that at the beam waist which is approximately equal to the square of the beam convergence, that is, $(r_K/r_M)^2\,[=\pi r_K^2/(\pi r_M^2)]$).

One may choose to write the beam convergence as $r_K/r_M = (r_K/r_A)(r_A/r_M)$, which in view of the relation

$$(\theta_0 \approx)\ \frac{r_K}{r_c} = \frac{r_A}{r_a} \quad \text{(Fig. 6.3.1)}$$

may be alternatively expressed as

$$\frac{r_K}{r_M} = \left(\frac{r_c}{r_a}\right)\left(\frac{r_A}{r_M}\right) . \tag{6.3.74}$$

Plugging (6.3.73) into (6.3.74) and taking θ_0 and hence θ'_0 (see (6.3.70)) to be small so that we can take $\tan\theta'_0 \approx \theta'_0$, we have:

$$\frac{r_K}{r_M} = \left(\frac{r_c}{r_a}\right)\exp\left(\frac{\theta_0'^2}{k^2}\right) . \tag{6.3.75}$$

Substituting θ'_0 from (6.3.72) into (6.3.75),

$$\frac{r_K}{r_M} = \left(\frac{r_c}{r_a}\right)\exp\left[\left(\frac{\theta_0^2}{k^2}\right)\left\{1-\left(\frac{1}{6}\right)\left(\frac{r_c}{r_a}\right)\left(\frac{1}{G_0^2}\right)\left(\frac{dG^2}{d(r_c/r)}\right)_0\right\}^2\right] . \tag{6.3.76}$$

remembering that the subscript 0 with G or with the derivative of G^2 refers to the quantity at $r = r_a$. Further it may be seen that

$$\frac{\theta_0^2}{k^2} = \left(\frac{9}{2}\right)G_0^2 , \tag{6.3.77}$$

which is obtained from (6.3.37) by replacing $\cos\theta_0$ by $1-\theta_0^2/2$ for small values of θ_0 and that

$$\left(\frac{dG^2}{d(r_c/r)}\right)_0 = 2G_0\left(\frac{dG}{d(r_c/r)}\right)_0 . \tag{6.3.78}$$

Making use of (6.3.77) and (6.3.78) in (6.3.76), we then get

$$\frac{r_K}{r_M} = \left(\frac{r_c}{r_a}\right) \exp\left[\frac{9}{2} G_0^2 \left\{1 - \frac{1}{6}\left(\frac{r_c}{r_a}\right)\left(\frac{1}{G_0^2}\right)\left(\frac{2}{G_0}\right)\left(\frac{dG}{d(r_c/r)}\right)_0\right\}^2\right] . \tag{6.3.79}$$

Finally, the substitution of (6.3.30) and its derivative both evaluated at $r = r_a$ into (6.3.79) yields the following explicit expression for the beam convergence:

$$\frac{r_K}{r_M} = \left(\frac{r_c}{r_a}\right) \exp\left[\frac{1}{2}\left\{-1 + \left(\frac{12}{5}\right)\ln\frac{r_c}{r_a} + \left(\frac{27}{40}\right)\ln^2\frac{r_c}{r_a} + \left(\frac{369}{2200}\right)\ln^3\frac{r_c}{r_a} + \ldots\right\}^2\right] . \tag{6.3.80}$$

One may generate data for r_K/r_M for various values of r_c/r_a, for instance, in the form of a plot between these quantities, and interpret the same to find r_c/r_a if r_K/r_M is known. Alternatively, one may like to invert the series (6.3.80) for the same purpose. One of the methods of doing this is outlined in appendix *a6.3.3*.

6.3.4 Output Design Parameters

Taking the beam voltage(V_0), the beam current (I_0), the beam radius, which is also equal to the beam-waist radius r_M, and the cathode operating current density as the four input design parameters we may proceed to find the output design parameters: the cathode radius of curvature r_c, the anode radius of curvature r_a, the cathode-disc radius r_K, the anode-aperture radius r_A, the interelectrode spacing d (= r_c - r_a) and the throw of the gun d_m through the following steps [9].

i) The cathode disc area is equal to $2\pi\, r_c^2 (1-\cos\theta_0)$ which may be approximated as $\pi\, r_k^2$ in view of the relations $\cos\theta_0 \approx 1 - \theta_0^2/2$ and $\theta_0 \approx r_K/r_c$, for small values of θ_0. Multiplying this area ($\pi\, r_k^2$) by the cathode operating current density one gets the beam current (I_0). This gives r_K as:

$$r_K = \left[\frac{I_0}{\pi\,(\text{operating cathode current density})}\right]^{1/2} .$$

ii) Since r_K is known from step (i), the beam convergence r_K/r_M is also known, r_M being one of the known input beam parameters, namely, the beam radius which is to be maintained beyond the throw of the gun with the help of a separate focusing structure.

iii) Now that r_K/r_M is known one may find r_c/r_a as discussed following (6.3.80). (see also appendix *a6.3.3* for a method to invert the series (6.3.80) for this purpose).

iv) Since r_c/r_a is known one may find G_0 with the help of (6.3.33). This value of $G_0\{u\}$ together with the given value of I_0 and V_0 may be put in (6.3.36) to find θ_0.

v) One may use the relation $\theta_0 \approx r_k/r_c$ to find r_c as $r_c = r_K/\theta_0$, where r_K is known from step (i) and θ_0 from step (iv).

vi) That r_c/r_a is known from step (iii) and r_c from step (v), r_a may be found as $r_a = r_c/(r_c/r_a)$.

vii) θ_0 being known from step (iv) and r_a from step (vi), r_A is found from the relation $\theta_0 \approx r_A/r_a$ as $r_A = \theta_0 r_a$.

viii) r_c and r_a being known from steps (v) and (vi), respectively, the interelectrode spacing d is found as $d = r_c - r_a$.

ix) The throw of the gun, measured as the distance of the beam waist from the anode, d_m, can be found from (6.3.56), since now r_A is known from step (vii) and r_M is known as one of the input parameters.

6.3.5 Conformal Mapping of Electrode Shapes

The conformal transformation is a method of finding the solution of a physical problem for a given configuration from the corresponding solution which is known apriori for a simple configuration [6]. In the present context, the solution for potential function at a point outside a strip beam derived from a flat cathode has already been obtained in a simpler rectangular system (see section *6.2*). Our objective is to exploit the solution for this configuration and use the conformal transformation technique to find the corresponding solution at a point outside a convergent conical beam derived from a spherical-cup cathode. Subsequently, the result could be interpreted to find the electrode shapes from equipotentials.

In this method a suitable complex analytic function is used to transform the cross-sectional geometry of the problem from one plane to another. Thus if the function is represented as: $W = f\{Z\} = f\{z + j\,y\} = u_r + j\,u_i$ (see (a6.2.1)), then corresponding to a point (z, y) on the Z-plane, with z and y as the real and imaginary axes, respectively, one would have a point (u_r, u_i) on the W-plane with u_r and u_i as the real and imaginary axes, respectively. So, for an elementary portion of the line on the Z-plane one would have a corresponding portion of the line on the W-plane as well. It may also be shown that an incremental region on the W-plane is changed in magnitude by a factor, and rotated through an angle, with respect to the corresponding incremental region on the Z-plane (see appendix *a6.3.4*). Thus by this kind of transformation from one plane to another the relative angles between the lines are preserved and the corresponding incremental areas enjoy being similar in shape (though they differ in scale). Hence the transformation is called *conformal*. It may be mentioned that the incremental area, be it on the Z or W plane, could be formed as being enclosed between the lines u_r = constant, $u_r + \Delta u_r$ = constant, u_i = constant and $u_i + \Delta u_i$ = constant. Further, obviously on the W-plane, a u_r = constant line, which is parallel to the imaginary axis, would intersect at a right angle with a u_i = constant line, which is parallel to the real axis. So, these lines when depicted on the Z-plane should also maintain these property of perpendicular intersection. This can also be independently appreciated by studying the properties of a complex analytic function (see appendix *a6.3.5*). It is worth mentioning here that in electrostatic problems if we choose u_r, the real part of f{Z}, as the potential function, as we have done in the case of a rectangular strip beam(section *6.2*), then we may take u_i, the imaginary part of f{Z}, to represent the electric flux function (see appendix *a6.3.6*). Therefore, the equipotential and electric flux lines should make perpendicular intersections on the Z-plane as do u_r = constant and u_i = constant lines .

In the present context, the configuration of the convergent conical beam problem is conformally mapped to the known configuration of the strip beam case. A suitable conformal transformation function is sought satisfying the conditions that the beam edge of the conical beam after transformation from Z- to W-plane coincides with the real axis of the W-plane above which the solution for the potential function is already known (section *6.2*). Since upon transformation the angles between orthogonal lines are preserved but not the magnitude of the areas enclosed between them as mentioned earlier (see also appendix *a6.3.4*), care must be taken to introduce a scale factor in the solution for potential at a point outside the beam by this method. This can be done ensuring that after transformation the beam-boundary values of potential remain invariant. In other words, the potential function obtained for the conical beam configuration obtained by this method should in the limit pass on to the potential function at the beam edge ($\theta = \theta_0$) which in turn could be known apriori from the following form of Langmuir and Blodgett's solution:

$$V\{r,\theta_0\} = \left(\frac{9 I_0}{8\pi\varepsilon_0 (2|\eta|)^{1/2} (1 - \cos\theta_0)}\right)^{2/3} G^{4/3}\{u\} \ , \qquad (6.3.81)$$

an expression which is obtainable from (6.3.36) replacing therein $V_0^{3/2}$ by $[G_0^2\{u\}/G^2\{u\}]V^{3/2}$, in view of the relation (6.3.68).

Let us try here two transformation functions, namely, logarithmic [10] and bi-linear [11] which, it will be of interest to see, would lead to the one and the same potential function.

6.3.5.A Logarithmic transformation

Let

$$W = f\{Z\} = \ln Z = u_r + j\, u_i \ , \qquad (6.3.82)$$

with

$$Z = z + j\, y = r \exp(j\theta) \ . \qquad (6.3.83)$$

Separating the real and imaginary parts of (6.3.83), we have $z = r\cos\theta$ and $y = r\sin\theta$ giving

$$r = (z^2 + y^2)^{1/2}, \qquad \theta = \tan^{-1} y/z\ , \tag{6.3.84}$$

for the location of a point outside the beam of conical configuration (Fig. 6.3.3(a)). Substituting (6.3.83) into (6.3.82), we get

$$W = u_r + j\,u_i = \ln r + j\,\theta \tag{6.3.85}$$

which then on equating the real and imaginary parts gives

$$u_r = \ln r\,, \quad u_i = \theta\ , \tag{6.3.86}$$

for the logarithmic transformation function (6.3.82) assumed.

Now let us choose to map on the W-plane (Fig. 6.3.3(b)) the cathode ($r = r_c$) at $u_r = u_{ro}$, say, and the beam edge ($\theta = \theta_0$) at $u_i = u_{io}$, say, where the subscript o has been used to indicate that the values are the reference values. For these reference values the relation corresponding to (6.3.86) may be written as:

$$u_{r0} = \ln r_c\ , \qquad u_{i0} = \theta_0\ . \tag{6.3.87}$$

In the next step let us transform the cathode and the beam edge from the plane W ($= u_r + j\,u_i$) to W' ($= u'_r + j\,u'_i$) such that the origin ($u'_r = 0$, $u'_i = 0$) of the W'-plane is located at a point corresponding to the point ($u_r = u_{ro}$, $u_i = u_{io}$) of the W-plane, and the cathode coincides with the imaginary axis $u'_r = 0$ and the beam edge with the real axis $u'_i = 0$ of the W'-plane (Fig. 6.3.3(c)). Clearly then u'_r is related u_r as $u'_r = u_r - u_{ro}$, and u'_i to u_i as $u'_i = u_i - u_{io}$, so that one may write the functional relationship between W' and W as

$$\begin{aligned} W' &= u'_r + ju'_i = (u_r - u_{ro}) + j\,(u_i - u_{io}) \\ &= (u_r + j\,u_i) - (u_{ro} + j\,u_{io}) \\ &= W - (u_{ro} + j\,u_{io})\ . \end{aligned} \tag{6.3.88}$$

Substituting (6.3.85) and (6.3.87) into (6.3.88) we then have

$$W' = u'_r + j\,u'_i = \ln r + j\,\theta - (\ln r_c + j\,\theta_0)$$

which simplifies to:

$$W' = u'_r + j\,u'_i = \ln r/r_c + j\,(\theta - \theta_0)\ . \tag{6.3.89}$$

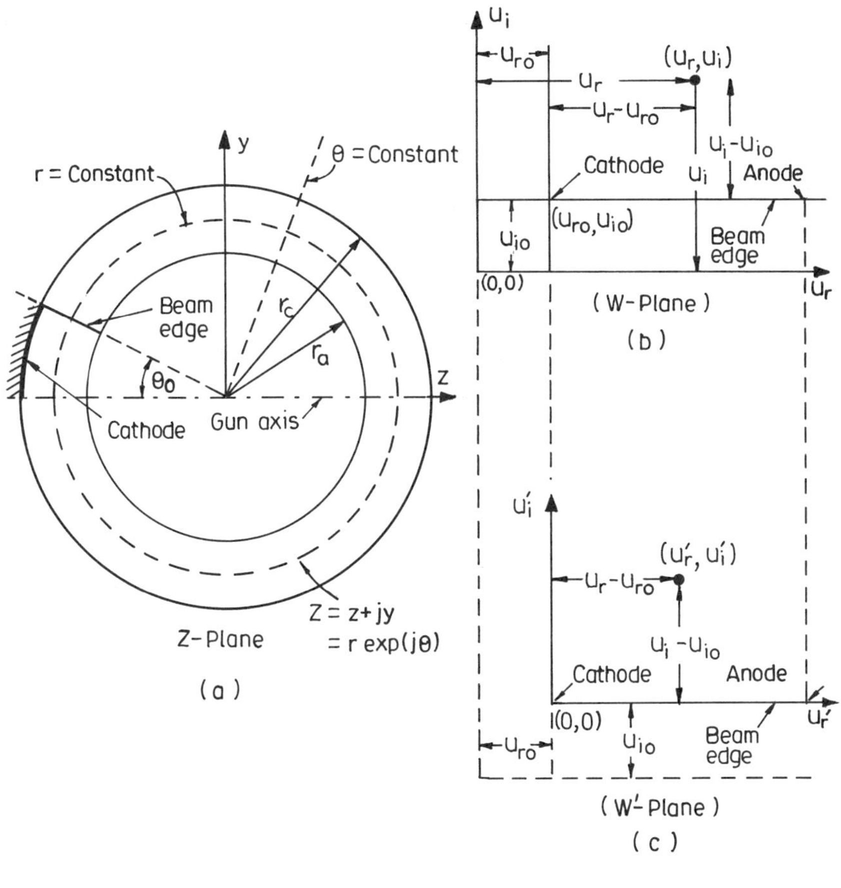

Fig. 6.3.3. Cross-sectional geometry of the conical beam configuration showing half-cone angle θ_0, half the emitting portion of the cathode, the beam edge, the cathode and the anode spheres and the gun axis on the Z-plane **(a)**, the first conformal transformation thereof showing the cathode at $u_r = u_{ro}$ and the beam edge at $u_i = u_{io}$ on the W-plane **(b)**, and the second conformal transformation thereof showing the cathode at u'_r $(= u_r - u_{ro}) = 0$ and the beam edge at u'_i $(= u_i - u_{io}) = 0$ on the W'-plane **(c)**.

Now that by the second stage of transformation the beam edge and the cathode have been made to coincide with the real and the imaginary axes of the W'-plane, respectively, one has now actually changed the configuration of a conical beam to that of a rectangular strip beam. Therefore, now potential function at a point outside the beam ($\theta > \theta_0$) can be found directly using the same method as followed in the rectangular strip beam case outlined in section *6.2*, as follows (see (6.2.4)):

$$V\{r,\theta\} = Re\,W' = Re\,f\{u'_r + j\,u'_i\} = Re\;f\{\ln r/r_c + j\,(\theta - \theta_0)\}\,, \qquad (6.3.90)$$

making use of (6.3.89). Here, the function f has to be so chosen that, as in the case of a rectangular strip beam, in the limit (i) $V\{r,\theta\}$ passes on at $\theta = \theta_0$ to the beam edge function $V\{r,\theta_0\}$ predicted by Langmuir and Blodgett's solution (6.3.81), a condition which is similar to (6.2.2), and that (ii) $(\partial V/\partial\theta)_{\theta = \theta_0} = 0$ making zero the azimuthal component of electric field intensity ($= (-1/r)\,\partial V/\partial\theta$) (see (2.5.11) through (2.5.12(c))) at the beam edge ($\theta = \theta_0$) and hence ensuring that the beam edge electrons are not subject to any force deviating their flow from the radial direction, a condition which is similar to (6.2.3).

It is convenient to express the series (6.3.30) for $G\{u\}$ occurring in (6.3.81) as

$$G\{u\} = \sum_{n=1,2,3,\ldots} a_n\,(\ln r/r_c)^n = \sum_{n=1,2,3,\ldots} b_n\,(\ln r_c/r)^n\,, \qquad (6.3.91)$$

where

$$a_1 = -1,\ a_2 = 3/10,\ a_3 = -3/40,\ a_4 = 63/4400,\ \text{and so on; and}\ b_n = (-1)^n a_n\,. \qquad (6.3.92)$$

Subject to the above conditions (i) and (ii), and remembering (6.3.91) through (6.3.92) one may easily identify the function in (6.3.90) to get $V\{r,\theta\}$ for $\theta > \theta_0$, *outside the beam*, as:

$$V\{r,\theta\} = Re\;f\{\ln\frac{r}{r_c} + j\,(\theta - \theta_0)\}$$

$$= Re\left(\frac{9\,I_0}{8\pi\varepsilon_0(2|\eta|)^{1/2}(1-\cos\theta_0)}\right)^{\frac{2}{3}}\left[\sum_{n=1,2,3,\ldots} a_n\left(\ln\frac{r}{r_c} + j\,(\theta-\theta_0)\right)^n\right]^{\frac{4}{3}}$$

$$= Re\left(\frac{9\,I_0}{8\pi\varepsilon_0(2|\eta|)^{1/2}(1-\cos\theta_0)}\right)^{\frac{2}{3}}\left[\sum_{n=1,2,3,\ldots} b_n\left(\ln\frac{r_c}{r} - j\,(\theta-\theta_0)\right)^n\right]^{\frac{4}{3}}. \qquad (6.3.93)$$

One may check that for $\theta = \theta_0$ (6.3.93) passes on to (6.3.81) through (6.3.91)(condition (i)) and $\partial V/\partial\theta$ evaluated at $\theta = \theta_0$ vanishes since $\partial f/\partial\theta$ becomes purely imaginary (condition(ii)). Since upon conformal transformation the incremental areas, though they maintain the similarity in shape, differ in scale(appendix *a6.3.4*), it is implied that the function f should include this scale factor. Such scale factor is automatically taken care of by condition (i).

6.3.5.B Bilinear transformation

In this transformation in addition to the intersection of the cathode and the anode with the beam edge, at C and A, respectively, let us depict another point P between C and A on the beam edge on the Z-plane (Fig. 6.3.4(a)). We seek for a suitable function which would transform the beam edge together with the points C, P and A of the Z-plane to the real axis ($u_i = 0$) of the W-plane taking care to place these points in the increasing order of u_r with C located at the origin ($u_r = 0$) (Fig. 6.3.4(b)) Hence, let us propose a bilinear transformation of the following cross-ratio relationship [6]

$$\frac{(Z_a - Z)(Z_c - Z_p)}{(Z_a - Z_p)(Z_c - Z)} = \frac{(W_a - W)(W_c - W_p)}{(W_a - W_p)(W_c - W)} . \tag{6.3.94}$$

Here the subscripts c, a and p refer to the beam edge points ($\theta = \theta_0$) at the cathode, the anode and a point in between, respectively, and the unscripted quantities refer to a point outside the beam. With the help of (6.3.82) and (6.3.83) the quantities in (6.3.94) are:

on the beam edge($\theta = \theta_0$, $u_i = 0$):

$$Z_c = r_c \exp(j\theta_0) \quad \text{(a)} \qquad W_c = 0\ (u_{rc} = u_{ic} = 0) \quad \text{(d)}$$

$$Z_a = r_a \exp(j\theta_0) \quad \text{(b)} \qquad W_a = u_{ra}\ (u_{ia} = 0) \quad \text{(e)}$$

$$Z_p = r \exp(j\theta_0) \quad \text{(c)} \qquad W_p = u_{rp}\ (u_{ip} = 0) \quad \text{(f)} \tag{6.3.95}$$

outside the beam($\theta > \theta_0$, $u_i > 0$):

$$Z = r \exp(j\theta) \qquad\qquad W = u_r + j\,u_i . \tag{6.3.96}$$

Using (6.3.95) one may express (6.3.94) as

$$\left[\frac{r_a \exp(j\theta_0) - r\exp(j\theta)}{(r_a - r)\exp(j\theta_0)}\right]\left[\frac{\exp(j\theta_0) - (r/r_c)\exp(j\theta_0)}{\exp(j\theta_0) - (r/r_c)\exp(j\theta)}\right] = \frac{(1 - W/u_{ra})(0 - W_P)}{(1 - u_{rP}/u_{ra})(0 - W)} . \tag{6.3.97}$$

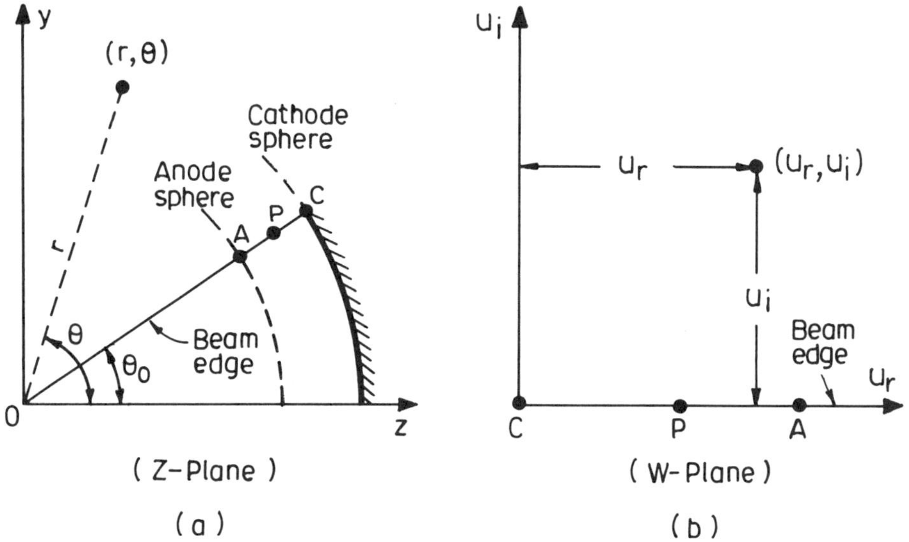

Fig. 6.3.4 Cross section of the conical beam through a half cone angle (θ_0) showing a point (r, θ) outside the beam edge ($\theta > \theta_0$) as well as three points on the beam edge — one at the cathode (C), the other at the anode (A) and another between C and A at P on the Z-plane **(a)** and their transformed location on the W-plane **(b)**.

Now, mathematically there is no restriction on the separation between the cathode and the anode. Therefore, let us, on the Z-plane, keeping the anode at a finite distance from the origin (the common center of curvature of the cathode and anode spheres), take the cathode to infinity: $r_c \to \infty$ (Fig. 6.3.4(a)), and, at the same time, on the W-plane, retaining the cathode at the origin take the anode to infinity: $u_{ra} \to \infty$. Substituting these conditions, namely, $r_c \to \infty$, $u_{ra} \to \infty$ into (6.3.97), one may obtain the following relation after simplification:

$$W = \frac{(u_{rp})(r/r_a - 1)}{(r/r_a)\exp j(\theta - \theta_0) - 1} . \qquad (6.3.98)$$

Now, let us choose P such that

$$u_{rp} = (r/r_a - 1)^{-1} . \qquad (6.3.99)$$

The relation (6.3.99) put in (6.3.98) makes

$$W = [(r/r_a) \exp j(\theta - \theta_0) - 1]^{-1} \quad (6.3.100)$$

At the beam edge, $\theta = \theta_0$ and $u_i = 0$, and for the point P on it we have from (6.3.85) and (6.3.100)

$$u_{rp} = (r/r_a - 1)^{-1} , \quad (6.3.101)$$

a relation that agrees with (6.3.99). (6.3.101) may be rearranged to give

$$r_a/r = (1 + 1/W)^{-1}_{\theta = \theta_0} \quad . \quad (6.3.102)$$

Now let us choose to express r_c/r occurring in the series (6.3.91) for G{u} with the help of (6.3.102) as:

$$\frac{r_c}{r} = \left(\frac{r_c}{r_a}\right)\left(\frac{r_a}{r}\right) = \left(\frac{r_c}{r_a}\right)\left(1 + \frac{1}{W}\right)^{-1}_{\theta=\theta_0} \quad . \quad (6.3.103)$$

Therefore Langmuir and Blodgett's potential distribution on the beam edge (6.3.81), through (6.3.91) and (6.3.103), may be expressed as

$$V\{r,\theta_0\} = \left(\frac{9I_0}{8\pi\varepsilon_0(2|\eta|)^{1/2}(1-\cos\theta_0)}\right)^{2/3}\left[\sum_{n=1,2,3..} b_n\left(\ln\frac{r_c}{r}\right)^n\right]^{4/3}$$

$$= \left(\frac{9I_0}{8\pi\varepsilon_0(2|\eta|)^{1/2}(1-\cos\theta_0)}\right)^{2/3}\left[\sum_{n=1,2,3...} b_n\left\{\ln\left(\left(\frac{r_c}{r_a}\right)\left(1+\frac{1}{W}\right)^{-1}_{\theta=\theta_0}\right)\right\}^n\right]^{4/3} \quad (6.3.104)$$

Now, that by the bilinear transformation the beam edge has been made to be coincident with the real axis on the W-plane, the potential at a point above this axis that is outside the beam ($\theta > \theta_0$) is obtainable as in the rectangular strip beam case as follows(see (6.2.4) and also (6.3.90)):

$$V\{r,\theta\} = Re\, f\{W\} \quad . \quad (6.3.105)$$

In (6.3.105), f{W} has to be so selected that, at $\theta = \theta_0$, (6.3.105) would pass on to (6.3.104). This motivates one to pick-up the function from (6.3.104) to write

$$V\{r,\theta\} = Re\left(\frac{9\,I_0}{8\,\pi\,\varepsilon_0\,(2|\eta|)^{1/2}\,(1-\cos\theta_0)}\right)^{2/3}\left[\sum_{n=1,2,3...} b_n\left\{\ln\left(\left(\frac{r_c}{r_a}\right)\left(1+\frac{1}{W}\right)^{-1}\right)\right\}^n\right]^{4/3} . \tag{6.3.106}$$

The logarithmic term occurring in (6.3.106) may be shown with the help of (6.3.100) to be:

$$\begin{aligned}\ln\,[(r_c/r_a)(1 + 1/W)^{-1}] &= \ln\,[(r_c/r)\exp -j\,(\theta - \theta_0)]\\ &= \ln r_c/r\ -\ j\,(\theta - \theta_0)\ .\end{aligned} \tag{6.3.107}$$

Plugging (6.3.107) into (6.3.106), we have the potential function at a point *outside the beam* $(\theta > \theta_0)$ as:

$$V\{r,\theta\} = Re\left(\frac{9I_0}{8\pi\varepsilon_0\,(2|\eta|)^{1/2}\,(1-\cos\theta_0)}\right)^{\frac{2}{3}}\left[\sum_{n=1,2,3..} b_n\left\{\ln\frac{r_c}{r} - j(\theta-\theta_0)\right\}^n\right]^{\frac{4}{3}} . \tag{6.3.108}$$

It may be mentioned that the scale factor upon conformal transformation is inherent in (6.3.108) as discussed following (6.3.93). Further one may see that (6.3.108) is identical with (6.3.93). In other words both the logarithmic and bilinear transformation have led us to the one and the same expression for potential at a point outside the convergent conical beam. The potential distribution thus obtained by conformal transformation in the conical beam case may be interpreted to find the equipotential lines on the cross-sectional geometry of the conical beam configuration as done in the case of a rectangular strip beam configuration. From the equipotentials one may find the shapes of the electrodes (here BFE and anode) also, as discussed in section *6.2*. The method is referred to as the conformal mapping of electrode shapes.

Appendix a6.3.1 *Definite integral in the expression for throw*

Putting

$$\xi = (\ln R)^{1/2}\,, \tag{a6.3.1}$$

we have

$$R = \exp \xi^2$$

which on differentiation gives

$$dR = 2\xi \exp \xi^2\, d\xi\ . \tag{a6.3.2}$$

Substituting (a6.3.1) and (a6.3.2) into the integral in the right hand side of (6.3.55), we get

$$\int \frac{dR}{(\ln R)^{1/2}} = 2\int \exp \xi^2 \, d\xi \ . \tag{a6.3.3}$$

Substituting the series expansion of $\exp\xi^2$ *in (a6.3.3), then*

$$\int \frac{dR}{(\ln R)^{1/2}} = 2\int \left(1+\frac{\xi^2}{1!}+\frac{\xi^4}{2!}+\frac{\xi^6}{3!}+\ldots\right) d\xi$$

which upon a simple integration yields

$$\int \frac{dR}{(\ln R)^{1/2}} = 2\left[\xi+\frac{\xi^3}{(3)(1!)}+\frac{\xi^5}{(5)(2!)}+\frac{\xi^7}{(7)(3!)}+\ldots\right] . \tag{a6.3.4}$$

Expressing ξ *in terms of R with the help of (a6.3.1), one may write (a6.3.4), that is, the definite integral of (6.3.55), as*

$$\int_1^{r_A/r_M} \frac{dR}{(\ln R)^{1/2}} = 2\left[\ln^{\frac{1}{2}} R+\frac{\ln^{\frac{3}{2}} R}{(3)(1!)}+\frac{\ln^{\frac{5}{2}} R}{(5)(2!)}+\frac{\ln^{\frac{7}{2}} R}{(7)(3!)}+\ldots\right]_1^{r_A/r_M} . \tag{a6.3.5}$$

Putting the limits in the right hand side of (a6..3.5), one gets the value of the definite integral that can be now substituted in (6.3.55) to obtain the expression for d_m *as given by (6.3.56).*

Appendix a6.3.2 *Focal length of an aperture-lens*

In conventional optics, the phenomenon of refraction takes place at the interface between two media of dissimilar refractive indices. In electron optics the quantity corresponding to the refractive index is the electron velocity [8]. Thus if the electron-optical ray deviates in passing from one region to another, it is implied that an electron does have different velocities in these two regions supposedly of two different uniform potentials. Although the normal component $v\cos\theta$ *of electron velocity* v*, say, in a region would change at the boundary between two regions, the tangential component* $v\sin\theta$ *would remain conserved, where* θ *is the angle of the electron optical ray in a medium measured from the normal to boundary. This can be identified with the well known Snell's law of conventional optics taking the electron velocity* v *in electron optics to be a quantity analogous to the refractive index in conventional optics,* v *being further related to the potential* V *in the region by the energy conservation relation* $v = (2|\eta|V)^{1/2}$ *(for instance, see (6.1.6)). Clearly, then in electron optics a uniform electric field intensity region in which potential varies uniformly is analogous to a medium of uniformly-varying refractive index in conventional optics.*

The present appendix is dedicated to deducing some fundamental relations and expressions which are of relevance to studying an electrostatic lens. First we take up the expression for potential in the cylindrical system of coordinates (r, θ, z) which may be obtained by solving the following Laplace's equation for an azimuthally symmetric situation (∂/∂θ) as is found in an electrostatic lens (see (2.6.5) through (2.6.4(b))):

(a6.3.6) $$\frac{1}{r}\frac{\partial}{\partial r}\left(\frac{r\,\partial V}{\partial r}\right) + \frac{\partial^2 V}{\partial z^2} = 0 .$$

One may try a series solution of (a6.3.6) as [8]:

(a6.3.7) $$V\{r,z\} = \sum_{n=0,1,2,\ldots\infty} a_n\{z\}\, r^n .$$

If (a6.3.7) is substituted in (a6.3.6), the relevant derivatives worked out, and the coefficients of the terms involving r^n are equated, we get the recurrence relation

(a6.3.8) $$a_{n+2} = -\frac{a_n''\{z\}}{(n+2)^2} .$$

Because of azimuthal symmetry, V{r, z} should have a value that is irrespective of whether r is positive or negative. This would then allow only the terms with an even power of r to be present in the series (a6.3.7), or in other words we should have

(a6.3.9) $$a_1 = a_3 = a_5 = \ldots = 0 .$$

The other non-zero coefficients may be written with the help of (a6.3.8) as:

(a6.3.10) $$a_2 = -\frac{a_0''\{z\}}{2^2}$$

$$a_4 = -\frac{a_2''\{z\}}{4^2} = \frac{a_0''''\{z\}}{2^2.4^2} ,$$

$$a_6 = -\frac{a_4''\{z\}}{6^2} = -\frac{a_0''''''\{z\}}{2^2.4^2.6^2} , \text{ and so on .}$$

We notice that the first term of the series (a6.3.7), corresponding to n = 0, is $a_0\{z\}$ which is the only term in the series that is independent of r. It may also be seen that for r = 0 all the terms of the series except this term, namely, $a_0\{z\}$ would become null since each of such terms involves r with a non-zero index of power. Thus we identify $a_0\{z\}$ as

(a6.3.11) $$a_0\{z\} = V_0\{z\} , \text{ say,}$$

the value of the potential at the axis (r = 0). Substituting (a6.3.9) and (a6.3.10) through (a6.3.11) in (a6.3.7), we have (a6.3.12) $$V\{r,z\} = V_0\{z\} - V_0''\{z\}\frac{r^2}{2^2} + V_0''''\{z\}\frac{r^4}{2^2.4^2} + \cdots \ .$$

Considerable amount of information about an electron optical path can be gathered if a differential equation could be formulated describing such path in terms of potential V as a function of r and z, which in turn could be obtained from (a6.3.12) if the potential function along the axis of a system of azimuthal symmetry is known. Obviously, the problem becomes quite simple if such differential equation is first formulated in a rectangular system of coordinates and then it is extended to a cylindrical system as relevant to the present problem as follows.

For this purpose, let us write the equations of motion of an electron subject to a two-dimensional electric field in the rectangular system of coordinates as follows:

(a6.3.13) $$\frac{d^2x}{dt^2} = \frac{e}{m}E_x \qquad \text{and}$$

(a6.3.14) $$\frac{d^2y}{dt^2} = \frac{e}{m}E_y \ .$$

In addition, we have the energy conservation equation in terms of V (see (6.1.6)):

(a6.3.15) $$\frac{1}{2}m\left[\left(\frac{dx}{dt}\right)^2 + \left(\frac{dy}{dt}\right)^2\right] = |e|V.$$

Two additional relations which are helpful for the purpose are:

(a6.3.16) $$\frac{dy}{dt} = \frac{dy}{dx}\frac{dx}{dt}$$

whence

(a6.3.17) $$\frac{d^2y}{dt^2} = \left(\frac{d^2y}{dx^2}\right)\left(\frac{dx}{dt}\right)^2 + \left(\frac{dy}{dx}\right)\left(\frac{d^2x}{dt^2}\right) \ .$$

Substituting (a6.3.13) and (a6.3.14) into (a6.3.17) one may obtain

(a6.3.18)
$$\left(\frac{dx}{dt}\right)^2 = \frac{\left(\frac{e}{m}\right)\left(E_y - E_x\frac{dy}{dx}\right)}{\frac{d^2y}{dx^2}} .$$

Further, putting (6.3.16) into (6.3.15) we have

(a6.3.19)
$$\frac{1}{2}m\left(\frac{dx}{dt}\right)^2\left[1+\left(\frac{dy}{dx}\right)^2\right] = |e|V .$$

Substituting (a6.3.18) into (a6.3.19), we then get

(a6.3.20)
$$2V\frac{d^2y}{dx^2} = \left(E_x\frac{dy}{dx} - E_y\right)\left[1+\left(\frac{dy}{dx}\right)^2\right] .$$

Restricting ourselves to an electron path close to the axis, we can continue to use (a6.3.20) also for a cylindrical system, however, changing y to r and x to z, enabling us to write:

$$2V\frac{d^2r}{dz^2} = \left(E_z\frac{dr}{dz} - E_r\right)\left[1+\left(\frac{dr}{dz}\right)^2\right]$$

which for $(dr/dz)^2 << 1$ (under paraxial-ray approximation) simplifies to:

(a6.3.21)
$$2V\frac{d^2r}{dz^2} = E_z\frac{dr}{dz} - E_r .$$

In view of the relations $E_z = -\partial V/dz$ and $E_r = -\partial V/dr$(see (2.5.11) through (2.5.12(b))), we then get from (a6.3.21):

(a6.3.22)
$$2V\frac{d^2r}{dz^2} = \frac{\partial V}{\partial r} - \frac{\partial V}{\partial z}\frac{dr}{dz} ,$$

the desired paraxial ray equation for an electron trajectory.

Now, one has from (a6.3.12) and its derivatives:

(a6.3.23)
$$V = V_0\{z\} - V_0''\{z\}\frac{r^2}{4} + \cdots \qquad \cong V_0\{z\}$$

$$\frac{\partial V}{\partial z} = V_0'\{z\} - V_0'''\{z\}\frac{r^2}{2} + \cdots \qquad \cong V_0'\{z\}$$

$$\frac{\partial V}{\partial r} = -V_0''\{z\}\frac{r}{2} + V_0''''\{z\}\frac{r^3}{16} + \cdots \qquad \cong -V_0''\{z\}\frac{r}{2} \ ,$$

for an electron path close to the axis (that is, for small r). (a6.3.23) may be substituted into (a6.3.22) to get

$$2V_0\{z\}\frac{d^2r}{dz^2} = -V_0''\{z\}\frac{r}{2} - V_0'\{z\}\frac{dr}{dz}$$

which may be re-arranged as :

(a6.3.24)
$$\frac{d^2r}{dz^2} + \left(\frac{V_0'\{z\}}{2V_0\{z\}}\right)\frac{dr}{dz} = -\left(\frac{V_0''\{z\}}{4V_0\{z\}}\right)r \ .$$

An alternative form of (a6.3.24) is:

(a6.3.25)
$$V_0^{1/2}\{z\}\frac{d}{dz}\left(V_0^{1/2}\{z\}\frac{dr}{dz}\right) = -V_0''\{z\}\frac{r}{4}$$

which upon integration gives

(a6.3.26)
$$\int_{z_1}^{z_2} d\left(V_0^{1/2}\{z\}\frac{dr}{dz}\right) = -\frac{r}{4}\int_{z_1}^{z_2} V_0^{-1/2}\{z\}V_0''\{z\}\,dz \ ,$$

taking r outside the integral of the right hand side by treating it as a constant for a short lens. Here, z_1 *and* z_2 *refer to the points in the regions to the left and to the right of the lens which represent the boundaries of appreciable potential variation.*

Evaluation of the integral of the left hand side of (a6.3.26) is rather straightforward so that (a6.3.26) now takes the form

(a6.3.27)
$$V_0^{1/2}\{z_2\}\left(\frac{dr}{dz}\right)_{z_2} - V_0^{1/2}\{z_1\}\left(\frac{dr}{dz}\right)_{z_1} = -\frac{r}{4}\int_{z_1}^{z_2} V_0^{-1/2}\{z\}V_0''\{z\}\,dz.$$

Now, if we choose a parallel ray to be incident on the lens from its left, then $[dr/dz]_{z_1} = 0$. *This simplifies (a6.3.27) as:*

(a6.3.28)
$$V_0^{1/2}\{z_2\}\left(\frac{dr}{dz}\right)_{z_2} = -\frac{r}{4}\int_{z_1}^{z_2} V_0^{-1/2}\{z\}V_0''\{z\}\,dz \ .$$

For such a parallel ray the ray will converge to or appear to diverge from the second focus of the lens, according as the lens is of converging type of positive focal length or diverging type of negative focal length. From simple geometrical considerations one then has the relation

$$\left(\frac{dr}{dz}\right)_{z_2} = -\frac{r}{f_2}\,, \tag{a6.3.29}$$

such that, for a positive value of f_2 (for a converging lens), $[dr/dz]_2$, the slope of the ray emerging from the lens, becomes negative, while, for a negative value of f_2 (diverging lens), the same(slope) becomes positive. Putting (a6.3.29) into (a6.3.28), and canceling r from both sides of the equation, we get the following expression for the focal length f_2 :

$$\frac{1}{f_2} = \frac{1}{4V_0^{1/2}\{z_2\}}\int_{z_1}^{z_2} V_0^{-1/2}\{z\}V_0''\{z\}dz\,. \tag{a6.3.30}$$

In order to evaluate the integral in (a6.3.30) let us put

$$p = V_0^{-1/2}\{z\} \quad \text{and} \tag{a6.3.31}$$

$$q = V_0'\{z\}\,, \quad \text{along with their differentials:}$$

$$dp = -\frac{1}{2}V_0^{-3/2}\{z\}\,V_0'\{z\}\,dz \quad \text{and}$$

$$dq = V_0''\{z\}\,dz\,.$$

Making use of (a6.3.31), we may write

$$\int V_0^{-1/2}\{z\}\,V_0''\{z\}\,dz = \int p\,dq = \int pq^0\,dq \tag{a6.3.32}$$

$$= pq - \int q\,dp \text{(evaluating the integral by parts).}$$

Substituting (a6.3.31) into (a6.3.32),

$$\int V_0^{-1/2}\{z\}\,V_0''\{z\}\,dz = V_0^{-1/2}\{z\}\,V_0'\{z\} + \frac{1}{2}\int V_0'^2\{z\}\,V_0^{-3/2}\{z\}\,dz\,. \tag{a6.3.33}$$

Making use of (a6.3.33) the expression for the focal length, (a6.3.30) becomes

$$\frac{1}{f_2} = \frac{1}{4V_0^{1/2}\{z_2\}}\left(V_0^{-1/2}\{z\}\,V_0'\{z\}\right)_{z_1}^{z_2} + \frac{1}{8V_0^{1/2}\{z_2\}}\int_{z_1}^{z_2} V_0'^2\{z\}V_0^{-3/2}\{z\}\,dz \tag{a6.3.34}$$

$$= \frac{1}{4V_0^{1/2}\{z_2\}}\left(V_0^{-1/2}\{z_2\}\,V_0'\{z_2\} - V_0^{-1/2}\{z_1\}\,V_0'\{z_1\}\right) + \frac{1}{8V_0^{1/2}\{z_2\}}\int_{z_1}^{z_2} V_0'^2\{z\}V_0^{-3/2}\{z\}\,dz\,.$$

In the case of a single aperture electrostatic lens presented by the anode at a positive potential V_0 with respect to the cathode, the variation of potential and its derivatives along the axis in the vicinity of the anode is such that one gets $V_0'\{z_2\}$ as substantially different from $V'_0\{z_1\}$ though one may take $V_0\{z_2\} \triangleq V_0(z_1) \triangleq V_0$ [8]. For such distributions of $V_0\{z\}$ and its derivatives, in the vicinity of the anode one may take $V_0\{z_2\} \triangleq V_0(z_1) \triangleq V_0$, the anode potential (represented by the symbol without the argument z), though $V_0'\{z_2\} \neq V'_0\{z_1\}$. This gives the expression for focal length f_2 from (a6.3.34) as:

(a6.3.35)
$$\frac{1}{f_2} = \frac{V_0'\{z_2\} - V_0'\{z_1\}}{4V_0} .$$

In writing (a6.3.35) we have dropped the term containing the definite integral (which may be found by finding the area under the plot of $V_0'^2\{z\}\, V_0^{-3/2}\{z\}$ versus z within the integration limits) as it is found to be much small compared to the remaining terms of the right hand side of (a6.3.34). Choosing to write $E_1 = -V_0'\{z_1\}$ and $E_2 = -V_0'\{z_2\}$ (see (2.5.11) through (2.5.12b)) we have from (a6.3.35):

(a6.3.36)
$$\frac{1}{f_2} = \frac{E_1 - E_2}{4V_0} ,$$

where E_1 and E_2 are the electric field intensities along the positive z direction on the two sides of the anode aperture, respectively. The reciprocal of (a6.3.36) may be seen to be identical with (6.3.62).

Appendix a6.3.3 *Inversion of the series expression for beam convergence*

One may invert the series (6.3.80) to obtain an expression for the ratio of the cathode-to-anode radii of curvature that can be used to find this ratio if the beam convergence is known. Let us take logarithm of (6.3.80) to get

(a6.3.37)
$$y = x + \frac{1}{2}\left(-1 + \frac{12}{5}x + \frac{27}{40}x^2 + \frac{369}{2200}x^3 + \ldots\right)^2 ,$$

where $y = \ln r_K/r_M$ and $x = \ln r_c/r_a$. In principle, expanding the square in its right hand side, (a6.3.37) could be expressed as

(a6.3.38)
$$y = a_0 + a_1 x + a_2 x^2 + a_3 x^3 + \ldots ,$$

where the coefficients a_0, a_1, a_2, a_3 etc. are the constants which may be found by comparing (a6.3.37) (with its square expanded) with (a6.3.38). Now, let us arbitrarily choose known x values: x_1, x_2, x_3 etc., say, and find with the help of (a6.3.38) the corresponding y-values: y_1, y_2, y_3 etc., say. Now let us propose to invert the series (a6.3.38) as:

(a6.3.39)
$$x = b_0 + b_1 y + b_2 y^2 + b_3 y^3 + \ldots ,$$

where b_0, b_1, b_2, b_3, etc. are the constants which have to be found out. Now that the values y_1, y_2, y_3, etc. of y for the corresponding values x_1, x_2, x_3, etc. of x are known, we may write from (a6.3.39)

(a6.3.40)
$$x_1 = b_0 + b_1 y_1 + b_2 y_1^2 + b_3 y_1^3 + \ldots$$
$$x_2 = b_0 + b_1 y_2 + b_2 y_2^2 + b_3 y_2^3 + \ldots$$
$$x_3 = b_0 + b_1 y_3 + b_2 y_3^2 + b_3 y_3^3 + \ldots ,$$

and so on. In (a6.3.40) except b_0, b_1, b_2, b_3 *etc. all other quantities, namely,* y_1, y_2, y_3, *etc. are known for the chosen values of* x_1, x_2, x_3 *etc. Thus one may form as many simultaneous equations like (a6.3.40) as there are number of terms which we would like to retain in the inverted series(a6.3.39), and solve it for as many unknown coefficients* b_0, b_1, b_2, b_3, *etc. The coefficients* b_0, b_1, b_2, b_3 *etc. being thus known they may be substituted in (a6.3.39) to get the series (a6.3.37) inverted. Thus if* $x = r_K/r_M$ *is known one can find with the help of (a6.3.39)* $y = \ln r_c/r_a$, *the antilogarithm of which gives the desired value of* r_c/r_a.

Appendix a6.3.4 *Infinitesimal-region transformation from one plane to another*

Let a point be represented at (y,z) on the complex Z-plane in which the real axis is z and the imaginary axis is y. In view of the relation $f\{Z\} = f\{z + j\,y\} = u_r + j\,u_i$ *(see a6.2.1) the same point could be located at* (u_r, u_i) *on the W-plane in which* u_r *and* u_i *are the real and imaginary axes, respectively.*

Let the derivative dW/dZ of the analytic function which is unique and independent of the direction of approach (see appendix a6.2.1) be expressed in terms of its magnitude $|dW/dZ|$ *and phase* ϕ *as*

$$dW/dZ = |dW/dZ|\,\exp(j\phi)$$

which gives

$$dW = |dW/dZ|\,dZ\,\exp(j\phi)\,. \tag{a6.3.41}$$

Expressing dZ in terms of its magnitude $|dZ|$ *and phase* ψ *as*

$$dZ = |dZ|\,\exp(j\psi) \tag{a6.3.42}$$

and plugging (a6.3.42) into (a6.3.41), we have

$$dW = |dW/dZ|\,|dZ|\,\exp j(\psi + \phi)\quad. \tag{a6.3.43}$$

By comparing (a6.3.42) with (a6.3.43) one may notice that the infinitesimal region on the W-plane is changed by a factor of $|dW/dZ|$ *and rotated by an angle of* ϕ *with respect to the corresponding infinitesimal region on the Z-plane. The rotation of the infinitesimal region suggests that the relative angle between* u_r = *constant and* u_i = *constant lines should not alter by transformation from one plane, say, W-plane to Z-plane. These lines being parallel to the imaginary and real axes, respectively, on the W-plane, the angle between these lines would obviously be a right angle. So, on the Z-plane, too, the same angle, namely, a right angle should be maintained between these lines. This has also been independently appreciated in appendix a6.3.5.*

Appendix a6.3.5 *Lines at right-angle intersection on the complex plane.*

If one combines (a6.2.3) and (a6.2.4) one obtains

$$\left(-\frac{\partial u_r/\partial z}{\partial u_r/\partial y}\right)\left(-\frac{\partial u_i/\partial z}{\partial u_i/\partial y}\right) = -1\;. \tag{a6.3.44}$$

The left hand side of (a6.3.44) is the product of two terms, the first term representing the slope of the u_r = constant curve while the second term representing the slope of the u_i = constant curve, both at a point (y, z) on the Z-plane. The product of the two terms being equal to unity but with a negative sign (see (a6.3.44)) immediately suggests that these two curves must intersect at the point (y,z) at a right angle.

Appendix a6.3.6 Representation of flux and potential functions

Let us write the element of electric flux $d\phi_E$ between the curves u_i = constant and $u_i + du_i$ = constant on the Z-plane (see Fig. a6.3.1) as

(a6.3.45) $$d\phi_E = \mathbf{E} \cdot \mathbf{ds} \;,$$

where **ds** *is an element of area enclosed between the curves. If we interpret $d\phi_E$ as the element of electric flux per unit length (the length being measured perpendicular to the Z-plane), then* **ds** *may be put as*

(a6.3.46) $$\mathbf{ds} = ds_z\, \mathbf{a}_z + ds_y\, \mathbf{a}_y \;,$$

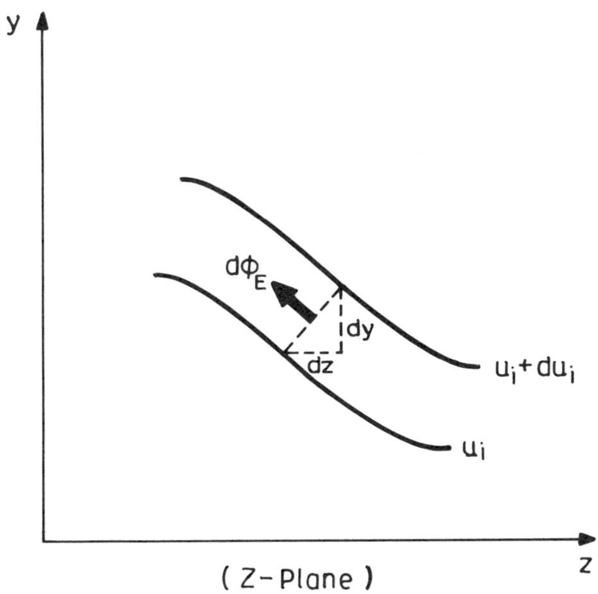

Fig. a6.3.1 Element of electric flux per unit length enclosed between u_i = constant and $u_i + du_i$ = constant lines.

where

(a6.3.47) $$ds_z = (dy)(1) = dy \quad and \quad ds_y = (dz)(1) = dz \;.$$

Substituting $\mathbf{E} = E_z\,\mathbf{a}_z + E_y\,\mathbf{a}_y$ *and (a6.3.46) through (a6.3.47) into (a6.3.45) one may write* (a6.3.48) $$d\phi_E = -E_z\, dy + E_y\, dz\ .$$

In view of (2.5.11) through (2.5.12a), (a6.3.48) becomes

(a6.3.49) $$d\phi_E = \partial V/\partial z\, dy - \partial V/\partial y\, dz\ .$$

Now if we choose u_r as the potential function, we have from (a6.3.49) replacing V by u_r :

(a6.3.50) $$d\phi_E = \partial u_r/\partial z\, dy - \partial u_r/\partial y\, dz\ .$$

Next let us express the incremental quantity du_i as :

$$du_i = \partial u_i/\partial z\, dz + \partial u_i/\partial y\, dy$$

which using (a6.2.3) and (a6.2.4) may be expressed as

(a6.3.51) $$du_i = -\partial u_r/\partial y\, dz + \partial u_r/\partial z\, dy\ .$$

Comparing (a6.3.51) with (a6.3.50), we see that

$$d\phi_E = du_i$$

which upon integration gives

(a6.3.52) $$\phi_E = u_i\ ,$$

taking $\phi_E = 0$ at $u_i = 0$ which makes the integration constant equal to zero. Thus u_i represents the flux function (see (a6.3.52)) corresponding to u_r representing the potential function.

6.4 Summary

In a practical linear beam device the electron beam of the desired cross section, current and velocity is thrown into the interaction region with the help of an electron gun. The source of electrons, if thermionic, could be cathodes of modern dispenser types. In a separate appendix, an overview on cathodes is given (see appendix *a6.0.1*). The cathode technology aims at achieving high emission density, low operating temperature, less susceptibility to poisoning and surface degradation, less warm-up time, etc.

The electron beam considered as a charge fluid may be treated by the basic laws of electrostatics. Child-Langmuir's law may be developed by solving the one-dimensional Poisson's equation for a rectilinear flow of electrons between a cathode and an anode, both taken as large and flat. The current density may be increased by increasing the potential difference between these electrodes and/or reducing their separation. For a given geometry the quantity beam current divided by the beam voltage raised to the index of power of 3/2, called the beam perveance, is a constant and gives an important gun parameter if such a gun could be realised by allowing the beam to pass through an aperture or a grid of the anode.

In a practical parallel-flow electron gun system both the electron beam and the cathode are of finite dimensions. In a method due to Pierce, a rectangular strip electron beam is derived from a finite-size flat cathode using an additional electrode called the beam forming electrode (BFE) close to the cathode outside the beam. The BFE and the anode external to the beam coincide with the equipotentials at their respective electrode potentials. The equipotentials may be formed from the potential distribution outside the beam which should satisfy Laplace's equation and at the same time pass on to the potential distribution at the beam edge which is known apriori from the solution of Poission's equation, for instance, in the form of Child-Langmuir's law for a one-dimensional rectilinear flow of electrons. In addition, the potential distribution outside the beam must have a zero gradient in a direction normal to the beam edge so that no force could be exerted transverse to the desired electron motion. The properties of an analytic function (appendix *a6.2.1*) have been used to suggest the potential distribution outside the beam in such a problem.

The solution to the simple rectangular cross section beam is approximately valid near the beam edge even for a circular-cylindrical beam. For example, in the rectangular system the BFE having the same zero potential as the cathode is in the form of a planar hat that makes an angle of 67.5° (see (6.2.6)) with the beam edge. Since the points near the beam edge of a circular cross section beam would fail to see the curvature of the beam edge, the BFE for such a beam would meet the beam edge at the same angle, namely, 67.5°. However, a more practical method of forming a circular cross section beam would be to first derive the beam from a large spherical-cup cathode and then converge it to a lower beam cross-sectional area. The solution corresponding to Child-Langumuir's law in the rectangular system is known in the spherical system as Langmuir-Blodgett's solution which is valid for a one-dimensional radial flow of electrons between a cathode, in the form of a complete sphere, and a concentric anode sphere.

In a practical Pierce-gun system forming a convergent conical beam, the cathode is part of a complete sphere in the form of a spherical cup. The problem is to find the basic output parameters, namely, the cathode radius of curvature, the cathode-disc radius, the distance between the cathode and the anode, the anode-aperture radius and the location of the waist of the beam beyond the aperture or grid of the anode. Such a beam waist would form when the beam is subject only to the space-charge field beyond the anode. The electron beam instead of converging to a common center of curvature of the cathode and anode spheres would under the space-charge field transverse to its motion pass through a minimum beam radius at the beam waist, beyond which it would diverge out unless it is prevented from doing so by a focusing structure that is separate from the electron gun system (chapter *7*). The beam spread beyond the anode under the space-charge field is modified due to the modification in the beam transit angle at the anode due to the anode acting as an electrostatic lens. In a separate appendix the basic principles of an electrostatic lens (appendix *a6.3.2*) are given. By combining the relevant relations pertaining to the space-charge flow between the cathode and the anode, mainly based on Langmuir and Blodgett's solution, and those beyond the anode under beam spread in the field-free region modified at the transit at the anode due to lens effects, one can find an explicit expression for the beam convergence in terms of a series involving the ratio of the cathode-to-anode radii of curvature, the beam convergence being taken as the ratio of the cathode disc-to- beam waist radius. One can subsequently invert such series to find the ratio of the cathode-to-anode radii of curvature when the beam convergence is known.

Once the basic output parameters have been found one can find the electrode shapes for the Pierce gun forming a convergent-flow beam. For this purpose the method of conformal transformation is used in conjunction with the solution for potential outside a simpler rectangular strip beam configuration which could be obtained rather easily (section *6.2*). Two transformations, namely, logarithmic and bilinear have been tried out. By such a transformation the actual configuration on a given complex plane is transformed onto a complex plane in which the real axis coincides with the beam edge and the imaginary axis with the cathode edge. The solution above the real axis on the transformed plane is then interpreted with the help of the transformation relation to predict the solution of potential outside the beam edge of the actual configuration. Interestingly, one and the same solution is obtained for both the transformations used. The solution then can be interpreted to find the equipotentials and hence predict the shapes of the BFE and the anode at their respective potentials.

The basic principles of forming an electron beam have been developed here starting from the fundamentals of electrostatics (chapter *2*). The method has been developed here from first principles to synthesize Pierce electron gun when the relevant beam and cathode parameters are known. In the case of a convergent-flow beam derived from a spherical-cup cathode one can find by the method described in this chapter both the basic output parameters and the electrode shapes of the gun. The Pierce guns are essentially low-perveance (~1 microperv) guns in which the anode aperture is well separated from the cathode. Special care must be made while designing a high-perveance gun to reduce the offsetting effect of the anode aperture if brought closer to the cathode with a view to increasing the beam perveance. In an innovation called the Müller gun not studied here, the anode has a tube that projects into the BFE region providing a higher beam-perveance.

From the purview of the present chapter we have excluded the analysis of the gun which gives one a picture of the beam trajectory for a given configuration of the gun. This gives one a scope to adjust the electrode potential, geometry and shapes till the desired beam trajectory is obtained. We have also not considered a grid-controlled gun in which it is possible to control the beam current by applying relatively small voltages on a control grid. Non-intercepting gridded gun (NIG) is an innovation that reduces the current intercepted by a control grid in the gun. A large-intercept current can cause excess grid heating to the point at which the grid could start to emit or ultimately fail [12]. In an NIG, an additional shadow grid is provided very close and, in fact, electrically connected to the cathode and very precisely aligned with respect to the control grid. In special-purpose applications, an electron gun may be required to provide a multi-perveance beam for multi-mode (power level) operation of a device (see appendix *a8.2.1*). The basic principles relevant to these state-of-the-art concepts in electron guns have, however, been amply provided in this chapter.

Problems

6.1 Find the beam current of a parallel-flow electron gun derived from a flat cathode of area 4mm^2 separated from the anode at a potential 2 kV and at a distance 2mm.

Answer: 73.7 mA.

6.2 Calculate G_0, Langmuir and Blodgett's function G evaluated at the anode, for the values of the ratio of the cathode-to-anode radii of curvature equal to 1.5, 2.0, 2.5, 5.0 and 10.0, respectively.

Answer: 0.46, 0.87, 1.24, 2.82 and 5.40, respectively.

6.3 (a) What is the anode current that could be obtained from a diode that consists of a complete cathode sphere of radius 22.5 cm as the emitter and a concentric inner anode sphere of radius 15cm at anode potential 10kV. (b) How would the anode current be modified if these radii of curvature are each doubled while keeping the anode potential unchanged.

Answer: (a) 134 amps (b) unchanged.

6.4 Show that a beam current ~1 amp could be obtained from a Pierce-gun derived from the basic diode configuration of problem 6.3 in which the cathode emits only through a half cone angle of 10° about the gun axis.

6.5 An electron beam of diameter 3.2 mm with beam voltage 2kV and beam current 20mA is formed by a Pierce-gun using a spherical-cup cathode. If the transit diameter of the beam at the anode is 1.63mm, calculate the throw of the gun.

Answer: 5.32 mm.

6.6 A Pierce gun using a spherical-cup cathode forms a 10kV, 1amp, 1mm(radius) electron beam. (a) Calculate the throw of the gun and (b) hence find at what distance beyond the beam waist would the beam double its radius if no separate focussing structure were used to confine it. Take the radius of the beam at the anode as 2mm.

Answer: (a) and (b) both 12mm.

Hint: The radius of the beam at the anode is double the beam-waist radius and the beam trajectory has a symmetry about the beam waist on its two sides.

6.7 A convergent Pierce gun that has the beam parameters of problem 6.5 operates at a cathode emission density of 50 mA/cm^2. Calculate its output parameters in the following sequence: cathode aperture radius (r_k), beam convergence (r_k/r_M), ratio of the cathode-to-anode radii of curvature (r_c/r_a), half beam-cone angle (θ_0), cathode radius of curvature (r_c), anode radius of curvature (r_a), anode aperture radius (r_A), interelectrode distance ($d = r_c - r_a$), ratio r_A/d (to check that its value is much less than unity), throw(d_m).

Answer: $r_k = 3.57$mm, $r_k/r_M = 2.23$, $r_c/r_a = 1.79$, $\theta_0 = 7.01°$, $r_c = 29.22$mm, $r_a = 16.33$mm, $d = 12.89$mm, $r_A/d = 0.15$, $d_m = 22.29$mm.

6.8 Show that the BFE of a spherical-cup cathode convergent Pierce gun meets the beam edge at an angle of 67.5°.

Hint: Retain for a point extremely close to the beam only the first term in the series expression (6.3.108) or (6.3.93) and show using deMoivre's theorem that the potential outside the beam at a point (r, θ) extremely close to the beam edge is proportional to $\cos(4\psi/3)$ where $\psi = \tan^{-1}[(\theta - \theta_0)/\ln r_c/r] = \tan^{-1}[(\theta-\theta_0)/\ln\{1+(r_c - r)/r\}] \cong \tan^{-1}[(\theta - \theta_0)/(r_c/r)] = \tan^{-1}[r(\theta - \theta_0)/r_c]$. Now by putting $\cos(4\psi/3) = 0$ corresponding to zero BFE potential get $4\psi/3 = \pi/2$ giving $\psi = 3\pi/8 = 67.5°$. Next appreciate from the geometrical consideration of the cross section of the configuration that the angle between the beam edge and the line joining the cathode edge and the close-by point (r, θ) on the BFE is also equal to $\tan^{-1}[r(\theta - \theta_0)/r_c]$ which has been found to be equal to ψ and which in turn has a value 67.5°.

References

[1] J. R. Pierce, "Rectilinear flow in beams," *J. Appl. Phys.* **11** (1940) 548-554.

[2] P. Zalm and A. J. A. Van Stratum, "Osmium dispenser cathodes," *Philips Tech. Rev.* **27** (1966) 69-75.

[3] J. L. Cronin, "Modern dispenser cathodes," *Proc. IEE* **128, pt. I** (1981) 19-32.

[4] C. D. Child, "Discharge from hot CaO," *Phys. Rev.* **32** (1911) 492-511.

[5] J. R. Pierce, *Theory and Design of Electron Beams* (Van Nostrand, New York, 1958).

[6] E. Kreyzig, *Advanced Engineering Mathematics* (John Wiley and Sons, Inc., New York,1967).

[7] I. Langmuir and K. B. Blodgett, "Currents limited by space charge between concentric spheres," *Phys. Rev.* **24** (1924) 49 - 59.

[8] K. R. Spangenberg, *Vacuum Tubes* (Mc Graw-Hill, New York, 1948).

[9] U. Tiwari and B. N. Basu, "Noniterative method for the synthesis of convergent Pierce electron guns," *IEEE Trans. Electron Dev.* **ED-34** (1987) 1218-1221.

[10] U. Tiwari and B. N. Basu, "Noniterative synthesis of convergent Pierce electron guns by the conformal mapping of electrode shapes using logrithmic transformation," *IEEE Trans. Electron Dev.* **ED-35** (1988) 1184-1187.

[11] H.Z. Sar-el, "Revised theory of Pierce-type electron guns," *Nuclear Instrumentation and Methods* **203** (1982) 21-33.

[12] *Hughes TWT and TWTA Handbook* (A bulletin for business circulation), Electron Dynamics Div., Hughes Aircraft Co., California.

CHAPTER 7

CONFINEMENT OF AN ELECTRON BEAM

7.0 Introduction

The electron beam once it is formed by an electron gun (see chapter *6*) has also to be confined beyond the gun — in the region where it could take part in the interaction with an electromagnetic wave. (The mechanism of such interaction has been dealt with later in chapter *8*). However, unless special care is taken, the beam of electrons left to themselves would diverge out due to Coulomb repulsion between them (see section *6.3.3*). In this chapter we take up the study of a focusing structure that does the function of constraining the electrons to move parallel to the beam axis beyond the gun to the interaction region. So it may be noted that the purpose of a *focusing* structure here is not actually to focus the electrons to a point. We consider here a magnetic focusing structure that provides a Lorentz force that counteracts the force responsible for the divergence of electrons. Actually, the focusing structure provides at its entry adequate radial component of magnetic flux density which would be responsible for an azimuthal component of electron velocity of the beam. Consequently, the interaction between the azimuthal component of electron velocity and the axial component of magnetic flux density provided by the structure would give rise to the required radial magnetic force to counter-balance the space-charge force plus the centrifugal force of the circular electronic motion.

The magnetic flux density and other relevant conditions which would ensure that the beam would retain the parameters set by the electron gun can be found by studying the motion of the beam electrons subject to the force of the space-charge field and to the opposite Lorentz force[1]-[5].

The space-charge field of an electron beam has already been estimated in section *2.2* (see also section *2.3* and *2.6*). Similarly, the Lorentz force has also been introduced in section *3.3*. Thus the problem of focusing or confinement of an electron beam taken up in the present chapter of part two of this volume is essentially the application of some of the basic concepts of electromagnetics developed in chapter *2* and *3* of part one.

7.1 Beam Space-Charge Field

The intensity of space-charge field **E** inside a circular cross-section electron beam as found by various methods in chapter *2* (see equations (2.2.2) or (2.3.7) or (2.6.8)) is given by

$$\mathbf{E} = \frac{\rho r}{2\varepsilon_0}\mathbf{a}_r \quad . \tag{7.1.1}$$

The force F_s, say, due to this space-charge field on an electron of charge e is given by

$$\mathbf{F}_s (= e\mathbf{E}) = \frac{e\rho r}{2\varepsilon_0}\mathbf{a}_r \quad . \tag{7.1.2}$$

It may be noted that since the volume charge density of electrons ρ carries a negative sign, **E** as can be seen from (7.1.1) is directed radially inward. But since e also carries a negative sign the product $e\rho$ occurring in the numerator (7.1.2) is a positive quantity, and consequently this makes F_s, given by (7.1.2), radially outward.

Example 7.1.1 Show that for a cylindrical beam of circular cross section the space-charge force would be more for a higher beam current, for a lower beam voltage, and if the beam is thinner; and also that the force will be maximum for edge electrons.

If v_z is the velocity of electrons all constrained to move parallel to the beam axis then the volume charge density of electrons ρ is related to the electron current density J_e as

$$J_e = \rho v_z$$

which may be alternatively expressed in terms of the electron current $I_e (= \pi a^2 J_e)$ as

$$\rho = \frac{I_e}{\pi a^2 v_z} , \tag{7.1.3}$$

where a is the beam radius. Putting in (7.1.3)

$$v_z = (2 | \eta | v_0)^{1/2} , \tag{7.1.4}$$

expressed in terms of beam voltage V_0 (see (6.1.6)), η being the charge-to-mass ratio of an electron, one gets

$$\rho = \frac{I_e}{\pi a^2 (2|\eta| v_0)^{1/2}} \quad . \tag{7.1.5}$$

Substituting (7.1.5) into (7.1.2) one obtains

$$\mathbf{F}_s = \frac{|e| I_0 r}{2\pi\varepsilon_0 a^2 (2|\eta| v_0)^{1/2}}\mathbf{a}_r , \tag{7.1.6}$$

where $I_0 = | I_e |$ is the beam current interpreted as a positive quantity. It directly follows from (7.1.6) that F_s will be more for a higher beam current, for electrons at a greater radial distance from the axis of the beam (for a larger value of r) being maximum for edge electrons, for a thinner beam (for a smaller value of a) and for a lower beam voltage V_0.

7.2 Elementary Magnetic Flux through a Beam Cross Section

In the present study of the dynamics of an electron beam supposedly of circular cross section, subject to the dc magnetic field of the focusing structure, we consider the axis of the beam to be essentially coincident with that of the focusing structure (Fig. 7.2.1). Let us now form [3] an element of area dS, say, by rotating an element of electron trajectory of infinitesimal length PQ inside the beam through an infinitesimal azimuthal angle $d\theta$ (Figs. 7.2.1 (a) and (b)). Let the coordinates of P be (r, θ, z) and those of Q be $(r+dr, \theta+d\theta, z+dz)$, respectively. The element of area dS is then given by the product of PP′ and PQ. Now, clearly PP′ = $rd\theta$ (Fig. 7.2.1(b)). One may further see from the right-angled triangle QPN, where N is the foot of the perpendicular drawn from Q on the radially outward line through P, then PQ = (PN) $\sec\psi$ = dr $\sec\psi$, putting PN = dr, where ψ is the angle between PQ and the radially outward line through P. It may be seen that the angle ψ also equals the angle between the unit vector inwardly normal to the area element, $\mathbf{a}_n$, and the unit vector parallel to the axis, $\mathbf{a}_z$. This can be better appreciated by considering $\mathbf{a}_n$ and $\mathbf{a}_z$ both on the element of trajectory PQ (Fig. 7.2.1(c)). Now, substituting PP′ = $rd\theta$ and PQ = dr $\sec\psi$ into dS = (PP′)(PQ) one has

$$dS = r \sec\psi \, d\theta \, dr . \tag{7.2.1}$$

The unit vector $\mathbf{a}_n$ may be expressed in terms of the axial and radial unit vectors $\mathbf{a}_z$ and $\mathbf{a}_r$, respectively, as:

$$\mathbf{a}_n = \cos\psi \, \mathbf{a}_z - \sin\psi \, \mathbf{a}_r . \tag{7.2.2}$$

Now, the applied dc magnetic flux density **B** in general has both the axial and radial components which, however, becomes predominantly axial down the focusing structure away from the entry plane at the gun end. **B** may be expressed as

$$\mathbf{B} = B_r \mathbf{a}_r + B_z \mathbf{a}_z . \tag{7.2.3}$$

Making use of (7.2.1)-(7.2.3) one thus obtains:

the element of magnetic flux through the area element (= ds)

$$= \mathbf{B}.\mathbf{a}_n \, ds = (B_r \mathbf{a}_r + B_z \mathbf{a}_z).(\cos\psi \, \mathbf{a}_z - \sin\psi \, \mathbf{a}_r) \, r \sec\psi \, d\theta \, dr$$

$$= - B_r \sin\psi \, r \sec\psi \, d\theta \, dr + B_z \cos\psi \, r \sec\psi \, d\theta \, dr$$

$$= - B_r r \tan\psi \, d\theta \, dr + B_z r \, d\theta \, dr . \tag{7.2.4}$$

Putting $\tan\psi = QN/PN = dz/dr$ (Fig. 7.2.1(c)) one may then write (7.2.4) as:

the element of magnetic flux through the area element (= dS)

$$= - B_r r \, (dz/dr) d\theta \, dr + B_z r \, d\theta \, dr = - B_r r \, dz \, d\theta + B_z r \, dr \, d\theta . \tag{7.2.5}$$

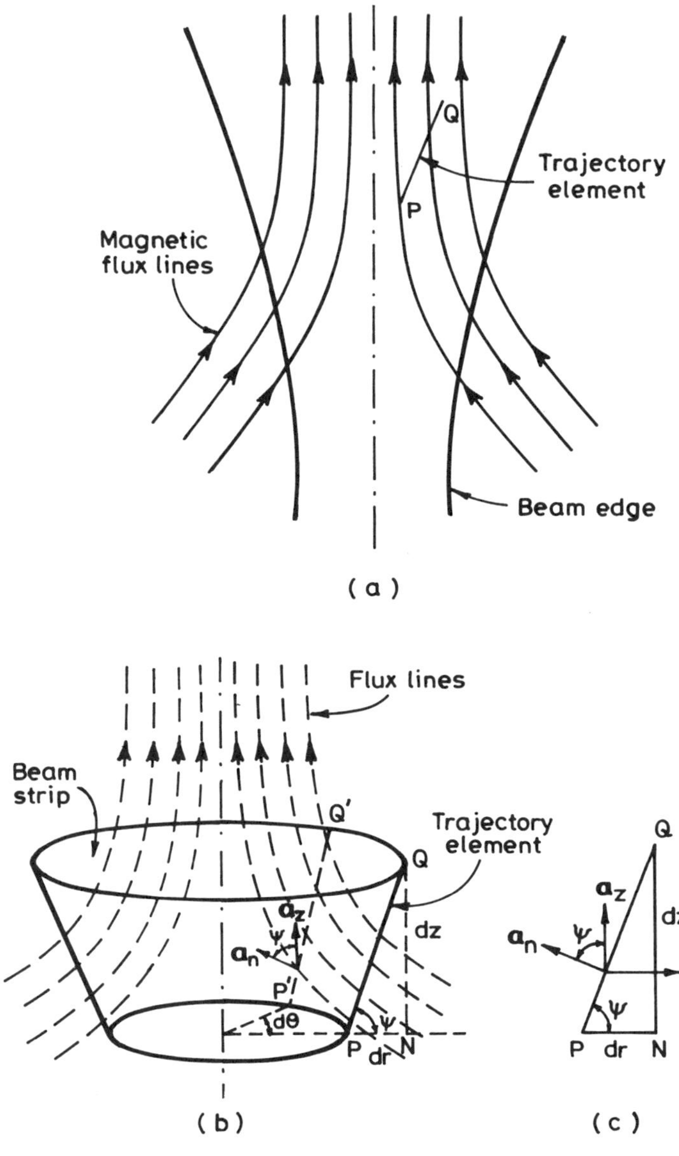

Fig. 7.2.1 Cross-section of the beam through its axis showing the magnetic flux lines cutting through the beam edge which down the axis become parallel to the latter, and also showing an element of electron trajectory PQ inside the beam **(a)**; an element of area dS (= PQQ′P′) generated by rotating the element of electron trajectory PQ through an infinitesimal angle dθ, for the estimate of magnetic flux through such an area element, and the unit vector $\mathbf{a}_n$ inwardly normal to the area element and the axial unit vector $\mathbf{a}_z$ **(b)** and the unit vectors $\mathbf{a}_n$ and $\mathbf{a}_z$ drawn on the element of trajectory **(c)**.

Integrating (7.2.5) between $\theta = 0$ and 2π one obtains $d\phi_B$, the element of magnetic flux through an element of strip of beam which is obtained by making a complete revolution of the element of electron trajectory PQ, as follows[3]:

$$d\phi_B = -2\pi r\,(B_r\,dz - B_z\,dr)\,. \qquad (7.2.6)$$

7.3 Angular Acceleration Equation for Electronic Motion

The angular acceleration equation of an electron subject to the azimuthal component of Lorentz force due to the magnetic field of the focusing structure may be written as:

$$\frac{1}{r}\frac{d}{dt}(r^2\frac{d\theta}{dt}) = \eta(\mathbf{v}\times\mathbf{B})_\theta\,,$$

which after expansion of the cross-product in the cylindrical system of coordinates and re-arrangement of terms becomes

$$d(r^2\frac{d\theta}{dt}) = \eta r(B_r v_z - B_z v_r)\,dt\,, \qquad (7.3.1)$$

where v_z and v_r are the axial and radial components of electron velocity, respectively. Putting v_z (= dz/dt) and v_r (= dr/dt) in (7.3.1) one obtains

$$d(r^2\frac{d\theta}{dt}) = \eta r(B_r\frac{dz}{dt} - B_z\frac{dr}{dt})\,dt = \eta r(B_r\,dz - B_z\,dr)\,. \qquad (7.3.2)$$

With the help of (7.2.6) one may then express (7.3.2) as:

$$d(r^2\frac{d\theta}{dt}) = -\eta r\frac{d\phi_B}{2\pi r} = -\frac{\eta}{2\pi}d\phi_B\,. \qquad (7.3.3)$$

The magnetic flux lines corresponding to $d\phi_B$ passing through the beam strip also cuts across the beam edge on the one hand and becomes parallel to the axis of the beam on the other hand. Therefore, if one integrates (7.3.3) one gets

$$r^2\frac{d\theta}{dt} = -\frac{\eta\phi_B}{2\pi} + K\,, \qquad (7.3.4)$$

where K is an integration constant. ϕ_B refers to the total flux lines entering the beam across its edge which become predominantly parallel to the beam axis, that is, perpendicular to the circular cross section of the portion of the beam of area πr^2, down the focusing structure away from the entry plane of the latter. ϕ_B may, therefore, be expressed as

$$\phi_B = \pi r^2 B\,, \qquad (7.3.5)$$

where B is the *predominantly-axial* magnetic flux density of the focusing structure (B_z = B), away from its entry plane.

The integration constant K may be evaluated by substituting into (7.3.4) $d\theta/dt = 0$ at the cathode where the electrons emit from as

$$K = \frac{\eta\phi_{BC}}{2\pi}, \tag{7.3.6}$$

where $\phi_B = \phi_{BC}$ is the magnetic flux at the cathode. Substituting (7.3.6) into (7.3.4) one then obtains

$$\frac{d\theta}{dt} = -\frac{\eta}{2\pi}\left(\frac{\phi_B - \phi_{BC}}{r^2}\right). \tag{7.3.7}$$

An alternative form of (7.3.7) may be obtained by putting in the latter the expression for ϕ_B given by (7.3.5) and its corresponding expression at the cathode

$$\phi_{BC} = \pi\, r_k^2\, B_C, \tag{7.3.8}$$

where r_k is the radius of the beam at the cathode referring to a convergent Pierce electron gun derived from a spherical-cup cathode (see Fig. 6.3.1). Thus one obtains for the angular velocity of an electron

$$\frac{d\theta}{dt} = -\frac{\eta B}{2}\left(1-(\frac{B_C}{B})(\frac{r_k}{r})^2\right). \tag{7.3.9}$$

Further, since one may write the electron cyclotron frequency ω_c as

$$\omega_c = -\eta B, \tag{7.3.10}$$

η carrying the negative sign (see problem 3.7), one may express (7.3.9) as:

$$\frac{d\theta}{dt} = \frac{\omega_c}{2}\left(1-\left(\frac{B_C}{B}\right)\left(\frac{r_k}{r}\right)^2\right). \tag{7.3.11}$$

The expression (7.3.7) or (7.3.11) is referred to as Busch's theorem[4].

The tangential electron velocity of the beam v_θ is given by

$$v_\theta = r\frac{d\theta}{dt} \tag{7.3.12}$$

which may be read with the help of the expression for $d\theta/dt$, (7.3.7), in terms of magnetic flux, or by (7.3.11), in terms of magnetic flux density.

A particular case of interest is the one in which no magnetic flux links up with the cathode, that is,

$$\phi_{BC} = 0, \tag{7.3.13}$$

which yields from (7.3.7) while using (7.3.5) and (7.3.10):

$$\frac{d\theta}{dt} = \frac{\omega_c}{2} = \omega_L \text{ (for a magnetically-shield cathode)}, \qquad (7.3.14)$$

where ω_L , half the cyclotron frequency, is called the Larmor frequency [1] (see also problem 7.3.). It is of interest to see from (7.3.14) that the angular velocity of a beam electron is independent of r which suggests that in this case, that is, for $\phi_{BC} = 0$, the electron beam would rotate like a rigid bar. It may be noted that the condition $\phi_{BC} = 0$ given by (7.3.13) would be satisfied:

i) if the cathode is shielded from the magnetic field, that is, for $B_C = 0$ giving $\phi_{BC} = 0$, as can be seen from (7.3.8);

ii) if the cathode is a point cathode, that is, $r_k = 0$, again giving $\phi_{BC} = 0$, as can be seen from (7.3.8); and

iii) if the magnetic flux lines are tangential to the cathode surface, that is, the magnetic flux density vector at the cathode is at right angle to the area vector over the cathode surface thereby making the dot product of these two vector equal to zero and hence $\phi_{BC} = 0$.

7.4 Radial Acceleration Equation for Electronic Motion

The radial acceleration of an electron down the focusing structure, subject to the radial space-charge field given by (7.1.1) as well as to the radial component of Lorentz force due to the magnetic field of the focusing structure, is given by

$$\frac{d^2r}{dt^2} - r\left(\frac{d\theta}{dt}\right)^2 = \eta[E_r + (\mathbf{v} \times \mathbf{B})_r] . \qquad (7.4.1)$$

Substituting $E_r = \rho r/2\varepsilon_0$ from (7.1.1) and expanding the cross-product $(\mathbf{v} \times \mathbf{B})_r$, remembering that here the magnetic flux magnetic flux density has no azimuthal component ($B_\theta = 0$) and $B_z = B$, one may express (7.4.1), now referring to a beam-edge electron, as follows:

$$\frac{d^2r}{dt^2} = \eta\left(\frac{\rho r}{2\varepsilon_0} + v_\theta B\right) + r\left(\frac{d\theta}{dt}\right)^2 , \qquad (7.4.2)$$

where ρ is the volume charge density of electrons. From (7.1.3), replacing a = r therein and remembering that r refers here to the beam-edge electron, we get

$$\rho = \frac{I_e}{\pi r^2 v_z} , \qquad (7.4.3)$$

where I_e is the electron current constituted by the electrons across the entire cross section of the beam of area πr^2, all supposedly moving in the z-direction with a velocity v_z. Let us now substitute (7.3.12) and (7.4.3) into (7.4.2) and use (7.3.14) and (7.3.10), to obtain for a magnetically shield cathode:

$$\frac{d^2 r}{dt^2} = \frac{\eta I_e}{2\pi\varepsilon_0 v_z r} - \frac{\eta^2 B^2 r}{2} + \frac{\eta^2 B^2 r}{4}$$

which on putting $\eta = -|\eta|$ and $I_e = -I_0$ becomes

$$\frac{d^2 r}{dt^2} = \frac{|\eta| I_0}{2\pi\varepsilon_0 v_z r} - \frac{|\eta|^2 B^2 r}{2} + \frac{|\eta|^2 B^2 r}{4}, \tag{7.4.4}$$

where I_0 is the beam current interpreted as positive. Thus in (7.4.4) we obtain the equation of force per unit mass on an edge electron.

It may be noted that the first, second and third terms of the right hand side of (7.4.4), which in fact have followed from the corresponding terms of (7.4.2), represent the three components of radial force per unit mass on an edge electron. The first of these terms is the radially outward space-charge force per unit mass on an edge electron, the second term is the radially-inward Lorentz centripetal force per unit mass, while the third term is the radially-outward centrifugal force per unit mass on an edge electron due to the circular motion of the electron. Also, it may be seen that the magnetic-field dependent second and third terms, more precisely the 'centripetal' second term reduced by the 'centrifugal' third term, may be interpreted as the restoring force part opposing the space-charge force part represented by the first term. Combining the second and third terms of (7.4.4) one may express the latter as:

$$\frac{d^2 r}{dt^2} = \left(\frac{|\eta| I_0}{2\pi\varepsilon_0 v_z}\right)\frac{1}{r} - \left(\frac{|\eta|^2 B^2}{4}\right) r\,. \tag{7.4.5}$$

Now, in order to obtain an equation for the trajectory of the edge electron let us choose to express dr/dt as (remembering that $v_z = dz/dt$):

$$\frac{dr}{dt} = \frac{dr}{dz}\frac{dz}{dt} = v_z \frac{dr}{dz},$$

which on differentiation second-time gives

$$\frac{d^2 r}{dt^2} = v_z \frac{d^2 r}{dz^2}\frac{dz}{dt} = v_z^2 \frac{d^2 r}{dz^2}. \tag{7.4.6}$$

Substituting (7.4.6) into (7.4.5) one then obtains

$$\frac{d^2r}{dz^2} = \frac{C_1}{r} - C_2 r \,, \tag{7.4.7}$$

where C_1 and C_2 are the constants given by

$$C_1 = \frac{|\eta| I_0}{2\pi\varepsilon_0 v_z^3} \quad \text{and} \quad C_2 = \frac{|\eta|^2 B^2}{4 v_z^2} \,,$$

which may be expressed in terms of beam voltage V_0 corresponding to v_z , given by (7.1.4), as follows:

$$C_1 = \frac{|\eta| I_0}{2\pi\varepsilon_0 (2|\eta| V_0)^{3/2}} \qquad \text{(a)}$$

and

$$C_2 = \frac{|\eta|^2 B^2}{4(2|\eta| V_0)} \,. \qquad \text{(b)} \tag{7.4.8}$$

The trajectory of an edge electron is described by (7.4.7) which in turn depends upon the beam voltage V_0 and current I_0 and the magnetic flux density B of the focusing structure. It will be of interest to find the solution of (7.4.7) that would give the position of an edge electron with respect to the axis of the electron beam as the electron moves down the focusing structure. This has been taken up in the following section.

7.5 Brillouin Focusing Conditions for Minimum Beam Scalloping

Let us now find by studying the solution of (7.4.7) what conditions need to be satisfied to ensure that the beam cross-sectional area does not vary with the axial distance from its value at the entry plane of the focusing structure [3]. One may thus desire, referring to a conical convergent-flow beam derived from a curved cathode (section *6.3*), that the beam radius r remains constant at a value equal to the beam-waist radius r_M = a, say (see Fig. 6.3.1). Let

$$r = a + \delta \,, \tag{7.5.1}$$

where $\delta \ll a$ represents the variation of the beam radius over the desired constant beam radius a (= r_M). Putting (7.5.1) in (7.4.7) one obtains:

$$\frac{d^2\delta}{dz^2} = \frac{C_1}{a+\delta} - C_2(a+\delta)$$

$$= \frac{C_1}{a}\left(1+\frac{\delta}{a}\right)^{-1} - C_2 a - C_2\delta$$

which, after expanding binomially and ignoring higher powers of δ/a (<< 1), takes the following form:

$$\frac{d^2\delta}{dz^2} = \frac{C_1}{a}\left(1-\frac{\delta}{a}\right) - C_2 a - C_2\delta$$

$$= \left(\frac{C_1}{a} - C_2 a\right) - \left(\frac{C_1}{a^2} + C_2\right)\delta \ . \tag{7.5.2}$$

If now the beam voltage V_0 and current I_0 as well as the magnetic flux density are so chosen that the following relation:

$$\frac{C_1}{a} = C_2 a \tag{7.5.3}$$

is satisfied, then the first term of (7.5.2) would vanish while its second term simplifies to $C_1/a^2 + C_2 = 2\,C_2$. This makes (7.5.2) read as

$$\frac{d^2\delta}{dz^2} = -2C_2\delta \ . \tag{7.5.4}$$

Putting

$$2C_2 = m^2 \tag{7.5.5}$$

in (7.5.4), one thus gets

$$\frac{d^2\delta}{dz^2} = -m^2\delta \ , \tag{7.5.6}$$

an equation that has the well-known solution [3]

$$\delta = A\sin(mz) + B\cos(mz)\ , \tag{7.5.7}$$

where A and B are the constants.

If it is assumed that the electron beam enters the focusing magnetic field at $z = 0$ with a radius $r = a$, then one may put in (7.5.7) the following boundary condition (see equation (7.5.1)):

$$\delta = 0 \quad \text{at} \quad z = 0\ . \tag{7.5.8}$$

This makes

$$B = 0\ . \tag{7.5.9}$$

In view of (7.5.9), one gets from (7.5.7) the following expression for the radius perturbation δ of the beam [3] :

$$\delta = A\sin(mz)\ . \tag{7.5.10}$$

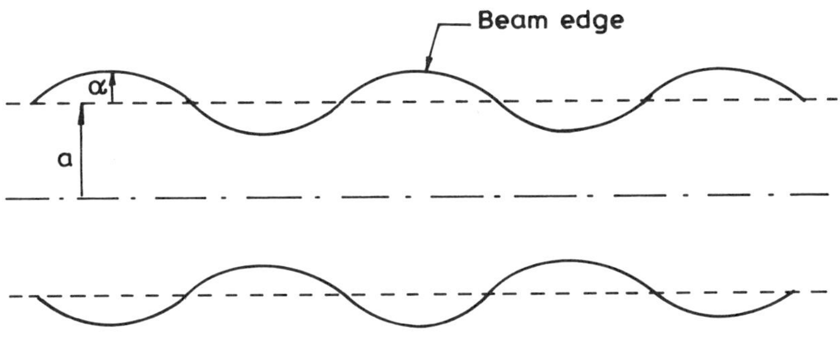

Fig. 7.5.1 Sinusoidal variation of the beam cross-sectional radius (beam-scalloping) along the structure if the cathode is magnetically shielded ($\phi_{BC} = B_c = 0$) and the beam radius at the entry plane ($z = 0$) is the value that is desired to be maintained constant, without reference to the radial velocity of the beam at the entry plane.

Substituting (7.5.10) in (7.5.1) one may then write

$$r = a + A\sin(mz)\,, \tag{7.5.11}$$

which shows that the beam scallops (see Fig. 7.5.1) with its radius varying sinusoidally over its mean radius 'a' — the value which the focusing structure is supposed to maintain as constant with the axial distance. From (7.5.11) one may easily find the wavelength of the scallop (see problem 7.4) in terms of beam parameters with the help of (7.5.5) and (7.4.8(b)).

One must now seek for an additional boundary condition to ensure that the beam does not scallop at all. For this purpose let us differentiate (7.5.10) to get

$$\frac{d\delta}{dz} = A\,m\cos(mz)\,. \tag{7.5.12}$$

It follows from (7.5.12) that if we satisfy the boundary condition [3]

$$\frac{d\delta}{dz} = 0 \quad \text{at} \quad z = 0\,, \tag{7.5.13}$$

then we have from (7.5.12) $Am = 0$, giving $A = 0$ since $m \neq 0$ (see the definition of m in terms of C_2 following (7.5.4), C_2 in turn being given by (7.4.8(b))). This ($A = 0$) now gives from (7.5.10) $\delta = 0$ and hence from (7.5.1):

$$r = a\,. \tag{7.5.14}$$

Thus the additional boundary condition (7.5.13) now ensures that the beam radius is constant, remaining independent of the axial distance z from the entry plane of the focusing structure, given by (7.5.14).

Thus at this stage it is worth recalling all the conditions which have led to the result of no beam scalloping as given by (7.5.14) and stated following it. These conditions may be put together as follows.

i) The cathode is shielded from the magnetic flux as stated by (7.3.13).

ii) The relation (7.5.3) is satisfied. Substituting (7.4.8) into (7.5.3) and rearranging terms this condition may be expressed as:

$$B_B^2 = \frac{\sqrt{2}\, I_0}{\pi\varepsilon_0 |\eta|^{3/2} V_0^{1/2} a^2} = 0.69 \times 10^{-6} \frac{I_0}{V_0^{1/2} a^2} \text{ (Tesla)}^2 . \qquad (7.5.15)$$

The above relation (7.5.15) relating the magnitude of magnetic flux density to the beam parameters V_0 , I_0 and a is called the Brillouin relation. The subscript B with the magnitude of the magnetic flux density B thus stands for the Brillouin value to be provided by the focusing structure. According to (7.5.15), a larger magnetic flux density would be necessary to confine an electron beam of a higher beam current, lower beam voltage and smaller beam radius. In fact, this is also expected since the space-charge force to be counterbalanced would also be more for such a beam, as discussed following (7.1.6).

iii) The boundary condition (7.5.8) is valid. This requires that the beam should enter the magnetic field of the focusing structure with its radius equal to the desired value ($r = a$) which the structure is supposed to maintain as constant.

iv) The boundary condition (7.5.13) is valid. This demands that the beam should enter the magnetic field of the focusing structure parallel to the axis. In other words, there should be no radial component of beam velocity at the entry of the focusing structure.

The above conditions (i)-(iv) are known as the Brillouin conditions. A scheme that ensures that these conditions for the confinement of an electron beam is realised is called the Brillouin focusing and an electron beam so formed is known as the Brillouin beam [1]. Such a beam may be produced [2], [3] (Fig. 7.5.2) by screening the cathode from the magnetic field by a pole piece to ensure condition (i). The precise value of the axial magnetic flux density (Brillouin value) given by (7.5.15) as per the desired beam parameters (condition ii) is to be attained abruptly at the entry of the focusing structure. Such a beam is usually derived from a curved cathode in a convergent-flow electron gun (section *6.3*). The beam waist-radius (= r_M) is precisely set equal to the beam radius a, desired to be maintained constant, in order to realise condition (iii). Further, the beam-waist (Fig. 6.3.1) is made to coincide with the abrupt beginning of the magnetic field region that ensures condition (iv), since at the beam waist the beam is parallel to the axis and the electrons of the beam have no transverse velocities.

Thus to summarise, in a Brillouin focusing scheme that is to be compatible with a convergent Pierce gun derived from a spherical-cup cathode, a magnetic field of precise value of flux density determined by the required beam parameters should abruptly begin at the beam waist (entry of the focusing structure) where the electrons have no component of transverse velocity and have a radius value equal to the one that is required to be maintained as constant with the axial distance, it being further ensured that the cathode is essentially shielded magnetically.

In a practical arrangement [2], [3] to realise Brillouin conditions one has to take care in shielding the cathode from magnetic field by employing a screen in the form of a pole piece, made of a magnetic material, with a hole through which the electron beam can pass (Fig. 7.5.2). It has been found that such an arrangement limits the value of the beam perveance to a lower value for a large value of the hole radius of the pole piece relative to the beam radius so that the values of the magnetic flux density and the beam parameters remain reasonably close to those required to satisfy the Brillouin conditions [2].

In the foregoing treatment of Brillouin focusing we have assumed uniform emission from the cathode. We have also ignored thermal velocity effects. In practice therefore the required magnetic flux density would be greater than the predicted Brillouin value B_B given by (7.5.15). Further, Brillouin conditions are prone to being offset by the unpredictable space-charge forces due to the formation of positive ions in the tube. Also it may be recalled that in the derivation of Brillouin conditions a dc beam current density is taken. However, in the actual operation of the tube, the beam current density changes locally due to rf bunching, more so in the large-signal regime. This, it has been found in Brillouin focusing, causes a large change in the beam radius [2]. In the following section is considered an alternative focusing scheme, called the confined-flow, which is less sensitive to the variation of beam parameters, which though would, as we will see, call for a greater magnetic flux density than is needed in Brillouin focusing.

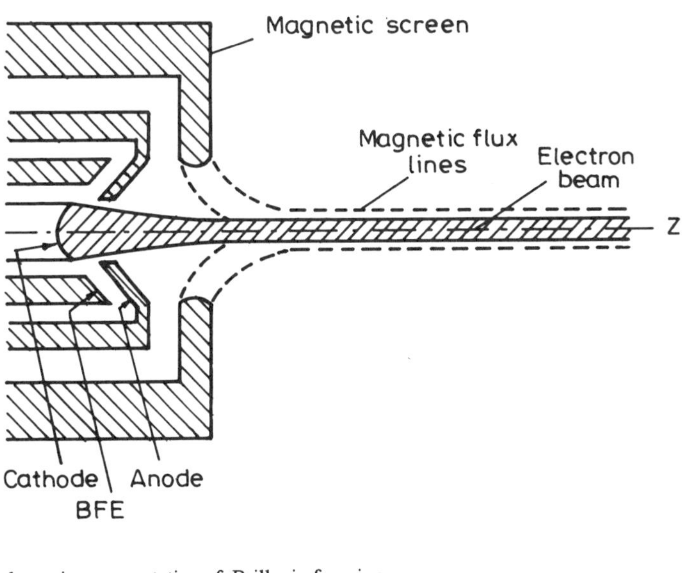

Fig. 7.5.2 Schematic representation of Brillouin focusing.

<u>*Example 7.5.1*</u> For an edge electron of a circular cross-section electron beam in a Brillouin magnetic field region, verify that (a) the magnitudes of the radially outward space-charge and centrifugal forces are equal to each other; that (b) the magnitude of the Lorentz-centripetal force is twice that of the centrifugal force; and that (c) the sum of the magnitude of the space-charge and centrifugal forces is equal to the magnitude of the Lorentz-centripetal force.

(a) The magnitude of the radially outward space-charge force on an edge electron (r = a), F_s, may be written from (7.1.6) as:

$$F_S = \frac{|e| I_o}{2\pi\varepsilon_0 a(2|\eta| v_0)^{1/2}} . \qquad (7.5.16)$$

The magnitude of the radially outward centrifugal force on an edge electron (r = a) due to its circular motion, F_{cf}, may be written with the help of (7.3.12), (7.3.14) and (7.3.10) as:

$$F_{cf} = \left(\frac{m v_\theta^2}{a}\right)_{r=a} = m\left(\frac{d\theta}{dt}\right)^2_{r=a} a$$

$$= \frac{m\omega_c^2 a}{4} = \frac{|e||\eta| aB^2}{4} . \qquad (7.5.17)$$

Putting Brillouin magnetic flux density from (7.5.15) in (7.5.17) one obtains

$$F_{cf} = \frac{|e| I_o}{2\pi\varepsilon_0 a(2|\eta| V_0)^{1/2}} . \qquad (7.5.18)$$

Comparing (7.6.16) and (7.6.18) one may thus easily see:

$$F_{cf} = F_s . \qquad (7.5.19)$$

(b) The magnitude of the radially inward centripetal Lorentz force on an edge electron (r = a), F_{cp}, may be written with the help of (7.3.12), (7.3.14) and (7.3.10) as:

$$F_{cp} = |e\, v_\theta B| = |e|(\frac{d\theta}{dt})(a)(B)$$

$$= \frac{|e|\omega_c aB}{2} = \frac{|e||\eta| aB^2}{2} . \qquad (7.5.20)$$

Comparing (7.5.20) with (7.5.17), one may see

$$F_{cp} = 2 F_{cf} . \qquad (7.5.21)$$

The explicit expression for F_{cp} in terms of beam parameters may be then written by substituting (7.5.18) in (7.5.21) as:

$$F_{cp} = \frac{|e| I_o}{\pi \varepsilon_0 a (2|\eta| V_0)^{1/2}} \quad . \tag{7.5.22}$$

(c) One may then either combine (7.5.19) and (7.5.21) or compare (7.5.22) with (7.5.18) and (7.5.16) combined to obtain:

$$F_S + F_{cf} = F_{cp} \quad .$$

<u>Example 7.5.2</u> Verify that the radial force on an edge electron of a Brillouin beam is zero yielding a space-charge balanced situation.

The radial force on an edge electron is obtained by multiplying (7.4.5) by m and putting r = a and $B = B_B$ therein to obtain

$$m\left(\frac{d^2 r}{dt^2}\right)_{r=a} = \frac{m|\eta| I_0}{2\pi\varepsilon_0 v_z a} - \frac{|\eta|^2 B_B^2}{4} a \quad . \tag{7.5.23}$$

Substituting (7.1.4) and (7.5.15) in (7.5.23), the radial force on an edge electron becomes

$$m\left(\frac{d^2 r}{dt^2}\right)_{r=a} = \frac{m|\eta| I_0}{2\pi\varepsilon_0 a (2|\eta| V_0)^{1/2}} - \frac{m|\eta| I_0}{2\pi\varepsilon_0 a (2|\eta| V_0)^{1/2}} = 0 \quad .$$

7.6 Confined-Flow Focusing

Stringent Brillouin conditions derived in section *7.5* are somewhat difficult to realise in a practical design. The difficulties would be increased if in the device one employs a convergent-flow electron gun, derived from a spherical-cup cathode, instead of a simple parallel-flow gun, derived from a flat cathode. Further, it has been mentioned in the foregoing section that in Brillouin focusing the beam radius is quite sensitive to the variation of beam current density. In an alternative focusing scheme which is less prone to beam current density modulation, magnetic flux lines are allowed to thread into the cathode unlike in Brillouin focusing [2]-[5].

In Pierce's design of an electron gun (considered in the preceding chapter), however, no magnetic field is allowed for in the gun region unless magnetic flux lines are aligned with the beam trajectories. It, therefore, turns out that magnetic flux lines that cut through the beam waist should also extend over the entire cathode disc following the beam trajectories. This enables one to write the following relation between the flux densities at the cathode and the beam waist, B_c and B_W, respectively, by equating the flux through the beam waist with that through the cathode disc:

$$B_W \pi a^2 = B_c \pi r_k^2 , \tag{7.6.1}$$

where a (= r_M) is the beam-waist radius and r_k is the cathode-disc radius (Fig. 6.3.1). From (7.6.1) one sees that the cathode flux density B_c is compressed from its value B_W at the beam waist by a factor of the area convergence of the gun (see (6.3.3(d))) as:

$$B_c = \frac{B_W}{(r_k / a)^2} . \tag{7.6.2}$$

For confined-flow focusing let us now formulate an equation of the radial force per unit mass on an edge electron (r = a) similar to (7.4.5) which was obtained for Brillouin focusing. For this purpose let us start from (7.4.2) and adopt the same method as followed for Brillouin focusing in section *7.5*, however, taking (7.3.9) as the expression for dθ/dt instead of (7.3.14) read through (7.3.10). The method gives the following equation corresponding to (7.4.4):

$$\frac{d^2 r}{dt^2} = \frac{|\eta| I_0}{2\pi\varepsilon_0 v_z r} - \frac{|\eta|^2 r}{2}\left(B - B_c\left(\frac{r_k}{r}\right)^2\right)B + \frac{|\eta|^2 r}{4}\left(B - B_c\left(\frac{r_k}{r}\right)^2\right)^2 . \tag{7.6.3}$$

It may be seen that, putting $B_c = 0$, (7.6.3) passes on to (7.4.4). By a simple algebra, one may express (7.6.3) as

$$\frac{d^2 r}{dt^2} = \frac{|\eta| I_0}{2\pi\varepsilon_0 v_z r} - \frac{|\eta|^2 r}{4}\left(B^2 - B_c^2 \frac{r_k^4}{r^4}\right) . \tag{7.6.4}$$

Putting $B_c = 0$, (7.6.4) is seen to pass on to (7.4.5).

Now, if the confined-flow focusing has to simulate the same restoring force on an edge electron, (duly reduced by the centrifugal force) as the Brillouin focusing (see the discussion following (7.4.4)), one must set the magnetic-field dependent term of (7.6.4), which is valid for confined-flow focusing, equal to the corresponding term of (7.4.5), in which B is to be replaced by B_B , for Brillouin focusing. This gives remembering that (7.4.5) as well as (7.6.4) refers to an edge electron (r = a) [5]:

$$-\frac{|\eta|^2 a}{4}\left(B^2 - B_c^2 \frac{r_k^4}{a^4}\right) = -\frac{|\eta|^2 a}{4} B_B^2$$

whence we get

$$B^2 = B_B^2 + B_c^2 \frac{r_k^4}{a^4} . \tag{7.6.5}$$

It is clear from (7.6.5) that $B > B_B$ meaning thereby that, for an identical space-charge balanced condition, a confined-flow focusing structure is required to establish a larger magnetic flux density beyond the beam waist than an equivalent Brillouin focusing structure. Further, in view of (7.6.2) one may express (7.6.5) in terms of the magnetic flux density at the beam waist, B_W, as:

$$B^2 = B_B^2 + B_W^2. \quad (7.6.6)$$

Now, if we choose to take

$$B = p\,B_B \quad (p > 1), \quad (7.6.7)$$

then using (7.6.5)-(7.6.7), one may express B_W and B_c as

$$B_W = (1 - 1/p^2)^{1/2}\,B = (p^2 - 1)^{1/2}\,B_B. \quad (7.6.8)$$

and

$$B_c = (1 - 1/p^2)^{1/2}(a/r_k)^2\,B = (p^2 - 1)^{1/2}(a/r_k)^2\,B_B. \quad (7.6.9)$$

It may be seen from (7.6.7)-(7.6.9) that if the focusing magnetic flux density beyond the beam waist B were typically chosen as a factor (= p) 1.5 larger than the Brillouin value ($B_W \sim 1.5\,B_B$), then a relatively smaller magnetic flux density B_W would be required at the beam waist: $B_W \sim 1.1\,B_B$, and taking the area convergence of the gun as $(r_k/a)^2 \sim 50$, only a small fraction of the Brillouin flux density would be required to be linked up with the cathode: $B_c \sim 0.02\,B_B$.

7.7 Periodic Permanent Magnet (PPM) Focusing

The magnetic flux density required for confining an electron beam may be obtained either by a solenoid or a permanent magnet (PM) focusing structure. For a linear-beam electron beam device like the traveling-wave tube, a large magnetic flux density will be required over a large interaction length to confine *a large-current* beam for a high power or *a small-radius beam* for a high operating frequency (see Brillouin relation (7.5.15)). (The transverse dimension of the interaction structure and hence also the radius of the beam have to be reduced for a high operating frequency of the device.) The focusing structure, be it a PM or a solenoid, becomes rather heavy and bulky if the required magnetic flux density is high. Besides, the solenoid needs an external power supply. Particularly for air-borne and space applications, a solenoid is, therefore, inconvenient, though it is capable of providing a uniform magnetic flux density over a large interaction length of the device.

A PM structure is suitable for providing a uniform magnetic flux density over a small interaction length. In a device like the travelling-wave tube in which the focusing magnetic flux density to be provided has to extend over a large interaction length, the PM structure, too, becomes quite heavy and large. This is because if the length of a PM is increased by a factor of N, then all other dimensions of the structure have also to be increased in the same proportion to ensure that the magnetic flux density provided by the structure in the interaction region does not fall. For a tubular magnet, for instance, this can be done by increasing the length of the magnet by a factor of N and, at the same time, by increasing its inner and outer radii each by the same factor N. It can be easily then appreciated that this would amount to increasing the volume and hence the weight of the magnet to N^3 times its initial value. The increase in weight of a PM is attributed to the increase in magnetic field outside the magnet and hence to the increase in the external magnetic energy stored.

In an improvised version of a PM, called the periodic permanent magnet (PPM) [6]-[10], the length of the structure could be increased N times without requiring to increase its transverse dimensions as is necessary in a PM. Thus one gets the advantages of a PPM over its PM counterpart [6] in terms of weight by a factor of $N^3/N = N^2$. In a PPM instead of a single PM of its length increased by a factor of N, an array of N magnet cells of unchanged lengths are used. Furthermore, with the advent of light-weight magnetic materials like somarium-cobalt, ALNICO-5, etc. it has been possible for one to reduce the weight of a PPM to a considerable extent and render it suitable for air-borne and space applications. Two alternative schemes of magnetisation that use ring magnets can be conceived [11](Fig. 7.7.1). In one, the magnetisation is *axial* but its direction reverses from cell to cell which amounts to like poles of adjacent cells facing each other (Fig. 7.7.1(a)). In another, magnetisation is *radial*, becoming alternately inward and outward from cell to cell (Fig. 7.7.1(b)). It is easy to appreciate that, in each of these schemes, at a point far from a PPM, the magnetic flux density would be considerably small because of the cancellation of fields due to adjacent magnets. Consequently, the external magnetic energy stored in a PPM would be smaller than that in a PM. This also accounts for the reduction of weight of a PPM over its PM counterpart.

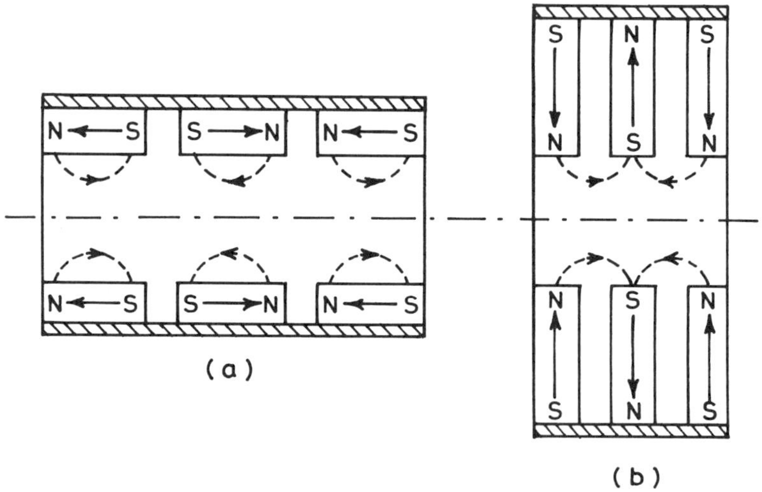

Fig. 7.7.1 Ring-magnet cells (shown typically, three, in number) in a PPM for the axial (a) and radial magnetisation (b) showing magnetic flux lines (dotted) inside the structure.

In a practical configuration magnetic fluxes are brought close to the axis of a PPM with the help of magnetic pole pieces arranged between magnet cells. A typical arrangement with reference to an axially-magnetised PPM is shown in Fig. 7.7.2. One may also conceive a similar arrangement for a radially-magnetised PPM (see problem 7.5).

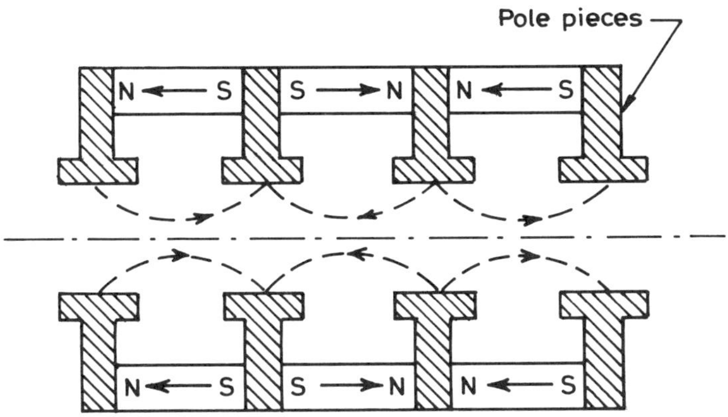

Fig. 7.7.2 A typical arrangement of magnetic pole pieces between magnets to bring magnetic flux density close to the axis of an axially magnetised PPM.

Let us now analyse the behaviour of an electron beam subject to the magnetic flux density of a PPM which is assumed to be uniform over any beam cross section but which varies from cross section to cross section, along z, periodically. Approximating this variation as 'sinusoidal' one may choose to express B in the following 'cosinusoidal' form [6]:

$$B = B_0 \cos\frac{2\pi z}{L} = B_0\left(\frac{1+\cos(4\pi z/L)}{2}\right)^{1/2}, \tag{7.7.1}$$

where B_0 is the peak magnetic flux density, L is the axial periodicity of variation in B, half of which, that is, L/2, it may be appreciated with reference to Fig. 7.7.2, is equal to the distance between two consecutive opposite (positive and negative) maxima in the axial variation of magnetic flux density. It is implied in the assumed axial variation of B given by (7.7.1) that, at z = 0, that is, at the entry of magnetic field of the structure, B has the peak value, B_0 .

Now that we have already described the electronic motion in the form of (7.4.7), we may use the latter for a PPM, remembering, however, that now B, instead of remaining constant with z, varies as (7.7.1). Thus one may express the beam-trajectory equation (7.4.7), with the help of (7.4.8(a)) and (7.4.8(b)), putting (7.7.1) in the latter:

$$\frac{d^2r}{dz^2}+\left(\frac{|\eta|^2 B_0^2\,(1+\cos(\frac{4\pi z}{L}))}{8(2|\eta|V_0)}\right)(r)-\left(\frac{|\eta|I_0}{2\pi\varepsilon_0(2|\eta|V_0)^{3/2}}\right)\left(\frac{1}{r}\right) = 0 . \qquad (7.7.2)$$

One may now express (7.7.2) in a convenient normalised form. For this purpose let us define [6]:

$$\sigma = \frac{r}{a}, \qquad (a)$$

and

$$T = \frac{2\pi z}{L}. \qquad (b) \qquad (7.7.3)$$

With the help of (7.7.3(a)) and (7.7.3(b)), it can be may easily shown that

$$\frac{d^2 r}{dz^2} = \left(\frac{4\pi^2 a}{L^2}\right)\left(\frac{d^2\sigma}{dT^2}\right). \qquad (7.7.4)$$

Substituting (7.7.3) and (7.7.4) into (7.7.2) one then obtains the following equation describing the beam trajectory in terms of the normalised beam edge radius σ and the normalised axial distance T:

$$\frac{d^2\sigma}{dT^2}+\alpha(1+\cos 2T)\sigma-\frac{\beta}{\sigma} = 0 , \qquad (7.7.5)$$

where the parameters α and β are given by:

$$\alpha = \frac{|\eta|L^2B_0^2}{64\pi^2 V_0} \qquad (a)$$

and

$$\beta = \frac{|\eta|L^2I_0}{8\pi^3a^2\varepsilon_0(2|\eta|V_0)^{3/2}} = \frac{|\eta|L^2(\text{Perv})}{8\pi^3a^2\varepsilon_0(2|\eta|)^{3/2}}, \qquad (b) \qquad (7.7.6)$$

where ‘Perv’ is the beam perveance as defined following (6.3.38).

<u>*Example 7.7.1*</u> Show that, for a PPM, $\alpha = \beta$ corresponds to (a) $B_{rms} = B_B$ and also to (b) $\omega_p = \omega_L$, where B_{rms} is the root mean square value of B which varies periodically with z as (7.7.1); ω_p is the beam-plasma frequency, and ω_L is the Larmor frequency corresponding to the peak magnetic flux density (= B_0).

(a) If $\alpha = \beta$, the right hand side of (7.7.6(a)) will be equal to that of (7.7.6(b)). This gives

$$\frac{|\eta| L^2 B^2}{64\pi^2 V_0} = \frac{|\eta| L^2 I_0}{8\pi^3 a^2 \varepsilon_0 (2|\eta| V_0)^{3/2}}$$

whence one gets

$$B_{rms}^2 = \frac{B_0^2}{2} = \frac{\sqrt{2}\, I_0}{\pi\varepsilon_0 |\eta|^{3/2} V_0^{1/2} a^2} \qquad (\alpha = \beta). \tag{7.7.7}$$

Comparing (7.7.7) with (7.5.15) one immediately sees that

$$B_{rms} = B_B \qquad (\alpha = \beta)\ . \tag{7.7.8}$$

(b) The angular beam-plasma frequency is given by (see section *8.1.1*):

$$\omega_p^2 = \frac{|\eta||\rho|}{\varepsilon_0}\ , \tag{7.7.9}$$

where ρ, here, is comprised of only the dc (time-independent) part. ω_p as given by (7.7.4) is a measure of the beam current I_0. In fact one may write (7.7.9) with the help of (7.1.3) and (7.1.4) as

$$\omega_p^2 = \frac{|\eta||I_e|}{\pi a^2 \varepsilon_0 v_z} = \frac{|\eta|\, I_0}{\pi a^2 \varepsilon_0 (2|\eta| V_0)^{1/2}}\ . \tag{7.7.10}$$

If I_0 from (7.7.7) is substituting in (7.7.10), one gets

$$\omega_p^2 = \frac{|\eta|^2 B_0^2}{4} \qquad (\alpha = \beta)\ . \tag{7.7.11}$$

Further, from (7.3.14) and (7.3.10) one may write ω_c corresponding to $B = B_0$ as

$$\omega_L^2 = \frac{\omega_c^2}{4} = \frac{|\eta|^2 B_0^2}{4}\ . \tag{7.7.12}$$

Comparing (7.7.11) with (7.7.12) one then obtains

$$\omega_p = \omega_L \qquad (\alpha = \beta)\ . \tag{7.7.13}$$

<u>*Example 7.7.2*</u> Obtain the expression for α and β in terms of ω_p, ω_L and Ω, the 'angular' frequency with which an electron of the beam moving down the PPM passes through a particular phase of magnetic flux density of the PPM.

Time an electron takes to pass between two successive phases of magnetic flux density as it moves along the axis is L/v_z, and hence the corresponding frequency will be the reciprocal of this time, that is, equal to v_z/L. Therefore, the corresponding angular frequency Ω is given by 2π times the frequency as follows:

$$\Omega = \frac{2\pi v_z}{L} . \tag{7.7.14}$$

which in view of (7.1.4) becomes

$$\Omega = \frac{2\pi (2|\eta| V_0)^{1/2}}{L} . \tag{7.7.15}$$

Now, from (7.7.12) and (7.7.15) one may obtain B_0^2 and $(2|\eta|V_0)^{1/2}$, respectively, and substitute them in (7.7.6(a)) to obtain

$$\alpha = \frac{1}{2} \frac{\omega_L^2}{\Omega^2} . \tag{7.7.16}$$

Similarly, substituting I_0 and $(2|\eta|V_0)^{1/2}$ obtained from (7.7.10) and (7.7.15), respectively, into (7.7.6(b)), one gets

$$\beta = \frac{1}{2} \frac{\omega_p^2}{\Omega^2} . \tag{7.7.17}$$

Incidentally, one may verify the findings in Example 7.7.1(b) by equating the right hand sides of (7.7.16) and (7.7.17), for $\alpha = \beta$, to obtain the relation $\omega_p = \omega_L$, a result that is in agreement with (7.7.13).

Now that we have examined in the preceding two examples the relation between the beam and magnetic field parameters corresponding to the condition $\alpha = \beta$, let us see next if this condition does have any relevance to maintaining the beam radius constant with the axial distance. For this purpose one has to solve the beam trajectory equation (7.7.5) for a PPM.

One may gain a sufficient insight into the PPM focusing if one ignores the last term of the left hand side of (7.7.5) by letting $\beta = 0$, which corresponds to a low beam current as can be seen from (7.7.6(a)). The trajectory equation would then reduce to a form of Maitheu's equation the solution of which is known to have alternative pass and stop bands [6], [12] on the scale of the parameter α which, as can be seen from (7.7.6(a)), is proportional to B_0^2. In the pass band the solution for the normalised beam radius σ, defined by (7.7.3(a)), is found to be stable and be periodic with the normalised axial distance T defined by (7.7.3(b)). The beam radius ripples in this region would depend upon the value of α relative to that of β. On the other hand, in the stop band, the solution is unstable corresponding to a cumulative divergence of the beam with the increasing axial distance. Thus the stop bands would correspond to no beam transmission through the structure, and the design must exclude these stop bands. If has been found that[5], [6], in the alternating pass and stop bands on the α-scale, the first band is essentially the pass band extending between $\alpha = 0$ and 0.66. The first stop band, which immediately follows the first pass band, extends between $\alpha = 0.66$ and 1.72. The next band to follow is the second pass band that extends between $\alpha = 1.72$ and 3.76. Following this one has the second stop band extending between $\alpha = 3.76$ and 6.1, and so on. Thus it may be seen that on the α-scale the stop bands are in general wider than the corresponding pass bands. For instance, the second stop band is wider than the second pass band. Also, the widths of both the pass and the stop bands progressively increase with the order of the band. For instance, the second pass band is wider than the first pass band, though the former calls for a larger value of magnetic flux density than the latter. It may be noted that the above findings for vanishingly small beam currents ($\beta = 0$) continue to be valid even for higher beam currents. The detailed computational study has revealed that the ripples in the radius of the beam are reduced to a minimum if the value of α is chosen around that of β. The relation between the beam and the magnetic structure parameters, $\alpha = \beta$, has already been studied in Examples 7.7.1 and 7.7.2. Thus for minimum ripples in the beam radius, theoretically one should choose B_{rms} ($= B_0/\sqrt{2}$) $= B_B$ corresponding to $\alpha = \beta$, as indicated by (7.7.8) which states that the root mean square value of magnetic flux density of the axially-alternating value of a PPM should be equal to the uniform Brillouin flux density. In other words one may choose the theoretical value of the peak magnetic flux density, B_0, from (7.7.7) as:

$$B_0 = \left(\frac{2\sqrt{2}\, I_0}{\pi\varepsilon_0 |\eta|^{3/2} V_0^{1/2} a^2} \right)^{1/2} = 1.17 \times 10^{-3} \frac{I_0^{1/2}}{a V_0^{1/4}} \text{ Tesla} . \qquad (7.7.18)$$

The period of the PPM, L, may be found by restricting the value of α to lie in the first pass band, that is, choosing $\alpha < 0.66$ (below the cutoff of the band). Since the condition $\alpha = \beta$ has to be maintained for minimum beam-radius ripples, this would also correspond to $\beta < 0.66$. It then follows from (7.7.6) that one should see that either of the following conditions are also satisfied in addition to (7.7.18):

$$\frac{B_0^2 L^2}{V_0} < \left(\frac{64\pi^2}{|\eta|} \right) (0.66), \text{i.e., } 2.4 \times 10^{-9} \text{ MKS} \qquad (7.7.19)$$

$$(\text{Perv}) \left(\frac{L}{a} \right)^2 < (16\sqrt{2}\, \pi^3 \varepsilon_0 |\eta|^{1/2})(0.66) \quad \text{i.e., } 1.7 \times 10^{-3} \text{ MKS} . \qquad (7.7.20)$$

Thus in a PPM, equation (7.7.18) and either (7.7.19) or (7.7.20) are obtained as the two basic design equations.

In this section the basic features of a PPM has been studied with reference to Brillouin focusing. It may be mentioned that the concept of a PPM can also be extended to the confined-flow focusing. In this way the light-weight advantage of a PPM could be combined with the ruggedness of the confined-flow focusing with respect to ion and rf defocusing. The first pass-band in such a case, however, becomes narrower as B_c (the flux density threading into the cathodes) increases [10].

7.8 Summary

The function of a focusing structure is to confine the electron beam of given parameters beyond the electron gun over a desired interaction length (where the electron beam is supposed to interact with electromagnetic waves in a practical electron beam device). In a microwave tube the focusing is usually done by a magnetic structure. A solenoid can provide a uniform magnetic field over a large interaction length. However, if the beam current is large, as it could be in a high-power microwave tube, the solenoid would be rather heavy. Morerover, the solenoid requires a separate power supply for current. Particularly for air-borne and space applications, therefore, the solenoid becomes very much inconvenient. A permanent magnet can also be used in place of a solenoid. But, though suitable for a small interaction length, a permanent magnet, too, becomes rather heavy for a large interaction length. This is because one has to increase the transverse dimensions of a permanent magnet in the same proportion as its length in order to retain in tact the required value of magnetic flux density. A periodic permanent magnet (PPM) is a light-weight substitute for a PM. Further, with the advent of light magnetic materials the weight of a PPM could be reduced.

In a scheme called Brillouin focusing, the cathode is magnetically shielded. From the solution of the angular acceleration equation of electronic motion one obtains Busch's theorem, an important corollary of which is that the beam rotates like a rigid bar for a shielded cathode. This also gives the concept of Larmor frequency and its relationship with the cyclotron frequency. The solution of the radial acceleration of electronic motion suggests the conditions which are required to be satisfied to make the beam radius constant with the axial distance. It turns out that a Brillouin focusing structure has to be strictly compatible with a convergent Pierce gun using a magnetically-shielded spherical-cup cathode. Thus the beam-waist radius as per electron gun design should be precisely equal to the value that is desired to be maintained constant by the focusing structure. Also, the magnetic field should abruptly start at the beam waist, where the electrons move parallel to the axis, having no transverse component of velocity. Finally, the flux density to be provided by the uniform magnetic-field structure should satisfy the Brillouin relation that involves the beam parameters namely the beam voltage, current and radius. It is rather difficult to satisfy exactly these ideal and stringent conditions of Brillouin focusing in practice.

In another scheme, called the confined-flow, a magnetic flux density larger than is required in Brillouin focusing is applied beyond the beam waist. In such scheme a small fraction of magnetic flux threads into the cathode, the flux lines following the beam trajectories in the cathode-axial region of the gun. In the confined-flow focusing the beam diameter is found to be less sensitive to the beam current density variation due to rf bunching, positive-ion formation etc. than in a corresponding Brillouin focusing.

The PPM can be used for both the Brillouin and the confined-flow focusing for the reduction of the size and weight of the structure. The PPM allows the increase in the length of the structure without having to increase its transverse dimensions. An array of small magnets with reverse polarities brought close together are used instead of a single magnet. In a cylindrical structure, the use of magnetic pole pieces between individual magnet cells makes it possible to bring the magnetic flux close to the axis of the structure. Both axial and radial magnetisation are possible in a PPM. The weight of a PPM is reduced by a factor of the square of the number of magnet cells over its uniform-field, single-magnet counterpart. A study that has been carried out with particular reference to Brillouin focusing suggests that the solution of the equation of electronic motion subjected to the axially-periodic (assumed to be sinusoidal) magnetic flux density of a PPM has alternate pass and stop bands over a scale that is proportional to the peak magnetic flux density. In the design, the axial period of a PPM should be so chosen as to avoid a stop band where the solution is unstable corresponding to a cumulative beam divergence. It also turns out that the beam-radius ripples in the pass band are reduced to a minimum, if the rms value of magnetic flux density of a PPM are chosen to be equal to the Brillouin value in terms of the beam parameters. The PPM which has the advantage of reducing the weight of the structure becomes of considerable relevance in the confined-flow focusing since the latter calls for more flux density than the Brillouin focusing.

Problems

7.1 Calculate the Brillouin magnetic flux required to confine a 10 kV, 1 A, 3 mm electron beam.

Answer : 0.036 Tesla .

7.2 In a confined-flow focusing arrangement compatible with a convergent Pierce gun the magnetic flux density at the beam waist is 66 % of the Brillouin value. If the area convergence of the gun is 25, what percentage of the Brillouin flux density does thread into the cathode ?

Answer : 2.7 % .

7.3 Ignoring the space-charge force, that is, on putting $I_0 = 0$ in the basic acceleration equation (7.4.5) for an edge electron, develop the concept of Larmor frequency. Hence also appreciate by making reference to an appropriate radial coordinate system that the angular frequency of rotation of an electron about its centre is $\omega_C = 2\omega_L$ rather than ω_L .

7.4 Show that the wavelength of the beam scallop, λ_s, is given by

$$\lambda_s = \frac{\pi\,(4|\eta|v_0)^{1/2}}{\omega_L} .$$

Hint: Study (7.5.10) and find λ_s in terms of m which is given by (7.5.5), in terms of C_2, which, in turn, is given by (7.4.8(b)).

7.5 Suggest an arrangement of magnetic pole pieces between magnets to bring the magnetic flux density close to the axis of a 'radially-magnetised' PPM and indicate magnetic flux lines therein, in a manner similar to what has been shown in the case of an 'axially-magnetised' PPM in Fig. 7.7.2.

7.6 Choosing the design of a PPM at 30 % below the cutoff of the first stable band on the α-scale, find its length period (L) for a (V_0 =) 1.5 kV beam, taking the peak magnetic field density (B_0) as 800 gauss.

Answer: L = 1.3 mm.

7.7 Appreciate a) that at the edge of the stop band, $\alpha = 0.66$, the rotating electrons of the beam become synchronous with the periodic magnetic impulse of the PPM and b) that the latter also becomes synchronous with the electrons oscillating at the natural electron plasma frequency, for $\beta \approx \alpha$, corresponding to the minimum beam scalloping, each of these cases, (a) and (b) corresponding to a diverging beam [5].

Hint : a) $\Omega \approx \omega_L$ b) $\Omega \approx \omega_p$.

References

[1] L. Brillouin, "A theorem of Larmor and its importance for electrons in magnetic fields," *Phys. Rev.* **67** (1945) 260-266.

[2] J. F. Gittins, *Power Traveling-Wave Tubes* (English University Press, London, 1964).

[3] V. N. Shevchik, G. N. Shvedov, and A. V. Soboleva, *Wave and Oscillatory Phenomenon in Electron Beams at Microwave Frequencies* (Pergamon Press, Oxford, 1966).

[4] M. Chodorow and C. Susskind, *Fundamentals of Microwave Electronics* (Mc Graw-Hill, New York, 1964).

[5] A. S. Gilmour, Jr., *Principles of Traveling Wave Tubes* (Artech House, Boston, 1994).

[6] J. T. Mendel, CF. Quate, and W. H. Yocum, "Electron beam focusing with periodic permanent magnetic fields," *Proc. IRE* **42** (1954) 800-810.

[7] J. T. Mendel, "Magnetic focusing of electron beams", *Proc IRE* **43** (1955) 327-331.

[8] J. E. Sterrett and H. Heffner, "The design of periodic magnetic focusing structures," *IRE Trans. Electron Dev.* **ED-5** (1958) 35-42.

[9] K. J. Harker, "Periodic focusing of beams from partially shielded cathodes," *IRE Trans. Electron Dev.* **ED-2** (1955) 13-19.

[10] K. Ura and M. Terader, "Equivalence of periodic magnetic field to uniform magnetic field in electron beam focusing," *IEEE Trans. Electron Dev.* **ED-13** (1966) 930-934.

[11] *Hughes TWT and TWTA Handbook*, Electron Dynamics Division, Hughes Aircraft Company, Torrance, California.

[12] N. W. Mac. Lachlan, *Theory and Applications of Mathieu Function* (Oxford University Press, New York, 1947).

CHAPTER 8

GROWING-WAVE ELECTRON BEAM DEVICES

8.0 Introduction

Now that some concepts have been developed on the formation of a linear electron beam (chapter *6*) and its confinement (chapter *7*) as well as on the behaviour of an interaction structure (chapter *5*), it remains then to be seen what happens if an electron beam so formed and confined is allowed to interact with an electromagnetic wave supported by such a structure. Such an interaction, as we will see (section *8.2*), provides the mechanism of amplification in a practical electron beam device called the travelling-wave tube (TWT) which finds extensive applications in satellite communication and electronic warfare systems. The principles of double-stream and beam-plasma amplifiers — the devices which do away with a slow-wave structure (SWS) (chapter *5*) but use another medium instead (for instance, an electron beam or a plasma) — have also been touched upon (section *8.3*). All these electron beam devices under study belong to a category of microwave tubes called the linear or the O-type ("O" has been derived from the last letter of TPO which is an acronym for "tubes à propagation des ondes") in which the kinetic energy of the electron beam is converted into electromagnetic energy. Finally, the study has been extended to another device under this category, namely, the gyro-TWT — so named as it is based on the principle of gyro-resonance between an electron beam gyrating in a helical path and an electromagnetic wave travelling down a conventional hollow-pipe metal waveguide — a device that has emerged out as a potential high power amplifier in the millimeter-and sub-millimeter-wave range (section *8.4*).

8.1 Space-Charge Waves on an Electron Beam

Let us consider a long linear beam of electrons constrained to move with a uniform velocity along the positive z direction; for the sake of simplicity let the beam have infinite dimensions perpendicular to the z-direction. Further, we assume that the negative charges of the electrons of the beam are neutralised by the positive charges of the ions which are always formed in a practical (electron beam) vacuum-tube device. Let now such a beam pass through a narrow gap at $z = 0$ across which a high-frequency voltage is applied to modulate the beam velocity. The average beam velocity beyond the gap, however, is assumed to remain constant in the space-charge neutralised beam. This will be consistent with the assumption of the neutralising presence of the positive ions but for which the negative space charge of the electrons would depress the potential in the region and cause the electrons to slow down. Obviously in this problem the motion of the ions, which are much heavier than the electrons, is ignored, and it is assumed that their presence causes little effects on the electronic motions caused by the modulator [1], [2].

Subject to the beam velocity modulation at z = 0 as mentioned above, and owing to the inherent mechanism in a space-charge neutralised electron beam to restore its neutrality if it is disturbed, the charge-density waves are set up on the beam propagating along z. In fact, viewed in a coordinate system moving with the average velocity of electrons of the beam, these waves, called the Hahn and Ramo Space-charge waves [3], as discussed following (8.1.22), turn out to be simply oscillations at a frequency called the plasma frequency — the natural frequency with which a cloud of electrons in equilibrium will oscillate, if perturbed [2], [4], [5] (see appendix *a8.1.1*).

In order to predict the behaviour of the space-charge waves on an electron beam let us first recall the longitudinal current density relation (6.1.7) which, on putting $J = J_0+J_1$, $\rho = \rho_0+\rho_1$ and $v = v_0 + v_1$, takes the following form:

$$J_0 + J_1 = (\rho_0 + \rho_1)(v_0 + v_1) = \rho_0 v_0 + \rho_0 v_1 + \rho_1(v_0 + v_1) \,, \tag{8.1.1}$$

where the subscript '0' refers to the unperturbed (dc) part of a quantity that remains constant with time and space coordinates, and the subscript '1' refers to the perturbation (rf) part which varies with z and time t but remains constant over the cross section of the beam considered to be of infinite extent. In view of the relation $J_0 = \rho_0 v_0$, and under the small-signal approximation $v_1 << v_0$, one may get from (8.1.1)

$$J_1 = \rho_0 v_1 + v_0 \rho_1 \,. \tag{8.1.2}$$

Taking the partial derivative of J_1 with respect to z, one may write

$$\frac{\partial J_1}{\partial z} = \rho_0 \frac{\partial v_1}{\partial z} + v_0 \frac{\partial \rho_1}{\partial z} \,. \tag{8.1.3}$$

It is also of relevance to write a relation between J_1 and ρ_1 in the form of the continuity equation (4.1.8) which, with the help of (2.3.5), and noting that here J_1 varies only with z and remains independent of the transverse coordinates, may be written as:

$$\frac{\partial J_1}{\partial z} = -\frac{\partial \rho_1}{\partial t} \,. \tag{8.1.4}$$

Substituting (8.1.4) into (8.1.3) and re-arranging terms one obtains

$$(\frac{\partial}{\partial t} + v_0 \frac{\partial}{\partial z})\rho_1 = -\rho_0 \frac{\partial v_1}{\partial z}$$

which may be expressed as

$$\mathcal{D}\,\rho_1 = -\rho_0\,\mathcal{D}_z\,v_1\ , \tag{8.1.5}$$

where
$$\mathcal{D}_z = \partial/\partial z \quad \text{and} \quad \mathcal{D} = \partial/\partial t + v_0\,\partial/\partial z \tag{8.1.6}$$

are the operators involving partial derivatives.

Next it is of interest to write the following force equation of an electron subjected to the longiludinal space-charge electric field E_z that comes into play when the equilibrium charge density of the beam is disturbed due to the longitudinal disturbance of electrons (see appendix *a8.1.1*):

$$m\frac{dv_1}{dt} = m\left(\frac{\partial v_1}{\partial t} + \frac{dz}{dt}\frac{\partial v_1}{\partial z}\right) = eE_z\ ,$$

which on putting $dz/dt = v = v_0 + v_1$ takes the following form

$$\frac{dv_1}{dt} = \frac{\partial v_1}{\partial t} + (v_0 + v_1)\frac{\partial v_1}{\partial z} = \eta E_z\ , \tag{8.1.7}$$

where $\eta(= e/m)$ is the charge-to-mass ratio of an electron, m being its mass and e its charge carrying a negative sign. Under the small-signal approximation, $v_1 << v_0$, the equation of acceleration of an electron, (8.1.7), becomes

$$\left(\frac{\partial}{\partial t} + v_0\frac{\partial}{\partial z}\right)v_1 = \eta E_z\ ,$$

which may be written using (8.1.6) as

$$\mathcal{D}\,v_1 = \eta\,E_z\ . \tag{8.1.8}$$

Now, operating $\mathcal{D}$ on (8.1.5) one obtains

$$\mathcal{D}^2\rho_1 = -\rho_0\,\mathcal{D}\,\mathcal{D}_z v_1 = -\rho_0\,\mathcal{D}_z\,\mathcal{D}\,v_1\ . \tag{8.1.9}$$

Plugging (8.1.8) into the right hand side of (8.1.9) one then gets

$$\mathcal{D}^2\rho_1 = -\rho_0\,\mathcal{D}_z(\eta\,E_s) = -\eta\rho_0\,\mathcal{D}_z E_z\ . \tag{8.1.10}$$

The longitudinal rf space-charge electric field E_Z developed may be related to the peturbation in the volume charge density, ρ_1, by one of the Maxwell's equations, namely, (4.4.1(a)) which may be spelt out in the present context with the help of (2.3.5) by taking $D_z = \varepsilon_0 E_z$ and remembering that the latter varies only with z, remaining independent of the transverse coordinates. Thus one obtains

$$\frac{\partial E_z}{\partial z} = \frac{\rho_1}{\varepsilon_0} ,$$

which using (8.1.6) becomes

$$\mathscr{D}_z E_z = \rho_1 / \varepsilon_0 . \tag{8.1.11}$$

Substituting (8.1.11) into the right-hand side of (8.1.10) one then gets

$$\mathscr{D}^2 \rho_1 = -\left(\frac{\eta\rho_0}{\varepsilon_0}\right)\rho_1 . \tag{8.1.12}$$

Expecting and looking for wave-like solutions for the rf quantities relevant to the beam that has been disturbed by the modulator one may take ρ_1 to vary as $\exp(j\omega t - \Gamma z)$, in which Γ in general is complex, a positive imaginary part in the latter corresponding to a wave propagating in the positive z direction, while the positive and negative real parts corresponding to exponentially attenuating and growing waves. In view of this rf dependence the operator $\mathscr{D}$ given by (8.1.6) becomes

$$\mathscr{D} = j\omega - v_0\Gamma . \tag{8.1.13}$$

Putting (8.1.13) in (8.1.12), one obtains

$$(j\omega - v_0\Gamma)^2 = -\left(\frac{\eta\rho_0}{\varepsilon_0}\right)^2 . \tag{8.1.14}$$

The terms of (8.1.14) have the dimension of angular frequency squared. So, choosing to put

$$\omega_p^2 = \frac{\eta\rho_0}{\varepsilon_0} \tag{8.1.15}$$

in the right hand side of (8.1.14), one may write

$$(j\omega - v_0\Gamma)^2 = -\omega_p^2 . \tag{8.1.16}$$

It may be noted that η and ρ_0 each carry a negative sign, and so one may alternatively express (8.1.15) as

$$\omega_p^2 = \frac{|\eta||\rho_0|}{\varepsilon_0} . \tag{8.1.17}$$

The physical significance of the angular frequency ω_p can be appreciated by putting in (8.1.16) $v_0 = 0$, corresponding to electrons having no dc component of velocity, to get [6]

$$\omega^2 = \omega_p^2 \ . \tag{8.1.18}$$

In fact, it can be shown that, if a stationary cloud of electrons in the neutralising background of positive ions, as they are in a state of matter called the plasma, is disturbed, then it will oscillate with the frequency given by (8.1.18), where ω_p , which is given by (8.1.17), is known as the angular plasma frequency (see appendix *a8.1.1*). ω, as can be seen from (8.1.18), is independent of β and as such the group velocity v_g ($= \partial\omega/\partial\beta$) of disturbance (see appendix *a5.5.1*) would be zero as is expected of such a localised, non-propagating disturbance. Now, turning back to (8.1.16) we find that its square root gives

$$j\omega - v_0\Gamma = \pm j\omega_p$$

whence we get

$$\Gamma = \frac{j(\omega \mp \omega_p)}{v_0} \ .$$

If we choose to write

$$\Gamma = j\beta \ , \tag{8.1.19}$$

then we see that the rf dependence $\exp(j\omega t - \Gamma z)$ discussed following (8.1.12) becomes $\exp j(\omega t - \beta z)$, where β by comparing the two expressions for Γ — (8.1.19) and the one immediately preceding it — is found as:

$$\beta = \frac{\omega \mp \omega_p}{v_0} = \beta_e \mp \beta_p \ , \tag{8.1.20}$$

where β_e ($= \omega/v_0$) and β_p ($= \omega_p/v_0$) are called the beam and the plasma propagation constants, respectively. The dispersion relation (8.1.20) (between ω and β) has been obtained here with reference to a beam of infinite cross-section. In a practical configuration the beam may be of a finite cross-section and enclosed, say, in a metal structure. It can be shown that, for such a bounded beam, (8.1.20) continues to hold, however, with ω_p (or β_p) reduced by a factor called the plasma frequency reduction factor (see appendix *a8.1.2*). Also from this general case one may derive as a special case the dispersion relation for a beam of infinite cross section which, as is expected, turns out to be the same as (8.1.20) (see appendix *a8.1.2*).

Now it emerges out from (8.1.20) that there are two space-charge waves corresponding to the plus and minus signs, respectively. The phase velocity v_p ($= \omega/\beta$) of these waves may be written from (8.1.20) as

$$v_p = \frac{\omega}{\beta} = \frac{\omega}{\omega \mp \omega_p} v_0 = \frac{\omega}{\beta_e \mp \beta_p} . \tag{8.1.21}$$

The group velocity v_g (see appendix *a5.5.1*) of both the space-charge waves, as can be seen from (8.1.20), is given by v_g $(= \partial\omega/\partial\beta) = v_0$. It may be mentioned that, for conditions typical in microwave tubes, $\omega > \omega_p$ in (8.1.21). The phase velocities of space-charge waves, as can be seen from (8.1.21), are then different from the dc beam velocity v_0. The upper sign in (8.1.21) corresponds to $v_p > v_0$ and refers to the so-called fast space-charge wave, and similarly the lower sign in it corresponds to $v_p < v_0$ and refers to the so-called slow space-charge wave on the beam. The dispersion relation between ω and β has been plotted in Fig. 8.1.1 with the help of (8.1.20) as two straight lines (shown as solid lines) each of slope v_0.

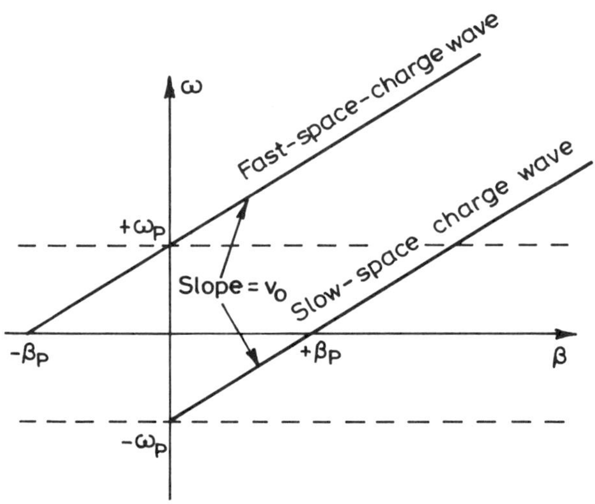

Fig. 8.1.1 ω-β dispersion plot of space-charge waves on an electron beam (solid lines) and their representation in a reference frame moving with the dc beam velocity (broken lines).

It may be noted that, in a frame of reference which moves with the dc beam velocity v_0, an observer 'sees' a Doppler-shifted frequency ω' given by [7]

$$\omega' = \frac{v_p - v_0}{v_p}\omega = \left(1 - \frac{v_0}{v_p}\right)\omega$$

into which one may substitute (8.1.21) to get

$$\omega' = \left(1 - \frac{\omega \mp \omega_p}{\omega}\right)\omega = \pm\omega_p \;, \tag{8.1.22}$$

the upper and the lower signs corresponding to the fast and the slow space-charge waves, respectively. It is of interest to note that (8.1.22) is identical with the natural frequency of oscillation of a stationary cloud of electrons in the background of neutralising positive ions — the electron plasma frequency, which is given by (8.1.18) through (8.1.17) (appendix *a8.1.1*). In other words, viewed in a coordinate system moving with the electron beam, the space-charge waves on a one-dimensional electron beam are oscillations at the electron plasma frequency which are independent of the wavelength (= $2\pi/\beta$) of disturbance[7] (Fig. 8.1.1).

Since the two space-charge waves on the beam have different phase velocities given by (8.1.21), they will slip by each other along the direction of propagation, down the electron steam, and hence should at alternate axial-distance intervals reinforce and cancel each other. This will thus set up standing waves characterised by a series of alternate nodes and antinodes in the relevant rf beam quantities. The type of the narrow-gap velocity modulator at the input end of the beam, z = 0, that has been assumed for the excitation of these space-charge waves demands that the first antinode of the beam velocity standing wave should coincide with the first node of the beam current density standing wave, and both should form at z = 0 (see (8.1.38) and (8.1.39)). One may model such a modulator as a pair of closely-spaced parallel grids across the beam between which an alternating voltage is applied. The transit time of the electron in the stream between the grids is considered to be small such that the number of electrons leaving the grids per unit area per second equals the number per unit area per second arriving on the other side. This makes zero the alternating component of beam convection current density ($J_1\{0\}= 0$) at the position of the grids (z = 0), though the application of the altenating voltage across the grids makes both the beam velocity and the volume charge density vary at z = 0, thus enabling one to write with the help of (8.1.2)

$$J_1\{0\} = \rho_0 v_1\{0\} + v_0 \rho_1\{0\} = 0 \;, \tag{8.1.23}$$

where the argument '0' refers to the quantities at z = 0.

Considering the rf dependence $\exp j(\omega t - \beta z)$ as discussed following (8.1.19) one may express v_1 as

$$v_1 = v_{1f} + v_{1s} = \hat{v}_{1f}\{0\}\exp j(\omega t - \beta_f z) + \hat{v}_{1s}\{0\}\exp j(\omega t - \beta_s z) \;, \tag{8.1.24}$$

where $\hat{v}_{1f}\{0\}$ and $\hat{v}_{1s}\{0\}$ represent the amplitudes of the fast (v_{1f}) and the slow (v_{1s}) space-charge wave components, v_1 at z = 0, respectively, and similarly one may express J_1 as

$$J_1 = J_{1f} + J_{1s} = \hat{J}_{1f}\{0\}\exp j(\omega t - \beta_f z) + \hat{J}_{1s}\{0\}\exp j(\omega t - \beta_s z) \;, \tag{8.1.25}$$

in terms of the amplitudes of the corresponding current density components. Here, β_f and β_s are the fast and the slow space-charge wave propagation constants given by (8.1.20) (see also the discussion following (8.1.21)) which may be re-written as follows:

$$\beta_f = \beta_e - \beta_p \quad \text{(a)}$$

and

$$\beta_s = \beta_e + \beta_p \, . \quad \text{(b)} \qquad (8.1.26)$$

The condition (8.1.23) may then be written with the help of (8.1.25) as

$$(\hat{J}_{1f}\{0\} + \hat{J}_{1s}\{0\}) \exp j\omega t = 0$$

whence one has

$$\hat{J}_{1f}\{0\} = -\hat{J}_{1s}\{0\} \, . \qquad (8.1.27)$$

The amplitude of the velocity modulation at z = 0 may also similarly be written with the help of (8.1.24) as

$$\hat{v}_1\{0\} \exp j\omega t = (\hat{v}_{1f}\{0\} + \hat{v}_{1s}\{0\}) \exp j\omega t$$

whence one gets

$$\hat{v}_1\{0\} = \hat{v}_{1f}\{0\} + \hat{v}_{1s}\{0\} \, . \qquad (8.1.28)$$

The rf velocity v_1, however, may be related to the beam convection current density J_1 and hence, as we will see, the condition (8.1.28) can be expressed in a new form in view of the condition (8.1.27). For this purpose, considering the rf dependence as $\exp j(\omega t - \beta z)$ as before, one may first using (8.1.4) express ρ_1 as

$$\rho_1 = \frac{\beta}{\omega} J_1 \qquad (8.1.29)$$

and then substitute it in (8.1.2) to obtain by a re-arrangement of terms:

$$v_1 = \frac{(\omega - v_0 \beta) J_1}{\omega \rho_0} \, . \qquad (8.1.30)$$

In terms of the fast and the slow space-charge wave components one may then write from (8.1.30):

$$v_{1f} = \frac{(\omega - v_0 \beta_f) J_{1f}}{\omega \rho_0} \quad \text{(a)}$$

and

$$v_{1s} = \frac{(\omega - v_0 \beta_s) J_{1s}}{\omega \rho_0} \, . \quad \text{(b)} \qquad (8.1.31)$$

Substituting (8.1.26) in (8.1.31) and remembering the definitions of the beam propagation constant $\beta_e = \omega/v_0$ and the plasma propagation constant $\beta_p = \omega_p/v_0$ (as given following (8.1.20)), one may write

$$v_{1f} = \frac{\omega_p}{\omega\rho_0} J_{1f} \qquad \text{(a)}$$

and

$$v_{1s} = \frac{-\omega_p}{\omega\rho_0} J_{1s} \ . \qquad \text{(b)} \qquad (8.1.32)$$

Making use of (8.1.27) and (8.1.32) one may then choose to express the amplitudes of the rf velocity components as follows:

$$\hat{v}_{1f}\{0\} = \frac{\omega_p}{\omega\rho_0} \hat{J}_{1f}\{0\} \qquad \text{(a)}$$

and

$$\hat{v}_{1s}\{0\} = \frac{-\omega_p}{\omega\rho_0} \hat{J}_{1s}\{0\} = \frac{\omega_p}{\omega\rho_0} \hat{J}_{1f}\{0\} \ . \qquad \text{(b)} \qquad (8.1.33)$$

Then from (8.1.28) and (8.1.33) one may obtain the following relation at the location of the beam velocity modulator ($z = 0$):

$$\hat{v}_{1f}\{0\} = \hat{v}_{1s}\{0\} = \frac{\hat{v}_1\{0\}}{2} \ . \qquad (8.1.34)$$

With the help of (8.1.33) and (8.1.34) one then obtains the corresponding relation for the amplitudes of the rf beam convection current density components:

$$\hat{J}_{1f}\{0\} = -\hat{J}_{1s}\{0\} = \frac{\omega\rho_0 \hat{v}_1\{0\}}{2\omega_p} \ . \qquad (8.1.35)$$

Now that we have obtained the fast and the slow space-charge wave components of the amplitudes of the rf beam velocity and current density as (8.1.34) and (8.1.35), respectively, we may write from (8.1.24) and (8.1.25) the following expression for the beam velocity and current density at a point down the electron stream (away from the input end) in terms of $\hat{v}_1\{0\}$, the amplitude of the input beam velocity modulation at $z = 0$, as follows:

$$v_1 = \frac{\hat{v}_1\{0\}}{2} \exp j(\omega t - \beta_f z) + \frac{\hat{v}_1\{0\}}{2} \exp j(\omega t - \beta_s z) \qquad (8.1.36)$$

and

$$J_1 = \frac{\omega\rho_0 \hat{v}_1\{0\}}{2\omega_p} \exp j(\omega t - \beta_f z) - \frac{\omega\rho_0 \hat{v}_1\{0\}}{2\omega_p} \exp j(\omega t - \beta_s z) \ . \qquad (8.1.37)$$

Substituting (8.1.20) in (8.1.36) and (8.1.37), and after a little trigonometry, one then obtains

$$v_1 = \hat{v}_1\{0\} \cos\beta_p z \ \exp j(\omega t - \beta_e z) \tag{8.1.38}$$

and

$$J_1 = \frac{j\omega\rho_0 \hat{v}_1\{0\}}{\omega_p} \sin\beta_p z \exp j(\omega t - \beta_e z) \ . \tag{8.1.39}$$

The rf charge density ρ_1 may similarly be expressed, with the help of (8.1.4) and remembering the time dependence $\exp j\omega t$, as:

$$\rho_1 = -\frac{1}{j\omega}\frac{\partial J_1}{\partial z}$$

into which one may substitute $\partial J_1/\partial z$ obtainable from (8.1.39) to get

$$\rho_1 = \frac{\rho_0 \hat{v}_1\{0\}}{\omega_p}(j\beta_e \sin\beta_p z - \beta_p \cos\beta_p z) \exp j(\omega t - \beta_e z)$$

which making use of the relations $\beta_e = \omega/v_0$ and $\beta_p = \omega_p/v_0$ becomes

$$\rho_1 = \frac{\rho_0 \hat{v}_1\{0\}}{v_0}(j\frac{\omega}{\omega_p} \sin\beta_p z - \cos\beta_p z) \exp j(\omega t - \beta_e z) \ . \tag{8.1.40}$$

If one chooses to express the rf beam velocity, current density and volume charge density standing waves by the respective real parts of their phasor representation, then one obtains from (8.1.38)-(8.1.40)

$$v_1 = \hat{v}_1\{0\} \cos \beta_p z \cos(\omega t - \beta_e z) \qquad \text{(a)}$$

$$J_1 = \frac{-\omega\rho_0 \hat{v}_1\{0\}}{\omega_p} \sin\beta_p z \sin(\omega t - \beta_e z) \qquad \text{(b)}$$

$$\rho_1 = \frac{\rho_0 \hat{v}_1\{0\}}{v_0}[-\frac{\omega}{\omega_p} \sin\beta_p z \sin(\omega t - \beta_e z) - \cos\beta_p z \cos(\omega t - \beta_e z)] \qquad \text{(c)} \tag{8.1.41}$$

<u>*Example 8.1.1*</u> Show that the standing beam current density and velocity waves are spatially apart by a quarter of the plasma wavelength and the distance between consecutive nodes (or antinodes) of either of the waves is equal to half the plasma wavelength.

It can be easily appreciated by examining (8.1.41) that the nodes of the beam velocity standing wave and the antinodes of the beam current density standing wave are given by $\cos \beta_p z = 0$ and

$\sin \beta_p z = \pm 1$, respectively, each giving $\beta_p z = (2n+1)\pi/2$, where n is an integer, suggesting that the nodes of the former coincide with the antinodes of the latter. Similarly, one may see again from (8.1.41) that the antinodes of the beam velocity standing wave corresponding to $\cos \beta_p z = \pm 1$ coincide with the nodes of the beam current density standing wave corresponding $\sin \beta_p z = 0$, each giving $\beta_p z = (2n)\pi/2$. Clearly, then it can be seen that β_p $(=2\pi/\lambda_p)$ times the distance between a node (or antinode) of one wave from the corresponding node (or antinode) of the other is equal to $(2n+1)\pi/2 - (2n)\pi/2 = \pi/2$ whence this distance is $\pi/(2\beta_p) = \lambda_p/4$, where λ_p is called the plasma wavelength. Thus it is by this distance, namely, $\lambda_p/4$ that the beam velocity and current density waves are spatially apart with respect to each other. Similarly, β_p $(= 2\pi/\lambda_p)$ times the distance between consecutive nodes (or antinodes) can be seen to be equal to $2\,(n+1)\pi/2 - (2n)\pi/2 = \pi$ and hence this distance becomes $\pi/\beta_p = \lambda_p/2$.

Example 8.1.2 Appreciate that the volume charge density variation of the beam dominants over the beam velocity variation in contributing to the beam current density variation in the space-charge standing waves, provided one takes $\omega >> \omega_p$, in most of the practical situations.

Taking $\omega >> \omega_p$, one gets from (8.1.41(c))

$$\rho_1 v_0 \approx \rho_1 \hat{v}_1\{0\}\left(\frac{-\omega}{\omega_p} \sin \beta_p z \sin(\omega t - \beta_e z)\right) . \tag{8.1.42}$$

Comparing (8.1.42) with (8.1.41(b)) we see that

$$J_1 \approx \rho_1 v_0 . \tag{8.1.43}$$

Then in view of (8.1.43) one may easily see that, in this case ($\omega >> \omega_p$), the rf current density, which in general comprises two terms given by (8.1.2) representing the beam velocity and volume charge density fluctuations, respectively, is predominantly contributed by the latter.

8.1.1 Kinetic Power of Space-Charge Waves

For a one-dimensional electron beam one may define [1], [8]-[12] a beam quantity having the dimension of potential called the kinetic voltage V_K

$$V_K = \frac{-v_0 v_1}{|\eta|} , \tag{8.1.44}$$

and also correspondingly another quantity, namely the kinetic power density $\mathcal{P}_K$ for the flow of electrons along z:

$$\mathcal{P}_k = \mathscr{P}_k \mathbf{a}_z = \frac{1}{2} Re V_K J_1^* \mathbf{a}_z . \tag{8.1.45}$$

The electromagnetic power P_{em} delivered by a portion of the linear electron beam between two cross-sectional plane (1 and 2) perpendicular to the flow of electrons (Fig. 8.1.2) may be found as the following surface integral over a volume enclosing the portion of the beam (see section *4.8*):

$$P_{em} = \frac{1}{2} Re \oint_S (\mathbf{E} \times \mathbf{H}^*) . \mathbf{a}_n \, dS \, . \tag{8.1.46}$$

Using Chu's power conservation theorem[8]-[12] (see appendix *a8.1.3* for its derivation) one may further express (8.1.46) as:

$$P_{em} = - \oint_S \mathcal{P}_K . \mathbf{a}_n \, dS \, . \tag{8.1.47}$$

It follow from (8.1.47) that the flux of the beam kinetic power density $\mathcal{P}_K$ through the surface of the volume enclosing the beam must be negative for P_{em} to be positive, that is, for a net electromagnetic power to flow out of the surface.

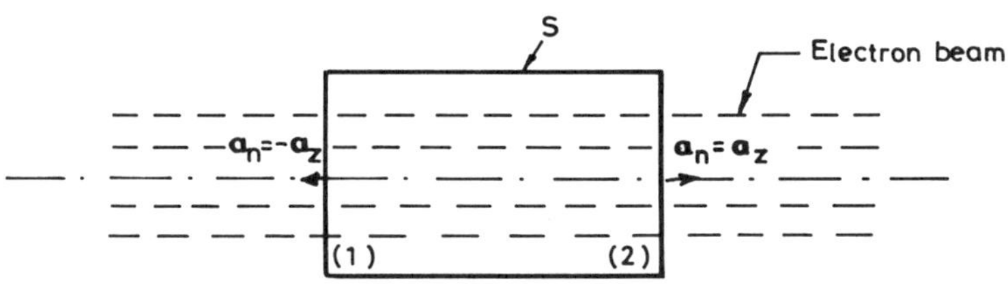

Fig. 8.1.2 Closed surface (S) of integration enclosing a portion of a linear beam between two reference cross sections 1 and 2.

Since $\mathcal{P}_K$, as can be seen from (8.1.45), is directed along z and its magnitude is zero outside the beam, the surface area over which the integration in (8.1.47) extends resolves down to the areas intercepted on the surface enclosure (S) by the beam at the two cross-sectional planes (1 and 2) considered. Thus one may write (8.1.47) as:

$$P_{em} = - \left(\oint_\alpha \mathcal{P}_k . \mathbf{a}_n \, d\alpha \Big|_{(1)} + \oint_\alpha \mathcal{P}_k . \mathbf{a}_n \, d\alpha \Big|_{(2)} \right) \, , \tag{8.1.48}$$

where α is the cross-sectional area of the beam, perpendicular to the flow of electrons, that remains constant, along the length of the beam. Clearly, the unit vectors normal to the surface at the cross-sections 1 and 2 are $\mathbf{a}_n = - \mathbf{a}_z$ and $\mathbf{a}_n = \mathbf{a}_z$, respectively (Fig. 8.1.2). Thus (8.1.48) becomes

$$P_{em} = -[-(\mathcal{P}_K \cdot \mathbf{a}_z)|_{(1)}\alpha + (\mathcal{P}_K \cdot \mathbf{a}_z)|_{(2)}\alpha]. \tag{8.1.49}$$

Putting (8.1.45) in (8.1.48) one then obtains

$$P_{em} = -\alpha\,(-\mathcal{P}_{K(1)} + \mathcal{P}_{K(2)}) = \alpha\,(\mathcal{P}_{K(1)} - \mathcal{P}_{K(2)}) . \tag{8.1.50}$$

Let us now see how the average kinetic energy of the beam changes over an rf cycle corresponding to a transfer of power from the beam to electromagnetic waves. In order to appreciate this let the excitation be such that the outward flux through the cross section (2) down the stream contributes more significantly to the electromagnetic power than the inward flux at the cross section (1) (Fig. 8.1.2). Then one may see from (8.1.50):

$$P_{em} \cong -\alpha\,\mathcal{P}_{K(2)} .$$

which, using (8.1.45) and (8.1.44), becomes

$$P_{em} \cong \frac{v_0}{2|\eta|} Re\, v_1 J_1^* \alpha . \tag{8.1.51}$$

It follows from (8.1.51) that, for P_{em} to be positive, corresponding to a kinetic power delivered from the beam to electromagnetic waves, v_1 and J_1 should be in phase. This means that, when v_1 is positive, J_1 is also positive, and similarly when v_1 is negative, J_1 is also negative. Therefore, during the positive rf half cycle, when v $(= v_0+v_1) > v_0$, then $|J|$ $(= |J_0 + J_1| = |-|J_0| + J_1| = |J_0| - J_1) < |J_0|$, similarly, during the negative rf half cycle, when v $(= v_0 + v_1) < v_0$, then $|J|$ $(= |J_0| - J_1) > |J_0|$ (corresponding to v_1 and J_1 each having negative values). In other words, over an rf cycle, the rate of flow of electron through a beam cross-section down the stream will be more (corresponding to $|J| > |J_0|$) for those electrons which have a kinetic energy less than their unperturbed value in the absence of modulation ($v < v_0$) than that (corresponding to $|J| < |J_0|$) for those which have a kinetic energy more than their unperturbed value ($v > v_0$). Thus, taking an average over all the electrons passing the cross section, the beam kinetic energy becomes smaller than it would be in the absence of modulation — a result that has essentially been found consequent upon an assumed electromagnetic power delivered from the beam (corresponding to a positive value of P_{em}).

At this stage, it is worth examining the contributions to the magnitude of the beam kinetic power density $\mathcal{P}_K$ from the fast and the slow space-charge waves on the beam. For this purpose, let us express $\mathcal{P}_K$ substituting (8.1.44) in (8.1.45) as follows:

$$\mathcal{P}_K = Re \frac{-v_0 v_1 J_1^*}{2|\eta|} . \tag{8.1.52}$$

Further, substituting (8.1.24) and (8.1.25) into (8.1.52) one may express $\mathcal{P}_K$ in terms of the slow and fast space-charge wave components as:

$$\mathcal{P}_K = Re\,\frac{-v_0(v_{1f}+v_{1s})(J_{1f}^*+J_{1s}^*)}{2|\eta|}\;. \tag{8.1.53}$$

Substituting v_{1f} and v_{1s} from (8.1.32) into (8.1.53) one gets

$$\mathcal{P}_K = Re\,\frac{-v_0\omega_p}{2|\eta|\omega\rho_0}(J_{1f}J_{1f}^*+J_{1f}J_{1s}^*-J_{1s}J_{1f}^*-J_{1s}J_{1s}^*)$$

which may be expressed in terms of the amplitudes of the rf beam current densities with the help of equation (8.1.25) as

$$\mathcal{P}_K = Re\,\frac{-v_0\omega_p}{2|\eta|\omega\rho_0}(\hat{J}_{1f}^2\{0\}-\hat{J}_{1s}^2\{0\}+2j\sin\beta_p z)$$

$$= \frac{-v_0\omega_p}{2|\eta|\omega\rho_0}(\hat{J}_{1f}^2\{0\}-\hat{J}_{1s}^2\{0\})\;. \tag{8.1.54}$$

Making use of the relations $\rho_0 = -|\rho_0|$, $v_0^2 = 2|\eta|V_0$, where V_0 is the dc beam voltage corresponding to the dc beam velocity v_0 (see equation (6.1.6)), $J_0 = -|J_0|$, $\omega_p^2 = |\eta|\,|\rho_0|/\varepsilon_0$ (see equation (8.1.17)) and $|J_0| = |\rho_0|\,v_0$ (from the $J_0 = \rho_0 v_0$ as mentioned following (8.1.1)), one may easily express (8.1.54) as

$$\mathcal{P}_K = \frac{\omega_p V_0}{\omega\,|J_0|}(\hat{J}_{1f}^2\{0\}-\hat{J}_{1s}^2\{0\}). \tag{8.1.55}$$

Thus one may see from (8.1.55) that the amplitude of the fast and the slow space-charge components of the rf current density contributes, respectively, to the positive and the negative parts of the value of the magnitude of the kinetic power density and that their relative values determine the sign of $\mathcal{P}_K$. In fact, if the amplitudes of these components are equal, $\mathcal{P}_K = 0$ and, consequently, one obtains from (8.1.47) $P_{em} = 0$. In practical electron beam devices, one may thus predominantly excite the slow space-charge wave component which carries a negative kinetic power density (as can be seen from (8.1.55)) and which, in turn, could be responsible, (as can be appreciated from (8.1.47)), for the delivery of electromagnetic power from the beam (corresponding to a positive value of P_{em}). This also, as a corollary discussed following (8.1.51), suggests that the excitation of such a slow space-charge wave component can be effected only by extracting the kinetic energy from the beam. A more quantitative description of the delivery of power from the beam can be made referring back to the problem of electromagnetic power flow through the surface of the volume enclosing a portion of the electron between two cross sectional planes at 1 and 2 (Fig. 8.1.2). Thus substituting (8.1.55) in (8.1.50) with reference to this problems one obtains

$$P_{em} = \frac{\omega_p V_0}{\omega |J_0|}[(\hat{J}_{1s}^2\{0\}_{(2)} - \hat{J}_{1s}^2\{0\}_{(1)}) - (\hat{J}_{1f}^2\{0\}_{(2)} - \hat{J}_{1f}^2\{0\}_{(1)})] . \qquad (8.1.56)$$

It is clear from (8.1.56) that in a growing-wave linear beam device (like a travelling-wave tube, see section *8.2*) in which the amplitudes of the space-charge wave components grow ($\hat{J}_{1s,f\,(2)} > \hat{J}_{1s,f\,(1)}$), one must couple the slow space-charge wave component of the beam to a circuit wave in order to realise a positive value of P_{em}, that is, for the delivery of power from the beam to the circuit.

<u>*Example 8.1.3*</u> Show that the electron beam should be faster than the phase of the circuit wave for a power transfer from the beam to the circuit.

In order to couple the slow space-charge wave on the beam to the circuit waves, one has to make these waves synchronous. This requires that the phase velicity $v_{p\,(circuit)}$ of the circuit be equal to the phase velocity of the slow space charge wave given by (8.1.21) taking the lower sign in its right hand side. In other words, one has to have

$$v_{p(circuit)} = \frac{\omega}{\omega + \omega_p} v_0$$

which may also be put as

$$v_0 = \frac{\omega + \omega_p}{\omega} v_{p(circuit)} ,$$

which shows that $v_0 > v_{p\,(circuit)}$, the difference between the two velocities increasing with ω_p, that is, with the beam charge density $|\rho_0|$ (see (8.1.17)).

Appendix a8.1.1 Plasma oscillations of a cloud of electrons

Let us consider an ensemble of electrons and positive ions, as they usually are in a plasma, specified by an overall change neutrality. If the electrons in the ensemble are displaced to a small extent from their equilibrium positions relative to the heavier stationary ions, a space-charge electric field will come into play between the region of positive concentration of charges wherefrom the electrons are depleted and the region of negative concentration of charges where the electrons are bunched. This space-charge electric field acts in the direction of the displacement of electrons and hence provides a restoring force in the opposite direction on them. This compels the displaced electrons to return to their original equilibrium positions and restore the neutrality. But the electrons because of their stored kinetic energy overshoot though the restoring force again comes into play and makes them return to their equilibrium positions, and thus the electrons oscillate about their mean positions. The natural angular frequency of oscillation of electrons is called the plasma frequency or Langmuir frequency [13]. Here, we give two methods of finding the space charge field that is developed due to electronic motions.

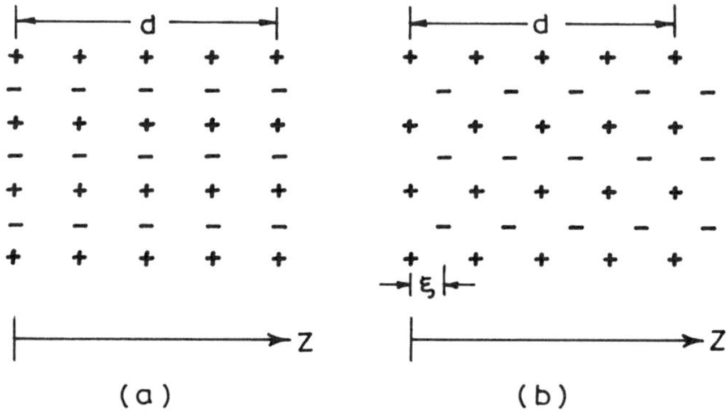

Fig. a8.1.1 Alternate layers of electrons and positive ions in a charge-neutralised slab in equilibrium **(a)** and the same when the equilibrium is disturbed by a small displacement of electron layers relative to the positive-ions layers **(b)**.

In the first of these methods [4] let us consider a thin slab of infinite area transverse to its thickness comprising alternate layers of electrons and positive ions of equal concentrations (Fig. a8.1.1). The net charge of electrons on either faces of the slab is zero when the electrons are in equilibrium, that is, when they are not displaced relatively to the positive ions (Fig. a8.1.1(a)). However, when they are displaced, say, by a small distance ξ along z, a positive charge layer appears on one face of the slab while a negative charge layer does on the other face (Fig. a8.1.1(b)). Since ξ is small, the positive surface charge appearing on an arbitrary area A on an end face of the slab may be taken as owing to the depletion of electrons within a volume of a right cylinder of length ξ and base area A lying on the end face. This charge will obviously be equal to $|\rho_0|\xi A$, where $|\rho_0|$ is the magnitude of the equilibrium volume charge density of electrons. The corresponding surface charge density $|\rho_s|$ on the end face of this slab is then obtained by dividing this charge by the area A to obtain

$$|\rho_s| = \frac{|\rho_0|\xi A}{A} = |\rho_0|\xi \ . \qquad \text{(a8.1.1)}$$

*Similarly, one can find the negative surface charge density on the other face of the slab caused by the accumulation of electrons as equal to $-|\rho_0|\xi A/A = -|\rho_0|\xi = -|\rho_s|$. For a small thickness of the slab one may find [14] the space-charge field intensity within it by adding the electric field intensities due to these positive and negative planar charge distributions using the concepts of electrostatics developed in chapter 2. (See equation (2.1.18); see also problem 2.6). The space-charge electric field intensity **E** thus becomes*

$$\mathbf{E} = \left(\frac{|\rho_s|}{2\varepsilon_0}\right)(\mathbf{a}_z) + \left(\frac{-|\rho_s|}{2\varepsilon_0}\right)(-\mathbf{a}_z) \ . \qquad \text{(a8.1.2)}$$

One notes here while finding the contributions to **E** *by the individual sheets of charge that the unit vector normal to the sheet of charge on one face of the slab reverses its sign on the other face of the slab of an opposite surface charge density. Substituting (a8.1.1) in (a8.1.2) one then gets*

$$\mathbf{E} = \frac{|\rho_s|}{\varepsilon_0}\mathbf{a}_z = \frac{|\rho_0|\xi}{\varepsilon_0}\mathbf{a}_z \; . \tag{a8.1.3}$$

It is clear from (a8.1.3) that, corresponding to a positive displacement ξ *, an electric field intensity is developed within the slab in the positive z direction, the magnitude,* E_z *, of which may be written from (a8.1.3) as*

$$E_z = \frac{|\rho_0|\xi}{\varepsilon_0} \; . \tag{a8.1.4}$$

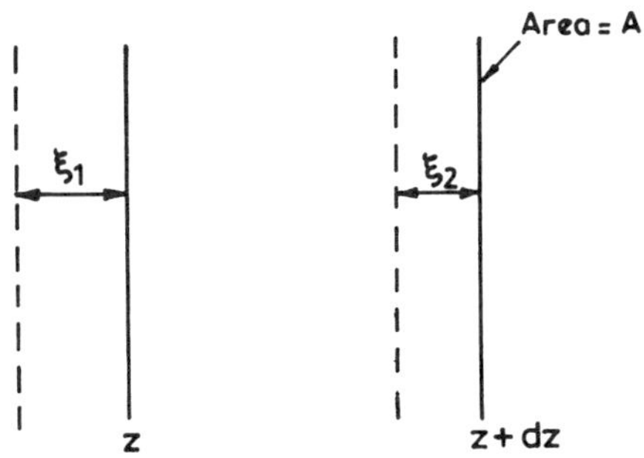

Fig. a8.1.2 An infinitesimal slab of overall-neutral ensemble of electrons and positive ions in which to find the charge imbalance due to displacements of electrons.

In an alternative method [5] E_z *is found by considering a thin slab of infinitesimal thickness lying between* z *and* $z+dz$ *within the ensemble of electrons and positive ions. In equilibrium the net charge within the slab (of thickness dz) is zero. However, when the equilibrium is disturbed due to electronic motion there would be an imbalance of charges flowing in and out of the volume of the slab (Fig. a8.1.2). Letting* ρ_1 *be the varying (rf) part of volume charge density developed within the slab at any instant of time, one may express the corresponding charge* q_1 *developed within it as*

$$q_1 = \rho_1 \, A \, dz \; , \tag{a8.1.5}$$

where A is an arbitrary area of the slab. Further, at this instant, let the electronic charges ($=\rho_0 \, \xi_1 \, A$) *originally outside the slab, lying within the length* ξ_1 *from the edge of the slab at z and extending over the area A, enter the slab, and let at the same time the electronic charges* ($=\rho_0 \, \xi_2 \, A$) *originally inside the slab, lying within the length* ξ_2 *from the edge of the slab at* $z+dz$ *and extending over the same area A, leave the slab (Fig. a8.1.2). Here,* ρ_0 ($= -|\rho_0|$) *is the unperturbed volume charge density of electrons.* ξ_1 *is the distance which the electrons outside the slab, and situated farthest from its edge at z, move to reach the edge.* ξ_2 *is the distance which the electrons inside the slab, and situated farthest from the other edge at* $z + dz$*, move to reach the edge. Then one may put* (a8.1.6) $$\xi_2 = \xi_1 + \frac{\partial \xi}{\partial z} dz \ .$$

From the above considerations, the unbalanced charge q_1 *developed within the slab may be found by subtracting the charge leaving the slab from the charge entering it as follows:*

(a8.1.7) $$q_1 = \rho_0 \xi_1 A - \rho_0 \xi_2 A = -\rho_0 A(\xi_2 - \xi_1) \ .$$

Comparing (a8.1.5) and (a8.1.7) one may then write

(a8.1.8) $$\rho_1 dz = -\rho_0 (\xi_2 - \xi_1) \ .$$

Substituting (a8.1.6) in (a8.1.8) one then obtains the following expression for the rf volume charge density ρ_1 *:*

(a8.1.9) $$\rho_1 = -\rho_0 \frac{\partial \xi}{\partial z} \ .$$

The space-charge electric field intensity **E** *may be related to the rf volume charge density* ρ_1 *with the help of one of Maxwell's equations (4.4.1(b)) (expressed in terms of electric displacement* **D** $= \varepsilon_0$ **E** *) as:*

(a8.1.10) $$\nabla . \mathbf{E} = \frac{\rho_1}{\varepsilon_0} \ .$$

The slab of a cloud of electrons, in the neutralising background of positive ions considered here, is a part of an ensemble which has been assumed to have infinite transverse dimensions. Therefore, one may take here E_z *to be independent of the transverse coordinates and vary only with z in the longitudinal direction in which the electrons move when the equilibrium is disturbed, and, consequently, one may spell out the divergence of* **E** *in (a8.1.10), with the help of (2.3.5), as:*

(a8.1.11) $$\nabla . \mathbf{E} = \frac{\partial E_z}{\partial z} \ .$$

In view of (a8.1.11) one may then write (a8.1.10) as

(a8.1.12) $$\frac{\partial E_z}{\partial z} = \frac{\rho_1}{\varepsilon_0} \ .$$

Substituting (a8.1.9) in (a8.1.12) one then obtains

$$\frac{\partial E_z}{\partial z} = -\frac{\rho_0}{\varepsilon_0} \frac{\partial \xi}{\partial z}$$

which on integration gives

(a8.1.13) $$E_z = -\frac{\rho_0}{\varepsilon_0} \xi = \frac{|\rho_0|}{\varepsilon_0} \xi \ ,$$

the constant of integration being zero since no space-charge field is generated, $E_z = 0$ *, when the electrons are in equilibrium,* $\xi = 0$ *. Thus we have by an alternative approach arrived at the same expression for the space charge field as (a8.1.4).*

In order to study the electronic motion, subject to the space-charge field thus obtained as either (a8.1.4) or (a8.1.13), let us now write the force equation along z of an electron as follows:

$$m \frac{d^2 \xi}{dt^2} = e E_z$$

whence

$$\frac{d^2 \xi}{dt^2} = \eta E_z = \frac{\eta |\rho_0|}{\varepsilon_0} \xi = -\frac{|\eta||\rho_0|}{\varepsilon_0} \xi , \tag{a8.1.14}$$

η *(= e/m = - | e | /m = -| η |) being the charge-to-mass ratio of an electron. Plugging (8.1.17) in (a8.1.14) one then obtains the following equation of acceleration of an electron:*

$$\frac{d^2 \xi}{dt^2} = -\omega_p^2 \xi . \tag{a8.1.15}$$

The equation of motion given by (a8.1.15) may be identified as the simple-harmonic motion of angular frequency ω_p *. Taking the time dependence as* $\exp j\omega t$*, one then gets from (a8.1.5)*

$$\omega^2 = \omega_p^2 , \tag{a8.1.16}$$

which has been used in the text as (8.1.18). The general solution of (a8.1.15) comprising two parts corresponding to two solutions of (a8.1.16), namely, $\omega = \pm \omega_p$ *is :*

$$\xi = A \exp(j\omega_p t) + B \exp(-j\omega_p t) , \tag{a8.1.17}$$

where A and B are constants. An alternative form of (a8.1.17) is:

$$\xi = C \cos\omega_p t + D \sin\omega_p t , \tag{a8.1.18}$$

where $C = A + B$ *and* $D = j(A - B)$ *are constants.*

In most practical situations the ensemble of charged particles are of a finite transverse cross section and it may be enclosed in an envelope which could be metallic. As a result the space-charge field E would not be purely radial and, therefore, the axial electric field intensity, E_z *, and hence the restoring force on electrons would reduce as compared to their values in the case of an infinite transverse cross section of the ensemble. Obviously, this is manifested in the reduction in the value of* ω_p *occurring in the acceleration equation (a8.1.15). In view of this, further, it becomes necessary to replace, in (a8.1.16)-(a8.1.18),* ω_p *by*

$$\omega_q = F \omega_p , \tag{a8.1.19}$$

the reduced angular plasma frequency, F (< 1) being called the plasma-frequency reduction factor (see appendix a8.1.2) [14], [15].

Appendix a8.1.2 Plasma-frequency reduction factor

The dispersion relation (8.1.20) has been deduced in the text for an electron beam of infinite cross section. Now let us take up a more general case in which a cylindrical electron beam of circular cross section is enclosed in a coaxial conducting cylindrical envelope. Let the radius of the beam be 'a' and that of the metal enclosure be 'b', there being in general a clearance between the beam and the metal enclosure.

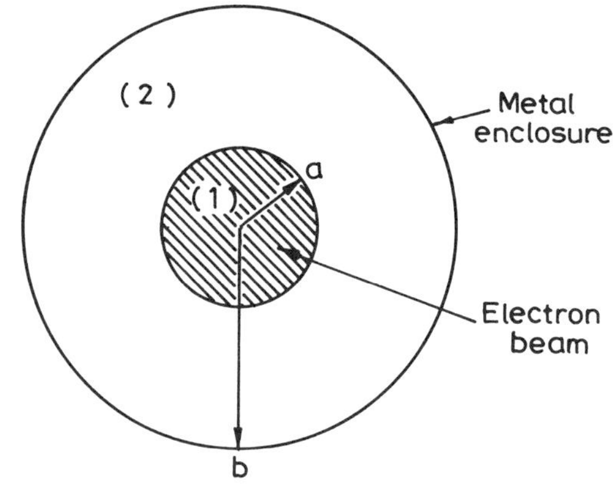

Fig. a8.1.3 Cross section of a cylindrical electron beam in a coaxial conducting envelope showing two regions (1) , inside the beam, and (2), between the beam and the envelope.

In order the analyse the problem let us recall the inhomogeneous wave equation (4.5.1) which may be read through (4.5.2) in the present context by interpreting $\boldsymbol{J}_c$ (that involves in (4.5.1)) as the beam convection current density. Thus we get

$$\nabla^2 \boldsymbol{E} - \mu_0 \varepsilon_0 \frac{\partial^2 \boldsymbol{E}}{\partial t^2} = \mu_0 \frac{\partial \boldsymbol{J}}{\partial t} + \nabla\left(\frac{\rho}{\varepsilon_0}\right) \tag{a8.1.20}$$

in which the subscript 'c' in the symbol for the convection current density has been dropped. One may then write from (a8.1.20) the following wave equation for the axial component of electric field intensity E_z, inside the beam, which provides the restoring force on the electrons when the latter are axially disturbed from equilibrium (see appendix a8.1.1):

$$\nabla^2 E_z - \mu_0 \varepsilon_0 \frac{\partial^2 E_z}{\partial t^2} = \mu_0 \frac{\partial J_1}{\partial t} + \frac{1}{\varepsilon_0} \nabla \rho_1 \Big|_z , \tag{a8.1.21}$$

where the longitudinal beam convection current density J and the volume charge density ρ of the beam have been taken as $J = J_0 + J_1$ and $\rho = \rho_0 + \rho_1$, respectively, as defined preceding (8.1.1), giving $\partial J/\partial t = \partial J_1 /\partial t$ and $\nabla\rho = \nabla\rho_1$. Spelling out the Laplacian ∇^2 and $\nabla\rho_1 |_z$, the z-component $\nabla\rho_1$, with the help of (2.6.4(b)) and (2.5.12(b)), respectively, in the cylindrical system of coordinates, while treating the problem essentially as azimuthally symmetric ($\partial /\partial\theta = 0$) , one may write from (a8.1.21):

(a8.1.22)
$$\frac{\partial^2 E_z}{\partial r^2} + \frac{1}{r}\frac{\partial E_z}{\partial r} + \frac{\partial^2 E_z}{\partial z^2} - \mu_0 \varepsilon_0 \frac{\partial^2 E_z}{\partial t^2} = \mu_0 \frac{\partial J_1}{\partial z} + \frac{1}{\varepsilon_0}\frac{\partial \rho_1}{\partial z}.$$

Taking the rf dependence as exp $j(\omega t - \beta z)$ (see, for instance, the discussions following (8.1.19)), one obtains from (a8.1.22)

(a8.1.23)
$$\frac{\partial^2 E_z}{\partial r^2} + \frac{1}{r}\frac{\partial E_z}{\partial r} - (\beta^2 - k^2) E_z = j\omega \mu_0 J_1 - \frac{j\beta \rho_1}{\varepsilon_0},$$

where $k = \omega (\mu_0 \varepsilon_0)^{1/2} = \omega / c$ is the free-space propagation constant. Substituting ρ_1 from (8.1.29) in (a8.1.23), one then obtains

(a8.1.24)
$$\frac{\partial^2 E_z}{\partial r^2} + \frac{1}{r^2} - (\beta^2 - k^2) E_z = \frac{-j(\beta^2 - k^2)}{\omega \varepsilon_0} J_1 .$$

Now, with the help of (8.1.30) one may express J_1 in terms of v_1, which in turn can be expressed in terms of E_z with the help of (8.1.8) through (8.1.6), as follows:

(a8.1.25)
$$J_1 = -\frac{j\eta \rho_0 E_z}{(\omega - v_0 \beta)^2} .$$

Substituting (a8.1.25) in (a8.1.24) and re-arranging terms one then gets

(a8.1.26)
$$\frac{\partial^2 E_z}{\partial r^2} + \frac{1}{r}\frac{\partial E_z}{\partial r} - (\beta^2 - k^2)\left[1 - \frac{\omega_p^2}{(\omega - v_0 \beta)^2}\right] E_z = 0 ,$$

in writing which the relation $\omega_p^2 = \eta \rho_0 / \varepsilon_0 = |\eta| |\rho_0| / \varepsilon_0$ has been used (see equation (8.1.15) and (8.1.17)). If E_z does not vary across the beam cross section perpendicular to the axis, as if the beam were of infinite cross section, then one may put $\partial^2 E_z / \partial r^2 = \partial E_z / \partial r = 0$ in (a8.1.26) to get

$$(\beta^2 - k^2)\left[1 - \frac{\omega_p^2}{(\omega - v_0 \beta)^2}\right] E_z = 0$$

whence one obtains, obviously since $\beta^2 \neq k^2$ (that is, $\beta^2 - k^2 \neq 0$),

$$1 - \frac{\omega_p^2}{(\omega - v_0 \beta)^2} = 0$$

giving on simplification

(a8.1.27)
$$\beta = \frac{\omega \mp \omega_p}{v_0} = \beta_e \mp \beta_p$$

which, as is expected, becomes identical with (8.1.20), the expression for the axial propagation constants of space-charge waves earlier derived in the text considering right from the beginning the electron beam to be of infinite cross section. However, when the effects of the finite transverse dimensions of the electron beam are considered, (a8.1.27) is expected to be modified in view of the reduction in the value of the plasma frequency, as discussed following (8.1.20) and preceding (a8.1.19). Let us therefore in this case choose to put β *as (a8.1.27) with* ω_p *(or* β_p *) replaced by* $\omega_q = F\omega_p$ *(or* $\beta_q = F\beta_p$ *) giving*

$$\beta = \frac{\omega \mp F\omega_p}{v_o} = \beta_e \mp F\beta_p \quad , \tag{a8.1.28}$$

whence one gets

$$(\omega - v_0 \beta)^2 = F^2 \omega_p^2 \quad . \tag{a8.1.29}$$

Substituting (a8.1.29) in (a8.1.26) one may then express the wave equation in the presence of the beam as:

$$\frac{\partial^2 E_z}{\partial r^2} + \frac{1}{r}\frac{\partial E_z}{\partial r} + G^2 E_z = 0 \tag{a8.1.30}$$

where

$$G^2 = (\beta^2 - k^2)\left(\frac{\omega_p^2}{(\omega - v_0 \beta)^2} - 1\right) = \gamma^2 \left(\frac{1}{F^2} - 1\right) \quad \text{and} \tag{a8.1.31}$$

$$\gamma^2 = \beta^2 - k^2 \quad . \tag{a8.1.32}$$

Since for a particular situation, $\beta^2 >> k^2$, $F < 1$ *one may write the following approximate relations, respectively, from (a8.1.31) and (a8.1.32) :*

$$G^2 \cong \beta^2 \left(\frac{1}{F^2} - 1\right) \tag{a8.1.33}$$

and

$$\gamma^2 \cong \beta^2 \quad . \tag{a8.1.34}$$

Similarly, one may write from (a8.1.28), for $\omega >> \omega_p$ *and* $F < 1$,

$$\beta^2 \cong \beta_e^2 \quad . \tag{a8.1.35}$$

Substituting (a8.1.35) in (a8.1.33) and (a8.1.34) one then may write

$$G^2 \cong \beta_e^2 \left(\frac{1}{F^2} - 1\right) \tag{a8.1.36}$$

and

$$\gamma^2 \cong \beta_e^2 \quad . \tag{a8.1.37}$$

It may be appreciated from (a8.1.36) that G^2 *is positive. In view of this, one identifies (a8.1.18) as the zeroth-order ordinary Bessel equation which has following solution (see (5.2.8)) :*

$$E_{z1} = A_1 \, J_0\{Gr\} + B_1 \, Y_0\{Gr\} \quad , \tag{a8.1.38}$$

where A and B are the field constants. The subscript 1 refers to the region 1, inside the beam (Fig. a8.1.3). J_0 *and* Y_0 *are the zeroth-order Bessel functions of the first and second kinds, respectively. At*

$r = 0$, which refers to the axis of the beam, $Y_0\{Gr\} = Y_0\{0\} \to -\infty$. Therefore, it follows from (a8.1.38) that the electric field intensity E_{z1} at the beam axis would blow up to have infinite values if one had not put $B_1 = 0$. Thus the finiteness of the electric field at the beam axis demands that one must put $B_1 = 0$ in (a8.1.38) to write

(a8.1.39) $$E_{z1} = A_1 \; J_0\{Gr\} \; .$$

Outside the electron beam ($\omega_p = 0$), that is, in region 2 outside the beam (Fig. a8.1.3), the wave equation (a8.1.26) takes the following form

(a8.1.40) $$\frac{\partial^2 E_z}{\partial r^2} + \frac{1}{r}\frac{\partial E_z}{\partial r} - \gamma^2 E_z = 0 \; .$$

Since γ^2 is positive (see equation (a8.1.37)), (a8.1.28) may be identified with the zeroth-order modified Bessel equation which has the following solution (see (5.2.9)):

(a8.1.41) $$E_{z2} = A_2 \; I_0\{\gamma r\} + B_2 \; K_0\{\gamma r\} \; ,$$

where the subscript 2 refers to the region 2. I_0 and K_0 are the zeroth order modified Bessel function of the first and second kinds, respectively.

Further, with the help of Maxwell's equations one may express $H_{\theta 1}$ and $H_{\theta 2}$, the azimuthal magnetic field intensity in the region 1 and 2, as follows (see problem 4.5 in which one has to take $\partial/\partial\theta = 0$ for the present azimuthally symmetric case):

(a8.1.42) $$H_{\theta_{1,2}} = \frac{j\omega\varepsilon_0}{\beta^2 - \omega^2 \mu_0 \varepsilon_0} \frac{\partial E_{z_{1,2}}}{\partial r} \; .$$

Now one may relate, as we will see, the field constants A_1, A_2 and B_2 with the help of following electromagnetic boundary conditions (see section 4.7):

(a8.1.43) $$E_{z1}\big|_{r=a} = E_{z2}\big|_{r=a} \; ,$$

the tangential electric field intensities being continuous at the interface between the regions 1 (inside the beam) and 2 (outside the beam), that is, at the beam boundary ($r = a$);

$$H_{\theta 1}\big|_{r=a} = H_{\theta 2}\big|_{r=a} \; ,$$

the surface current density at the beam boundary ($r = a$) being 0, which in view of (a8.1.42) gives

(a8.1.44) $$\left.\frac{\partial E_{z1}}{\partial r}\right|_{r=a} = \left.\frac{\partial E_{z2}}{\partial r}\right|_{r=a} \; ; \text{ and}$$

(a8.1.45) $$E_{z2}\big|_{r=b} = 0 \; ,$$

the tangential electric field intensity at the conducting boundary $r = b$ being zero.

Substituting (a8.1.39) and (a8.1.41) into the three boundary conditions (a8.1.43)-(a8.1.45), one then respectively obtains three simultaneous equations in three field constants A_1, A_2 and B_2. The first of these equations is:

(a8.1.46) $$A_1 \; J_0\{Ga\} \; - \; A_2 \, I_0\{\gamma a\} \; - B_2 \; K_0\{\gamma a\} \; = 0 \, ,$$

The second equation is :

$$A_1 G J_0'\{Ga\} - A_2 \gamma I_0'\{\gamma a\} - B_2 \gamma K_0'\{\gamma a\} \; = \; 0 \, ,$$

which, since $J_0'\{x\} = -J_1\{x\}, \;\; I_0'\{x\} = I_1\{x\}$ *and* $K_0'\{x\} = -K_1\{x\}$,

becomes

(a8.1.47) $$-A_1 \, G \, J_1\{Ga\} - A_2 \; \gamma I_1\{\gamma a\} \; + B_2 \; \gamma K_1\{\gamma a\} \; = 0 \; ,$$

J_1 , I_1 *and* K_1 *being the relevant first order functions. The third equation is :*

(a8.1.48) $$A_2 \;\; I_0\{\gamma b\} \; + B_2 \; K_0\{\gamma b\} \; = 0 \; .$$

For a non-trivial solution one obtains from (a8.1.46) - (a8.1.48) the following condition:

(a8.1.49) $$\begin{vmatrix} J_0\{Ga\} & -I_0\{\gamma a\} & -K_0\{\gamma a\} \\ -GJ_1\{Ga\} & -\gamma I_1\{\gamma a\} & \gamma K_1\{\gamma a\} \\ 0 & I_0\{\gamma b\} & K_0\{\gamma b\} \end{vmatrix} = 0 \; .$$

One may then easily simplify (a8.1.49) to obtain the following dispersion relation :

(a8.1.50) $$\frac{J_0\{Ga\}}{Ga \;\, J_1\{Ga\}} = \frac{K_0\{\gamma a\} I_0\{\gamma b\} \; - \; I_0\{\gamma a\} K_0\{\gamma b\}}{K_1\{\gamma a\} I_0\{\gamma b\} \; + \; I_1\{\gamma a\} K_0\{\gamma b\}} \; .$$

One may solve the transcedental dispersion relation (a8.1.50) by numerical methods. It may be noted that near $\beta \approx \beta_e$ *, one may take the right hand side of (a8.1.50) to be more or less constant for a given value* $\gamma a \approx \beta_e a$ *(see equation (a8.1.34) and (a8.1.35)). However, near* $\beta \approx \beta_e$ $((\omega - v_0 \beta) \; \approx \; 0)$*, the left hand side of (a8.1.50) is prone to large variations as can be seen from the dependence of G on* $(\omega - v_0 \beta)$ *from (a8.1.31). Thus, for a given value of* $\beta_e a$*, one may take, for solution, the following equation by putting* $\gamma a \approx \beta_e a$ *and* $\gamma b \approx (\beta_e a)(b/a)$ *in the right hand side of (a8.1.50):*

(a8.1.51) $$\frac{J_0\{Ga\}}{Ga \;\, J_1\{Ga\}} = \frac{K_0\{\beta_e a\} I_0\{(\beta_e a)(b/a)\} \; - \; I_0\{\beta_e a\} K_0\{(\beta_e a)(b/a)\}}{K_1\{\beta_e a\} I_0\{(\beta_e a)(b/a)\} \; + \; I_1\{\beta_e a\} K_0\{(\beta_e a)(b/a)\}} \; .$$

For given values $\beta_e a$ *and b/a, the value of the right hand side of (a8.1.51) becomes fixed, say, at a value* $R\{\beta_e a, b/a\}$*. Then what one may do is to plot the left hand side of (a8.1.51) as the ordinate against Ga as the abcissa to get a plot that intersects the constant R-line parallel to the abcissa at infinite number of points. The values of Ga corresponding to all such points of intersection give a series of solutions of different orders for Ga. The plasma-frequency reduction factor may then be obtained for a given order of solution using(a8.1.36) as*

(a8.1.52) $$F \;\; = \;\; \left(1 + \frac{(Ga)^2}{(\beta_e a)^2} \right)^{-1/2} .$$

Substituting (a8.1.52) in (a8.1.28) the axial propagation constants are found as

(a8.1.53)
$$\beta = \frac{\omega \mp \left(1+\dfrac{(Ga)^2}{(\beta_e a)^2}\right)^{-1/2} \omega_p}{v_0} .$$

From the numerical results one finds [15] that the value of F, which is, in general, less than unity, increases with b/a, the ratio of the envelope-to-beam radii, and tends to unity at large values of $\beta_e a$. Also the value of F becomes progressively less as the order of the solution is increased. In other words, for the lowest-order space-charge waves, the reduced plasma frequency has the highest value. The axial electric field intensity of the lowest-order wave does not have a zero over the beam cross section. The next higher-order wave has a single zero over the beam cross section, and the next to this order has two zeros, and so on [15]. Further, since for a higher-order of solution, the value of F becomes less, one concludes that the value of the effective plasma frequency ω_q (= $F\omega_p$) and that of the effective plasma propagation constant β_q (= $F\beta_p$) will become each less while that of the effective plasma wavelength λ_q (= $2\pi/\beta_q$) will become more, as the order of solution is increased. Consequently, then it follows from (8.1.32), in which one has to replace ω_p by ω_q, that, with the increase of the order of the solution, the ratio of the beam-current-to-velocity modulation increases. It also consequently follows from Example 8.1.1, in which one has to interpret λ_p as λ_q, that as the order of solution is increased, the distance between the nodes (or antinodes) of the velocity and current density space-charge standing waves increases.

It is instructive to take up the special case of b/a = 1, corresponding to the beam completely filling the region inside the conducting enclosure, which yields from (a8.1.50), $J_0\{Ga\} = 0$ giving a series of solutions: Ga = 2.405, 5.520, 8.645, ... For numerical appreciation taking $\beta_e a = 1$, typically, one may obtain for the lowest order wave (Ga = 2.405), F = 0.384 from (a8.1.52). Then from (a8.1.53) one obtains, for this case, $\beta = (\omega \mp 0.384\omega_p)/v_0$ showing how the propagation constants in the typical case considered ($\beta_e a = 1$ and b/a = 1) change from those given by (a8.1.27) for an infinite beam, that is, from $\beta = (\omega \mp \omega_p)/v_0$.

Appendix a8.1.3 *Chu's power conservation theorem*

The problem of the electron beam considered in section 8.1 has been considered as one-dimensional; as such in the theoretical analysis no vectorial expressions were involved. Here, we continue to take the excitation to be longitudinal, but choose to express the relevant quantities in the vectorial form as follows:

(a8.1.54) (a) $\mathbf{J_1} = J_1\, \mathbf{a}_z$

(b) $\mathbf{v_0} = v_0\, \mathbf{a}_z$, and

(c) $\mathbf{v_1} = v_1\, \mathbf{a}_z$.

In the light of this, one may re-write[1] the following vectorial expressions from (8.1.2), (8.1.4) (originating from (4.1.8)), and the equation of motion (8.1.8), respectively, involving the time dependence exp (jωt):

(a8.1.55)
$$\mathbf{J_1} = \rho_0\, \mathbf{v_1} + \mathbf{v_0}\, \rho_1$$

(a8.1.56) $$\nabla . \boldsymbol{J}_1 = -\frac{\partial \rho_1}{\partial t} = -j\omega\rho_1 \, , \qquad \text{and}$$

(a8.1.57) $$(\partial/\partial t + \boldsymbol{v_0} \, . \nabla)\, \boldsymbol{v_1} = j\omega \boldsymbol{v_1} + (\boldsymbol{v_0} \, . \nabla)\, \boldsymbol{v_1} = \eta \boldsymbol{E} \, .$$

Starting from (a8.1.54)-(a8.1.57), one may proceed to arrive at the power conservation theorem[12] as follows. Equation (a8.1.56) is multiplied by v_0 and use is made of (a8.1.55) to get

$$\boldsymbol{v_0} \; \nabla . \boldsymbol{J_1} = -j\omega \boldsymbol{J_1} + j\omega \rho_0 \, \boldsymbol{v_1} \, ,$$

the complex conjugate of which gives

(a8.1.58) $$\boldsymbol{v_0} \; \nabla . \boldsymbol{J}_1^* = j\omega \boldsymbol{J}_1^* - j\omega \rho_0 \, \boldsymbol{v}_1^* \, .$$

Now, taking the dot product of $\boldsymbol{J}_1^*$ with $\boldsymbol{E}$ obtainable from (a8.1.57) one may write, putting η as $-|\eta|$, the following expression:

(a8.1.59) $$\boldsymbol{J}_1^* . \boldsymbol{E} = -\frac{j\omega}{|\eta|} \boldsymbol{J}_1^* . \boldsymbol{v}_1 - \boldsymbol{J}_1^* . \frac{(\boldsymbol{v}_0 . \nabla)\boldsymbol{v}_1}{|\eta|} \, .$$

Further, taking the dot product of $\boldsymbol{v_1}/|\eta|$ with (a8.1.58) and re-arranging terms, one obtains

(a8.1.60) $$\frac{j\omega}{|\eta|} \boldsymbol{J}_1^* . \boldsymbol{v}_1 = \frac{\boldsymbol{v}_1 . \boldsymbol{v}_0 \, \nabla . \boldsymbol{J}_1}{|\eta|} + \frac{j\omega \rho_0 \boldsymbol{v}_1 . \boldsymbol{v}_1^*}{|\eta|} \, .$$

Substituting (a8.1.60) in (a8.1.59) one gets

(a8.1.61) $$\boldsymbol{J}_1^* . \boldsymbol{E} = \frac{-\boldsymbol{v}_1 . \boldsymbol{v}_0 \nabla . \boldsymbol{J}_1^*}{|\eta|} - \frac{j\omega\rho_0 \boldsymbol{v}_1 . \boldsymbol{v}_1^*}{|\eta|} - \boldsymbol{J}_1^* . \frac{(\boldsymbol{v}_0 . \nabla)\boldsymbol{v}_1}{|\eta|} \, .$$

Now, using the vector indentity $\nabla . (\psi \boldsymbol{A}) = \boldsymbol{A} . \nabla\psi + \psi \nabla . \boldsymbol{A}$, where ψ is a scalar and $\boldsymbol{A}$ is a vector, and interpreting $\boldsymbol{A}$ as $\boldsymbol{J}_1^*$ and ψ as $\boldsymbol{v_1} . \boldsymbol{v_0}$, let us write [12]

(a8.1.62) $$\boldsymbol{v}_1 \, . \, \boldsymbol{v_0} \;\; \nabla . \boldsymbol{J} = \nabla . (\boldsymbol{v}_1 \, . \, \boldsymbol{v}_0 \;\boldsymbol{J}_1^*) - \boldsymbol{J}_1^* . \; \nabla (\boldsymbol{v}_1 \, . \, \boldsymbol{v_0}) \, .$$

Substituting (a8.1.62) in (a8.1.61) one thus obtains

(a8.1.63) $$\boldsymbol{J}_1^* . \boldsymbol{E} = -\frac{j\omega \rho_0 \boldsymbol{v}_1 . \boldsymbol{v}_1^*}{|\eta|} - \frac{\nabla . (\boldsymbol{v}_1 . \boldsymbol{v}_0 \boldsymbol{J}_1^*)}{|\eta|} - \frac{\boldsymbol{J}_1^* . [(\boldsymbol{v}_0 . \nabla)\boldsymbol{v}_1 - \nabla(\boldsymbol{v}_1 . \boldsymbol{v}_0)]}{|\eta|} \, .$$

The last term in (a8.1.63) is:

$$-\frac{\boldsymbol{J}_1^*}{|\eta|} . [(\boldsymbol{v}_0 . \nabla)\boldsymbol{v}_1 - \nabla(\boldsymbol{v}_1 . \boldsymbol{v}_0)] \, .$$

which, using (a8.1.54) and expanding the gradient according to (2.5.12) and remembering that the rf quantities have been assumed to vary only with z, simply becomes (a8.1.64) $$-\frac{\boldsymbol{J}_1^*}{|\eta|}\left(v_0\frac{\partial v_1}{\partial z}-v_0\frac{\partial v_1}{\partial z}\right)=0 \ .$$

In view of (a8.1.64), and putting $\rho_0 = -|\rho_0|$, *one may simplify (a8.1.63) as*

(a8.1.65) $$\boldsymbol{J}_1^*.\boldsymbol{E} = \frac{j\omega|\rho_0|\boldsymbol{v}_1.\boldsymbol{v}_1^*}{|\eta|}-\frac{\nabla.(\boldsymbol{v}_1.\boldsymbol{v}_0\,\boldsymbol{J}_1^*)}{|\eta|} \ .$$

Now, we can also obtain the following expression for $\boldsymbol{J}_1^* \,.\, \boldsymbol{E}$ ($= \boldsymbol{E} \,.\, \boldsymbol{J}_1^*$) *using a different approach by interpreting* $\boldsymbol{J}_c$ *as* $\boldsymbol{J}_1$, *the rf beam convection current density, in (4.8.54) which was derived earlier as an intermediate step towards developing the concepts of the complex Poynting vector:*

(a8.1.66) $$\boldsymbol{J}_1^*.\,\boldsymbol{E} = -\,\nabla.\,(\boldsymbol{E}\times\boldsymbol{H}^*) - j\omega\mu_0\,\boldsymbol{H}.\,\boldsymbol{H}^* + j\omega\varepsilon_0\,\boldsymbol{E}.\,\boldsymbol{E}^* .$$

Comparing (a8.1.66) with (a8.1.65) one may then easily see

$$-\nabla.(\boldsymbol{E}\times\boldsymbol{H}^*)-j\omega\mu_0\,\boldsymbol{H}.\boldsymbol{H}^*+j\omega\varepsilon_0\,\boldsymbol{E}.\boldsymbol{E}^* = \frac{j\omega|\rho_0|\boldsymbol{v}_1.\boldsymbol{v}_1^*}{|\eta|}\frac{-\nabla.(\boldsymbol{v}_1.\boldsymbol{v}_0\,\boldsymbol{J}_1^*)}{|\eta|}$$

which may be re-arranged as

(a8.1.67) $$\nabla.\left((\boldsymbol{E}\times\boldsymbol{H}^*)-\frac{\boldsymbol{v}_1.\boldsymbol{v}_0\boldsymbol{J}_1^*}{|\eta|}\right) = -\frac{j\omega|\rho_0|\boldsymbol{v}_1.\boldsymbol{v}_1^*}{|\eta|}-j\omega\mu_0\boldsymbol{H}.\boldsymbol{H}^*+j\omega\varepsilon_0\boldsymbol{E}.\boldsymbol{E}^* \ .$$

Now, one may see from (a8.1.67) that half the real part of (a8.1.67) is

(a8.1.68) $$\frac{1}{2}Re\,\nabla.\left((\boldsymbol{E}\times\boldsymbol{H}^*)-\frac{\boldsymbol{v}_1.\boldsymbol{v}_0\,\boldsymbol{J}_1^*}{|\eta|}\right)=0 \ .$$

Therefore, integrating (a8.1.68) over a closed volume (τ), *enclosing a portion of a linear beam between two reference cross sections (Fig. 8.1.2), one obtains*

(a8.1.69) $$\frac{1}{2}Re\oint_\tau\nabla.\left((\boldsymbol{E}\times\boldsymbol{H}^*)-\frac{\boldsymbol{v}_1.\boldsymbol{v}_0\,\boldsymbol{J}_1^*}{|\eta|}\right)d\tau = 0 \ .$$

Applying Gauss's theorem (see appendix a4.8.1) to (a8.1.69) one may change the volume integral occurring in the latter to the following surface integral:

(a8.1.70) $$\frac{1}{2}Re\oint_S(\boldsymbol{E}\times\boldsymbol{H}^*).\boldsymbol{a}_n\,ds = \frac{1}{2}Re\oint\frac{\boldsymbol{v}_1.\boldsymbol{v}_0\boldsymbol{J}_1^*}{|\eta|}.\boldsymbol{a}_n\,ds$$

$$= -\frac{1}{2}Re\oint_s V_k\,\boldsymbol{J}_1^*.\boldsymbol{a}_n dS \ ,$$

where

(a8.1.71) $$V_k = -\frac{v_1 . v_0}{|\eta|} .$$

In view of (a8.1.54), one may further express (a8.1.70) as

(a8.1.72) $$P_{em} = \frac{1}{2} Re \oint_S (\boldsymbol{E} \times \boldsymbol{H}^*).\boldsymbol{a}_n dS = -\frac{1}{2} Re \oint_S V_k J_1^* \boldsymbol{a}_z . \boldsymbol{a}_n dS = -\oint_S \mathcal{P}_k . \boldsymbol{a}_n dS ,$$

where P_{em} *is defined in the text following (8.1.45). Also,* V_k*, given by (a8.1.71), becomes*

(a8.1.73) $$V_k = -\frac{v_1 \boldsymbol{a}_z . v_0 \boldsymbol{a}_z}{|\eta|} = -\frac{v_0 v_1}{|\eta|} ;$$

and $\mathcal{P}_k$ *is given by:*

(a8.1.74) $$\mathcal{P}_k = \mathcal{R}_k \boldsymbol{a}_z = \frac{1}{2} Re\, V_k J_1^* \boldsymbol{a}_z .$$

Here, we see that (a8.1.73) and (a8.1.74) are identical with (8.1.44) and (8.1.45), respectively. Also, we have found in (a8.1.72) Chu's power conservation theorem that has been used in the text as (8.1.47).

8.2 Amplification in a Travelling-Wave Tube

We have seen in the preceding section that an electron beam treated as a separate entity supports space-charge waves which are not of amplifying type. In fact Γ, which occurs in the dependence of the rf space-charge wave quantities as exp ($j\omega t$ - Γz), turned out to be purely imaginary corresponding to waves which are of neither growing nor attenuating type (see equation (8.1.19) and (8.1.20)). Further, it was also appreciated that power could be transferred from the beam to a circuit if the latter be coupled to the slow space-charge wave supported by the former (see the discussion following (8.1.56)). In this section we develop the theory of a practical growing wave electron beam device, called the travelling-wave tube (TWT), in which the rf power is fed as an input to one end of a circuit, which is essentially a slow-wave structure (see chapter *5*), while the rf power which spatially grows at the expense of the beam power is coupled out of the circuit at its other end as the output power. The TWT has a unique combination of bandwidth and gain which makes it unparallel in the family of electron devices — amongst both solid-state and electron-beam types. Moreover, the present-day technology, which is being continuously updated as well as the use of the state-of-the-art design concepts has made a major breakthrough in enhancing the power, efficiency, life and reliability of a TWT (see appendix *a8.2.3*); the device has thus found extensive applications in communication — both in civil and in defence.

The TWT consists of an electron gun, a slow-wave structure, which also includes a built-in attenuator, a collector, an input coupler, an output coupler and a magnetic focusing structure (Fig. 8.2.1). The principles of the slow-wave structure, the electron gun and the magnetic focusing structure have been described at length in chapters *5, 6 and 7*, respectively. A linear electron beam, usually of circular cross section, is formed with the help of the electron gun and is confined by the focusing structure in the interior and along the length of the slow-wave structure. The rf power is injected, with the help of the input coupler, into the slow-wave circuit at its end that is closer to the electron gun, while the rf power that spatially grows at the expense of the beam power is coupled out of the circuit from its other end.

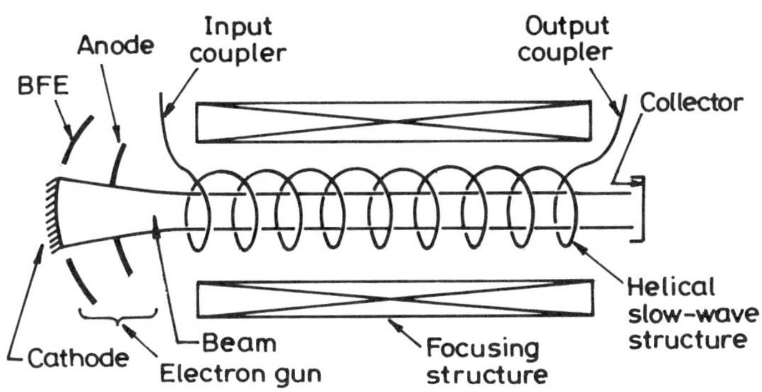

Fig. 8.2.1 Schematic of a TWT showing its major parts.

The cold slow-wave circuit, that is, the slow-wave structure when considered isolated without a beam, supports waves which like the space-charge waves on an isolated beam (section *8.1*) are of neither growing nor attenuating (the circuit being assumed lossless) type. Therefore, it may be intuitively guessed that one forward cold circuit wave plus two forward fast and slow space-charge waves and one backward cold circuit wave, the latter being generated if the circuit is not terminated in a match, will combine in a beam-circuit wave-coupled system of the TWT to generate three forward waves and one backward wave. It will be of interest to see if, in such a system, $-\Gamma$ occurring in the rf dependence $\exp(j\omega t - \Gamma z)$ referring to any of these four waves turns out to be complex comprising a negative imaginary and a positive real part, thus corresponding to a forward, growing wave.

The dispersion relation of a TWT has been found by Pierce [16] by an ingenious method. In this method the expressions for the ratio of the circuit voltage-to-rf beam convection current are found in the form of the so-called circuit (section *8.2.1*) and electronic equations (section *8.2.2*), respectively. These expressions are then equated to obtain the dispersion relation of the device which is essentially a relation between Γ and ω (see section *8.2.3*). In fact, the dispersion relation, it will be seen, will turn out to be a fourth degree equation which has four solutions corresponding to four waves that were intuitively expected to be generated in the beam-circuit coupled system as mentioned above. Out of these four solutions, the one corresponding to -Γ containing a positive real part subsequently becomes relevant to the derivation of the gain formula of the TWT (sections *8.2.3 and 8.2.4*).

8.2.1 Circuit Equation

Let us first study how the behaviour of the slow-wave circuit could be affected by the proximity of a modulated electron beam. The slow-wave circuit, irrespective of its type (see appendix *a5.0.1*), may be considered as equivalent to a semi-infinite transmission line defined by a set of distributed line parameters (see section *5.4*). We then make here two approaches to find the circuit equation for the ratio of the circuit voltage-to-beam convection current. In the first of these, the effect of the modulated electron beam manifests itself in the modification of the current telegraphist's equation [16] and, consequently, as it will be shown, the circuit equation is obtained in a form that involves the characteristic impedance of the transmission line equivalent of the slow-wave structure (section *8.2.1A*). In the second approach we start from how to the circuit electric field intensity, originally due to an input rf power injected into the circuit, is added the field intensities contributed by a set of infinitesimal current generators distributed all along the circuit to simulate the presence of the rf beam current density [1]. Consequently, in this second approach the circuit equation, as we will see (section *8.2.1B*), involves the interaction impedance of the slow-wave structure (section *5.5.2*) instead of the characteristic impedance of the transmission line equivalent of the structure, unlike in the first approach. Let us take up these two approaches now one by one.

8.2.1A Equation involving the characteristic impedance

Let us consider an element of length dz of the beam-circuit coupled system as shown in Fig. 8.2.2(a) in which the circuit, which is an arbitrary slow-wave structure, though shown typically as a helix here, is considered to be in the close vicinity of the beam. Now, with the help of (4.2.13)-(4.2.15) one may write for this coupled system;

$$\nabla .(\mathbf{J}_1 + \mathbf{J}_d) = 0 , \qquad (8.2.1)$$

where $\mathbf{J}_1$ and $\mathbf{J}_d$ are the convection and the displacement current densities, respectively. Taking the closed volume integral of the left hand side of (8.2.1) over a surface barely enclosing a volume 'τ' containing the beam portion of the element of the coupled system (see Fig. 8.2.2(a)), one gets

$$\oint_{\tau} \nabla.(\mathbf{J}_1 + \mathbf{J}_d)d\tau = 0 \ , \qquad (8.2.2)$$

since, here the integrand itself according to (8.2.1) vanishes. One may now using Gauss's theorem convert the volume integral of (8.2.2) into the surface integral as follows:

$$\oint_{S} (\mathbf{J}_1 + \mathbf{J}_d).\boldsymbol{a}_n dS = 0, \qquad (8.2.3)$$

where S represents the surface area of the enclosure (Fig. 8.2.2a). Thus, according to (8.2.3), the flux of the sum of the convection and the displacement current densities through the surface enclosing the beam element becomes zero.

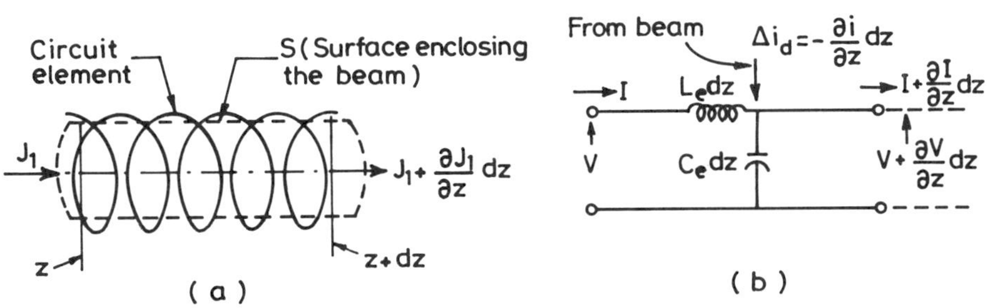

Fig. 8.2.2 Element of length (dz) of the beam-circuit coupled system and a surface enclosing its beam portion over which to take the closed integral to find the flux of current density as in (8.2.2) **(a)** and its transmission-line analogue showing the displacement current linked up from the beam to the circuit element **(b)**.

Let us now choose to separate out the convection and the displacement current density parts of (8.2.3) as:

$$\oint_{S} \mathbf{J}_1.\mathbf{a}_n \, dS = -\oint_{S} \mathbf{J}_d.\mathbf{a}_n \, dS \ . \qquad (8.2.4)$$

Further in the present one-dimensional problem, the portion of the surface close to the curved surface of the beam element does not contribute to the surface integral of the left hand side. Also, if the beam were not modulated, the axial flux of the current density through the portion of the surface close to and cutting across the beam at z would be equal and opposite to that through the portion of the surface close to the beam at z + dz (Fig. 8.2.2(a)). However, if we consider a modulated beam then these two quantities will be different; the former (at z) will be $\mathbf{J}_1.\alpha\, \mathbf{a}_n = \mathbf{J}_1\, \mathbf{a}_z \,.\, \alpha\, (-\mathbf{a}_z) = -J_1\alpha$ and the latter (at $z + dz$) will be $[\mathbf{J}_1 + (\partial \mathbf{J}_1/\partial z)dz]\,.\,\alpha\, \mathbf{a}_n = [J_1 + (\partial J_1/\partial z)dz]\mathbf{a}_z \,.\, \alpha\, \mathbf{a}_z = [J_1 + (\partial J_1/\partial z)dz]\alpha$, where α is the cross-sectional area of the beam. Adding these quantities in the left hand side of (8.2.4) one may then express the latter as

$$\alpha \frac{\partial J_1}{\partial z} dz = -\Delta i_d \,, \tag{8.2.5}$$

where

$$\Delta i_d = \oint_S \mathbf{J}_d . \mathbf{a}_n \, dS$$

represents the elements of the displacement current being equal to the flux of the displacement current density through the enclosure which may be considered to be predominantly contributed by the flux through the portion of the closed surface that is close to the curved surface of the beam rather than through the portion of the surface close to and cutting across the beam cross sections near z and z+dz, respectively (Fig. 8.2.2(b)). This is a valid assumption as long as the beam is thin. Moreover, considerable simplification in the analysis can be achieved if it is assumed that the element of the displacement current Δi_d thus obtained is completely linked up with the closeby circuit. Therefore, putting

$$J_1 \alpha = i \,, \tag{8.2.6}$$

where i is the rf beam current, one may write from (8.2.5)

$$\Delta i_d = -\frac{\partial i}{\partial z} dz \tag{8.2.7}$$

as the displacement current that is linked up with the circuit. In view of this, the transmission line analogue of the element of the beam-circuit coupled system may be represented as in Fig. 8.2.2(b). Considering the structure to be lossless one may take $R_e = G_e = O$, where R_e is the series resistance per unit length and G_e is the shunt conductance per unit length of the equivalent transmission line of the slow-wave structure (see appendix *a5.4.1* and compare Fig. 8.2.2(b) with Fig. a5.4.1). (The equivalent circuit analysis of a helical slow wave structure has been given in section *5.4* .) We notice that the element of the displacement current Δi_d that links up with the circuit will generate an equivalent conduction current in the circuit [1], and, consequently, this will modify the usual current telegraphist's equation. Thus one can write, applying Kirchoff's current law, to the element of the beam-circuit coupled system of Fig. 8.2.2(b), the following equation (which passes on to (a5.4.2) in the absence of the beam and taking $G_e = 0$):

$$I + \Delta i_d = I + \frac{\partial I}{\partial z} dz + C_e dz \frac{\partial V}{\partial t} \,, \tag{8.2.8}$$

where V and I are the circuit voltage and current, respectively, and C_e is the shunt capacitance per unit length of the transmission line equivalent of the slow-wave structure, G_e being the shunt conductance per unit length of the line (taken as zero for the lossless structure).

Substituting (8.2.7) in (8.2.8) and simplifying one then obtains

$$\frac{\partial I}{\partial z} = -C_e \frac{\partial V}{\partial t} - \frac{\partial i}{\partial z} \tag{8.2.9}$$

which is obtained as the modified form of the current telegraphist's equation. One may notice that (8.2.9) is identical with (a5.4.4) taking $G_e = 0$ for a lossless circuit, except for the addition of the term ($= -\partial i/\partial z$) due to the presence of the modulated beam.

The voltage telegraphist's equation of the equivalent transmission line however, remains unaffected by the presence of the modulated and may be written as (see equation (a5.4.3) in which one has to take $R_e = 0$ for the lossless circuit):

$$\frac{\partial V}{\partial z} = -L_e \frac{\partial I}{\partial t} , \tag{8.2.10}$$

where L_e is the series inductance per unit length of the transmission line equivalent of the slow-wave structure, R_e being the series resistance per unit length of the line (taken as zero for the lossless structure).

Assuming the rf quantities to vary as $\exp(j\omega t - \Gamma z)$ (see the discussion following (8.1.12)), one may write from (8.2.9) and (8.2.10), respectively:

$$-\Gamma I = -j\omega C_e V + \Gamma i \tag{8.2.11}$$

$$-\Gamma V = -j\omega L_e I . \tag{8.2.12}$$

Obtaining I from (8.2.12) and substituting it in (8.2.11) one may easily show

$$\frac{V}{i} = -\frac{\Gamma \Gamma_0 Z_0}{\Gamma^2 - \Gamma_0^2} , \tag{8.2.13}$$

where

$$\Gamma_0 = j\beta_0 , \tag{8.2.14}$$

β_0 being the cold axial phase propagation constant of the circuit (in the absence of the beam) which, considering the structure to be lossless ($R_e = G_e = 0$), is given by (see (a5.4.16)):

$$\beta_0 = \omega (L_e C_e)^{1/2} ; \tag{8.2.15}$$

Z_0 is the characteristic impedance of the transmission-line equivalent of the slow-wave structure given by (see (a5.4.19) in which one has to take $R_e = G_e = 0$ for the lossless circuit considered):

$$Z_0 = (Le/C_e)^{1/2} \quad . \tag{8.2.16}$$

Thus in (8.2.13) we have obtained an expression for the ratio of the circuit voltage-to-rf beam current, called the *circuit equation*, in a form that involves the characteristic impedance (Z_0) of the transmission line equivalent of the slow-wave circuit.

8.2.1B Equation involving the interaction impedance

The circuit equation (8.2.13) that has been derived in section *8.2.1A* is far from being exact since it refers to a rather over-simplified picture of the interaction between a thin beam and a closeby circuit in which the entire displacement current = $(-(\partial i/\partial z)dz)$ from the beam has been considered to be linked up with the element of length dz of the circuit. In an actual beam-circuit configuration it is extremely difficult to find the exact current linkage precisely. In an alternative approach the effect of the element of a modulated beam at a point on the circuit is simulated by an infinitesimal current generator at that point that 'sees' half the characteristic impedance of the transmission line, being equivalent to two such characteristic impedances in parallel, corresponding to two halves of the supposedly matched line. Such an infinitesimal generator sends two circuit waves in opposite directions, one to the left and one to the right such that the amplitudes of the circuit electric field intensity associated with these waves are equal [1]. Instead of bothering to find the exact current linkage from the beam to the circuit what we do in this approach is to find the contribution to the circuit field intensity at a circuit point from all such infinitesimal current generators distributed along the circuit both to the left and to the right of the point [1]. These contributions are then added to the circuit field intensity at the point caused by the power injected at the input end of the circuit to find the total circuit field intensity at the point in the presence of the modulated beam. Starting from the circuit field intensity thus obtained one may proceed to deduce the circuit equation by a method to be described as follows.

Let dE_R and dE_L be the amplitudes of the electric field intensity at a point on the circuit associated with two waves, one travelling to the right and the other to the left of the point, respectively, both launched by an infinitesimal current generator. It is easy to appreciate then that the electric field intensity at a point on the circuit caused by an infinitesimal current generator to the left of the point is $dE_R \exp -j\beta_0(z - x)$, which is associated with a wave travelling to the right from the generator, where β_0 is the axial phase propagation constant; z is the distance of the point and x is the distance of the infinitesimal current generator, both measured from the input end of the circuit (Fig. 8.2.3). Similarly, the electric field intensity at the point caused by a current generator to the right of the point is $dE_L \exp -j\beta_0(x - z)$, which is associated with a wave travelling to the left from the generator (Fig. 8.2.3). Integrating the contributions from all such infinitesimal generators to the left and to the right of the point and also taking into account the contribution from the power injected at the input end of the circuit, one may write the electric field intensity E{z} at a circuit point distant z from the input end as follows:

$$E\{z\} = E_i \exp(-j\beta_0 z) + \int_0^z \zeta_R\{x\} \exp - j\beta_0(z-x)dx + \int_z^\ell \zeta_L\{x\} \exp - j\beta_0(x-z)dx \ , \quad (8.2.17)$$

where E_i is the amplitude of the circuit electric field intensity injected at the input end (z = 0) of the circuit. Here, we have defined

$$dE_R = \zeta_R\{x\}\, dx \qquad \text{(a)}$$

and

$$dE_L = \zeta_L\{x\}\, dx \ . \qquad \text{(b)} \qquad (8.2.18)$$

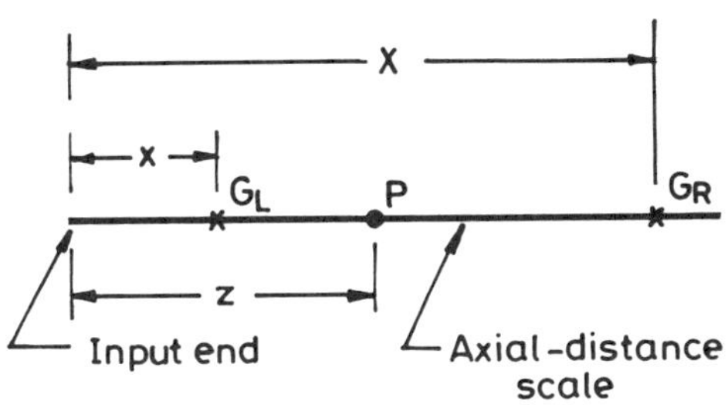

Fig. 8.2.3 The scale to measure the axial distances of the point (z) where to find the electric field intensity, shown as a dot (P), as well as of the points of locations of two arbitrarily chosen infinitesimal current generators (x), shown as crosses, one (G_L) to the left and the other (G_R) to the right of P.

However, here,

$$dE_R = dE_L = dE \ , \text{ say} \qquad \text{(a)}$$

and

$$\zeta_R\{x\} = \zeta_L\{x\} = \zeta\{x\} \ , \text{ say} \ , \qquad \text{(b)} \qquad (8.2.19)$$

since the current generators sees identical halves of the matched transmission line both to its left and to its right. Combining (8.2.18) and (8.2.19), one may write

$$dE_R = dE_L = dE = \zeta\{x\}dx \ . \qquad (8.2.20)$$

In view of (8.2.19), one may write the electric field intensity expressed as a function of z as:

$$E\{z\} = E_i \exp(-j\beta_0 z) + \int_0^z \zeta\{x\} \exp - j\beta_0 (z-x)dx + \int_z^\ell \zeta\{x\} \exp - j\beta_0 (x-z)dx \ . \quad (8.2.21)$$

Putting (8.2.21) as

$$E\{z\} = E_i \exp(-j\beta_0 z) + \mathcal{J}_1 + \mathcal{J}_2 \quad (8.2.22)$$

and differentiating it with respect to z we get

$$\frac{dE}{dz} = - j\beta_0 E_i \exp(-j\beta_0 z) + \frac{d\mathcal{J}_1}{dz} + \frac{d\mathcal{J}_2}{dz} \ , \quad (8.2.23)$$

where $\mathcal{J}_1$ and $\mathcal{J}_2$ are the integrals

$$\mathcal{J}_1 = \int_0^z \zeta\{x\} \exp - j\beta_0 (z-x)\, dx \quad (a)$$

and

$$\mathcal{J}_2 = \int_z^1 \zeta\{x\} \exp - j\beta_0 (x-z)\, dx \ . \quad (b) \quad (8.2.24)$$

$d\mathcal{J}_1/dz$ and $d\mathcal{J}_2/dz$ appearing in (8.2.23) may be obtained by differentiating (8.2.24) with respect to z as (see appendix *a8.2.1*):

$$\frac{d\mathcal{J}_1}{dz} = - j\beta_0 \mathcal{J}_1 + \zeta\{z\} \quad (a)$$

and

$$\frac{d\mathcal{J}_2}{dz} = j\beta_0 \mathcal{J}_2 - \zeta\{z\} \ . \quad (b) \quad (8.2.25)$$

Substituting (8.2.25) in (8.2.23) one gets

$$\frac{dE}{dz} = - j\beta_0 E_i \exp(-j\beta_0 z) - j\,\beta_0(\mathcal{J}_1 - \mathcal{J}_2)$$

which on differentiation gives

$$\frac{d^2E}{dz^2} = - \beta_0^2 E_i \exp(-j\beta_0 z) - j\beta_0 \left(\frac{d\mathcal{J}_1}{dz} - \frac{d\mathcal{J}_2}{dz}\right)$$

into which one may substitute (8.2.25) to get

$$\frac{d^2E}{dz^2} = -\beta_0^2 E_i \exp(-j\beta_0 z) - j\beta_0[-j\beta_0(\vartheta_1 + \vartheta_2) + 2\zeta\{z\}]$$

$$= -\beta_0^2 E_i \exp(-j\beta_0 z) - \beta_0^2(\vartheta_1 + \vartheta_2) - 2j\beta_0\zeta\{z\}$$

$$= -\beta_0^2[E_i \exp(-j\beta_0 z) + \vartheta_1 + \vartheta_2] - 2j\beta_0\zeta\{z\} \quad . \tag{8.2.26}$$

Interpreting $\zeta\{x\}$ as dE/dx (see (8.2.20)) and hence $\zeta\{z\}$ as dE/dz and using (8.2.22), one may express (8.2.26) as:

$$\frac{d^2E}{dz^2} = -\beta_0^2 E - 2j\beta_0 \frac{dE}{dz} \quad . \tag{8.2.27}$$

Now let us proceed to express (8.2.27) in a form that explicitly involves a beam parameter. For this purpose let us find the increment of circuit power dP at a point due to an infinitesimal current generator simulating the effect of a modulated beam element of length dz coupled to the circuit at the point. dP may be put as

$$dP = dP_R + dP_L \ , \tag{8.2.28}$$

where dP_R and dP_L are the increments of circuit power associated with two waves sent by the infinitesimal current generator to the left and to the right of the point, respectively, which may be expressed, with the help of the differential of (5.5.4), in which β has to be interpreted as β_0

$$2\beta_0^2 K dP_R = d(E_R^2) = 2\, E_R\, dE_R$$

and

$$2\beta_0^2 K\, dP_L = d(E_L^2) = 2\, E_L\, dE_L$$

whence one gets (see (8.2.28))

$$dP = dP_R + dP_L = \frac{E_R + E_L}{\beta^2 K} dE \ , \tag{8.2.29}$$

remembering $dE_R = dE_L = dE$ (see (8.2.19(a))), where K is the interaction impedance of the structure. Since the increment of circuit power dP at the point given by (8.2.29) owes to the rf modulation of the beam element of length dz, dP may also be equated to the power lost by the beam element of length dz subjected to the circuit field at the point. In order to find the latter, the beam element is divided into two halves, each of length dz/2, which experience the electric field intensities E_L and E_R associated with the power propagating to the left and to the right of the circuit point coupled to the beam element, respectively.

The power lost, that is the energy lost per second, by an electron belonging to that half of the modulated beam element which is subjected to the electric field intensity E_L may be found as $-eE_Lv_1$, being equal to the product of $-eE_L$, the force on the electron, and v_1, the distance moved by it per second under rf modulation. The power lost per electron thus found becomes positive, since e carrier a negative sign, v_1 and E_L being interpreted to have the same direction. Multiplying this quantity by $n\,\alpha\,dz/2$, the number of electrons in the element half considered, one may find the power lost by this half as$(-eE_Lv_1)(n\,\alpha\,dz/2)$, where n is the rf number density of electrons of the perturbed beam element and α is the beam cross-sectional area. Similarly, the power lost by the remaining half which is subjected to the circuit electric field intensity E_R may be found as $(-eE_R\,v_1)(n\,\alpha\,dz/2)$. Adding these quantities one may then find the power lost by the complete beam element of length dz which may be equated to dP as discussed following (8.2.29). Thus one gets

$$dP = -eE_L\,v_1\,n\alpha\,dz/2 - eE_R\,v_1\,n\,\alpha\,dz/2 \;. \tag{8.2.30}$$

Putting $i = J_1\alpha$ see (8.2.6) as the rf beam current and with $J_1 = nev_1$ as the rf beam current density (see appendix *a6.1.1*), one may express (8.2.30) as

$$dP = -i(E_R + E_L)\frac{dz}{2} \;. \tag{8.2.31}$$

The negative sign in (8.2.31) indicates that the power lost is positive if i and $(E_R + E_L)$ are directed oppositely [1]. Comparing (8.2.29) with (8.2.31) then one obtains

$$\frac{dE}{dz} = -\frac{\beta_0^2\,Ki}{2} \;. \tag{8.2.32}$$

Plugging (8.2.32) in (8.2.27) one may then write

$$\frac{d^2E}{dz^2} = -\beta_0^2E + j\beta_0^3Ki \;. \tag{8.2.33}$$

Assuming rf quantities to vary with z as exp $(-\Gamma z)$ as discussed preceding (8.2.11), one may express (8.2.33) as

$$\Gamma^2E = -\beta_0{}^2E + j\,\beta_0{}^3\,Ki \;. \tag{8.2.34}$$

Under a quasistatic assumption one may take (see equation (a8.2.10))

$$E = -\frac{\partial V}{\partial z}$$

which, in view of the rf dependence exp $(-\Gamma z)$, becomes

$$E = \Gamma V . \tag{8.2.35}$$

Substituting (8.2.35) in (8.2.34) and re-arranging terms one then obtains

$$\frac{V}{i} = \frac{j\beta_0^3 K}{(\Gamma^2 + \beta_0^2)\Gamma} . \tag{8.2.36}$$

Using the relation $\Gamma_0 = j\beta$ as introduced following (8.2.13), one may easily show that (8.2.36) becomes

$$\frac{V}{i} = -\frac{\Gamma_0^3 K}{(\Gamma^2 - \Gamma_0^2)\Gamma} . \tag{8.2.37}$$

Since we do not expect Γ to deviate much from Γ_0, we may put $\Gamma \approx \Gamma_0$ in (8.2.37) and choose to express the latter as

$$\frac{V}{i} = -\frac{\Gamma\Gamma_0 K}{\Gamma^2 - \Gamma_0^2} . \tag{8.2.38}$$

It is of interest to note that in (8.2.38) we have derived a new circuit equation which is identical with the circuit equation (8.2.13) except that now the interaction impedance K has replaced the characteristic impedance Z_0. Obviously, the circuit equation (8.2.38) is more accurate and more general than the circuit equation (8.2.13) in the sense that in deriving the former no such approximation has been made as that a thin beam is coupled to a nearby circuit as has been done in deriving the latter. Consequently, the interaction impedance turns out to be a more realistic parameter than the characteristic impedance in the circuit eequation, that is, in the expression for the ratio of the circuit voltage-to-rf beam current.

8.2.2 *Electronic Equation*

Let us now find the ratio of the circuit voltage-to-rf beam current by studing the motion of the electrons subjected to the circuit electric field intensity E plus the space-charge electric field intensity E_s. For this purpose, one may recall (8.1.8) which may be read with the help of (8.1.6) (involving the rf dependence exp ($j\omega t$ - Γz)) and replacing E_z by $E + E_s$ therein, as follows:

$$(j\omega - v_0\Gamma)v_1 = \eta(E + E_s) . \tag{8.2.39}$$

From (8.2.39) one may find

$$v_1 = \frac{\eta(E + E_s)}{j\omega - v_0\Gamma} . \tag{8.2.40}$$

Similarly, one may find from (8.1.11) interpreting E_z as E_s :

$$-\Gamma E_s = \frac{\rho_1}{\varepsilon_0}$$

which gives

$$E_s = -\frac{\rho_1}{\Gamma \varepsilon_0} \quad . \tag{8.2.41}$$

Further ρ_1 occuring in (8.2.41) may be expressed in terms of J_1 with the help of (8.1.4) as:

$$-\Gamma J_1 = -j\omega \rho_1$$

whence one gets

$$\rho_1 = \frac{\Gamma J_1}{j\omega} \quad . \tag{8.2.42}$$

Substituting (8.2.42) in (8.2.41) one obtains

$$E_s = -\frac{\Gamma J_1}{j\omega}\frac{1}{\Gamma \varepsilon_0} = -\frac{J_1}{j\omega\, \varepsilon_0}$$

which may in turn be substituted in (8.2.40) to get

$$v_1 = \frac{\eta(E - \dfrac{J_1}{j\omega \varepsilon_0})}{j\omega - v_0 \Gamma} \quad . \tag{8.2.43}$$

Substituting ρ_1 and v_1 from (8.2.42) and (8.2.43) in (8.1.2) one then obtains

$$J_1 = \rho_0 \frac{\eta(E - \dfrac{J_1}{j\omega \varepsilon_0})}{j\omega - v_0 \Gamma} + v_0 \frac{\Gamma J_1}{j\omega} \quad . \tag{8.2.44}$$

One may easily simplify (8.2.44) to get

$$J_1 \left((j\omega - v_0 \Gamma)^2 + \omega_p^2 \right) = j\omega \eta \rho_0 E \ , \tag{8.2.45}$$

where $\omega_p^2 = \eta\rho_0/\varepsilon_0 = |\eta||\rho_0|/\varepsilon_0$ (see equation (8.1.15) and (8.1.17)). Using the relation $J_0 = \rho_0\, v_0$ (see equation (8.1.1) and the discussion following it) and $\beta_e = \omega/v_0$ (as defined following (8.1.20)) and $E = \Gamma V$ (see (8.2.35) and also (a8.2.10)), one may express (8.2.45) as

$$J_1\left((j\omega - v_0\Gamma)^2 + \omega_p^2\right) = j\beta_e \eta J_0 \Gamma V . \tag{8.2.46}$$

Multiplying (8.2.46) by α and putting $J_1\alpha = i$ and $J_0\alpha = i_0$, one obtains from (8.2.46)

$$\frac{V}{i} = \frac{(j\omega - v_0\Gamma)^2 + \omega_p^2}{j\beta_e \eta i_0 \Gamma} . \tag{8.2.47}$$

Using the relations $\beta_p = \omega_p / v_0$ and $\beta_e = \omega / v_0$ (as defined following (8.1.20)), one may express (8.2.47) as

$$\frac{V}{i} = \frac{\left((j\beta_e - \Gamma)^2 + \beta_p^2\right) v_0^2}{j\beta_e \eta i_0 \Gamma} . \tag{8.2.48}$$

Putting $\eta i_0 = |\eta|\, I_0$, where $\eta = -|\eta|$ and $i_0 = -|\, i_0| = -I_0$, say, and $v_0^2 = 2\,|\eta| V_0$ (see the relation (6.1.6)), where V_0 is the dc beam voltage corresponding to dc beam velocity v_0, one may express (8.2.48) as:

$$\frac{V}{i} = \left(\frac{2\,V_0}{I_0}\right)\left(\frac{(j\beta_e - \Gamma)^2 + \beta_p^2}{j\beta_e \Gamma}\right) . \tag{8.2.49}$$

The expression for the ratio of the circuit voltage-to-beam current thus obtained in the form of (8.2.49) is known as the electronic equation.

8.2.3 TWT Dispersion Relation and Its Solutions

The dispersion relation of the beam-circuit coupled system of a TWT is obtained as follows by equating the right hand sides of the circuit and the electronic equations (8.2.38) and (8.2.49), respectively, both representing the one and the same quantity, namely, the ratio of the circuit voltage-to-beam covnection current:

$$-\frac{\Gamma \Gamma_0 K}{\Gamma^2 - \Gamma_0^2} = \frac{2\,V_0}{I_0} \frac{(j\beta_e - \Gamma)^2 + \beta_p^2}{j\beta_e \Gamma} . \tag{8.2.50}$$

The dispersion relation (8.2.50) comes out to be essentially a fourth-degree equation which, as instutively guessed in the beginning of the section, has four solutions — three forward waves and one backward wave. In fact, it is easy to see from (8.2.50) that for a zero beam current ($I_0 = 0$), one has

$$(\Gamma^2 - \Gamma_0^2)[(j\beta_e - \Gamma)^2 + \beta_p^2] = 0$$

which gives circuit waves and beam space-charge waves decoupled yieding $\Gamma = \pm\Gamma_0 = \pm j\beta_0$ (see (8.2.14)), for the two forward and backward circuit waves, and $\Gamma = j(\beta_e \mp \beta_p)$ for the two forward, fast and slow, space-charge waves (see (8.1.20) and (8.1.21)).

One may solve (8.2.50) numerically. But it is of interest to review Pierce's method of solving it [16]. Expecting that the propagation constant β of the beam-circuit-coupled system, defined as $\Gamma = j\beta$, changes from the beam propagation constant β_e only to a small extent, that is, taking $\beta \approx \beta_e$, let us put, for a *forward wave*:

$$-\Gamma\,(= -j\beta) = -j\beta_e + \xi\,, \tag{8.2.51}$$

where $\xi << j\beta_e$ remembering that the rf quantities have been assumed here to vary with z as exp $(-\Gamma z)$. Let us now take a case in which the circuit phase velocity $v_p = \omega/\beta_0$ is equal to the dc beam velocity $v_0 = \omega/\beta_e$, that is, $\beta_0 = \beta_e$, which in turn gives from (8.2.14)

$$\Gamma_0 = j\beta_0 = j\beta_e\,, \tag{8.2.52}$$

for a lossless circuit. (The nature of solutions of the dispersion relation for situations in which the dc beam velocity differs from the circuit phase velocity (see, for instance, Example 8.1.3), and the circuit, in general, is lossy has been discussed later in section *8.2.3A*). Since ξ has been assumed to be small as compared to $j\beta_e$, one may, in view of (8.2.51) and (8.2.52), substitute in the dispersion relation (8.2.50): $\Gamma \cong \Gamma_0 = j\beta_e$ in the numerator and $\Gamma^2 - \Gamma_0^2 = (\Gamma+\Gamma_0)(\Gamma-\Gamma_0) \cong -2\,j\beta_e\,\xi$ in the denominator of its left hand side; $j\beta_e - \Gamma = \xi$ in the numerator and again $\Gamma \cong \Gamma_0 = j\,\beta_e$ in the denominator of its right-hand side. This will then make (8.2.50) read as:

$$\xi^3 = -j\beta_e^3\left(\frac{KI_0}{4V_0}\right), \tag{8.2.53}$$

putting, however, $\beta_p = 0$ which actually would amount to ignoring the space-charge electric field intensity compared to the circuit field intensity (see section *8.2.2*). (The effects of including the space-charge effects have been discussed later in section *8.2.3A*).

Let us now choose to define two dimensionless quantities C and δ as follows[16]:

$$\xi = \beta_e C\,\delta\,, \tag{8.2.54}$$

where

$$C = \left(\frac{KI_0}{4V_0}\right)^{1/3}, \tag{8.2.55}$$

such that $C\delta << 1$. The condition $C\delta << 1$ would ensure the assumption that ξ is small compared to $j\beta_e$ as can seen from (8.2.51) which may be read now with the help of (8.2.54) as

$$-\Gamma = -j\beta_e + \beta_e C\,\delta \quad (8.2.56)$$

Substituting (8.2.54) and (8.2.55) in (8.2.53) one obtains a simple cubic equation:

$$\delta^3 = -j. \quad (8.2.57)$$

The solutions of (8.2.57) may be obtained by first expressing (8.2.57) as [18]

$$\delta^3 = \exp -j(\pi/2)\,,\ \exp -j(\pi/2+2\pi)\,,\ \exp -j(\pi/2-2\pi)\,,$$

and then taking the cube-root thereof giving

$$\delta_1 = \exp -j(\frac{\pi}{6}) = \frac{\sqrt{3}}{2} - j\frac{1}{2},$$

$$\delta_2 = \exp -j(\frac{5\pi}{6}) = -\frac{\sqrt{3}}{2} - j\frac{1}{2}$$

and

$$\delta_3 = \exp j(\pi/2) = j, \quad (8.2.58)$$

where the subscripts 1, 2 and 3, respectively, represent the three solutions. Substituting (8.2.58) in (8.2.56) one may write the corresponding expression for propagation constant as:

$$-\Gamma_1 = -j\beta_e + \beta_e C\left(\frac{\sqrt{3}}{2} - j\frac{1}{2}\right) = \beta_e C\frac{\sqrt{3}}{2} - j\beta_e\left(1+\frac{C}{2}\right) \quad \text{(a)}$$

$$-\Gamma_2 = -j\beta_e + \beta_e C\left(-\frac{\sqrt{3}}{2} - j\frac{1}{2}\right) = -\beta_e C\frac{\sqrt{3}}{2} - j\beta_e\left(1+\frac{C}{2}\right) \quad \text{(b)}$$

and

$$-\Gamma_3 = -j\beta_e + \beta_e C j = -j\,\beta_e(1-C)\,. \quad \text{(c)}\ (8.2.59)$$

On examining the real and imaginery parts of the right hand side of (8.2.59), and remembering the rf dependence exp $(-\Gamma z)$, one may notice that Γ_1 and Γ_2 correspond to two forward waves which exponentially grow and decay, respectively, with the same rate and which propagate with the same phase velocity $\omega/[\beta_e(1+C/2)] = v_0/(1+C/2)$ which is less than the dc beam velocity v_0. Similarly, Γ_3 corresponds to a forward wave that is neither growing nor decaying which propagates with the phase velocity $\omega/\beta_e(1-C) = v_0/(1-C)$ which is greater than the dc beam velocity v_0.

The fourth solution that has not been obtained so far must correspond to a backward wave as was intutively guessed in the beginning of this section. In order to find it let us now put instead of (8.2.51):

$$-\Gamma = j\beta_e + \xi , \tag{8.2.60}$$

corresponding to a *backward* wave. Consequently, instead of (8.2.56) now one has to take

$$-\Gamma = j\beta_e + \beta_e C\delta . \tag{8.2.61}$$

We continue to take ξ to be small compared to $j\beta_e$ and now, in view of (8.2.52) and (8.2.60), substitute in the dispersion relation (8.2.50): $-\Gamma \cong \Gamma_0 = j\beta_e$ in the numerator and $\Gamma^2 - \Gamma_0^2 = (\Gamma + \Gamma_0)(\Gamma - \Gamma_0) \cong 2j\beta_e \xi$ in the denominator of its left hand side; $j\beta_e - \Gamma \cong 2j\beta_e$ in the denominator of its right hand side. This yields again ignoring the space-charge effects ($\beta_p = 0$):

$$\xi = -\left(\frac{j\beta_e}{4}\right)\left(\frac{KI_0}{4V_0}\right). \tag{8.2.62}$$

Substituting (8.2.54) and (8.2.55) in (8.2.62), one then gets

$$\delta = -\frac{jC^2}{4}. \tag{8.2.63}$$

Plugging (8.2.63) in (8.2.61), one may write

$$-\Gamma_4 = j\beta_e\left(1 - \frac{C^3}{4}\right) \tag{8.2.64}$$

in which the subscript 4 is used to indicate that the solution thus obtained refers to the fourth wave. The fourth solution given by (8.2.64) corresponds to a wave propagating in the backward (negative z) direction, with a phase velocity $\omega/[\beta_e(1 - C^3/4)] = v_0/(1 - C^3/4)$ which is greater in magnitude than the dc beam velocity v_0.

If the structure is perfectly matched one may assume that the fourth wave, the propagations constant Γ_4 of which is given by (8.2.64), is not excited to a significant extent. Also, out of the three forward waves, the propagation constants Γ_1, Γ_2 and Γ_3 which are given by (8.2.59), the one of propagation constant Γ_1 given by (8.2.59(a)) becomes more significant at a reasonable distance from the input end of the structure since $-\Gamma_1$, unlike $-\Gamma_2$ and $-\Gamma_3$, contains a positive real part corresponding to an exponentially growing wave.The exponential growth rate, corresponding to the rf dependence exp $(-\Gamma_1 z)$, is then $\beta_e C \sqrt{3}/2$ per unit axial distance, as can be seen from (8.2.59(a)). The spatial growth rate of the relevant rf wave can be suitably interpreted, as we will see in section *8.2.4*, to derive the gain equation of a TWT.

8.2.3 A *Effects of lack of synchronism, circuit losses and space charge*

So far we have considered a *synchronous* beam in which the circuit phase velocity has been put equal to the beam velocity ($v_p = v_0$) enabling us to take $\beta_0 = \beta_e$ and $\Gamma_0 = j\beta_e$ (see equation (8.2.52) and the discussion preceding it). Let us now take a non-synchronous beam for which $v_p \neq v_0$ and put the cold circuit axial-propagation constant as [16],[17]

$$\beta_0 = \beta_e (1 + bC) , \tag{8.2.65}$$

where b is called the Pierce's velocity parameter. Since $\beta_e = \omega/v_0$ and $\beta_0 = \omega/v_p$, one may express b with the help of (8.2.65)as:

$$b = \frac{\beta_0 - \beta_e}{\beta_e C} = \frac{v_0 - v_p}{v_p C} . \tag{8.2.66}$$

Further, if we take the structure to be lossy we may modify (8.2.14) as

$$\Gamma_0 = \alpha_0 + j\beta_0 , \tag{8.2.67}$$

where α_0 is the cold circuit attenuation constant. One may choose to express α_0 as [16]

$$\alpha_0 = \beta_e C d , \tag{8.2.68}$$

where d is called the Pierce's loss parameter. Substituting (8.2.65) and (8.2.68) in (8.2.67) we may write the cold circuit propagation constant as [16],[17]:

$$\Gamma_0 = \beta_e C d + j\beta_e (1 + bC) . \tag{8.2.69}$$

Now, let us see how, in the light of the change in the cold propagation constant from (8.2.52) to (8.2.69), the solution of the dispersion relation (8.2.50) will change. In order to add more practical relevance to the problem let us further include the space-charge effects by taking $\beta_p \neq 0$ in (8.2.50). (It may be noted that in deriving (8.2.57) we took $b = 0$ (that is, considered a synchronous beam: $v_0 = v_p$; see (8.2.66)); $d = 0$ (that is, ignored circuit losses); and $\beta_p = 0$ (that is, ignored space-charge effects). In view of (8.2.51) and (8.2.69), let us now substitute in the dispersion relation (8.2.50): $\Gamma \cong \Gamma_0 = j\beta_e$ in the numerator and $\Gamma^2 - \Gamma_0^2 = (\Gamma + \Gamma_0)(\Gamma - \Gamma_0) \cong -2j\beta_e(\xi + \beta_e Cd + j\beta_e bC)$ in the denominator of its left hand side; $j\beta_e - \Gamma = \xi$ in the numerator and again $\Gamma \cong \Gamma_0 = j\beta_e$ in the denominator of its right hand side. This yields

$$j(\xi^2 + \beta_p^2)\left(\frac{\xi}{\beta_e} + Cd + jbC\right) = \frac{K I_0}{4 v_0} \beta_e^2 . \tag{8.2.70}$$

It may be seen that, putting $\beta_p = b = d = 0$ in (8.2.70) the latter passes on to (8.2.53). Substituting (8.2.54) and (8.2.55) in (8.2.70), one obtains

$$j\,(\beta_e^2 C^2 \delta^2 + \beta_p^2)(\delta + d + jb) \;=\; C^2 \beta_e^2 \;. \qquad (8.2.71)$$

Now, let us define a space-charge parameter as follows [16], [17]:

$$QC \;=\; \frac{1}{4}\left(\frac{\beta_p}{\beta_e C}\right)^2 \qquad (8.2.72)$$

and substitute $\beta_p^2 = 4\,QC\,(\beta_e C)^2$ obtained therefrom into (8.2.71) to get

$$(\delta^2 + 4\,QC)\,(j\,\delta + j\,d - b) \;=\; 1\;. \qquad (8.2.73)$$

It may be seen that if one puts $QC = 0$ (which also corresponds to $\beta_p = 0$) and $b = d = 0$ (for a synchronuous beam coupled to a lossless circuit), (8.2.73) passes on to (8.2.57). However, in a practical case in which $QC \neq 0$, $b \neq 0$, $d \neq 0$, one must solve (8.2.73) for δ numerically, and the solution would then be different from (8.2.58). It is convenient to express δ in terms of its real and imaginery parts as:

$$\delta \;=\; x \;+\; j\,y\;. \qquad (8.2.74)$$

The three forward-wave propagation constants may then be written with the help of (8.2.56) and (8.2.74) as:

$$-\Gamma_{1,2,3} = -j\beta_e + \beta_e C\,\delta_{1,2,3} = \beta_e C\,x_{1,2,3} - j\beta_e\,(1 - C\,y_{1,2,3})\;. \qquad (8.2.75)$$

As an example, for a synchronous beam ($b = 0$) that is coupled to a lossless circuit ($d = 0$), and also when the space-charge effects are ignored ($\beta_p = 0$), one may take in (8.2.75) the following values of $x_{1,2,3}$ and $y_{1,2,3}$ which may be identified by comparing (8.2.74) with (8.2.58) which, in fact, was obtained by solving the simple cubic equation (8.2.57):

$$x_1 = \sqrt{3}/2, \quad y_1 = -1/2\,, \qquad \text{(a)}$$

$$x_2 = -\sqrt{3}/2, \quad y_2 = -1/2\,, \qquad \text{(b)}$$

and

$$x_3 = 0, \quad y_3 = 1 \quad (b = d = QC = 0)\;. \quad \text{(c)} \qquad (8.2.76)$$

<u>*Example 8.2.1*</u> Starting from the dispersion relation of a TWT obtain the dispersion relation of space-charge waves on an electron beam if the latter is widely out of synchronism with the circuit wave.

It follows from (8.2.66) that b will take on large values when v_0 is largely different from v_p. In view of this one may obtain from (8.2.73)

$$\delta^2 + 4QC = \frac{1}{j\delta + jd - b} \rightarrow 0$$

for large values of b, whence one obtains

$$\delta = \pm\, j\, 2\, \sqrt{QC}\,. \tag{8.2.77}$$

Substituting (8.2.72) in (8.2.77), one then gets

$$\delta = \pm \frac{j\beta_p}{\beta_e C}\,. \tag{8.2.78}$$

Substitution of (8.2.78) in (8.2.56) yields

$$\Gamma = j(\beta_e \mp \beta_p)\,. \tag{8.2.79}$$

If one chooses to take $\Gamma = j\beta$, as it is in (8.1.19), one obtains from (8.2.79)

$$\beta = \beta_e \mp \beta_p\,. \tag{8.2.80}$$

One now immediate identifies the propagation constants given by (8.2.80) as identical with those of two space-charge waves on an electron beam considered as a separate entity, given by (8.1.20).

Pierce [16] has discussed the results of solving the cubic equation (8.2.73) for δ and hence for x and y (see equation (8.2.74)). It may be noted that the positive and negative values of x, respectively, correspond to the spatial growth and decay of rf waves in the beam-circuit coupled system. Let, out of the three solutions x_1, x_2 and x_3, one say, x_1 be positive. Then the spatial growth rate, as can be seen from (8.2.75), is $\beta_e C\, x_1$. It also follows from (8.2.75) that according as whether y_1, y_2 and y_3 are positive or negative, the beam will be slower or faster than the wave components which y_1, y_2 and y_3 refer to. The phase velocity of the wave components are $\omega/(\beta_e(1 - C\, y_{1,2,3})) = v_0/(1 - C\, y_{1,2,3})$. Some of Pierce's results may be summarised as follows.

(i) The results for the case in which b = d = QC = 0 have been already shown in (8.2.76) from which we notice that $x_2 = -\, x_1$. Thus x_2 is negative, x_1 being positive. In other words, the exponentially-attenuating wave decays with the same rate as that with which the exponentially-growing wave grows. Also, y_1 and y_2 are each equal to the same negative value meaning thereby that the growing and decaying waves have the same phase velocity that is less than the dc beam velocity.

(ii) If one takes, in general, a non-synchronous beam, that is, $b \neq 0$, but continues to ignore circuit losses and space-charge effects (d = QC = 0), then upto a positive limiting value $b = (3/2)(2)^{1/3}$, one finds x_2 as the mirror image of x_1 about the abcissa of the plot of x and y (as the ordinates) versus the velocity parameter b (as the abcissa), throughout the range of b covering both the negative and positive values of b, that is, for both the beams which are slower and faster than the cold circuit wave, respectively (see equation (8.2.66)). Thus, in this range, upto the limiting value of b $(= (3/2)(2)^{1/3})$, one gets $x_2 = -x_1$ giving the same rate of growth and attenuation for the growing and attenuating-wave components. Beyond this limiting value of b, one gets $x_1 = x_2 = 0$, corresponding to neither a growing nor an attenuating-wave component. It also emerges out that x_1 shows a positive peak corresponding to the maximum growth rate of the growing-wave component for a value of $b \approx (>)0$, that is, for a beam that is slightly faster than the cold circuit wave. Also, for all the values of b, positive and negative, one gets $x_3 = 0$. Further, one finds that upto the same limiting value of b, one gets y_1 and y_2 each equal to the same negative value at a given value of b; this means that for a given value of b, both the growing and attenuating-wave components have the same phase velocity which is less than the dc beam velocity. Beyond this limiting value, one gets $y_1 \neq y_2$. Also, one finds y_3 take on positive values for all values of b, thereby meaning that the wave component which y_3 corresponds to travels faster than the beam; this component for which $x_3 = 0$, as mentioned earlier, neither grows nor attenuates, that is, it has a constant amplitude, throughout the range of b.

(iii) Next if one takes a non-synchronous beam ($b \neq 0$) and a lossy circuit ($d \neq 0$) but continues to ignore space-charge effects ($QC = 0$), one notices that the value of x_1 decreases from its value for the no-loss case ($d = 0$). Also, the limiting value of b beyond which x_1 becomes zero is now extended. Further, unlike in the no-loss case ($d = 0$), x_2, though it has a negative value, is no longer a mirror image of x_1 about the abcissa in the plot of x_1 and x_2 against b; y_1 and y_2, though both of them are negative, are no longer equal to each other over any range of values of b; and x_3, in general, has a non-zero, negative value.

(iv) Finally, in order to study the space-charge effects one has to take $QC \neq 0$; for the sake of simplicity in this study one may take $d = 0$. It is found that the value of x_1 in this case ($QC \neq 0$) decreases from that in the situation in which the space-charge effects are ignored ($QC = 0$). For lower values of QC, both positive and negative values of b can yield non-zero values of x_1 which in turn is responsible for the generation of a growing-wave component; however, the range of b over which x_1 has a non-zero value is now restricted. With the further increase of the value of QC, this range is further restricted, that is, narrowed down, and shifted towards the positive values of b, and this shift could occur to such an extent that it might not allow any negative values of b for a non-zero value of x_1. Thus for large values of QC, in order to ensure the spatial growth of rf waves, the electron beam should be made faster than the circuit wave, however, precisely to a very limited extent. The optimum value of b corresponding to the peak value of x_1, that is, to the maximum growth rate, increases with the value of QC.

8.2.4 TWT Gain Equation

Let us consider the slow-wave structure of the TWT to be perfectly matched so that the backward wave component of an rf wave is not generated. This enables one to write the circuit voltage $V\{z\}$ to be expressed in terms of three forward-wave components as follows:

$$V\{z\} = V_1\{z\} + V_2\{z\} + V_3\{z\} , \qquad (8.2.81)$$

where the subscripts 1, 2 and 3 refer to the forward wave components of propagation constants Γ_1, Γ_2 and Γ_3, respectively, the latter being given by (8.2.75). In terms of these propagation constants one may express (8.2.81) in the following alternative form:

$$V\{z\} = V_1\{0\}\exp(-\Gamma_1 z) + V_2\{0\}\exp(-\Gamma_2 z) + V_3\{0\}\exp(-\Gamma_3 z) . \qquad (8.2.82)$$

Substituting (8.2.75) into (8.2.82), the latter becomes:

$$V\{z\} = V_1\{0\}\exp(\beta_e C x_1 z) \exp -j\beta_e(1-Cy_1)z + V_2\{0\}\exp(\beta_e C x_2 z) \exp -j\beta_e(1-Cy_2)z$$
$$+ V_3\{0\}\exp(\beta_e C x_3 z) \exp -j\beta_e(1-Cy_3)z . \qquad (8.2.83)$$

The input circuit voltage V_{in}, at the input end of the device $z = 0$, may be written from (8.2.83) as:

$$V_{in} = V\{0\} = V_1\{0\} + V_2\{0\} + V_3\{0\} , \qquad (8.2.84)$$

in terms of the amplitudes of the three forward-wave components. Similarly, from (8.2.83) one may write the circuit voltage V_{out} at the output end of the device, $z = \ell$, say, where ℓ is the interaction length of the device, as follows:

$$V_{out} = V\{\ell\} = V_1\{0\}\exp(\beta_e C x_1 \ell) \exp -j\beta_e(1-Cy_1)\ell . \qquad (8.2.85)$$

In writing (8.2.85), however, we have considered the contributions from the last two terms of the right hand side of (8.2.83) to be negligibly small compared to that from the first term which grows as $\exp(\beta_e C x_1 \ell)$ over a large interaction length ℓ of the device, that is, taking the output end to be far enough from the input end of the interaction structure [16].

Now, if we can find a relation between $V_1\{0\}$ and V_{in}, and that is what we will do here, then using (8.2.85) we can express V_{out} in terms of V_{in} whence the gain of the device can be found. One may notice that, in (8.2.84), three unknowns, namely $V_1\{0\}$, $V_2\{0\}$ and $V_3\{0\}$ are involved. So, let us proceed to form two more equations in these unknowns so that including (8.2.84) we would have three equations which may be solved to find the unknown of interest, here, $V_1\{0\}$.

For this purpose let us invoke the condition that at the entry of the interaction structure, $z = 0$, the beam is unmodulated so that the input rf beam velocity $v_{1,in}$ and beam current density $J_{1,in}$ are each equal to zero:

$$\mathrm{v}_{1,\mathrm{in}} = 0 \tag{8.2.86}$$

and

$$J_{1,\mathrm{in}} = 0\,. \tag{8.2.87}$$

Now, we can express $\mathrm{v}_{1,\mathrm{in}}$ and $J_{1,\mathrm{in}}$ as follows in terms of the amplitudes of the three forward-wave components at $z = 0$, just as we have expressed V_{in} in the form of (8.2.84):

$$\mathrm{v}_{1,\mathrm{in}} = \mathrm{v}_1\{0\} = \mathrm{v}_{1,1}\{0\} + \mathrm{v}_{1,2}\{0\} + \mathrm{v}_{1,3}\{0\}\,, \tag{8.2.88}$$

and

$$J_{1,\mathrm{in}} = J_1\{0\} = J_{1,1}\{0\} + J_{1,2}\{0\} + J_{1,3}\{0\}\,. \tag{8.2.89}$$

Substituting (8.2.88) and (8.2.89) in (8.2.86) and (8.2.87) respectively, we can then write

$$\mathrm{v}_{1,1}\{0\} + \mathrm{v}_{1,2}\{0\} + \mathrm{v}_{1,3}\{0\} = 0 \tag{8.2.90}$$

and

$$J_{1,1}\{0\} + J_{1,2}\{0\} + J_{1,3}\{0\} = 0\,. \tag{8.2.91}$$

Now, let us show, by using some of the findings in section 8.2.1, that both (8.2.90) and (8.2.91) can be expressed in terms of $V_1\{0\}$, $V_2\{0\}$ and $V_3\{0\}$. Substituting the expression for E_s in terms of J_1 given following (8.2.42) into (8.2.39), one gets

$$(j\omega - \mathrm{v}_0\Gamma)\mathrm{v}_1 = \eta\left(E - \frac{J_1}{j\omega\varepsilon_0}\right)\,. \tag{8.2.92}$$

Obtaining J_1 from (8.2.45) and substituting it into (8.2.92), one obtains

$$(j\omega - \mathrm{v}_0\Gamma)\mathrm{v}_1 = \eta E\left(1 - \frac{\omega_p^2}{(j\omega - \mathrm{v}_0\Gamma)^2 + \omega_p^2}\right)$$

which simplifies to

$$\mathrm{v}_1 = \frac{\eta E(j\omega - \mathrm{v}_0\Gamma)}{(j\omega - \mathrm{v}_0\Gamma)^2 + \omega_p^2}\,. \tag{8.2.93}$$

Putting $E = \Gamma V$ from (8.2.35) in (8.2.93) and using the relations $\beta_e = \omega/\mathrm{v}_0$ and $\beta_p = \omega_p/\mathrm{v}_0$ (as defined following (8.1.20)), one then obtains

$$\mathrm{v}_1 = \frac{\eta\Gamma V}{\mathrm{v}_0}\,\frac{j\beta_e - \Gamma}{(j\beta_e - \Gamma)^2 + \beta_p^2}\,,$$

which in view of the relations $\Gamma \approx j\beta_e$ and $j\beta_e - \Gamma = \beta_e C\delta$ (see equation (8.2.56)) becomes

$$v_1 = \frac{j\eta\beta_e V}{v_0} \frac{\beta_e C\delta}{\beta_e^2 C^2 \delta^2 + \beta_p^2} . \tag{8.2.94}$$

Substituting $\beta_p^2 = 4QC\ \beta_e^2 C^2$ from (8.2.72) into (8.2.94), we get

$$v_1 = \frac{j\eta V}{v_0 C\delta\left(1 + \dfrac{4QC}{\delta^2}\right)} . \tag{8.2.95}$$

In terms of a voltage V′ defined as

$$V' = \frac{V}{1 + \dfrac{4QC}{\delta^2}} , \tag{8.2.96}$$

one may choose to express (8.2.95) as

$$v_1 = \frac{j\eta V'}{v_0 C\delta} . \tag{8.2.97}$$

The relation between J_1 and V has already been obtained in the form of (8.2.46) which, using the relation $\Gamma \approx j\beta_e$, $j\beta_e - \Gamma = \beta_e C\delta$ and $\beta_p^2 = 4\ QC\ \beta_e^2\ C^2$, as we have done in obtaining (8.2.95), becomes

$$J_1 = -\frac{\eta J_0 V}{v_0^2 C^2 \delta^2 (1 + \dfrac{4QC}{\delta^2})} . \tag{8.2.98}$$

Substitution of (8.2.96) in (8.2.98) gives

$$J_1 = -\frac{\eta J_0 V'}{v_0^2 C^2 \delta^2} . \tag{8.2.99}$$

From (8.2.97) and (8.2.99), one may then write

$$v_{1,\upsilon}\{0\} = \frac{j\eta V'_\upsilon\{0\}}{v_0 C \delta_\upsilon} \tag{8.2.100}$$

and

$$J_{1,\upsilon}\{0\} = -\frac{\eta J_0 V'_\upsilon}{v_0^2 C^2 \delta_\upsilon^2} , \tag{8.2.101}$$

respectively, where υ = 1, 2 and 3 correspond to the three forward-wave components. Substituting (8.2.100) and (8.2.101) in (8.2.90) and (8.2.91), respectively, one then obtains

$$\frac{V_1'\{0\}}{\delta_1} + \frac{V_2'\{0\}}{\delta_2} + \frac{V_3'\{0\}}{\delta_3} = 0 \tag{8.2.102}$$

and

$$\frac{V_1'\{0\}}{\delta_1^2} + \frac{V_2'\{0\}}{\delta_2^2} + \frac{V_3'\{0\}}{\delta_3^2} = 0 \ . \tag{8.2.103}$$

Also, one may write with the help of (8.2.98)

$$V_\upsilon\{0\} = V_\upsilon'\{0\}\left(1 + \frac{4QC}{\delta_\upsilon^2}\right) = V_\upsilon'\{0\} + 4QC\frac{V_\upsilon'\{0\}}{\delta_\upsilon^2} \ , \tag{8.2.104}$$

υ = 1, 2 and 3 referring to the three forward-wave components. Therefore, substituting (8.2.104) into (8.2.84) we may write

$$V_{in} = V_1'\{0\} + V_2'\{0\} + V_3'\{0\} + 4QC\left(\frac{V_1'\{0\}}{\delta_1^2} + \frac{V_2'\{0\}}{\delta_2^2} + \frac{V_3'\{0\}}{\delta_3^2}\right). \tag{8.2.105}$$

In view of (8.2.103), the fourth term of (8.2.105) vanishes enabling one to write

$$V'_1\{0\} + V'_2\{0\} + V'_3\{0\} = V_{in} \ . \tag{8.2.106}$$

Solving (8.2.102), (8.2.103) and (8.2.106) one then obtains

$$V_1'\{0\} = \frac{V_{in}}{\left(1 - \dfrac{\delta_2}{\delta_1}\right)\left(1 - \dfrac{\delta_3}{\delta_1}\right)} \ , \tag{8.2.107}$$

$$V_2'\{0\} = \frac{V_{in}}{\left(1 - \dfrac{\delta_3}{\delta_2}\right)\left(1 - \dfrac{\delta_1}{\delta_2}\right)} \ , \tag{8.2.108}$$

and

$$V_3'\{0\} = \frac{V_{in}}{\left(1 - \dfrac{\delta_1}{\delta_3}\right)\left(1 - \dfrac{\delta_2}{\delta_3}\right)} \ . \tag{8.2.109}$$

With the help of (8.2.104) and (8.2.107) one may write

$$V_1\{0\} = \left(1+\frac{4QC}{\delta_1^2}\right)\left(\frac{1}{(1-\frac{\delta_2}{\delta_1})(1-\frac{\delta_3}{\delta_1})}\right)V_{in} . \qquad (8.2.110)$$

Finally, in order to get a relation between V_{out} and V_{in} let us substitute (8.2.110) in (8.2.85) to get

$$V_{out} = \left(1+\frac{4QC}{\delta_1^2}\right)\left(\frac{1}{(1-\frac{\delta_2}{\delta_1})(1-\frac{\delta_3}{\delta_1})}\right)[\exp -j\beta_e(1-C\,y_1)\,\ell][\exp(\beta_e C\,x_1\,\ell)]\,V_{in} . \qquad (8.2.111)$$

From (8.2.11), the gain of the device in decibels, G, can be found as:

$$G = 20\log_{10}\left|\frac{V_{out}}{V_{in}}\right| = A + 20\log_{10}[\exp(\beta_e C x_1 \ell)] , \qquad (8.2.112)$$

where

$$A = 20\log_{10}\left|\left(1+\frac{4QC}{\delta_1^2}\right)\left(\frac{1}{(1-\frac{\delta_2}{\delta_1})\,(1-\frac{\delta_3}{\delta_1})}\right)\right| . \qquad (8.2.113)$$

The second term of the right hand side of (8.2.112) may be put as

$$20\log_{10}[\exp(\beta_e C\,x_1\,\ell)] = 20\log_e[\exp(\beta_e C\,x_1\,\ell)]\log_{10}e = 20\,\beta_e C\,x_1\,\ell\,\log_{10}e$$

$$= (20\,C\,x_1\log_{10}e)(\beta_e\ell) . \qquad (8.2.114)$$

Now, if we express the interaction length ℓ in terms of N, the number of beam wavelengths, $\lambda_e = 2\pi/\beta_e$, then we may put $\ell = N\lambda_e = 2\pi N/\beta_e$, whence we get

$$\beta_e\,\ell = 2\pi N . \qquad (8.2.115)$$

Plugging (8.2.115) in (8.2.114) and then substituting the latter in (8.2.113), we obtain

$$G = A + BCN , \qquad (8.2.116)$$

where

$$B = (40\pi\log_{10}e)\,x_1 \approx 54.6\,x_1 . \qquad (8.2.117)$$

The gain of the TWT in decibels may be found using the formula (8.2.116), if the velocity parameter b, the loss parameter d and the space-charge parameter QC are known. For this purpose, one has to solve (8.2.73) for the three values δ_1, δ_2, δ_3 which are required to calculate A using (8.2.113); the real part of δ_1 gives x_1 (see (8.2.74)) which is required to calculate B using (8.2.117).

<u>*Example 8.2.2*</u> Ignoring space-charge effects, structure losses and taking a synchronous beam (QC = d = b = 0), show (a) that the input signal is distributed evenly among the three forward wave components generated in the beam-circuit coupled system of the TWT and (b) that its gain is given by $G = -9.54 + 47.3\ CN$.

(a) In this case, QC = d = b = 0, (8.2.73) passes on to (8.2.57) for which the solution for $\delta_{1,2,3}$ are given by (8.2.58) and, correspondingly, the values of $x_{1,2,3}$ and $y_{1,2,3}$ by (8.2.76). Taking QC = 0 in (8.2.110) one gets

$$V_1\{0\} = \frac{V_{in}}{(1-\frac{\delta_2}{\delta_1})\,(1-\frac{\delta_3}{\delta_1})} \quad . \tag{8.2.118}$$

In fact, for QC = 0, one may interpret $V'_1\{0\}$ as $V_1\{0\}$, as can be seen from (8.2.104), and with this interpretation one finds (8.2.118) as identical with (8.2.107). Similarly, interpreting $V'_2\{0\} = V_2\{0\}$ and $V'_3\{0\} = V_3\{0\}$ for QC = 0, one may, respectively, write from (8.2.108) and (8.2.109):

$$V_2\{0\} = \frac{V_{in}}{(1-\frac{\delta_3}{\delta_2})\,(1-\frac{\delta_1}{\delta_2})} \tag{8.2.119}$$

and

$$V_3\{0\} = \frac{V_{in}}{(1-\frac{\delta_1}{\delta_3})\,(1-\frac{\delta_2}{\delta_3})} \quad . \tag{8.2.120}$$

Substituting the values of δ_1, δ_2 and δ_3 from (8.2.58) in (8.2.118),

$$V_1\{0\} = \frac{\delta_1^2\, V_{in}}{(\delta_1-\delta_2)\,(\delta_1-\delta_3)} = \frac{\left(\frac{\sqrt{3}}{2}-\frac{j}{2}\right)^2 V_{in}}{(\sqrt{3})\left(\frac{\sqrt{3}}{2}-\frac{j3}{2}\right)} = \frac{\left(\frac{\sqrt{3}}{2}-\frac{j}{2}\right)^2 V_{in}}{(\sqrt{3})(\sqrt{3})\left(\frac{1}{2}-\frac{j\sqrt{3}}{2}\right)}$$

$$= \frac{\left(\frac{3}{4}-2(\frac{\sqrt{3}}{2})(\frac{j}{2})-\frac{1}{4}\right)V_{in}}{(\sqrt{3})(\sqrt{3})\left(\frac{1}{2}-\frac{j\sqrt{3}}{2}\right)} = \frac{\left(\frac{1}{2}-\frac{j\sqrt{3}}{2}\right)V_{in}}{(\sqrt{3})(\sqrt{3})\left(\frac{1}{2}-\frac{j\sqrt{3}}{2}\right)} = \frac{V_{in}}{3} \quad . \tag{8.2.121}$$

Similarly, substituting the values of δ_1, δ_2 and δ_3 in (8.2.119) and (8.2.120),

$$V_2\{0\} = \frac{V_{in}}{3} \tag{8.2.122}$$

and

$$V_3\{0\} = \frac{V_{in}}{3}. \tag{8.2.123}$$

Thus, from (8.2.121)-(8.2.123), we conclude that the input signal is evenly distributed among the three forward-wave components.

(b) If we take QC = 0 and substitute the value of δ_1, δ_2 and δ_3 from (8.2.58) in (8.2.113) we get

$$A = 20\log_{10}\left|\frac{1}{(1-\frac{\delta_2}{\delta_1})(1-\frac{\delta_3}{\delta_1})}\right|$$

which as per calculations already done in (a) becomes

$$A = 20\log_{10}\left|\frac{1}{3}\right| = -20\log_{10}3 = -20\times 0.4771 \cong -9.54 . \tag{8.2.124}$$

Substituting $x_1 = \sqrt{3}/2$ from (8.2.76(a)) in (8.2.117),

$$B = 54.6 \times \sqrt{3}/2 = 54.6 \times 0.866 = 47.3 . \tag{8.2.125}$$

Putting (8.2.124) and (8.2.125) in (8.2.116),

$$G = -9.54 + 47.3\ CN . \tag{8.2.126}$$

For the simple case of QC = d = b = 0, it is found that the value of A becomes negative while that of B positive. (A = –9.54 decibels and B = 47.3 decibels, see Example 8.2.2). It then follows from (8.2.116) that for a positive value of gain, G, the magnitude of BCN should exceed that of A. This means that the initial part of the interaction length (measured in terms of N) would be eaten up to compensate for the negative value of A. For instance, from (8.2.116), it can be seen that one must have CN > 0.2 in order to make G > 0. Thus A is called the launching loss of the TWT.

For a general case, QC ≠ 0, d ≠ 0, b ≠ 0, one should calculate A using (8.2.113). It may be thus seen that the value of A depends upon the value of δ which in turn can be obtained by solving (8.2.73). The nature of solution of δ and its dependence on the parameters QC, d and b has been already discussed earlier following Example 8.2.1. This dependence, therefore, should have a correlation with the dependence of A on these parameters. Since B also depends upon the nature of solution δ (see equations (8.2.117) and (8.2.74)), a similar conclusion can be made also for B. Some interesting aspects of the dependence of A and B on these parameters may be summarised as follows [19].

With the variation of the velocity parameter b, the value of A fluctuates exhibiting one or two peaks, for given values of d and QC, and, interestingly, for some combinations of these parameters, A may even exhibit a positive value at or near such a peak value of A. Also, for a given value of QC, A is found to decrease (that is, the magnitude of its value, if it is negative, increase) with the increase of the loss parameter d. It is also found that A can take on the same value, for the same value of d, but for the two different values of QC. The value of B also decreases with the increase in loss parameter d, however, upto a certain value of b, where it has the same value irrespective of d. Beyond this value, B is found to increase with d. This allows for a relatively larger departure from synchronism (larger values of b) for a lossy structure, a result that is similar to what was noticed for the growth parameter x_1 (as discussed following Example 8.2.1). The loss parameter d remaining the same, the value of B is found to increase with the decrease of the space-charge parameter QC.

Finally, it may be noted that the gain of the TWT given by (8.2.116) depends upon C and N. C, called the Pierce's gain parameter given by (8.2.55), in turn, depends upon the interaction impedance K of the slow-wave structure (see section *5.5*). Therefore, any design of the slow-wave structure must ensure a high value of K so that the value of G does not deteriorate. This aspect has a special significance while inhomogeneously or/and anisotropically loading a helix, with a view to broadbanding a TWT, (see section *5.8*), since the loading of the helix is associated with the reduction in the value of K (see section *5.5.2*). Also, it follows from (8.2.116) that, with the increase of N, the normalised interaction length of the device given by (8.2.115) representing the number of beam wavelengths occupying this length (as mentioned following (8.2.114)), the gain of the device should increase. However, this increase should not take place indefinitely as the gain formula (8.2.116) simply predicts. In fact, with the increase of the interaction length or/and with the increase of the input signal level, the nonlinear effects come into play which have been ignored while formulating the electronic equation of the device in section *8.2.2*; it may be recalled that in this formulation the acceleration equation (8.1.8) was used which in turn was obtained ignoring the nonlinear term $v_1\partial v_1/\partial z$ of (8.1.7), under small-signal approximation ($v_1 << v_0$).

Finally, it may be of interest to study the nature of the gain-frequency response from the simple gain formula (8.2.116) derived in this section. If we ignore the launching loss part (A), then the small-signal gain G ~ BCN is seen to be proportional to C and N which in turn are, respectively, proportional to $K^{1/3}$ (see (8.2.55)) and frequency ($N = \beta_e \ell/2\pi = \omega\ell/(2\pi v_0)$; see (8.2.115)). Now we may define a normalised frequency γa ($\approx \beta a = \omega a/v_p$, v_p being the phase velocity of the slow-wave structure; see (5.2.7)), and see that K, which is a function of γa, decreases with frequency as discussed in section *5.5.2*, while γa itself increases with frequency. Therefore, if we plot a normalised quantity proportional to gain, $(K/(\mu_0/\varepsilon_0)^{1/2})^{1/3}(\gamma a)$ against a normalised frequency γa, we expect that the normalised gain first increases with γa, reaches a peak value, and then decreases with γa. The value of γa where the gain of the device peaks becomes an important design parameter. For instance, one may start from this value of γa to design the helical slow-wave structure of the device (as illustrated in Example 5.4.1). Also, clearly then in order to realise a particular value of γa ($\approx \omega a/v_p$) for the desired gain one must decrease the transverse dimension of the structure (a) at high values of frequency (ω). This sets a basic limitation of a TWT in the millimeter/sub-millimeter-wave regime (see the discussion in the beginning of section *8.4;* for further elaboration see also problem 5.4).

8.2.4 A Attenuator effects: active attenuation

In the foregoing analysis leading to the TWT gain equation it has been assumed that the slow-wave structure is perfectly matched at the input and output ends. However, in actual practice, there would be reflections at these ends due to imperfect matching. Thus a forward wave reflected at the output end will travel to the input end unattenuated, the structure being taken as lossless. As a result of mismatch at the input end, this wave will be reflected at this end and set up three forward waves one of which spatially grows and travels to the output end where it is reflected again, and the repetition of the process would eventually cause a regenerative oscillation in the device. In order to prevent oscillations in the device, therefore, a lossy section is placed along the slow-wave structure, called the *attenuator*, roughly 1/3 to 1/2 way down the structure [16],[18],[20]. The attenuator also isolates the systems connected to the input and the output ends of the device. The attenuator may typically be in the form of an absorbing layer on the dielectric supports for the helix, if the latter is used as a slow-wave structure; in the case of a coupled-cavity structure, this may be in the form of a lossy ceramic button, loading a spacer cavity wall in the structure (see appendix *a8.2.3* for other relevant concepts in attenuators).

The attenuator not only does the function of absorbing the reflected energy and thus prevents oscillations but also absorbs the energy flowing in the forward direction. The length of the attenuator, therefore, does not contribute to the gain of the TWT, and, correspondingly, the interaction length of the device has to be increased to compensate for the attenuating portion of the interaction length. For a large-gain tube it may be required to increase the length of the interaction structure by providing more than one attenuator in the structure; the rule of thumb is to provide one attenuator section per about 20 dB gain of the device, so that a tube of 30 dB gain would require one section while one of 50 dB requires two sections [20]. For a high-gain, high-power tube, often the structure is severed into parts with proper care to match the structure at severs.

Interestingly an attenuator, although it provides a cold-circuit attenuation of say, ~ 70 dB, thereby rendering the amplitude of the circuit field of the forward wave to a negligibly small value beyond the attenuator, yet it does not render zero the overall gain of the device which is, say ~ 40 dB. This is because, even in the presence of the attenuator, the rf modulation on the beam remains, so that, in the forward direction, the spatially growing wave is excited again beyond the attenuator region. In the reverse direction, however, the growing wave interaction does not take place, and, consequently, the circuit field in the wave traveling in the reverse direction is reduced to zero beyond the attenuator and does not spatially grow further in this direction to reach the input end of the structure.

In order to estimate the order of hot attenuation caused by the attenuator let us proceed as follows [16], [18]. For the sake of simplicity the cold-circuit attenuation is assumed to be infinite so that the circuit voltage V beyond the attenuator is reduced to zero; also the length of the attenuator is considered to be negligibly small so that the time of passage of the beam electrons through the attenuator can be ignored, and in such a passage the rf beam velocity and current density remain conserved. This enables one to write:

$$V^b = 0\,, \tag{8.2.127}$$

$$v_1^b = v_1^a\,, \tag{8.2.128}$$

and

$$J_1^b = J_1^a\,, \tag{8.2.129}$$

where the superscripts a and b refer to the quantities at the entry and the exit of the attenuator, respectively.

Let us, for the sake of simplicity, ignore the space-charge effects (QC = 0). Then using (8.2.95), we get

$$v_{1,\upsilon}^b = \frac{j\eta V_\upsilon^b}{v_0 C\delta_\upsilon}\,, \tag{8.2.130}$$

where $\upsilon = 1$, 2 and 3, respectively, refer to the three forward-wave components. Similarly, using (8.2.98),

$$J_{1,\upsilon}^b = -\frac{\eta J_0 V_\upsilon^b}{v_0^2 C^2 \delta_\upsilon^2}\,, \tag{8.2.131}$$

for the three forward-wave components corresponding to $\upsilon = 1$, 2 and 3, respectively. In terms of the three forward-wave components, (8.2.127)-(8.2.129) may be expressed as:

$$V_1^b + V_2^b + V_3^b = 0, \tag{8.2.132}$$

$$v_{1,1}^b + v_{1,2}^b + v_{1,3}^b = v_{1,1}^a + v_{1,2}^a + v_{1,3}^a\,, \tag{8.2.133}$$

and

$$J_{1,1}^b + J_{1,2}^b + J_{1,3}^b = J_{1,1}^a + J_{1,2}^a + J_{1,3}^a\,. \tag{8.2.134}$$

Substituting (8.2.130) and (8.2.131) in (8.2.133) and (8.2.134), respectively, one may write:

$$\frac{V_1^b}{\delta_1} + \frac{V_2^b}{\delta_2} + \frac{V_3^b}{\delta_3} = \frac{V_1^a}{\delta_1} + \frac{V_2^a}{\delta_2} + \frac{V_3^a}{\delta_3} \tag{8.2.135}$$

and
$$\frac{V_1^b}{\delta_1^2}+\frac{V_2^b}{\delta_2^2}+\frac{V_3^b}{\delta_3^2} = \frac{V_1^a}{\delta_1^2}+\frac{V_2^a}{\delta_2^2}+\frac{V_3^a}{\delta_3^2} . \qquad (8.2.136)$$

Let the attenuator be placed at a distance $z = \ell_1$ from the input end ($z = 0$). Then with the help of (8.2.82), one may write

$$V^a_{1,2,3} = V_{1,2,3}\{0\} \exp(-\Gamma_{1,2,3}\ell_1) . \qquad (8.2.137)$$

Now, let us take, for the sake of simplicity, in addition to $QC = 0$ that has already been assumed, $d = b = 0$. This in turn allows us to put $V_1\{0\} = V_2\{0\} = V_3\{0\} = V_{in}/3$ (see equation (8.2.121)-(8.2.123)) in (8.2.137) giving

$$V^a_{1,2,3} = \frac{V_{in}}{3}\exp(-\Gamma_{1,2,3}\ell_1) . \qquad (8.2.138)$$

Putting (8.2.75) in (138),

$$V^a_{1,2,3} = \frac{V_{in}}{3}\exp(-j\beta_e+\beta_e C\delta_{1,2,3})\ell_1 . \qquad (8.2.139)$$

Substituting (8.2.139) in (8.2.135) and (8.2.136) one, respectively, gets:

$$\frac{V_1^b}{\delta_1}+\frac{V_2^b}{\delta_2}+\frac{V_3^b}{\delta_3} = \frac{V_{in}}{3}\exp(-j\beta_e\ell_1)\left(\frac{\exp(\beta_e C\delta_1)\ell_1}{\delta_1}+\frac{\exp(\beta_e C\delta_2)\ell_1}{\delta_2}+\frac{\exp(\beta_e C\delta_3)\ell_1}{\delta_3}\right) \qquad (8.2.140)$$

and

$$\frac{V_1^b}{\delta_1^2}+\frac{V_2^b}{\delta_2^2}+\frac{V_3^b}{\delta_3^2} = \frac{V_{in}}{3}\exp(-j\beta_e\ell_1)\left(\frac{\exp(\beta_e C\delta_1)\ell_1}{\delta_1^2}+\frac{\exp(\beta_e C\delta_2)\ell_1}{\delta_2^2}+\frac{\exp(\beta_e C\delta_3)\ell_1}{\delta_3^2}\right) . \qquad (8.2.141)$$

Thus we have now formed three equations, namely, (8.2.132), (8.2.140) and (8.2.141) in three unknowns V_1^b, V_2^b and V_3^b, which on solving for V_1^b for the case of $b = QC = d = 0$ for which one may use the values of δ_1, δ_2 and δ_3 from (8.2.58), yield[16], [18] (see problem 8.4 for a general form):

$$V_1^b = \frac{V_{in}}{3}\exp-(j2\pi N_1)\left(\frac{2}{3}\exp(2\pi CN_1\delta_1)-\frac{1}{3}\exp(2\pi CN_1\delta_2)-\frac{1}{3}\exp(2\pi CN_1\delta_3)\right), \qquad (8.2.142)$$

where we have put (see (8.2.115)):

$$\beta_e\,\ell_1 = 2\pi\,N_1 . \qquad (8.2.143)$$

Also with the help of (8.2.139) and using (8.2.143) we may write (see problem 8.4 for a general form):

$$V_1^a = \frac{V_{in}}{3}\exp-(j2\pi N_1)\exp(2\pi C N_1\delta_1)\,, \tag{8.2.144}$$

for the forward-wave growing wave component of the circuit voltage at the entry of the attenuator corresponding to δ_1 (= $x_1 + jy_1$) having a positive real part. Thus the amplitude ratio of the growing wave component of the circuit voltage at the exit to that the entry of the attenuator is obtained from (8.2.142) and (8.2.144) for the case of b = QC = d = 0 as:

$$\left|\frac{V_1^b}{V_1^a}\right| = \left|\frac{2}{3}-\frac{1}{3}\exp-2\pi CN_1(\delta_1-\delta_2)-\frac{1}{3}\exp-2\pi CN_1(\delta_1-\delta_3)\right|\,, \tag{8.2.145}$$

where δ_1, δ_2 and δ_3 have to be taken from (8.2.58). (For general form of (8.2.145) suitable for the case of b ≠ 0, QC ≠ 0, d ≠ 0, see problem 8.4 for the expressions of V_1^b and V_1^a). If the attenuator is located sufficiently far from the input end such that $CN_1 > \sim 0.2$, thereby ensuring that the launching loss is overcome (see the discussion following (8.2.126)), then the first term of the right hand side of (8.2.145) dominates over the remaining two giving

$$\left|\frac{V_1^b}{V_1^a}\right| \cong \frac{2}{3}\,.$$

Therefore, the 'hot' attenuation in decibels due to the presence of the attenuator is only ~ 20 $\log_{10}3/2$ = 3.52 dB, surprisingly though in the analysis the 'cold' attenuation has been assumed to be infinite! (For a general expression for hot attenuation in terms of δ_1, δ_2 and δ_3 suitable for the case b ≠ 0, QC ≠ 0, d ≠ 0, see problem 8.4.) It may be mentioned that if a finite time of transit of beam electrons through the attenuator section were considered a somewhat greater value of hot attenuation would be obtained [16], [18].

Appendix a8.2.1 *Differentiation of relevant definite intetegrals using Leibnitz formula*

First let us take the integral $\mathcal{I}_1$ given by (8.2.24(a)) and find its derivative with respect to z using the formula of Leibnitz for evaluating a definite integral when the limits of integration are functions of a parameter appering in the integrand, as follows

$$\frac{d}{dz}\int_a^b f\{x,z\}\,dx = \int_a^b \frac{\partial f\{x,z\}}{\partial z}dx + f\{b,z\}\frac{\partial b}{\partial z} - f\{a,z\}\frac{\partial a}{\partial z}\,. \tag{a8.2.1}$$

Now from (8.2.24(a)):

$$\mathscr{I}_1 = \int_0^z \zeta\{x\} \exp - j\beta_0 (z-x)\, dx = \int_a^b f\{x,z\}\, dx , \quad \text{say.}$$

Here, the quantities relevant to (a8.2.1) are:

(a8.2.2)
$$f\{x,z\} = \zeta\{x\} \exp - j\beta_o (z-x) \; ; \quad a = 0, \quad b = z \; ; \quad \frac{\partial a}{\partial z} = \frac{\partial (0)}{\partial z} = 0 \; ,$$

$$\frac{\partial b}{\partial z} = \frac{\partial z}{\partial z} = 1 \; ; \quad \frac{\partial f\{x,z\}}{\partial z} = - j\,\beta_0\, f\{x,z\} ; \quad f\{b,z\} = \zeta(b) \exp - j\,\beta_0 (z-b)$$

$$= \zeta(z) \exp - j\,\beta_0 (z-z) = \zeta(z) \; , \quad f\{a,z\} = \xi(a) \exp - j\,\beta_0 (z-a)$$

$$= \zeta(0) \exp - j\beta_0 (z-0) = \zeta(0) \exp(-j\,\beta_0\, z) \; .$$

Therefore, using (a8.2.1) , (a8.2.2) one may write

(a8.2.3)
$$\frac{d\,\mathscr{I}_1}{dz} = \int_{a=0}^{b=z} - j\,\beta_0\, f\{x,z\}\, dx + (\zeta(z))(1) - (\zeta(0) \exp(-j\,\beta_0\, z))(0)$$

$$= -j\,\beta_0\, \mathscr{I}_1 + \zeta(z) \; .$$

Similarly, from (8.2.24(b)):

$$\mathscr{I}_2 = \int_z^l \zeta(x) \exp - j\,\beta_0 (x-z)\, dx = \int_a^b f\{x,z\}\, dx \; , \quad \text{say}$$

The quantities relevant to (a8.2.1) in this case are:

(a8.2.4)
$$f\{x,z\} = \zeta(x) \exp - j\,\beta_0 (x-z) ; \quad a = z, \quad b = l \; ;$$

$$\frac{\partial a}{\partial z} = \frac{\partial z}{\partial z} = 1, \quad \frac{\partial b}{\partial z} = \frac{\partial l}{\partial z} = 0 ; \quad \frac{\partial f\{x,z\}}{\partial z} = j\,\beta_0\, f\{x,z\} ;$$

$$f\{b,z\} = \zeta(b) \exp - j\beta_0 (b-z) = \zeta(l) \exp - j\,\beta_0 (l-z) ,$$

$$f\{a,z\} = \zeta(a) \exp - j\,\beta_0 (a-z) = \zeta(z) \exp - j\,\beta_0 (z-z) = \zeta(z) \; .$$

Using (a8.2.1) and (a8.2.4) one may write

(a8.2.5) $$\frac{d\,\mathcal{I}_2}{dz} = \int_{a=0}^{b=z} j\,\beta_0\, f\{x,z\}\,dx \;+\; (\zeta(l)\exp - j\,\beta_0\,(l-z))(0) \;-\; \zeta(z)$$

$$= j\,\beta_0\,\mathcal{I}_2 - \zeta(z)\,.$$

In the text (a8.2.3) and (a8.2.5) have been used as (8.2.25(a)) and (8.2.25(b)), respectively.

Appendix a8.2.2 Relation between the circuit potential and field under quasi-static assumption

Let us assume that the electron beam is close enough to the circuit. With particular reference to a helical slow-wave structure we have already related the axial electric field intensity at the circuit with the circuit potential as (5.4.9) which is, however, valid for a general circuit and may be re-written here as

(a8.2.6) $$E = j\beta_0 V - j\omega A\,.$$

In (a8.2.6) we have dropped the subscripts used in (5.4.9) to suit the nomenclature used in this section and used the symbol β_0 instead of β for the cold propagation constant of the circuit. In fact, it is also worth recalling the more basic equation, from which (5.4.9) was obtained, as follows:

(a8.2.7) $$E = -\frac{\partial V}{\partial z} - \frac{\partial A}{\partial t}\,.$$

In the time -independent case, from (a8.2.6) and (a8.2.7), E simply becomes

(a8.2.8) $$E = -\frac{\partial V}{\partial z} = j\beta_0 V\,.$$

It may be now appreciated under what situation one may take the electrostatic relation (a8.2.8) as a fairly approximate relation instead of (a8.2.7) which is more accurate for a time-varying situation. For this purpose, let us examine (5.4.13) which may be written in the present context as

(a8.2.9) $$E = \frac{j(\beta_0^2 - \omega^2\,\mu\,\varepsilon)V}{\beta_0}\,.$$

In a slow-wave circuit $\beta_0^2 >> \omega^2\,\mu\,\varepsilon$ (corresponding to $\omega^2/\beta_0^2 << 1/(\mu\,\varepsilon)$, that is, to the square of the phase velocity of electromagnetic waves through the circuit being much smaller than the square of the velocity of electromagnetic waves in the unbounded medium in which the circuit is placed). So, one gets from (a8.2.9) the following approximate relation

$$E = j\,\beta_0 V$$

which remembering that the rf quantities in a lossless cold circuit vary as $\exp(-j\beta_0 z)$ may be put as

(a8.2.10) $$E = -\partial V/\partial z\,.$$

Appendix a8.2.3 TWT sub-assembly design for performance improvement

TWTs find a wide range of communication system applications —from electronic warfare (EW) to space exploration to the relaying of video signals [20]. This is because no other device can rival the TWTs unique combination of bandwidth and gain. Moreover, with the advent of new technology and innovative concepts, the power and efficiency of the TWT have also been enhanced considerably. In this appendix some of the concepts in the design of TWT sub-assemblies vis-à-vis the improvement of various performance characteristics have been briefly outlined.

The trend in electronic counter measure (ECM) has been towards wide-band systems. Competitively, an electronic counter counter measure system (ECCM) has to have a wide-band frequency agility. Special considerations are, therefore, required to broadband a TWT if it is to be used in such an EW system. Further, since the inception of television broadcasting satellite (TVBS) communication system the TWT has played a significant role in their development [21]. The TWT constitutes the major satellite power-consuming element, and, therefore, a TVBS communication TWT should be designed for high efficiency. The life and reliability is also a vital factor to be considered in the design of such a tube.

The sub-assembly which needs most attention in design for increasing the bandwidth of the TWT is the slow-wave structure. The helix has the widest bandwidth of all the slow-wave structures. The dispersion characteristics of a helix are required to be shaped by either anisotropically or inhomogeneously loading the helix in order to broadband a TWT, as discussed in section 5.7. One must ensure, however, that the method of dispersion-shaping does not cause a reduction in the value of the interaction impedance of the structure which in turn reduces the gain and efficiency of the device. Also it should be noted that, by loading [22]the phase velocity of the rf wave decreases which in turn reduces the pi-point frequency which is the potential backward -mode oscillation frequency of the structure [22]. The pi-point frequency is certainly a bandwidth limiting factor to be considered in the design of a wide-band TWT. It may be seen from (8.2.116) that one of the factors on which the device gain depends is the value of $N = \beta_e l / 2\pi = \omega l / 2\pi v_0$ *(see (8.2.115)),* l *being the interaction length,* ω *the operating angular frequency and* v_0 *the beam velocity. At lower frequencies (*ω*), therefore, the value of N would be less unless l is increased. However, if l is increased to increase the value of N, then in a multi-octave TWT, at higher frequencies, N would become so much large, and hence the gain of the device so much high, that the device becomes prone to oscillations[23],[24]. One may combat this problem [24] by using a two-section output helix— the section closer to the collector end being dispersion-shaped, say, by anisotropic loading and the remaining section subjected to no such loading to give a conventional positive dispersion. The loaded dispersion-shaped section contributes to the gain of the device over the entire band but the unloaded section becomes synchronous with the beam only at lower frequencies, going progressively farther from synchronism as the frequency increases. Therefore, the technique effectively reduces the interaction length at higher frequencies as required to prevent oscillations. Hence multi-octaves are achievable using helix TWTs if they are properly designed [24].*

Thus, the control of helix dispersion plays a significant role in wide-band TWTs. It may be noted that the electron bunching and consequent energy transfer from the beam to the electromagnetic wave depends greatly upon the relative value of the phase velocity (v_p*) of the slow-wave circuit with respect to the dc electron beam velocity (*v_0*), that is, on the velocity parameter b (*$=(v_0 - v_p)/(v_p C)$*; see (8.2.66)). The interaction impedance of the structure as discussed following (5.5.13), and hence the gain parameter C (see (8.2.55)), is less at higher frequencies. Therefore, it is possible to make b fairly independent of frequency using a negative-dispersion circuit which has a higher value of* v_p *at higher frequencies, so that, when C decreases (or increases) at higher (or lower) frequencies, then at the same time the quantity* $(v_0 - v_p)/v_p$ *also decreases (or increases) (thereby making b fairly constant with frequency). Therefore, the bunching process and hence most of the operating parameters can be made relatively constant by using a negative dispersion or a constant - b circuit[25] . It has been found that dispersion-shaping not only widens the bandwidth of the TWT but also can bring about performance improvement with respect* *to the reduction in harmonic content, AM/PM distortion, etc., matching of gain-frequency response between modes in dual-mode tubes, linearity, constancy of the difference between the small-signal and saturated gains, susceptibility to beam-voltage changes, etc. [25].*

The efficiency of a TWT which becomes a serious design concern for a TVBS system can be increased by two methods — by compensating for the lack of synchronisation between the electron beam and the rf wave when the former delivers its kinetic energy to the latter, and by collecting the electrons after they have delivered kinetic energy to electromagnetic waves, at somewhat a depressed potential[20].

In a helix-TWT the re-synchronisation can be done in the first of the above methods by tapering the pitch of the helix (obviously by decreasing its value)near the output[21], [26]; in a coupled-cavity TWT this can be done by providing a velocity-jump cavity near the output that provides a boost in the operating beam voltage[20]. The second method, that is, of depressed collection can be implemented independent of the slow-wave circuit design. The method provides the soft-landing of electrons at the collector and thus it partially recovers the kinetic energy of the beam. The depressed collector may be of single-staged, in its simplest form, or multi-staged type. In an idealised situation, had all the electrons after they have taken part in energy exchange in the interaction region possessed the same velocity, they could have been retarded to a halt at the collector by applying a suitable potential that is depressed with respect to the interaction structure. This could have thus given a hundred percent collector efficiency. In a practical situation, however, the electrons do emerge out from the interaction region with a velocity spread. This is because the electron emission from the cathode is non-uniform and also because the emitted electrons are subsequently subjected to magnetic focusing and rf modulation. Therefore, since there exist electrons of various energy classes to be collected, the collector need to be segmented into various stages each biased at an appropriate potential for the collection of each energy class of electrons in the spent-beam. This is implemented in a multi-stage depressed collector (MDC) system [27], [28] in which the slowest electrons are deflected away from the axis first and collected, at the highest-potential stage, nearest to the interaction structure, while the fastest electrons are collected by the collector segment remotest from the interaction structure, at the lowest potential stage, which could be depressed typically to as low a value as that of the cathode potential. Besides giving an improvement in the device efficiency, an MDC reduces the collector heating and associated cooling problems; causes a reduction in hardness of X-rays due to reduced velocities of electrons impinging on the collector; and also relaxes the required value of the gain parameter C, due to enhanced efficiency achieved, and hence simplifies the design of the device.

There are some more aspects in collector design that need to be given special attention. For instance, one must take into account the slow spent-beam electrons reversing their flow and constituting a back streaming of electrons. Such electrons are likely to be collected by a higher potential than what is just necessary for the purpose and this offsets the power saving purpose of MDC; they also produce an excessive thermal load on the circuit, degrade rf performance and cause instabilities. The problem is augmented also by the fast electrons in the spent beam which impinge on the collector to generate secondary electrons. It has been found that the collector efficiency can be increased by increasing the number of collector stages, the improvement in incremental efficiency, however diminishing as the number of stages increases, and actually not much improvement is found beyond three to four numbers of collector stages. Further, one has to make a trade-off since the increase of number of stages invites more complexity in the design of power supply. Also, one must note that the optimum design of an MDC is a function of the input drive which controls the distribution of various collector currents. The choice for material for the collector is also an important consideration. For instance, from various considerations one may typically choose pyrolytic graphite which is characterised by its high emissivity, excellent thermal conductivity in a plane perpendicular to the deposition direction, light weight, excellent mechanical strength and a very low secondary emission coefficient. These collectors may be assembled simply by force-fitting without any brazing resulting in no risk of any outgassing or long-term degradation problems. The collector may also be projected beyond the satellite's outer surface for cooling.

The average rf output power of a TWT can be controlled by thermal management of the interaction structure. If thermal dissipation is not proper, the interaction structure may get heated to its melting point, and moreover if PPM focusing is used, the magnetic material may reach its Curie temperature, thus taking the tube close to destruction [20]. From thermal considerations a helix-derived structure such as a ring-and-bar has a more average power capability than a simple helix as discussed in appendix a5.0.1, similarly, a coupled cavity structure has a more average power capability than a ring-and-bar structure. With special care in the thermal management of the helical structure, however, one may increase the average power capability of a helix TWT. One must use good thermally-conducting dielectric materials such as isotropic boron nitride, anisotropic pyrolytic boron nitride (APBN), beryllia, alumina, diamond etc. for this purpose. More importantly, however, one must take care in reducing the thermal resistance at the interface between the helix and the supports and that between the supports and the envelope which become enhanced in vacuum. From thermal considerations a brazed-helix assembly considerably increases the average power capability of a TWT as compared to a force-fitted helix assembly. In this technology [29], [30], the helix is made out of a copper tape and is supported by beryllia rods in a copper envelope. The beryllia is chosen as the support materials because it has a good thermal conductivity and moreover it can be easily brazed to copper. The hot rf losses of the structure are also reduced due to a lower helix temperature and to a higher electrical conductivity and lower temperature coefficient of resistivity of the helix material (copper, instead of usual tungsten etc.). Further, another average-power limiting factor to be considered in the TWT design is the grid heating in a system operating with a high pulse repetition frequency[20]. For an increased average power of the device in such a system, therefore, the grid-intercept current should be reduced to a minimum by employing a non-intercepting gridded gun (NIG) in the device (see section 6.4).

The peak rf output power of a TWT is determined by the dc beam power ($V_0 I_0$) which is proportional to $V_0^{5/2}$, since the dc beam current I_0 is proportional to $V_0^{3/2}$ for a constant beam perveance, where V_0 is the dc beam voltage (see for instance, (6.1.16) and (6.3.36)-(6.3.38)). With respect to dc beam voltage then a centipede structure has more peak power capability than a coupled-cavity structure while the latter has more such capability then a helix (appendix a5.0.1). The limit of the peak power of a helix TWT is set by the dc beam voltage above which the circumference of the helix becomes an appreciable fraction of the free-space wavelength. This is actually set by the value of γa taken in the design. For instance, we have found (see the discussion preceding section 8.2.4A) that, with the increase of γa, the gain of the TWT first increases, till it reaches a peak value, and then decreases, where γ is the radial propagation constant of the structure which is approximately equal to its axial propagation constant β under the slow-wave assumption (as discussed following (5.3.8)); a is the helix radius. One may then note that $\gamma a \cong \beta a = \omega a/v_p \cong \omega a/v_0 = 2\pi f a/(2\,|\eta|V_0)^{1/2} = (2\pi a/\lambda)\, c/(2\,|\eta|V_0)^{1/2}$. This will give the value of $2\pi a/\lambda \sim 0.3$, for $V_0 = 10$ kV and taking $\gamma a = 1.5$, typically, at which value the gain becomes maximum. For such an appreciable ratio of the circumference-to-free space wavelength, $2\pi a/\lambda$, the first backward space-harmonic mode of the helix gets accenuated, and backward-wave oscillations (BWO) at pi-point frequencies which correspond to axial phase shift per turn, βp, close to π becomes a major source of instability in the device. In fact, close to the pi-point frequency both the forward and the backward modes are coupled to the electron beam and the backward mode gain is enhanced by the forward-wave gain by coupling via the beam. Also, the forward-wave gain is relatively reduced and any attempt to increase it by increasing the interaction length will make the device prone to oscillations. Thus BWO at high beam voltage operation can cause in-band power holes, drive induced oscillations and prevent the achievement of a high peak power [24].

Some of the methods that are commonly used for suppressing the above type of BWO are: the introduction of frequency sensitive attenuation[31], [32] into the circuit, the use of pitch taper for fast space-charge wave coupling[24] and the use of a conical helix[33].

In the first of the above methods [31], [32], frequency-sensitive attenuation can be achieved by printing a meander line whose total un-wrapped length (L) is equal to half the wavelength of BWO frequency so that the line resonates at this frequency. The BWO frequency is chosen above the upper edge of the band. Ideally, the meander line provides no loss in the operating band, and even if it does, it does not substantially affect the in-band tube performance (see section 8.2.4A).

The gold meander line with a titanium adhesive thin layer can be printed on one or more helix rods, say of boron nitride(substrate) between helix turns, using the standard microwave integrated circuit (MIC) lift-off technique[32]. It is found that, for a given pattern width (W), the resonant frequency almost linearly decreases with the ratio of the unwrapped length-to-pattern width (L/W). For a given value of L/W, the resonant frequency is lowered by an increase in W or in substrate thickness. The resonant loss is placed only on one or two helix support rods thus making the structure geometry asymmetric that causes a stop band centred at the pi-point frequency. However, since this stop band increases the loss at this frequency such asymmetric arrangement appears to be beneficial[32]. The resonator lowers the phase velocity and to compensate it the helix pitch has to be increased which, however, results into the lowering of the interaction impedance at the upper band edge. If both resonant and distributed losses are to be provided, one can do so either by placing them on opposite faces of the rod, say, rectangular or by placing resonant loss on one or two of the helix support rods and distributed loss on the remaining rod(s) [32].

In the second method of suppression of BWO based on fast space-charge wave coupling [24], an output helix is constructed with a pitch step. In the smaller pitch section closer to the output end, the forward circuit wave is synchronous with the slow space-charge wave providing the forward wave gain. However, the backward circuit wave is also nearly synchronous with the slow space-charge wave and the length of this section is enough to induce backward wave oscillations. This backward wave then moves into the increase-pitch section of increased phase velocity where it becomes synchronous with the fast space-charge wave. The method is much simpler than the first method of resonant attenuation. The drawbacks of the method are that the pitch step required is not the same as that is required for efficiency enhancement, as discussed before, and that some coupling between the forward circuit wave and the fast space-charge wave degrades the tube performance. In the third method of the suppression of BWO [33], a conical helix is used whose diameter varies along its length such that its pitch-to-diameter ratio remains constant. This ensures that the phase velocity of the forward wave remains constant along the length though the phase velocity of the space harmonics, including the first negative harmonic, changes along the length. As a result, useful interaction takes place only with the forward wave and hardly with other space harmonics.

Apart from BWO, another band-width limiting factor in a helix TWT is the band-edge oscillation [32], [33], [37] which is caused by the asymmetry in the positioning of the dielectric support rods around the helix or by different cross sections of these rods. As a result of this asymmetry, a stop band appears in the ω–β *diagram of the structure at or near which the circuit impedance can become quite high resulting into a low starting current for oscillations. It is found that a distributed loss removes the stop band due to asymmetry and thus stabilises the device against such band-edge oscillations.*

Besides BWO and band-edge oscillations in a TWT, the oscillations are also caused by imperfect matches at the input and output ends, and in order to suppress these oscillations the lossy section called the attenuator is used which becomes an integral part of the slow-wave structure of the device, as discussed in section 8.2.4A. The required loss is provided in the form of a lossy coating on dielectric helix support rods, in the case of a helix TWT and a lossy ceramic button on the spacer-cavity wall, in the case of a coupled-cavity structure.

Pyrolytic cracking of hydrocarbons [38] out of various methods now becomes a standard technique of depositing the attenuating layer on dielectric support rods. In this technique the cracking of a suitable material such as benzene, heptane, etc. takes place in a low pressure and inert, say, argon atmosphere, in the vicinity of the work piece which is uniformly heated all around, say, by resistance heating. The thickness of the coating is gradually tapered to zero at the ends of the attenuating region to obtain a good impedance match over a wide range of frequencies.

The attenuator not only does the function of absorbing reflected energy but also absorbs the useful forward energy. (The quantitative effect of the attenuator on the TWT gain has been estimated in section 8.2.4A). Therefore, the portion of the slow-wave structure dedicated to attenuation does not contribute to the gain, and so the interaction length has to be increased to campensate for this portion. The higher the gain, the more attenuator sections will be required; a rule of thumb is one section per about 20 dB gain.

For high-power, high-efficiency tube the most common technique of suppressing the backward wave is to severe the slow-wave structure by splitting it into, say, two parts roughly in the middle. The input section is terminated in approximately the characteristic impedance at the point where the structure is severed. The output section of the structure is also terminated similarly at the point of severe. Thus the severe loads of the two sections face each other. For high-power tubes, the power absorbed by severe loads must be carried out of the tube proper through matched transmission lines and subsequently dissipated in dummy loads that are external to the tube structure.

Two more concepts of providing attenuation to stabilise the device are coupled-helix[34] and ferrite [35], [36] attenuators. In a coupled-helix attenuator which can, however, be used with a non-metalic tube envelope, a helix concentric with and contrawound with respect to the main helix is wound over a lossy sleeve outside the envelope. In order to minimise reflections the coupler helix has to be terminated at both of its ends in matched loads, and the parameters such as the resistivity of the sleeve material, separation between the main and the coupler helices, pitch angles of the two helices, length of the coupler helix, etc. have to be optimised[34]. The ferrite attenuator which is essentially a non-reciprocal component [35], [36], may be in the form of a cylinder around the helix and may form a part of the tube envelope itself. The dc magnetic field required to magnetise the ferrite may be either axial or circumferential. The axial magnetising field may be obtained from the main focusing field used for beam confinement. The circumferential magnetising field may be produced by an axial current which can be removed after magnitisation. If a coupled-cavity slow-wave structure is used, the slots of the cavity wall may be ferrite- loaded to achieve the non-reciprocal attenuation.

We have already discussed some of the design considerations with respect to the TWT performance, namely, bandwidth, efficiency and average as well as peak power of a TWT. Next important considerations on which TWT designers may like to give attention pertain to the operating frequency of the device which with proper care can be increased to the millimeter-wave region. The transverse dimension of the interaction structure need to be reduced considerably in the millimeter-wave region. This demands precision and tight tolerances for the parts which become extremely small whether the device is a helix or a coupled- cavity TWT. The diameter of the electron beam to be positioned in the slow-wave structure has also to be reduced; this requires a large magnetic field for the confinement of the beam.

The problem of focusing becomes further aggravated by the presence of the thermal electrons in the beam formed by the usually low-perveance electron gun of the millimeter-wave tube. In order to achieve low noise and high reliability, a large permanent magnet focusing structure may be used, and to compensate for the small cathode diameter one has to operate the cathode at a high current density loading. This poses a reliability problem for a millimeter-wave tube[39]. Also, in a millimeter-wave tube, since the parts become miniaturised, heat dissipation becomes more severe, and thermal management requires more attention, and more so in the case of a helix TWT for which it becomes more difficult to implement the brazed helix technology with such miniaturised parts. In order to improve heat dissipation, one may use II-A diamond as the helix support material. For the same purpose, in another method, beryllia is plasma-sprayed over the outer surface of the helix and the whole pressure-fitted in the envelope [33], [40]. Circuits such as the ridge-loaded helical waveguides [41] have been tried out for their large size, good heat dissipation capabilities, tightly controlled tolerances on the circuit pitch at a reasonable cost, and for their ability to mitigate the effect of the backward space harmonic.

Coupling of rf power in and out of the slow-wave structure with minimum reflections at their ends poses a big challenge in the TWT design. The couplers have to be commensurate with the slow-wave structure with respect to its characteristics such as bandwidth. In general, the couplers will have a different characteristic impedance than the slow-wave structure and therefore the transitions between them have to be carefully designed for a minimum VSWR . Thus, for instance, if a waveguide is coupled to a helix, the shorter dimension of the former may be suitably tapered in the transition region; for a wide bandwidth a ridged waveguide may be used. Also, with a metal vacuum envelope, rf windows have to be provided in the tube body through which the input/output rf connections have to be made.

One method of coupling rf power in and out of a helix with a non-metallic envelope is to use a helical coupler[42]-[44] which uses a helix wound in opposite sense with respect to the main helix (that is, contra-wound) which is placed outside the tube. The parameters which control the performance of such a coupler are the ratio of the coupler-to-main helix radii, the ratio of the cotangents of the pitch angles of the two helices, etc. In such a system, when proper boundary conditions are satisfied, two forward waves on the individual helix produce a spatial beat phenomenon involving the longitudinal and transverse electric field modes, mid-way between the helices, in which the maxima of the stationary interference pattern of one helix coincide with the minima of the similar pattern on the other[42], [43]. Thus energy is built up periodically on one helix at the expense of energy propagating along the other. An optimum coupling length, which becomes equal to half the beat wavelength, exists for the coaxial system for complete power transfer from one helix to another, and vice-versa. For broadband coupling the parameter of the coupler should be optimised to get a constant optimum length for a wide range of frequencies [42], [44]. In a conventional coupler, called the synchronous coupler, the phase propagation constants on the two helices are the same. For low-noise TWT's the beam voltage is usually low, and correspondingly, the main helix inside the tube has a large number of turns. For such a structure the field outside the glass tube is weak, and there is a loose coupling between the two helices. Synchronous couplers fail to exhibit broadband characteristics for such a structure specially when the ratio of the coupler helix-to-main helix radii exceeds 2. In such a situation, the non-synchronous coupler[42] in which the two helices are proportioned to have different phase propagation constants, proves to be a better alternative than the synchronous coupler. For a non-synchronous coupler the spatial beat propagation constant is almost equal to the difference between the individual phase propagation constants of the waves on the isolated helices. In yet another concepts of coupled helices the relative propagation constants of the two helices are tapered by adjusting the pitch/diameter of the helices to obtain what is known as a tapered helical coupler which exhibits wide-band characteristics[42].

There is yet another type of broadband coupler in which an innovative concept of elliptical cavity coupling is utilised[45]. At one of the focii of a right-elliptic cylindrical cavity is positioned a conical antenna with its base fixed on one of the flat walls of the cavity of elliptical cross section. The tip of the cone is connected to the central conductor of the input/output coaxial connector through the remaining flat wall. At the other focus is connected a wire antenna which is connected to the helix at one end and is terminated in a quarter-wave cylinder at the other. Coupling of power takes place as a result of transfer of power between the wire and the conical antenna via reflection from the curved side wall of the elliptical cavity. The transition from the coaxial line to the cone is designed for best matching; the cone angle determines the radiation resistance. In order to keep the side wall area low and hence reduce losses the eccentricity of the ellipse has to be of a low value. However, it should not be so low that there is no space left for the placement of the conical antenna and the helix[45].

Finally, it may be mentioned that a very simple method of implementing wide-band coupling will be to connect the helix directly to the central conductor of a coaxial connector, however, taking due care in impedance matching. This requires the tapering of one or more dimensions of the structure such as the pitch of the helix which,in fact, in a typical arrangement, becomes an extension of a tapered-pitch slot on a metal cylinder which in turn is the extension of the central conductor of the coaxial cable; in this case the diameter of the metal tube envelope can also be simultaneously tapered away[17]. Such an arrangement, however, utilises much of the interaction length of the tube. In another arrangement the diameter of the central conductor of the coaxial cable may be tapered, that is, in fact reduced as it approaches the helix. In this method, however, a relatively long unsupported section of the central conductor of the coaxial line has to withstand mechanical shock and vibration conditions and, similarly the unsupported portion of the helix has to be provided proper cooling [17]. The design must, therefore, take care with respect to these considerations while implementing this method of tapered-dimension coupling [33]. Also in an alternative method in which the tapering of the dimension such as the helix pitch or envelope diameter is not needed, a simple multi-step quarter wavelength transformer is used for impedance matching [46].

The impedance matching at the input/output couplers is of utmost importance from the standpoint of the variation of the gain and phase of the TWT with frequency. If the slow-wave structure is split into a number of sections, the mismatches at the ends of each section also contributes to the gain and phase ripples. Besides reducing the mismatches, one may also reduce the gain of each section to reduce the gain and phase ripples; and finally also by increasing the cold distributed loss of the circuit, which increases the attenuation of the regenerative feedback and backward waves, the gain and phase ripples can be minimised[20], [47]. Typically, for instance, this can be implemented in a coupled cavity TWT by providing tapered loss with the help of small cavities, adjacent to the circuit cavity, which are loaded with lossy ceramic buttons the protrusion or re-entrancy of which may be adjusted to control in-band loss which is tapered for a low mismatch[20]. It may be mentioned that the phase ripple is correlated with the gain ripple, the mechanism of their production being a feedback loop of the amplified signal; and thus reducing the gain ripple by minimising mismatches etc. tends to minimise the phase ripple as well [47].

The phase of the rf output signal with respect to the input depends on the factors responsible for changing the beam velocity which in turn depend upon the cathode/anode voltage, grid drive, filament voltage etc. and is measured by the phase pushing factor expressed in terms of degree per volt of the relevant electrode voltage. The power supply designer must take into consideration the phase pushing factor. The amplitude modulation (AM) -to-phase modulation (PM) conversion is a measure of the change in phase angle between the input and output signals as the input drive varies, in terms of the degrees per dB of the input power-level change at a specified output power level which in turn is measured in dB with respect to the saturated power level. The AM/PM conversion becomes maximum at a point below saturation and is small either at low power levels or at a level above the saturated power level. The AM/PM conversion is frequency dependent and becomes higher at the high frequency end of the tube's pass band[20]. The phase linearity of the device is a measure of the deviation from the linear phase-versus-frequency characteristics of a TWT which also manifests itself in a constant time delay that it takes for a signal to pass through the device, being equal to the derivative of the phase delay-versus-frequency characteristics [5]. The design of the slow-wave structure, therefore, needs attention with regard to phase compensation that will improve the phase linearity of the tube. The phase compensation in a helix TWT can be accomplished by a helix section of increased pitch which has the function of partially compensating for the phase shift caused by the deceleration of the beam as it surrenders energy to the rf field. Under the increased pitch section the slow-wave structure is actually coupled to the fast space-charge wave of the beam, and thus the beam revives its energy from the rf wave (section 8.1). Clearly, the increased pitch section in the output section causes efficiency degradation. Hence simultaneous improvements in the efficiency and linearity can be made by using a double taper helix. For the improvement of gain and phase characteristics, the dynamic velocity taper that needs a precise pitch profiling has also been successfully implemented [48].

We have already discussed earlier in this section the importance of improving the efficiency of a TWT for TVBS/space applications; we have also briefly outlined the efficiency improvement methods. In designing a space-qualified TWT we have to take into consideration other factors as well. Thus in order to reduce the weight of the space TWT a light-weight PPM focusing structure should be used (chapter 7). Further, the device should be rugged and be able to withstand vibration and shock conditions, say, during the launching of the satellite. Of course it should be designed for a long life. The key to getting a long life of the device is to operate the cathode at an optimum temperature; lightly load the cathode; use a potted heater for an optimum heater-cathode heat exchange, provide for the expansion and contraction of the heater wire during turn-on and turn-off cycles, use adequate emitter coating material if an oxide cathode is used; provide optimum porosity of the emitter if a dispenser cathode is used; subject the tube to low residual pressure tube processing; make adequate cooling for the tube; use an electronic power conditioner in which the power supply constitutes an integral part of the tube that provides protective measures against alarming extremes such as high circuit intercept-current, voltage surge, load mismatch, system arc and inadequate cooling, and so on.

Further, it may be required to modify the tube design as per any specific requirement. For instance, a single tube may be designed such that it can deliver CW power output with a pulse-up capability in a TWT called the dual-mode TWT [51]. In a typical dual perveance non-intercepting gridded (NIG) electron gun which can be used for this purpose, the control grid is split into an inner circular section and an outer annular section flushing with each other. For high current, high power-mode operation, both the sections are kept at a positive potential with respect to the cathode, while for low current, low power-mode operation, the potential of the outer section is reduced to zero or to some negative value. Thus the solid electron beams of two different cross-sectional areas are derived from the same cathode in the two modes such that the beam currents in the two modes become different. However, in both the modes the beam current densities remain unaltered as the beam current and the beam cross-sectional areas change by the same proportion. The operation of a TWT in more than two modes can also be implemented using an NIG tetrode electron gun [20].

Finally, it is worth reviewing methods to improve the TWT performance by external means. For example, a number of pulsed TWT's can be combined using 3dB magic tees and phase shifters to attain higher pulse power over a wide bandwidth. Such an operation also relaxes the thermal management which is shared by all the tubes. Also, the system can be operated in the condition of a graceful output power degradation and with higher intrinsic reliability because there is more redundancy in the system and each tube operates at a low output power level. Similarly, the problem of harmonic generation, in particular, the second harmonic component, which is the most dominant component in a TWT [51], and which baffles the broadbanding of a TWT particularly at the lower end of the operating frequency band, can be reduced and thus the electronic efficiency of the device increased by injecting, along with the signal to be amplified, another signal at the second harmonic frequency of the former. The relative amplitude and phase of the harmonically related signals can be adjusted to reduce the second harmonic output power and enhance the efficiency at the fundamental [20], [53], [54].

In this appendix various basic design considerations relevant to the performance improvement of a TWT have been briefly outlined. The discussions have, however, been made in a qualitative manner without going into the details of theoretical analysis and quantitative estimates and in no way do the present discussions claim to include the up-to-date, and often proprietory, state-of-the-art.

8.3 Double-Stream and Beam-Plasma Amplifiers

There exists a family of growing-wave electron beam devices which are based on the principle of space-charge wave amplification in which the slow-wave structure does not constitute a building block as it does in a TWT. One class amongst them uses structures that do not support electromagnetic waves at the operating frequencies of interest [19]. The resistance-wall amplifier in which an electron beam is enclosed in a resistive envelope belongs to this class. In one out of two more examples of the same class an electron beam of constant diameter is enclosed in a conducting wall of varying diameter, while in the other, a conducting wall of constant diameter encloses an electron beam of varying diameter. In another class of this family of devices an electron beam continues to constitute a major component and also the slow-wave structure continues to remain absent, though the role of the latter is effectively played by a medium rather than a physical structure, which could be an additional electron beam, as it is in a double-stream amplifier (DSA), or a stationary plasma, as it is in a beam-plasma amplifier (BPA). The principle of space-charge wave amplification in these two devices is briefly outlined in this section.

8.3.1 Double-Stream Amplifier

A double-stream amplifier consists of two intermingled streams or beams of electrons with slightly different dc velocities[55]-[60] (Fig. 8.3.1). In this device an electromagnetic circuit is used only to impress signal on the electron flow at the input end or to couple out amplified signal present in the flow at the output end [16]. DSA's enjoy some advantages over TWT's. In a TWT it is desirable to have an electron flow very close to the slow-wave structure in order to have a satisfactory interaction. The confinement of the electron flow close to the structure is, however, difficult without the interception of electrons by the structure, particularly, at very high values of the operating frequency in which case the transverse dimensions of the structure become very small; this entails both loss of efficiency and heating of the circuit. In a DSA the electrons interact with one another and

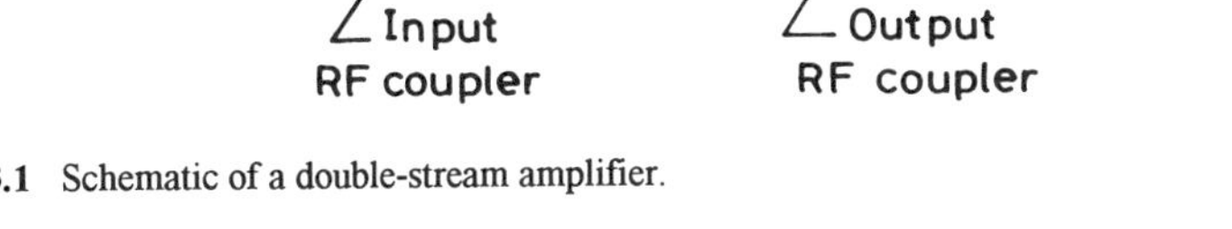

Fig. 8.3.1 Schematic of a double-stream amplifier.

they need not be close to a structure for interaction; also in such device a precise relation between the beam velocity and the phase velocity of the slow-wave structure need not be maintained as it is required in a TWT. The analysis of the DSA to be presented in this section will show that one may increase the operating frequency of the device by increasing the charge densities of the beams and/or by reducing the difference between their dc velocities. A dc magnetic field is used as in a TWT to constrain the electrons to move in the longitudinal direction. The electron gun of the device has to ensure the formation of interpenetrating electron beams which may be two coaxial annular beams. In an interesting arrangement some of the electrons from the faster stream are used for heating the cathode which emits the slower stream [59].

The interaction in a DSA ensures such a phasing of space-charge forces on one beam that it increases the bunching in the other beam. The space-charge force on an electron of an isolated beam tends to restore the electron to its equilibrium position when it is disturbed, the restoring force changing its direction each time the electron overshoots the equilibrium position. However, with the two beams in a DSA, the second beam always exerts a force on the electrons in the beam that is unidirectional moving them away from the equilibrium position [1]. This results in space-charge wave amplification which can be quantitatively understood only by studying the dispersion relation of the DSA which can be derived by a simple mathematical analysis to follow in this section.

A simple approximate condition of synchronism between the two beams may be obtained if it is assumed that the slow space-charge wave of the faster beam is coupled to the slower beam. The condition may be written by equating the phase velocity of the slow space-charge wave of the faster beam (see expression (*8.1.21*) which has to be taken with the lower sign in its right hand side) with the dc beam velocity of the slower beam as follows:

$$\frac{\omega}{\omega+\omega_{p1}} v_{01} = v_{02} , \qquad (8.3.1)$$

where the subscripts 1 and 2 refer to the faster and the slower beams, respectively. In terms of the volume charge density ρ_{01} of the faster beam, ω_{p1} is given by (see (8.1.15) and (8.1.17)):

$$\omega_{p1} = \left(\frac{\eta\rho_{01}}{\varepsilon_0}\right)^{1/2} = \left(\frac{|\eta|\,|\rho_{01}|}{\varepsilon_0}\right)^{1/2} . \qquad (8.3.2)$$

The corresponding expression for the slower beam is:

$$\omega_{p2} = \left(\frac{\eta\rho_{02}}{\varepsilon_0}\right)^{1/2} = \left(\frac{|\eta|\,|\rho_{02}|}{\varepsilon_0}\right)^{1/2} . \qquad (8.3.3)$$

Let us now define the following parameters which will prove to be useful in the analysis of a DSA:

$$\omega_p = (\omega_{p1}\,\omega_{p2})^{1/2} , \qquad (a)$$

$$r = \omega_{p1}/\omega_{p2} , \qquad (b)$$

$$v_{01} = v_0 + \Delta v , \qquad (c)$$

$$v_{02} = v_0 - \Delta v , \qquad (d)$$

$$v_0 = (v_{01} + v_{02})/2 , \qquad (e)$$

and

$$\Delta v = (v_{01} - v_{02})/2 , \qquad (f) \qquad (8.3.4)$$

the last two following, respectively, from the addition and the subtraction of (8.3.4(c)) and (8.3.4(d)).

If we take the volume charge densities of the two beams to be approximately equal, that is, $\rho_{01} \approx \rho_{02}$, then from (8.3.2)-(8.3.4) it follows that one may take $\omega_{p1} \approx \omega_{p2} \approx \omega_p$ enabling one to write from (8.3.1)

$$\frac{\omega}{\omega + \omega_p} v_{01} = v_{02}$$

which may be written with the help of (8.3.4(c)) and (8.3.4(d)) as

$$\frac{\omega + \omega_p}{\omega} = \frac{v_0 + \Delta v}{v_0 - \Delta v},$$

whence one easily gets

$$\frac{\omega_p}{\omega} = \frac{2\Delta v}{v_0 - \Delta v} \cong \frac{2\Delta v}{v_0},$$

which gives

$$X \cong \frac{1}{2}, \qquad (8.3.5)$$

where X is an inhomogeneity parameter of the DSA defined as follows:

$$X = \frac{\omega}{\omega_p} \frac{\Delta v}{v_0} . \qquad (8.3.6)$$

The optimum value of X corresponding to the maximum growth rate of signal in the device, however, will be found later to be different from what is approximately predicted by (8.3.5). The optimum value of X as well as the range of amplification, measured on the scale of X, it will be shown, depends upon the value of $r\ (= \omega_{p1}/\omega_{p2})$.

The analysis of the DSA is developed here under the following simplifying assumptions: the electron flow is essentially one-dimensional and the effective beam boundaries are neglected; the rf electric field in the direction of the stream is constant over the cross section of a beam; the effects of potential depression caused by space-charge as well as the collisional and thermal velocity effects are negligible; and the difference between the dc velocities of the two beams is much less than the dc velocities of these beams.

The electrons in either of the beams are subjected to the space-charge electric field intensities of both the beams thus enabling one to write (see (8.1.8)):

$$\mathcal{D}_1 v_1 = \eta(E_1 + E_2) \qquad (8.3.7)$$

and

$$\mathcal{D}_2 v_2 = \eta(E_1 + E_2), \qquad (8.3.8)$$

where $\mathcal{D}_1$ and $\mathcal{D}_2$ are the operators given by (see (8.1.6)):

$$\mathcal{D}_1 = \partial/\partial t + v_{01}\, \partial/\partial z \tag{8.3.9}$$

and

$$\mathcal{D}_2 = \partial/\partial t + v_{02}\, \partial/\partial z\,. \tag{8.3.10}$$

E_1 and E_2 are the space-charge electric field intensities, and v_1 and v_2 are the rf beam velocities, of the faster and the slower beams, respectively.

Now, with the help of one of the Maxwell's namely, (4.4.4(b)), one may write for the rf magnetic field intensity $\mathbf{H}_k$ over the cross section of the k*th* beam:

$$(\nabla \times \mathbf{H}_k)_z = J_k + \varepsilon_0 \frac{\partial E_k}{\partial t}\,, \tag{8.3.11}$$

where k = 1 and 2 refer to the faster and the slower beams, respectively. Taking the beams to be of large cross sections so that one may consider the rf fields to remain constant over them, one may write, with the help of (3.2.18), taking $\partial/\partial r = \partial/\partial\theta = 0$:

$$(\nabla \times H_k)_z = 0 \quad (k = 1, 2)\,. \tag{8.3.12}$$

Substituting (8.3.12) in (8.3.11), one may then separately write for the two beams, taking k = 1 and 2:

$$J_1 + \varepsilon_0 \frac{\partial E_1}{\partial t} = 0 \tag{8.3.13}$$

and

$$J_2 + \varepsilon_0 \frac{\partial E_2}{\partial t} = 0. \tag{8.3.14}$$

For the two beams, also one may write with the help of (8.1.2):

$$J_1 = \rho_{01} v_1 + v_{01} \rho_1 \tag{8.3.15}$$

and

$$J_2 = \rho_{02} v_2 + v_{02} \rho_2\,, \tag{8.3.16}$$

and with the help of (8.1.11):

$$\frac{\partial E_1}{\partial z} = \frac{\rho_1}{\varepsilon_0} \tag{8.3.17}$$

and

$$\frac{\partial E_2}{\partial z} = \frac{\rho_2}{\varepsilon_0}\,, \tag{8.3.18}$$

where ρ_1 and ρ_2 are the rf volume charge densities of the faster and the slower beams, respectively. Substituting J_1 and ρ_1 from (8.3.13) and (8.3.17), respectively, into (8.3.15), and similarly J_2 and ρ_2 from (8.3.14) and (8.3.18) respectively, into (8.3.16), we obtain:

$$-\varepsilon_0 \frac{\partial E_1}{\partial t} = \rho_{01} v_1 + v_{01}\varepsilon_0 \frac{\partial E_1}{\partial z} \tag{8.3.19}$$

and

$$-\varepsilon_0 \frac{\partial E_2}{\partial t} = \rho_{02} v_2 + v_{02}\varepsilon_0 \frac{\partial E_2}{\partial z} \; . \tag{8.3.20}$$

With the help of (8.3.9) and (8.3.10), one may easily express (8.3.19) and (8.3.20), respectively as:

$$v_1 = -\left(\frac{\varepsilon_0}{\rho_{01}}\right)\mathcal{D}_1 E_1 \tag{8.3.21}$$

and

$$v_2 = -\left(\frac{\varepsilon_0}{\rho_{02}}\right)\mathcal{D}_2 E_2 \; . \tag{8.3.22}$$

Substituting (8.3.21) in (8.3.7), and (8.3.22) in (8.3.8), we get

$$-\left(\frac{\varepsilon_0}{\rho_{01}}\right)\mathcal{D}_1^2 E_1 = \eta E_1 + \eta E_2 \; ,$$

and

$$-\left(\frac{\varepsilon_0}{\rho_{02}}\right)\mathcal{D}_2^2 E_2 = \eta E_1 + \eta E_2 ,$$

which, by making use of the relations (8.3.2) and (8.3.3) and after a little arrangement, respectively, become

$$(\mathcal{D}_1^2 + \omega_{p_1}^2) E_1 = -\,\omega_{p_1}^2 E_1 \tag{8.3.23}$$

and

$$(\mathcal{D}_2^2 + \omega_{p_2}^2) E_2 = -\,\omega_{p_2}^2 E_2 \; . \tag{8.3.24}$$

Taking the product of (8.3.23) and (8.3.24) we get

$$(\mathcal{D}_1^2 + \omega_{p_1}^2)(\mathcal{D}_2^2 + \omega_{p_2}^2) = \omega_{p_1}^2 \omega_{p_2}^2$$

which after simplification may be expressed as:

$$\frac{\omega_{p_1}^2}{\mathscr{D}_1^2} + \frac{\omega_{p_2}^2}{\mathscr{D}_2^2} = -1 . \tag{8.3.25}$$

As in the preceding sections let us assume the rf dependence as exp ($j\omega t - \Gamma z$). Further, considerable simplification results, as we will see while deriving the dispersion relation from (8.3.25), if we choose to express Γ as

$$\Gamma = \gamma + j\frac{\omega}{v_0} , \tag{8.3.26}$$

where v_0, the average of the dc velocities of the two beams, is given by (8.3.4(e)). Then in view of the assumed rf dependence, and invoking (8.3.26), one may express E_1 and E_2 as

$$\begin{aligned} E_{1,2} &= E_{1,2}\{z\} \exp j\omega t = E_{1,2}\{0\} \exp -\Gamma z \ \exp j\omega t \\ &= E_{1,2}\{0\} \exp -\gamma z \exp -j\omega z/v_0 \exp j\omega t . \end{aligned} \tag{8.3.27}$$

Now, operating $\mathscr{D}_1$ and $\mathscr{D}_2$ on E_1 and E_2 (see (8.3.9) and (8.3.10)), respectively, one obtains

$$\mathscr{D}_1 E_1 = \frac{\partial E_1}{\partial t} + v_{01}\frac{\partial E_1}{\partial z} \tag{8.3.28}$$

and

$$\mathscr{D}_2 E_2 = \frac{\partial E_2}{\partial t} + v_{02}\frac{\partial E_2}{\partial z} . \tag{8.3.29}$$

Substituting E_1 and E_2 from (8.3.27) and v_{01} and v_{02} from (8.3.4(c)) and (8.3.4(d)), respectively, into the right hand sides of (8.3.28) and (8.3.29), one gets

$$\mathscr{D}_1 E_1 = j\omega E_1 + (v_0 + \Delta v)(-\gamma E_1 - j\frac{\omega}{v_0} E_1) = \left(-\gamma(v_0 + \Delta v) - j\omega\frac{\Delta v}{v_0}\right) E_1 \tag{8.3.30}$$

and $$\mathscr{D}_2 E_2 = j\omega E_2 + (v_0 - \Delta v)(-\gamma E_2 - j\frac{\omega}{v_0} E_2) = \left(-\gamma(v_0 - \Delta v) + j\omega\frac{\Delta v}{v_0}\right) E_2 . \tag{8.3.31}$$

Now, since we have assumed the difference between the dc velocities of the two beams to be small compared to their dc velocities (see (8.3.4(c)-(f)), we may ignore Δv compared to v_0 in (8.3.30) and (8.3.31) so that we may now write the latter, respectively, as

$$\mathcal{D}_1 E_1 = \left(-\gamma v_0 - j\omega \frac{\Delta v}{v_0} \right) E_1$$

and

$$\mathcal{D}_2 E_2 = \left(-\gamma v_0 + j\omega \frac{\Delta v}{v_0} \right) E_2$$

whence $\mathcal{D}_1$ and $\mathcal{D}_2$ may be identified as

$$\mathcal{D}_1 = -\left(\gamma v_0 + j\omega \frac{\Delta v}{v_0} \right) \tag{8.3.32}$$

and

$$\mathcal{D}_2 = -\left(\gamma v_0 - j\omega \frac{\Delta v}{v_0} \right) . \tag{8.3.33}$$

Substituting (8.3.32) and (8.3.33) in (8.3.25) we then obtain

$$\frac{\omega_{p_1}^2}{\left(\gamma v_0 + j\omega \dfrac{\Delta v}{v_0} \right)^2} + \frac{\omega_{p_2}^2}{\left(\gamma v_0 - j\omega \dfrac{\Delta v}{v_0} \right)^2} = -1 . \tag{8.3.34}$$

Dividing the numerator and the denominator of each term of the left hand side of (8.3.34) by ω_p^2, we get

$$\frac{\omega_{p_1}^2 / \omega_p^2}{\left(\dfrac{\gamma v_0}{\omega_p} + j \dfrac{\omega}{\omega_p} \dfrac{\Delta v}{v_0} \right)^2} + \frac{\omega_{p_2}^2 / \omega_p^2}{\left(\dfrac{\gamma v_0}{\omega_p} - j \dfrac{\omega}{\omega_p} \dfrac{\Delta v}{v_0} \right)^2} = -1 . \tag{8.3.35}$$

Making use of (8.3.4(a) and (b)), one may then express (8.3.35) as

$$\frac{r}{(X - jY)^2} + \frac{r^{-1}}{(X + jY)^2} = 1 , \tag{8.3.36}$$

where

$$Y = \frac{\gamma v_0}{\omega_p} . \tag{8.3.37}$$

Thus we have deduced the dispersion relation (8.3.36) for the DSA in terms of a normalised propagation constant given by (8.3.37) and an inhomogeneity parameter given by (8.3.6). The dispersion relation (8.3.36) is a fourth-degree equation corresponding to four forward waves set up in the device in which the two beams are coupled, each beam, when isolated, supporting two forward space-charge waves.

One may easily study the nature of waves for the simple case in which the dc beam charge densities ρ_{01} and ρ_{02} of the beams are considered as equal giving $\omega_{p1}=\omega_{p2}$ (see (8.3.2) and (8.3.3)), that is r = 1 (see (8.3.4(b)). In this case the dispersion relation (8.3.3(b)) becomes

$$\frac{1}{(X-jY)^2}+\frac{1}{(X+jY)^2}=1\,. \tag{8.3.38}$$

Simplifying (8.3.38) to obtain a quadratic equation in Y^2 and then solving it one gets

$$Y = \pm[-(X^2+1) \pm (4X^2+1)^{1/2}]^{1/2} \qquad (r=1)\,. \tag{8.3.39}$$

An examination of (8.3.39) shows that, for $X^2 + 1< (4X^2 + 1)^{1/2}$, that is, for $X< \sqrt{2}$, two out of the four solutions of Y, and hence of γ (see (8.3.37)), are purely imaginary, being equal and opposite, and the remaining two solutions are purely real, also being equal and opposite. Consequently, this gives (see the rf dependence given by (8.3.27) two non-growing, unattenuated forward waves, one exponetially growing forward wave and one exponentially- attenuated forward wave in the device. The phase velocity of one of the unattenuated non-growing waves is less then v_0 while that of the other is greater than v_0, and the phase velocity of each of the growing and the attenuated waves is equal to v_0. This may be easily appreciated by taking, typically, X = 0.8 which gives, from (8.3.39), $Y = j\,1.88, -j\,1.88, -0.50$ and 0.50. Corresponding to these values of Y, one may write the rf dependence $\exp(j\omega t-\Gamma z)$ of the four waves as $\exp j[\omega t-\{(\omega+1.88\omega_p)/v_0\}z]$, $\exp j[\omega t-\{(\omega-1.88\omega_p)/v_0\}z]$, $\exp(0.5\omega_p/v_0)z\ \exp j[\omega t-(\omega/v_0)z]$ and $\exp(-0.5\omega_p/v_0)z$ $\exp j[\omega t-(\omega/v_0)z]$. Clearly, then the first two are of neither growing nor attenuated type, the third is exponentially growing with z, and the fourth is exponentially decaying with z. Their axial propagation constants, are $(\omega+1.88\omega_p)/v_0$, $(\omega-1.88\omega_p)/v_0$, ω/v_0 and ω/v_0, respectively, and the corresponding phase velocities (= ω / phase propagation constants) are $[\omega/(\omega+1.88\omega_p)]v_0$, $[\omega/(\omega-1.88\omega_p)]v_0$, v_0 and v_0, respectively. At large distances from the input end, the exponentially growing wave will eventually dominate over the remaining three waves and will be responsible for an amplified signal at the output end. For r = 1, the amplification range of the device is upto $X = \sqrt{2}$. Beyond this value of X, all the four solutions become purely imaginary corresponding to all the four waves neither growing nor being attenuated. For instance, for X = 1.6, one gets from (8.3.38), $Y = j\,2.63, -j\,2.63, j\,0.46$ and $-j\,0.46$. It is also of interest to maximise Y given by (8.3.39), for r = 1, in the amplification range ($0 < X< \sqrt{2}$) to find the optimum value of X as $1/\sqrt{2}$ which is different from the value approximately predicted by (8.3.5).

For $r \neq 1$, one has however, to solve (8.3.36) numerically. Taking r = 2, for X = 0.8, one obtains $Y = j\,1.66, -j\,2.25, j\,0.30-0.48$ and $j\,0.30+0.48$. It may be shown that, in this case, the range of amplification and the optimum inhomogeneity of beams, both expressed in terms of X, become higher than those for r = 1. It may be noted that once an optimum value of X(= $(\omega/\omega_p)(\Delta v/v_0)$; see (8.3.6)) has been chosen for maximum amplification, the frequency of operation will be more sensitive to the value of $\Delta v/v_0$ than to the value of r (= ω_{p1}/ω_{p2}). For the device to operate at high values of frequencies, it is necessary to increase ω_p (= $(\omega_{p1}\,\omega_{p2})^{1/2}$; see (8.3.4(a))), that is, increase the volume charge densities of the beams (that is, increase the beam currents) and to reduce $\Delta v/v_0$, the relative dc velocity difference between the beams.

Among the advantages of the DSA is its significant gain and wide bandwidth. Also, since the device does not use an interaction structure the transverse dimensions of which need to be reduced in proportion to wavelengths, the device has a potential for millimeter-wave amplification. On the other hand the cathode system of a DSA becomes more complicated than that of a TWT. For high values of the operating frequency, the beam current densities have to be large. However, for such high current densities it is likely that the potential depression of the beam becomes so large that the current to the collector is stopped. Although high operating frequencies can be achieved by reducing the relative dc velocity difference between the beams, yet the scope of this method is limited by the velocity spread of electrons, since if the latter limit is reached, the two beams become one and the same [15]. Also, it is worth noting that since in this device the slower beam gains energy at the expense of the faster one, the maximum available output energy is at most this energy difference and this limits the efficiency of the DSA. Finally, it may be remembered that the performance of a DSA must be evaluated in the presence of an output stage which typically is a helix. Thus the device may be looked upon as a TWT with enhanced bunching due to two-stream interaction. So, in practice it can not supercede the performance of a TWT [60].

Example 8.3.1 Obtain the dispersion relation of the fast and slow space-charge waves on a single electron beam as a special case of the dispersion relation of a DSA.

One may write from (8.3.25), putting $\omega_{p2} = 0$,

$$\frac{\omega_{p1}^2}{\mathscr{D}_1^2} = -1 , \qquad (8.3.40)$$

if we take the beam of dc velocity v_{02} as absent while the beam dc velocity v_{01} as present. Taking the rf dependence exp $(j\,\omega t - \Gamma z)$, one gets from (8.3.9)

$$\mathscr{D}_1 = j\,\omega - v_{01}\,\Gamma$$

which substituted in (8.3.40) gives

$$(j\omega - v_{01}\Gamma)^2 = -\,\omega_{p_1}^2 \qquad (8.3.41)$$

whence one gets

$$j\,\omega - v_{01}\Gamma = \pm\, j\,\omega_{p1} \,. \tag{8.3.42}$$

One may then write from (8.3.42), dropping the subscript 1,

$$\Gamma = \frac{j(\omega \mp \omega_p)}{v_0} \,. \tag{8.3.43}$$

Interpreting

$$\Gamma = j\,\beta \tag{8.3.44}$$

one obtains from (8.3.43):

$$\beta = \frac{\omega \mp \omega_p}{v_0} \,. \tag{8.3.45}$$

Thus we find (8.3.45) as identical with (8.1.20) which had been formed earlier for the space-charge waves on an isolated electron beam.

8.3.2 Beam-Plasma Amplifier

A beam-plasma amplifier consists of an electron beam penetrating through a plasma (Fig. 8.3.2). The growing wave results from the excitation of oscillations of the plasma electrons by the beam and the interaction of the oscillating electrons back on the beam, in a typical arrangement, one may modulate the beam by a signal before it interacts with the plasma, say, by a helix element and then demodulate the beam, as it leaves the interacting plasma medium, by another helix element, the modulation and demodulation helices being in turn coupled, respectively, to the input and output waveguide couplers [61].

At frequencies larger than the plasma frequency of the plasma medium, the density modulation in the beam-plasma coupled-system is of the same nature as that on a beam only. The density modulation grows at frequencies less than the plasma frequency of the plasma. (This will also be clear by examining the dispersion relation of the BPA to follow in this section). The mechanism of interaction can be qualitatively understood by considering the oscillations of the plasma electrons under the influence of space-charge forces arising because of bunches on the beam. At frequencies less then the plasma frequency of the plasma, the plasma electrons oscillate in phase with the space-charge forces in the beam. This results in the accumulation of the plasma electrons between the bunches of the beam. The space-charge forces due to the plasma electrons consequently oppose those due to the beam. The total space-charge field in the presence of the plasma, therefore, aids the oscillations of the beam electrons leading to eventual space-charge wave amplification. Thus the plasma here takes the role of one of the beams of the DSA. In fact, by putting one of the beam velocities as zero in the dispersion relation of the DSA, one may find the dispersion relation of the BPA as follows.

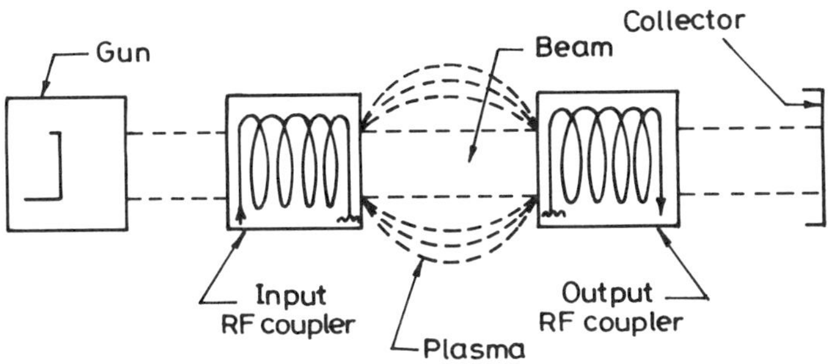

Fig. 8.3.2 Schematic of a beam-plasma amplifier.

Putting $v_{01} = v_{0b}$, the dc beam velocity of the BPA, and $v_{02} = 0$, corresponding to a stationary plasma, one may write respectively from (8.3.9) and (8.3.10):

$$\mathcal{D}_1 = \partial/\partial t + v_{0b}\,\partial/\partial z \tag{8.3.46}$$

and

$$\mathcal{D}_2 = \partial/\partial t \ . \tag{8.3.47}$$

Substituting (8.3.46), (8.3.47), $\omega_{p1} = \omega_{pb}$ and $\omega_{p2} = \omega_{pp}$, the last two being the plasma frequencies of the beam and the plasma, respectively, in (8.3.25) and remembering the rf dependence exp $(j\omega t - \Gamma z)$, one obtains

$$\frac{\omega_{pb}^2}{(j\omega - v_0\Gamma)^2} - \frac{\omega_{pp}^2}{\omega^2} = -1 \ , \tag{8.3.48}$$

which may alternatively be expressed as follows [62], in terms of the axial propagation constant β, which is related to Γ as (8.3.44):

$$\frac{\omega_{pb}^2}{(\omega - v_0\beta)^2} + \frac{\omega_{pp}^2}{\omega^2} = 1 \ . \tag{8.3.49}$$

From (8.3.48) it is easy to obtain

$$\Gamma = j\frac{\omega}{v_0} \mp \frac{\omega}{v_0}\frac{\omega_{pb}}{(\omega_{pp}^2 - \omega^2)^{1/2}} \ . \tag{8.3.50}$$

One may also choose to express the dispersion relation of a BPA in terms of the normalised parameters as we have done in the case of a DSA (see (8.3.39)). For this purpose let us define

$$\Omega = \frac{\omega}{\omega_p} ,$$

$$\omega_p = (\omega_{pb} \, \omega_{pp})^{1/2} ,$$

$$r = \omega_{pb} / \omega_{pp} ,$$

and

$$Y = \gamma v_0 / \omega_p , \tag{8.3.51}$$

where γ is related to Γ as (8.3.26), where v_0 has to be interpreted as the dc velocity of the beam (instead of the average of the dc beam velocities for the DSA). Using (8.3.51) one may then obtain from (8.3.50), invoking (8.3.26):

$$Y = \mp \frac{r\Omega}{(1 - r\Omega^2)^{1/2}} . \tag{8.3.52}$$

If follows from (8.3.50) that as long as the signal frequency is less than the plasma frequency of the plasma, $\omega < \omega_{pp}$, one of the values of Γ will contain a negative real part corresponding to the spatial growth of the signal; this corresponds to a negative real value of Y given by (8.3.52) as long as the normalised frequency $\Omega < r^{-1/2}$. Thus the amplification range of the device is $0 < \omega < \omega_{pp}$ (or $0 < \Omega < r^{-1/2}$).

The simple dispersion relation (8.3.50) or (8.3.52) predicts infinite growth rate in the BPA at $\omega = \omega_{pp}$ (or $\Omega = r^{-1/2}$). This is because in this deduction the thermal velocity of plasma electrons has been ignored compared to the dc beam velocity. However, if the finite temperature of the plasma were taken into consideration [15], the dispersion relation would be modified by the inclusion of the thermal velocity of plasma in the denominator of the second term of (8.3.48) or (8.3.49) such that the predicted growth rate now would become less and it would not become infinite at $\omega = \omega_{pp}$ or any other values of ω. Also, the growth rate does not become maximum at $\omega = \omega_{pp}$; rather it does so at a value $\omega < \omega_{pp}$ and also the amplification range extends beyond $\omega = \omega_{pp}$. It may be mentioned that the inclusion of the effects of collisions of plasma electrons [61] and those of finite geometry [62] of the system would cause a reduction of the predicted growth rate in a BPA.

The BPA would give millimeter wave amplification if a high density plasma were created to give a large value of ω_{pp} [62]. Also in such a device wide-band electronic tuning of frequency is available because of the possibility of varying the plasma electrodynamic characteristics electrically. Since the plasma is a highly reactive medium that is physically tenuous, rf heating is practically absent in the medium. Strong interaction is possible because of the intimate mixing of the electron beam with the interaction region, that is, plasma. No synchronism is required in such an interaction of reactive nature. This gives continuous bunching during the interaction resulting in high efficiency, and hence the device can be compact. The possibility of passing an electron stream of high density through a plasma as well as of isolating the plasma from the metal walls leads to an essential power increase of the BPA. In a BPA the scattering of the electron beam by the plasma and the gas background may be reduced by operating the system in a low pressure. Also, in such a device a beam diameter larger than ordinarily employed in a microwave tube may be used, because the gain is relatively independent of beam density. Interaction takes place effectively over the whole cross section of the beam that intimately mixes with the plasma and there need be no concern about the radial decay of fields as experienced with metal circuits. The use of large diameter beams relaxes the requirement of high current densities from the cathode and of large convergence in electron guns. However, in such a device special considerations should be made to overcome the problem of the bombardment of the cathode by ions, such as the employment of ion discharge holes and hollow cathodes. BPA's like the most-plasma devices have a large noise level. For the reduction of noise level, a negative-glow mode discharge which occurs at a rather high pressure in the tube and which has a low electron temperature may be used; the positive column could be removed by bringing the anode as close as possible to the cathode. Also, attention should be given to suppress plasma oscillations and fluctuations in the form of low-frequency modulation of the output signal such as by making a proper choice of geometries and other working parameters like pressure and magnetic field [64], [65].

8.4 Growing-Wave Interaction in a Gyro-TWT

As the operating frequency of a TWT (see section *8.2*) is increased it becomes necessary to proportionately reduce the transverse dimension of the slow-wave structure of the device (as explained preceding section *8.2.4A*). In fact, it can be shown that the relevant dimension would then become much less than the wavelength in free space (see problem 5.4). In the millimeter/sub-millimeter wave range, therefore, the slow-wave structure of the TWT would become rather tiny. Consequently, it becomes quite difficult, expensive and time-consuming to machine such a tiny structure which, moreover, being configured to have a periodic geometry (see appendix *a5.0.1*), does not have a simple form such as a hollow metal pipe which is typically used as a waveguiding structure at microwave frequencies (section *8.4.1*). The problem becomes all the more delicate if the structure, say, in the form of a helix has to be supported by a system of dielectric rods/bars in an envelope and becomes further complicated if the helix has to be anisotropically or inhomogeneously loaded for wide-band characteristics (section *5.8*; appendix *a8.2.3*). On the other hand the structure takes a simple form, say, a hollow metal cylinder of circular boundary — a cylindrical waveguide [4], [18](section *8.4.1*) — in a device called the gyro-TWT [66]-[68] — a growing-wave electron beam device that has been taken up for analysis in this section. It can be shown that the transverse dimension of the waveguide interaction structure of a gyro-TWT has also to be reduced in the millimeter/sub-millimeter wave band, however, not to as low a value as that of the slow-wave structure of a conventional TWT (see Example 8.4.1). The simplicity of the geometry as well as a larger transverse dimension of the interaction structure makes its fabrication easy in a gyro-TWT. Moreover, the structure being larger and more massive can handle more power and the problem of its thermal management also becomes easier.

In the case of a conventional TWT the radius of the electron beam to be placed inside the slow-wave structure has also to be reduced along with the transverse dimensions of the structure in the millimeter/sub-millimeter wave range. This will reduce the beam cross-sectional area and hence the dc beam current. Consequently, the dc beam power and hence the extractable rf output power of the device would also reduce. On the other hand, if any attempt is made to increase the dc beam current by increasing the cathode temperature, then the cathode life and hence the life of the device would also deteriorate. On the contrary, a larger allowable beam cross-sectional area enhances the extractable rf output power of a gyro-TWT.

Further, in a conventional TWT, for an effective interaction the electron beam has to be placed as close as possible to the slow-wave structure without the beam being intercepted by the structure. For a tiny structure the problem of confining the beam inside, including the alignment of the focusing structure, becomes quite difficult. Moreover, if the structure is a tiny helix, it is likely to be destroyed by beam interception. However, in the case of a gyro-TWT the electron beam need not be placed close to the interaction structure; in fact, for a better interaction, the beam has to be placed away from the structure — in a region where the former can experience the maximum rf electric field of the latter. This makes the problem of focusing and related beam interception by the structure much relaxed in a gyro-TWT. It may be mentioned here that the beam interception by the structure is also one of the factors responsible for the decrease of the device efficiency — a parameter which also degrades due to increased rf losses of the structure in the millimeter/sub-millimeter wave band. Interestingly, however, in some propagating modes of the interaction structure of a gyro-TWT, the attenuation of waves caused by rf losses decreases with increasing frequencies (for instance, in the TE_{01}-mode of a cylindrical waveguide [3], [4]).

In a gyro-TWT (Fig. 8.4.1) the electron beam is made periodic, instead of the interaction structure that supports an electromagnetic wave of phase velocity $>c$ ($= (\mu_0\varepsilon_0)^{-1/2}$), unlike in a conventional TWT, in which the beam is made non-periodic and the interaction structure periodic, the latter supporting an electromagnetic wave of phase velocity $< c$. Typically, the gyro-TWT consists of a hollow relativistic beam of mono-energetic electrons, inside a cylindrical waveguide, following helical trajectories about axes parallel to (or coinciding with) the axis of the waveguide, under the influence of a strong dc magnetic field. The interaction mechanism in such a device is based on the resonance between the gyrating electrons and the transverse rf electric field of the waveguide. The bunching of electrons in this device is essentially azimuthal and is caused by their relativistic change of mass with velocity. If the bunched electrons and the rf electric field are properly phase-synchronised,(section *8.4.2*), the net kinetic energy of the bunch will decrease resulting into an energy transfer form the electrons to the wave. Thus the gyro-TWT like a conventional TWT continues to be a kinetic energy conversion device, though both the bunching and the conversion of kinetic energy are azimuthal in a gyro-TWT, unlike in a TWT where they are axial.

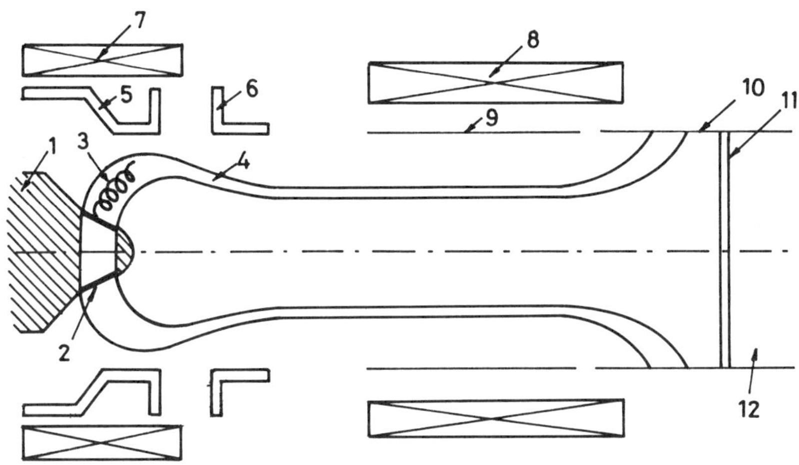

Fig. 8.4.1 Schematic of a gyro-TWT [1: Focusing electrode, 2:Cathode — emitting strip, 3: Helical electron trajectory, 4: Hallow electron beam, 5: First anode, 6: Second anode, 7: Gun solenoid, 8: Main solenoid, 9: Waveguide interaction structure, 10: Collector, 11: output window, 12: Output].

The helical electrons trajectories for a gyro-TWT can be formed by a magnetron injection gun (MIG) (so-named because of its appearance resembling a well-known microwave tube called the magnetron [67]). In an MIG (Fig. 8.4.1) the electrons are emitted by an annular-strip on the lateral face of a convex cathode[67], [69]. The beam is drawn off the cathode at an angle with the tube axis by the potential on the first (gun) anode into crossed electric and magnetic fields imparting a small (azimuthal) rotation of electrons. The required dc magnetic field for this purpose may be provided by the gun-solenoid. The second anode supplies a large amount of axial energy to the beam. Finally the beam is subjected to a slowly increasing magnetic field or adiabatic compression region thereby converting a large portion of its axial energy into rotational energy. It may be noted that the cathode used in the gun is operated temperature-limited rather than space-charge limited to minimise the velocity spread in the beam. The electrons in helical trajectories formed by the MIG take part in the cyclotron-resonance interaction with the transverse rf electric field in the waveguide interaction region. The required dc magnetic field for this purpose may be provided by the main solenoid. The beam after it takes part in interaction enters a magnetic decompression region and diverges off the axis to settle on the surface of the collector surrounding the beam. The rf output is coupled out from a region further down the axis of the tube beyond an output window which is vacuum-sealed to the tube. The output coupling arrangement has also to provide for the necessary waveguide-mode conversion since the mode of the waveguide interaction structure could, in general, be different form that of the output waveguiding system.

8.4.1 Cold Dispersion Relation of a Beam-Free Cylindrical Waveguide

Let us take a cylindrical waveguide in the transverse-electric (TE) mode ($E_Z = 0$; $H_Z \neq 0$) for which the axial component of magnetic field intensity is given by (see problem 5.14).

$$H_Z = C\, J_m\{k_t r\} \exp j(\omega t - \beta z - m\theta)\,, \tag{8.4.1}$$

where

$$k_t = (k_0{}^2 - \beta^2)^{1/2} \tag{8.4.2}$$

is the radial or transverse propagation constant and β is the axial propagation constant,

$$k_0\ (= \omega\, (\mu_0\varepsilon_0)^{1/2}) = \omega / c \tag{8.4.3}$$

being the free-space propagation constant. J_m is the mth order ordinary Bessel function of the first kind. m is the angular-harmonic number. C is a field constant which may be identified from (8.4.1) as the axial magnetic field intensity H_{Z0}, say, at the axis of the guide (r = 0) for the non-azimuthally-varying mode (m = 0) (noting that $J_0\{0\} = 1$ [4],[18]). Putting $C = H_{Z0}$ in (8.4.1) one may write

$$H_Z = H_{Z0}\, J_m\, \{k_t\, r\} \exp j(\omega t - \beta z - m\theta)\,. \tag{8.4.4}$$

The azimuthal and the radial components of electric field intensity may similarly be written as (problem 5.14):

$$E_\theta = \left(\frac{j\omega\mu_0}{k_t}\right) H_{z0}\, J'_m\{k_t\, r\} \exp j\,(\omega t - \beta z - m\theta) \tag{8.4.5}$$

and

$$E_r = -\left(\frac{\omega\mu_0}{k_t}\right)\left(\frac{m}{k_t r}\right) H_{z0}\, J_m\{k_t r\} \exp j\,(\omega t - \beta z - m\theta)\,. \tag{8.4.6}$$

At the wall of the waveguide (r = a) considered to be perfectly conducting one may put the electromagnetic boundary condition (see Table 4.7.1)

$$E_\theta\{a\} = 0\,, \tag{8.4.7}$$

where a is the wall radius of the guide. Substituting (8.4.5) into (8.4.7) one gets

$$J'_m\{k_t\, a\} = 0\,. \tag{8.4.8}$$

Equation (8.4.8) has an infinite number of roots[4], [18]. Let us put the n*th* root of (8.4.8) as

$$k_t\, a = X_{mn}\,, \tag{8.4.9}$$

where the subscripts m and n refer to the TE_{mn}-mode of the waveguide, the first (m) and the second (n) subscripts describing the circumferencial and the radial variations, respectively (see Example 8.4.1(b), for the radial variation). It is of interest to note some of the values of X_{mn} as[18]: $X_{01} = 3.832$, $X_{02} = 7.016$, $X_{03} = 10.174$, for TE_{01}, TE_{02} and TE_{03} modes, respectively; $X_{11} = 1.841$, $X_{12} = 5.331$ and $X_{13} = 8.536$, for TE_{11}, TE_{12} and TE_{13} modes, respectively; and $X_{21} = 3.054$, $X_{22} = 6.706$ and $X_{23} = 9.970$, for TE_{21}, TE_{22} and TE_{23} modes, respectively. Obtaining k_t^2 from both (8.4.2) and (8.4.9), and equating them, one gets

$$k_t^2 = \frac{X_{mn}^2}{a^2} = k_0^2 - \beta_{mn}^2\,, \tag{8.4.10}$$

where the subscripts m and n have also been used with the waveguide phase propagation constant β to indicate that the latter refers to the TE_{mn} mode guide. Putting (8.4.3) into (8.4.10),

$$\omega^2 - \beta_{mn}^2 c^2 - \frac{X_{mn}^2 c^2}{a^2} = 0\,. \tag{8.4.11}$$

The ω–β dispersion plot predicted by (8.4.11) is a hyperbola (Fig. 8.4.2) with a cut-off angular frequency $\omega_{cut,mn}$, obtained by putting $\beta_{mn} = 0$ in (8.4.11) as:

$$\omega_{cut,mn} = \frac{X_{mn} c}{a}\,. \tag{8.4.12}$$

Substituting (8.4.12) in (8.4.11), the dispersion relation becomes

$$\omega^2 - \beta_{mn}^2 c^2 - \omega_{cut,mn}^2 = 0\,, \tag{8.4.13}$$

which in terms of a cut-off wave number

$$k_{c,mn} = \frac{\omega_{cut,mn}}{c} \qquad (8.4.14)$$

may, alternatively, be expressed as

$$\omega^2 - \beta^2_{mn} c^2 - k^2_{c,mn} c^2 = 0 \; . \qquad (8.4.15)$$

The slope of the straight line joining the origin of the $\omega - \beta$ plot and any point on the waveguide hyperbola (Fig. 8.4.2) is found to be greater than that of the c-line (that is, the straight line whose slope is equal to c, the velocity of light in free space). This indicates that the phase velocity v_p ($= \omega/\beta$) of the wave supported by the waveguide is greater than c. Thus the waveguide — the interaction structure of a gyro-TWT — is a fast-wave structure. Hence, the gyro-TWT is classified as a fast-wave device unlike a conventional TWT which is essentially a slow-wave device.

The transverse propagation constant k_t which occurs in the field expressions (see (8.4.4)-(8.4.6)) may be expressed as

$$k_t = k_{c,mn} = \frac{X_{mn}}{a} \; . \qquad (8.4.16)$$

Thus substituting (8.4.16) in (8.4.5) one may see that the amplitude of the azimuthal component of electric field intensity E_θ varies as $J'_m\{X_{mn}\, r/a\}$.

<u>*Example 8.4.1*</u> (a) Find the radius of the cylindrical waveguide (interaction structure) of a gyro-TWT operated near the cut-off of the TE_{01}-mode, in terms of the free-space wavelength, and compare it with the radius of the helix of problem 5.4 of a conventional TWT. In what proportion will the waveguide radius increase from its value in the TE_{01}-mode if the device is operated in the TE_{02}-mode? (b) Find the location between the axis and the wall of a cylindrical waveguide operated in the TE_{02}-mode where the amplitude of electric field intensity becomes zero and reverses its sign (a parameter that becomes of relevance for the operation of the gyro-TWT at the second cyclotron-harmonic resonance (section *8.4.2*)).

(a) With the help of (8.4.14) and (8.4.16)

$$\omega_{cut} = \frac{X_{mn} c}{a} \qquad (8.4.17)$$

whence the wall radius in terms of the circular cut-off frequency f_{cut} ($= \omega_{cut}/2\pi$) $\cong f$, the operating frequency, is obtained as:

$$a = \frac{X_{mn} c}{2\pi f_{cut}} \cong \frac{X_{mn} c}{2\pi f} = \frac{X_{mn} \lambda_{free-space}}{2\pi} . \tag{8.4.18}$$

For TE_{01}-mode one has $X_{01} = 3.832$ (see the value of X_{mn} mentioned following (8.4.9)) giving

$$a = \frac{3.832}{2\pi} \lambda_{free-space} = 0.61\ \lambda_{free-space} \cong (3/5)\ \lambda_{free-space} .$$

Thus the waveguide radius in the TE_{01}-mode is found to be about 24 times as great as the helix radius ($\cong (1/40)\ \lambda_{free-space}$) of problem 5.4. Clearly, the waveguide radius will increase in the TE_{02}-mode (see (8.4.18)) by a factor $X_{02}/X_{01} = 7.016/3.832 = 1.83$ from that in the TE_{01} -mode.

(b) For TE_{02} -mode, $X_{mn} = X_{02} = 7.016$, the amplitude of E_θ will vary as $J'_0\{7.016\ r/a\}$ ($= -J_1\{7.016\ r/a\}$) and hence exhibit zeroes (see the values of roots of (8.4.8) mentioned following (8.4.9)) for the argument of the Bessel function 7.016 r/a = 7.016 and 3.832 which will correspond to r = a (waveguide wall) and r = 0.546a (between the axis and the wall of the guide) ($J'_0\{7.016\}$ ($= -J_1\{7.016\}$) $= J'_0\{3.832\}$ ($= -J_1\{3.832\}$) $= 0$). Also the nature of function $J_1\{x\}$ shows [4], [18] that its value reverse its sign at x = 3.832 where it also exhibits a zero. Therefore, in the present problem, it is easy to appreciate, there will be a reversal of the sign of the azimuthal electric field amplitude at r = 0.546a, that is, around the point where a zero is located inside the guide.

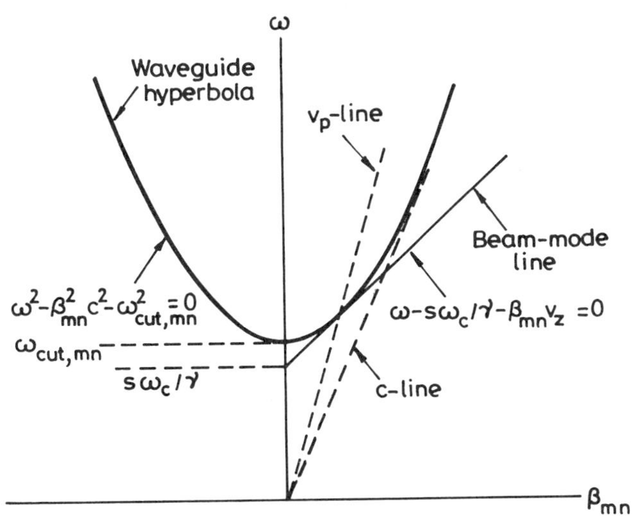

Fig. 8.4.2 The waveguide and beam-mode dispersion characteristics in a gyro-TWT.

8.4.2 Cyclotron Resonance, Relativistic Phase Bunching and Beam-Mode Dispersion Relation

The beam mode interaction in a gyro-TWT is essentially a cyclotron resonance phenomenon. An electron, when shot with a dc velocity in a region of uniform dc magnetic flux density B_0 in a direction perpendicular to that of the latter, gyrates in a circular orbit of radius r_L, called the Larmor radius, with an angular velocity ω_c (see (7.3.10)) given by (in this chapter we have used the symbol m_e for the electronics mass; the symbol m has been used for the angular harmonic number in rf dependence associated with propagation through a waveguide (see section *8.4.1*)):

$$\omega_c = \frac{|e| B_0}{m_e} \qquad \text{(a)}$$

and

$$\omega_c r_L = v_t , \qquad \text{(b)} \qquad (8.4.19)$$

where v_t is the orbital velocity, transverse to the direction of magnetic flux density. The relativistic interpretations of (8.4.19(a)) and (8.4.19(b)) have been discussed later (see equations (8.4.24) and (8.4.28)).

8.4.2A Fundamental and harmonic cyclotron resonance

In a gyro-TWT an electron follows a helical trajectory while interacting with the transverse electric field of a travelling electromagnetic wave supported by a waveguide (section *8.4.1*). However, considerable simplification, without loss of generality, is achieved in understanding the interaction mechanism if we assume the electron velocity in the axial direction to be zero [70]. In this simplified picture (Fig. 8.4.3), cyclotron resonance interaction between an electron gyrating with an angular frequency ω_c and a transverse rf electric field of angular frequency ω takes place if the following resonance relations are satisfied:

$$\omega = \omega_c \quad \text{(fundamental cyclotron resonance)} \qquad \text{(a)}$$

$$\omega = s\,\omega_c \quad (s\textit{th} \text{ harmonic cyclotron resonance}) . \qquad \text{(b)} \qquad (8.4.20)$$

The fundamental resonance condition $\omega = \omega_c$ corresponds to $T = T_c$, where T ($= 2\pi/\omega$) is the time period of one cycle of rf variation and T_c ($= 2\pi/\omega_c$) is the time of one gyration of an electron. Under this condition, if an electron is decelerated by the rf electric field at $t = 0$, and thus its kinetic energy is converted into the rf energy, it will continue to be decelerated at $t = T/2 = T_c/2$, that is, when the electron makes half an orbital gyration and the rf field reverses its sign of cycle at the same time (Fig. 8.4.3(a)). It can also be seen that the electron will continue to be decelerated at $t = T = T_c$ also, that is, when the electron orbits one complete gyration while the rf field completes one cycle of variation at the same time.

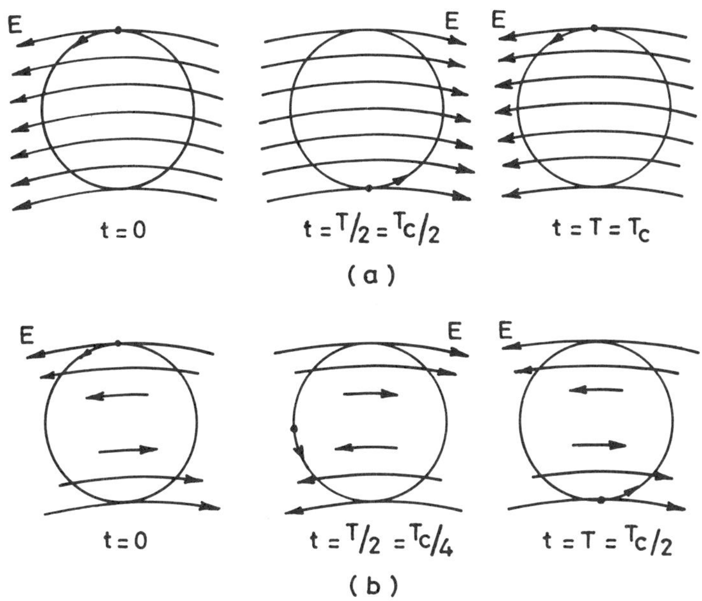

Fig. 8.4.3 Gyration of an electron over one period (T) of cycle of rf electric field (E) variation under fundamental **(a)** and second harmonic **(b)** resonance conditions.

The phenomenon of harmonic resonance can be easily understood with particular reference to the second harmonic ($s = 2$; $\omega = 2\omega_c$; see equation (8.4.20(b))) by considering an electron to orbit in an rf electric field that spatially reverses at the centre of the orbit (Fig. 8.4.3(b)). The condition $\omega = 2\omega_c$ corresponds to $T_c = 2T$ (since $\omega = 2\pi/T$ and $\omega_C = 2\pi/T_c$) giving $T/2 = T_c/4$ and $T = T_c/2$. It can be seen that under this condition if an electron is decelerated at $t = 0$, it is unaffected by the rf electric field that acts perpendicular to its motion at $t = T/2 = T_c/4$, that is, when the electron makes quarter of one orbital gyration and the rf field reverses its sign at the same time (Fig. 8.4.3(b)). The electron again finds itself in the decelerating rf electric field at $t = T = T_c/2$, that is, when the electron makes half an orbital gyration and the rf field completes a cycle of variation at the same time. It may be seen with the help of (8.4.19) and (8.4.20) that in cyclotron-harmonic operation the required magnetic flux density is reduced by a factor of s to:

$$B_0 = \frac{m_e \omega_C}{|e|} = \left(\frac{1}{s}\right)\left(\frac{m_e \omega}{|e|}\right) \tag{8.4.21}$$

from its value ($B_0 = m\omega/|e|$) at fundamental resonance (s = 1). Substituting the values of $|e|$ and m in (8.4.21) the dc magnetic flux density B_0 is estimated as (s = 1)

$$B_0 = 3.57 \times 10^{-11} \text{ Tesla per hertz of cyclotron frequency } (f_C = \omega_C/2\pi). \tag{8.4.22}$$

8.4.2B Relativistic effects and phase bunching

Under *relativistic* effects the mass of an electron with velocity v will increase from its rest mass by a factor of γ given by

$$\gamma = \frac{1}{(1 - v^2/c^2)^{1/2}}, \tag{8.4.23}$$

its value increasing from $\gamma = 1$ for v = 0 to $r \to \infty$ for $v \to c$. This gives the relativistic angular cyclotron frequency ω'_c with the help of (8.4.19) as:

$$\omega'_c = \frac{|e| B_0}{m_e} = \frac{|e| B_0}{\gamma m_{eo}} = \frac{\omega_c}{\gamma} \tag{8.4.24}$$

with

$$m_e = \gamma m_{eo} \tag{8.4.25}$$

interpreted as the relativistic electronic mass; and

$$\omega_c = \frac{|e| B_0}{m_{eo}} \tag{8.4.26}$$

interpreted as the non-relativistic (rest-mass) angular cyclotron frequency of an electron, m_{eo} being the rest mass of an electron. Thus under relativistic effects ($\gamma > 1$), the value of angular cyclotron frequency of an electron is reduced (see (8.4.24)). Also, consequently, the value of B_0 (see (8.4.21)) will be modified from that given by (8.4.22) by the relativistic mass factor γ as (s = 1):

$$B_0 = 3.57 \times 10^{-11} \gamma \text{ Tesla per hertz of cyclotron frequency}. \tag{8.4.27}$$

The Larmor radius r_L incorporating relativistic effects is obtained from (8.4.19) replacing ω_c by ω_c/γ (see equation (8.4.24)) as:

$$r_L = \frac{\gamma v_t}{\omega_c} . \qquad (8.4.28)$$

Therefore, it follows from (8.4.28) that under relativistic effects ($\gamma > 1$), the Larmor radius will increase.

If one defines a relativistic momentum p as

$$p = m_e v , \qquad (8.4.29)$$

where m_e is given by (8.4.25), then by a simple algebra it is possible to obtain another form of γ from (8.4.23) as

$$\gamma = \left(1 + \frac{p^2}{m_{eo}^2 c^2}\right)^{1/2} . \qquad (8.4.30)$$

For an electron following a helical trajectory one may resolve p into their axial and transverse components defined respectively as

$$p_z = m_e v_z \qquad \text{(a)}$$

and

$$p_t = m_e v_t , \qquad \text{(b)} \qquad (8.4.31)$$

where v_z and v_t are the axial and transverse components of electron velocity v. These components satisfy the following relations:

$$v^2 = v_z^2 + v_t^2 \qquad \text{(a)}$$

and

$$p^2 = p_z^2 + p_t^2 . \qquad \text{(b)} \qquad (8.4.32)$$

Using (8.4.32) one may choose to express (8.4.23) and (8.4.30), respectively, as:

$$\gamma = \left(1 - \frac{v_z^2 + v_t^2}{c^2}\right)^{-1/2} \qquad \text{(a)}$$

and

$$\gamma = \left(1 + \frac{p_z^2 + p_t^2}{m_{e0}^2 c^2}\right)^{1/2} . \qquad \text{(b)} \qquad (8.4.33)$$

For a relation that can be used to compute γ from the beam voltage, one may equate the relativistic kinetic energy $(\gamma m_{e0} - m_{e0})c^2$ with the potential energy $|e| V_0$ of the beam in view of the energy conservation to get the expression for γ in terms of beam voltage V_0 as:

$$\gamma = 1 + \frac{|e| V_0}{m_{e0} c^2} .$$

With the help of (8.4.31(b)) and (8.4.25) one may obtain an alternative form of the Larmor radius as:

$$r_L = \frac{p_t}{m_{eo} \omega_c} \tag{8.4.34}$$

which using (8.4.26) becomes

$$r_L = \frac{p_t}{|e| B_0} . \tag{8.4.35}$$

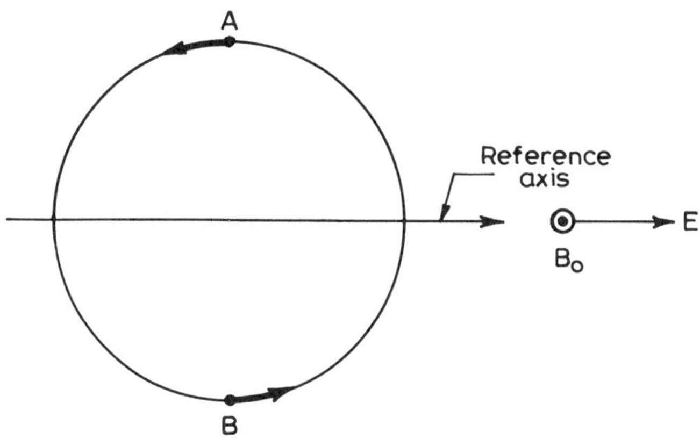

Fig. 8.4.4 Two reference electrons separated by 180^0 in phase taking part in relativistic bunching.

The relativistic effects take a significant role in the process of bunching of electrons in a gyro-TWT, unlike in a conventional TWT; further, the bunching is azimuthal rather than axial in a gyro-TWT. Although the trajectories of the hollow-beam electrons in this device are helical with their axes parallel to the axis of the waveguide, with their number distributed along each helical path, yet for the sake of understanding, as in section *8.4.2A*, let us assume that the axial velocity of electrons is zero. This gives a simplified picture in which the electrons are initially distributed uniformly over a single orbit of gyration; in order to avoid complexity let us take two representative electrons A and B initially separated by 180^0 in phase angle [71] (Fig. 8.4.4). For a dc magnetic flux directed towards the reader, the electrons orbit anticlockwise to provide the radially inward, centripetal Lorentz force required to maintain the circular motion. Under the influence of the rf electric field (E) with its direction shown, the electron A is subjected to the accelerating force and gains energy; its relativistic mass increases (the value of γ increases) and angular velocity decreases (see (8.4.24)). Thus the electron A rotates slower or accumulates a phase lag. On the other hand, the electron B is subjected to the decelerating force and loses energy; its relativistic mass decreases (the value of γ decreases) and angular velocity increases. Thus the electron B rotates faster or accumulates a phase lead. Consequently, the electrons A and B, while they continue to gyrate, do approach each other and thus tend to phase-bunch about a reference axis along the orbital diameter parallel to the rf electric field. These two electrons put together, however, do not contribute to the net transfer of energy to the rf wave, since while one of them gains energy the other loses it to the same amount. We can have a similar picture for the other electrons, too, apart form A and B, which are evenly distributed in the same orbit. As a result of the effect of relativistic angular velocity variations, after an integral number of wave-periods, these electrons will become bunched around the reference axis; again the energy gained by half of them above the reference axis will be equal to that lost by the remaining half below it. It may be noted that in the above description one may also allow for the change of radius of gyration of electrons under the influence of the rf field when the relativistic effects are considered. In fact, the electrons in the upper half space about the reference axis (Fig. 8.4.4) (for which the value of γ increases) will have a higher value of r_L (see (8.4.28)); and so they will spiral outward. Similarly, the electrons in the lower half-space about the reference axis (for which the value of γ decreases) will spiral inward. However, the effect of the change in radius of gyration will be rather small and one-time effect as compared to that of the angular velocity variation [66].

Now, for a net transfer of energy from the electrons to the rf wave it is necessary that the bunch of electrons should find themselves in the deccelerating field continuously. For this purpose, one must make ω slightly greater than ω_c/γ_t, say, for the fundamental-cyclotron resonance (see the relation (8.4.20(b)) in which the rest-mass angular cyclotron frequency ω_c is replaced (see (8.4.24)) by $\omega_c/\gamma = \omega_c/\gamma_t$, taking in the present simplified picture, $v_z = 0$ giving $\gamma = \gamma_t = (1-v_t^2/c^2)^{-1/2}$ (see (8.4.33(a))). If this slight-inequality condition $\omega \geq \omega_c/\gamma_t$ is satisfied, the particles on the average traverse an azimuthal coordinate space less than 2π, and the distribution of electrons after an integral number of wave periods will be bunched in the lower half space below the reference axis (Fig. 8.4.4) where the bunch finds itself in synchronism with the accelerating rf electric field to transfer its energy to the wave.

However, one must extend the above simplified picture of bunching developed for the case of $v_z = 0$ to that of $v_z \neq 0$, when the electrons actually follow helical paths. In this situation when the condition $\omega \geq \omega_c/\gamma$ is satisfied, the position where the bunch moves in the decelerating electric field (moving parallel to the latter) advances with a phase velocity approximately equal to the phase velocity of the waveguide. Thus in this case the bunch twists around the helix system with a pitch substantially greater than the electron pitch [66].

8.4.2C Beam-mode dispersion relation

In a gyro-TWT the periodicity of the interaction structure of a conventional TWT has been replaced by the periodicity of the beam of electrons which follow helical trajectories. This makes possible synchronism and interactions at phase velocities greater than c, the speed of light in free space. It may be seen from (8.4.5) that the rf field varies as exp $(j\psi)$, where ψ is the rf phase given by

$$\psi\{t, z, \theta\} = \omega t - \beta z - m\theta \,, \tag{8.4.36}$$

where β is the axial phase propagation constant of the waveguide. Let us find a condition of synchronism for a helical beam in cyclotron-harmonic resonance with the rf field of the wave propagating through the waveguide. However, before that let us review the phenomenon of second cyclotron harmonic resonance discussed in section *8.4.2A* for a simple case in which an electron gyrates a circular orbit on a plane, ignoring the axial electron velocity ($v_z = 0$). In this case (see Fig. 8.4.3(b)), the rf phase changes by 2π after a time $T_c/2$ when the electron makes half a complete rotation. The condition of second cyclotron harmonic resonance can then be mathematically described as

$$\psi\,\{(t_1 + T_c/2), z_1, \theta_1\} \;=\; \psi\,\{t_1, z_1, \theta_1\} + 2\pi \,, \tag{8.4.37}$$

where $t = t_1$ is the initial time (shown as $t_1 = 0$ in Fig. 8.4.3(b)), $z = z_1$ describes the plane of the orbit, $\theta = \theta_1$ is the initial angular position of the orbiting electron with respect to the waveguide coordinates. It may be noted that the angular position of the electron does not change after a time $T_c/2$ (see Fig. 8.4.3(b)). Putting (8.4.36) in (8.4.37), then one gets

$$\omega\,(t_1 + T_c/2) - \beta z_1 - m\theta_1 \;=\; \omega t_1 - \beta z_1 - m\theta_1 + 2\pi$$

whence the relation between ω and ω_c $(= 2\pi/T_c)$ is obtained as

$$\omega \;=\; 2\,\omega_c \,, \tag{8.4.38}$$

which, as expected, is found to be identical with (8.4.20(b)) in which, for the second-harmonic resonance, one has to interpret $s = 2$.

One may now easily extend the above treatment to the case of $v_z \neq 0$ for an electron following a helical trajectory. In this case, taking into account the axial advancement of the electron with the axial velocity v_z, one may put the condition of second cyclotron harmonic resonance, similar to (8.4.3), as

$$\psi\{(t_1 + T_c/2), (z_1 + v_z\,T_c/2), \theta_1\} \;=\; \psi\{t_1, z_1, \theta_1\} + 2\pi$$

which with the help of (8.4.36) yields the following relation between ω and ω_c $(= 2\pi/T_c)$:

$$\omega - \beta v_z - 2\omega_c = 0 . \tag{8.4.39}$$

Now, one may easily generalise (8.4.39) to the case of the sth cyclotron-harmonic resonance by replacing 2 by s as:

$$\omega - \beta v_z - s\omega_c = 0 . \tag{8.4.40}$$

If one assumes $v_z = 0$, (8.4.40) passes on to (8.4.20(b)). Putting $\beta = \omega/v_p$ and noting that the phase velocity of the waveguide $v_p > c$ and the axial beam velocity, $v_z < c$, the second term of the left hand side of (8.4.40), namely βv_z ($= \omega v_z/v_p$) called the classical Doppler shift becomes much smaller than ω, so that one gets $\omega = s\omega_c$ which is the same condition of synchronisation between ω and ω_c as predicted by (8.4.20(b)). If the beam-periodicity is ignored then dropping the third term in the left hand side of (8.4.40) one gets $\omega/\beta = v_p = v_z$, a condition which only a slow-waveguide structure of a conventional TWT can meet. On the other hand it is the beam-periodicity which becomes responsible in the case of a gyro-TWT to maintain a fast-wave synchronism between the beam and the rf waves ($\omega \approx s\omega_c$ ignoring the doppler shift βv_z in (8.4.40)). This corresponds to the intersection (see Fig. 8.4.2) between the waveguide dispersion hyperbola predicted by (8.4.13) and the beam-mode dispersion (straight) line predicted by

$$\omega - \beta_{mn} v_z - s\omega_c/\gamma = 0 \tag{8.4.41}$$

which is same as (8.4.40) in which the rest-mass angular cyclotron frequency ω_c is replaced by the corresponding relativistic quantity ω_c/γ (see (8.4.24)); also, we have used the subscripts m and n to indicate that the axial propagation constant refers to the TE_{mn}-mode of the guide considered here (section *8.4.1*).

Further, recalling the discussion at the end of section *8.4.2B*, one may put (8.4.41) as

$$\omega - \beta_{mn} v_z - s\omega_c/\gamma \sim (>)\, 0 \tag{8.4.42}$$

as the condition which will ensure that the bunch of electron do move in synchronism with the decelerating phase of the rf electric field of the waveguide.

<u>*Example 8.4.2*</u> Show that (a) the condition of synchronism between the axial-beam velocity and the velocity of the slow-wave of the rate of energy exchange between the helical beam and the rf wave and (b) the condition of synchronism between the phase velocity of the fast-cyclotron wave of the beam and that of the rf wave each correspond to the beam-mode dispersion relation.

(a) Let the electrons while following a helical path experience the magnitude of force F of amplitude F_0 due to the transverse rf electric field of the waveguide [72].

$$F = F_0 \exp j(\omega t - \beta_{mn} z) ,$$

at an angular position θ=0, with respect to the waveguide coordinates, due to the transverse rf electric field given by (8.4.5). The electrons moving with a transverse velocity v_t become parallel to the field with a time periodicity T_c (=$2\pi/\omega_c$) and exchange with the latter, with an

$$\text{energy exchange rate} = v_t F_0 \exp j(\omega t - \beta_{mn} z) . \tag{8.4.43}$$

Clearly, v_t will be periodic which may be represented in terms of its amplitude v_0 as:

$$v_t = v_0 \cos (2\pi/L_b)\, z , \tag{8.4.44}$$

where L_b is the periodicity or pitch of an electron given by

$$L_b = v_z T_c . \tag{8.4.45}$$

Putting (8.4.45) in (8.4.44),

$$v_t = v_0 \cos (\omega_c z/v_z) = v_0 \cos (\beta_c z) , \tag{8.4.46}$$

where β_c (= ω_c/v_z) is a cyclotron-wave number. Substituting (8.4.46) in (8.4.43) and choosing to put

$$\cos \beta_c z = [\exp (j\beta_c z) + \exp (-j\beta_c z)]/2 ,$$

one may get

$$\text{energy exchange rate} = \frac{v_0 F_0}{2} [\exp j(\omega t-(\beta_{mn} - \beta_c)z) + \exp j(\omega t -(\beta_{mn} + \beta_c)z)] . \tag{8.4.47}$$

If follows from (8.4.47) that the energy-exchange rate has two components of waves one travelling slower, with a phase velocity $\omega/(\beta_{mn} + \beta_c)$, than the other, with a phase velocity $\omega/(\beta_{mn}-\beta_c)$. Equating the axial velocity of the electrons with the phase velocity of the slower wave one may write

$$v_z = \frac{\omega}{\beta_{mn} + \beta_c} \tag{8.4.48}$$

which may be put as

$$\omega - \beta_{mn} v_z - \omega_c = 0 . \tag{8.4.49}$$

We thus find that (8.4.49) agrees with the beam-mode dispersion relation (8.4.41)(here, for $s = \gamma = 1$).

(b) When the electrons are perturbed to have rf transverse velocities under the influence of the Lorentz force of an axial dc magnetic field of flux density B_0, one may write the following equations of motion (ignoring relativistic effects):

$$m_{e0} \frac{dv_x}{dt} = e(\mathbf{v}_1 \times B_0\, \mathbf{a}_z)_x = v_{1y} B_0 \qquad \text{(a)}$$

and

$$m_{e0} \frac{dv_y}{dt} = e(\mathbf{v}_1 \times B_0\, \mathbf{a}_z)_y = - v_{1x} B_0 , \qquad \text{(b)} \tag{8.4.50}$$

choosing to express the electronic motion in the rectangular system of coordinates for an electron beam supposedly of infinite cross section. Assuming the rf quantities to vary as $\exp(j\omega t - \Gamma z)$, one may then express (8.4.50) as

$$(j\omega - v_z\Gamma)v_{1x} = -v_{1y}\,\omega_c \qquad \text{(a)}$$

and

$$(j\omega - v_z\Gamma)v_{1y} = v_{1x}\,\omega_c\,, \qquad \text{(b)} \qquad (8.4.51)$$

where we have put $\omega_c = -eB_0/m_{eo} = |e|\,B_0/m_{eo}$ (see (8.4.26)).

Also, v_z has been interpreted as the dc beam velocity, and the complete derivative d/dt expressed as

$$d/dt = \partial/\partial t + v_z\,\partial/\partial z \quad \text{(see for instance, (8.1.7))}\,.$$

Then the multiplication of (8.4.51(a)) and (8.4.51(b)) yields

$$(j\omega - v_z\Gamma)^2 = -\omega_c^2$$

whence one gets

$$\Gamma = j(\omega \mp \omega_c)v_z\,. \qquad (8.4.52)$$

It follows from (8.4.52) that the waves set up on the beam called the cyclotron waves have two components, one travelling faster and the other traveling slower than the beam in the axial direction. If the phase propagation constant of the fast cyclotron wave component is made synchronous with that of the waveguide mode β_{mn}, then

$$\beta_{mn} = \frac{\omega - \omega_c}{v_z}$$

giving

$$\omega - \beta_{mn}\,v_z - \omega_c = 0\,. \qquad (8.4.53)$$

We thus find (8.4.53) as identical with (8.4.49), again agreeing with the beam-mode dispersion relation (8.4.41) (here, for $s = \gamma = 1$).

8.4.3 Grazing Conditions

For strong and beam-wave interaction to occur, ω and β_{mn} must simultaneously satisfy [72], [73] the waveguide hyperbolic dispersion relation (8.4.13) and the beam-mode linear dispersion relation (8.4.41), as mentioned earlier in the discussion preceding (8.4.41). In general, the beam-mode line should intersect the wave-guide mode hyperbola at two points; however, a point of intersection corresponding to a negative value of β_{mn} should be avoided as it corresponds to a wave propagating in the backward direction. A particular point of interest is the *grazing* intersection, when the beam-mode line barely touches or grazes the waveguide-mode hyperbola, as shown in Fig. 8.4.2. At the grazing point, the slope of the hyperbola (= $d\omega/d\beta$), which also represents the group velocity of the wave (see appendix *a5.5.1*), becomes equal to the slope of the beam-mode line which is equal to v_z. Differentiating (8.4.13), one may find ($d\omega/d\beta$) as:

$$\frac{d\omega}{d\beta} = \frac{\beta_{mn} c^2}{\omega} ,$$

so that the above condition for grazing intersection may be put as:

$$v_z = \frac{\beta_{mn} c^2}{\omega} . \tag{8.4.54}$$

With the help of (8.4.41) and (8.4.54) one may show the grazing point ω as:

$$\omega = s\,\gamma_z^2\,\omega_c/\gamma \quad \text{(grazing)} , \tag{8.4.55}$$

where γ_z is defined as (see (8.4.33(a))):

$$\gamma_z = (1 - v_z^2/c^2)^{-1/2} . \tag{8.4.56}$$

Similarly, with the help of (8.4.13) and (8.4.54) one gets an alternative form of the grazing-point ω as:

$$\omega = \gamma_z\,\omega_{cut,\,mn} \quad \text{(grazing)} . \tag{8.4.57}$$

The grazing-point β_{mn} may now be found by substituting (8.4.57) in (8.4.54) to get

$$\beta_{mn} = \gamma_z\,\omega_{cut,\,mn}\,v_z/c^2 \quad \text{(grazing)} . \tag{8.4.58}$$

The grazing-point ω_c may be found by comparing (8.4.55) with (8.4.57) as

$$\frac{s\,\omega_c}{\gamma} = \frac{\omega_{cut,mn}}{\gamma_z} \quad \text{(grazing)} , \tag{8.4.59}$$

whence the grazing magnetic flux density may be found with the help of (8.4.19(a)) as:

$$B_0 = \frac{\gamma\, m_{eo}\,\omega_{cut,mn}}{|e|\, s\,\gamma_z} \quad \text{(grazing)} . \tag{8.4.60}$$

An alternative method to find the grazing conditions is as follows. ω is obtained from (8.4.41) and substituted in (8.4.13) to form a quadratic equation in β_{mn} which has a solution

$$\beta_{mn} = \frac{-v_z s\omega_c/\gamma \pm [v_z^2 s^2\omega_c^2/\gamma^2 - (v_z^2 - c^2)(s^2\omega_c^2/\gamma^2 - \omega_{cut}^2)]^{1/2}}{v_z^2 - c^2} . \tag{8.4.61}$$

Under grazing conditions, (8.4.61) has to have only one solution. This requires that the quantity under radical in (8.4.61) should become zero yielding a condition that is found to be identical with (8.4.59). With the term under radical put as zero, (8.4.61) yields the same expression for the grazing point β_{mn} as that given by (8.4.58). The latter substituted in (8.4.41) gives the same expression for the grazing point ω as (8.4.57).

8.4.4 Dispersion Relation of a Gyro-TWT

Let us now proceed to derive the dispersion relation of a gyro-TWT in which a periodic (helical) beam couples with a TE-mode of a non-periodic cylindrical waveguide [72]-[76]. We have obtained earlier a similar dispersion relation for a conventional TWT (see equation (8.2.50)) considering the coupling between a non-periodic electron beam and a periodic (helical) slow-wave structure. It is worth looking back into the dispersion relation to see how the individual beam mode and circuit mode dispersion relations are inherent in this relation of a conventional TWT. For this purpose let us re-arrange the terms of (8.2.50) as

$$(\Gamma^2 - \Gamma_0^2)\,[(j\beta_e - \Gamma)^2 + \beta_p^2] \cong -2\beta_e^4 C^3\,, \qquad \text{(conventional TWT)} \tag{8.4.62}$$

where we have put the approximation $\Gamma \cong \Gamma_0 \cong j\beta_e$. For a weak coupling, say, for a small beam current which also renders C (= $(KI_0/4V_0)^{1/3}$) have a small value, so much so that the right hand side of (8.4.62) may be set to zero [6], one gets

$$(\Gamma^2 - \Gamma_0^2)\,[(j\beta_e - \Gamma)^2 + \beta_p^2] = 0\,. \tag{8.4.63}$$

From (8.4.63) the decoupled circuit-mode relation is:

$$\Gamma^2 - \Gamma_0^2 = 0$$

which in terms of

$$\Gamma = j\beta \tag{8.4.64}$$

becomes

$$\beta = \pm\beta_0 \text{ (slow-wave structure mode),} \tag{8.4.65}$$

where β_0, the cold-circuit axial propagation constant of the slow-wave structure, is obtainable from the analysis of the structure (see, for instance, section *5.3* or *5.4* for the analysis of a helical slow-wave structure). Again, from (8.4.63) the decoupled beam-mode dispersion relation is:

$$(j\beta_e - \Gamma)^2 + \beta_p^2 = 0$$

which using (8.4.64) becomes

$$\beta = \beta_e \mp \beta_p$$

which is identical with (8.1.20), which ignoring space charges ($\beta_p = 0$) becomes

$$\beta = \beta_e , \tag{8.4.66}$$

which may also be put, remembering $\beta_e = \omega/v_o$, as

$$\omega - \beta v_0 = 0 \text{ (beam mode)} . \tag{8.4.67}$$

Our aim in this section is to derive the dispersion relation of a gyro-TWT similar to the conventional TWT relation (8.4.62) which also, in the case of weak coupling should decouple into the beam-mode and waveguide-mode dispersion relations (8.4.41) and (8.4.13), respectively. The decoupled waveguide mode relation should be obtained as

$$\beta = \pm \beta_{mn} \quad \text{(waveguide-mode)} , \tag{8.4.68}$$

similar to (8.4.65), where β_{mn} is given by the cold dispersion relation of the guide (8.4.15) which may be read with the help of (8.4.16). Similarly, the decoupled beam-mode relation should be obtained as

$$\beta = \beta_{e,gyro} \text{ (beam-mode)}, \tag{8.4.69}$$

similar to (8.4.66), where

$$\beta_{e,gyro} = \frac{\omega - s\omega_c / \gamma}{v_z} , \tag{8.4.70}$$

which substituted in (8.4.69) gives

$$\omega - \beta v_z - s\,\omega_c/\gamma = 0 , \tag{8.4.71}$$

that can be seen to be identical with (8.4.41) interpreting β as β_{mn}.

8.4.4A Wave equation for a beam-loaded cylindrical waveguide

Using (4.5.3) read with the help of (4.5.2), the wave equation in terms of the axial component of magnetic field intensity H_z for the TE_{mn}-mode in a cylindrical waveguide (section *8.4.1*), in the presence of an electron beam, may be written as

$$\nabla^2 H_z - \mu_0 \varepsilon_0 \frac{\partial^2 H_z}{\partial t^2} = -(\nabla \times \mathbf{J})_z . \tag{8.4.72}$$

Expanding the Laplacian, ∇^2, and the z-component of curl, $(\nabla\times\mathbf{H})_z$, with the help of (2.6.4(b)) and (3.2.18(b)), respectively, one may express (8.4.72) as:

$$\frac{\partial^2 H_z}{\partial z^2}+\frac{1}{r}\frac{\partial H_z}{\partial r}+\frac{1}{r^2}\frac{\partial^2 H_z}{\partial \theta^2}-\mu_0\varepsilon_0\frac{\partial^2 H_z}{\partial t^2}=-\frac{1}{r}\left(\frac{\partial}{\partial r}(rJ_\theta)-\frac{\partial J_r}{\partial \theta}\right). \tag{8.4.73}$$

We make here a simplifying assumption that the electron beam is sufficiently tenuous such that the spatial structure of the waveguide mode is unaffected by the presence of the beam [72], [75], [76]. Thus, in the beam-wave coupled system of the gyro-TWT, the radial and the azimuthal rf dependence continues to be of the form given by (8.4.4) as it is for a cold waveguide in the absence of the beam. However, the axial propagation constant β that appears in the rf axial dependence (given by (8.4.4)) will have to be interpreted now as different from that for a cold waveguide $\beta = \beta_{mn}$ that satisfies the cold dispersion relation (8.4.13). Consequently, we are looking here for a dispersion relation between ω and β for a beam-wave coupled system which will obviously be different from that between ω and β_{mn}, given by (8.4.13), for a cold waveguide.

Substituting (8.4.4) (interpreting β as the axial phase propagation constant of the beam-wave coupled system) into the first two terms of the left hand side of (8.4.73) one may show (appendix *a8.4.1*):

$$\frac{\partial^2 H_z}{\partial z^2}+\frac{1}{r}\frac{\partial H_z}{\partial r}=\left(-k_t^2+\frac{m^2}{r^2}\right)H_z\,. \tag{8.4.74}$$

A similar substitution of (8.4.4) into the remaining terms of the left hand side of (8.4.73) gives (appendix *a8.4.1*):

$$\frac{1}{r^2}\frac{\partial^2 H_z}{\partial \theta^2}+\frac{\partial^2 H_z}{\partial z^2}-\mu_0\varepsilon_0\frac{\partial^2 H_z}{\partial t^2}=\left(-\frac{m^2}{r^2}-\beta^2+k_0^2\right)H_z\,. \tag{8.4.75}$$

Plugging (8.4.74) and (8.4.75) into (8.4.73) one then gets the simplified form of wave equation for the cylindrical waveguide in the presence of an electron beam as:

$$(k_0^2-\beta^2-k_t^2)H_z=-\frac{1}{r}\left(\frac{\partial}{\partial r}(rJ_\theta)-\frac{\partial J_r}{\partial \theta}\right)$$

which when multiplied by μ_0 gives the equation in terms of B_z ($= \mu_0 H_z$) as follows:

$$(k_0^2-\beta^2-k_t^2)B_z=-\frac{\mu_0}{r}\left(\frac{\partial}{\partial r}(rJ_\theta)-\frac{\partial J_r}{\partial \theta}\right). \tag{8.4.76}$$

8.4.4B Dispersion relation from the wave equation for a beam-loaded waveguide

Let us consider a model of a thin hollow (annular) beam of monoenergetic electrons in helical trajectories all of the same Larmor radius r_L (see equation (8.4.35)) with their guiding centres uniformly distributed on a common guiding circle of radius r_0, equal to the average hollow beam radius, the radial thickness of the beam being $2r_L$ (Fig. 8.4.5).

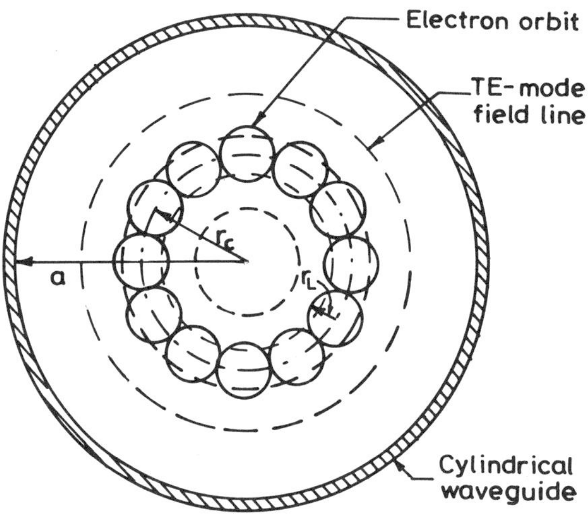

Fig. 8.4.5 Cross-sectional view of the gyro-TWT model showing the projection of electron orbits of the hollow beam on the cross-sectional plane of the waveguide in a uniform magnetic field (not shown) pointing towards the reader.

The dispersion relation of the gyro-TWT in the above model may be derived as follows. Substituting first B_z (= $\mu_0 H_z$) as (see equation (8.4.4))

$$B_z = B_{z0} J_m\{k_t r\} \exp j(\omega t - \beta z - m\theta) \tag{8.4.77}$$

in (8.4.76), where $B_{z0} = \mu_0 H_{z0}$; then multiplying it by $rJ_m\{k_t r\}$ and integrating over r between the axis of the waveguide r = 0 and its wall r = a, one obtains [72], [75], [76]:

$$(k_0^2 - \beta^2 - k_t^2)B_{z0} \exp j(\omega t - \beta z - m\theta)\int_0^a r\, J_m^2\{k_t r\}dr$$

$$= \int_0^a -\mu_0\left(\frac{\partial}{\partial r}(r J_\theta) - \frac{\partial J_r}{\partial \theta}\right)J_m\{k_t r\}\, dr\ . \tag{8.4.78}$$

The definite integral occurring in the left hand side of (8.4.78) is [72]:

$$\int_0^a r\, J_m^2\{k_t r\}\, dr = \frac{a^2}{2}(J_m'^2\{k_t a\} + K_{mn})$$

which, since for a TE_{mn}-mode $J'_m\{k_t a\}=0$ (see equation (8.4.8)), becomes[72]

$$\int_0^a r\, J_m^2\{k_t r\}\, dr = \frac{a^2 K_{mn}}{2}\,, \tag{8.4.79}$$

where
$$K_{mn} = \left(1 - \frac{m^2}{k_t^2 a^2}\right) J_m^2\{k_t a\}\,. \tag{8.4.80}$$

Then substituting (8.4.79) in (8.4.78) and re-arranging terms one may write a relation between k_0 ($=\omega/c$) and β as:

$$k_0^2 - \beta^2 - k_t^2 = \frac{-2\mu_0 \mathcal{J} \exp - j(\omega t - \beta z - m\theta)}{B_{z0}\, a^2 K_{mn}} \tag{8.4.81}$$

where the integral

$$\mathcal{J} = \int_0^a \left(\frac{\partial}{\partial r}(r J_\theta) - \frac{\partial J_r}{\partial \theta}\right) J_m\{k_t r\}\, dr\,. \tag{8.4.82}$$

The rf beam current densities J_θ and J_r occurring in the right side of (8.4.82) is determined by the dynamics of the beam particles (electrons) which is described by the relativistic Vlasov equation[72], [76], [77]:

$$\left(\frac{\partial}{\partial t} + \frac{\mathbf{p}}{\gamma m_{eo}}.\nabla - |e|\,(\mathbf{E} + \mathbf{v} \times \mathbf{B}).\nabla_p\right) f = 0\,, \tag{8.4.83}$$

where $f\{\mathbf{r}, \mathbf{p}, t\}$ is the electron distribution in $(\mathbf{r}, \mathbf{p})$ phase space, ∇f is the gradient of f in physical space and $\nabla_p f$ is its gradient in momentum space. The Vlasov equation applies to the tenuous beam that has been assumed here, as mentioned following (8.4.73), in which electron collisions can be ignored. If we represent f as

$$f = f_0 + f_1\,, \tag{8.4.84}$$

where f_0 is the dc or initial unperturbed part and f_1 is the perturbed part in the distribution function f_1 (with $f_1 \ll f_0$), then the beam current density puturbations J_r and J_θ can be evaluated through the following integral expressions:

$$J_r = -|e| \int f_1 v_r d^3p \qquad \text{(a)}$$

and

$$J_\theta = -|e| \int f_1 v_\theta d^3p \ , \qquad \text{(b)} \qquad (8.4.85)$$

where

$$d^3p = dp_t \ p_t d\phi \ dp_z \qquad (8.4.86)$$

is the element of volume in the momentum space, and v_r and v_θ, in terms of the transverse component v_t or p_t (see (8.4.31(b)) and (8.4.25)), are (Fig. 8.4.6):

$$v_r = v_t \cos(\phi - \theta) = \frac{p_t}{\gamma m_{eo}} \cos(\phi - \theta)$$

and

$$v_\theta = v_t \sin(\phi - \theta) = \frac{p_t}{\gamma m_{eo}} \sin(\phi - \theta), \qquad (8.4.87)$$

respectively, where r and θ are the real-space polar coordinates and p_t and ϕ are the momentum-space polar coordinates of the instantaneous position of the electron on the projection of the electron orbit on the cross-sectional plane of the waveguide (Fig. 8.4.6).

Let us now examine the integral $\mathcal{I}$ which occurs in the right hand side of (8.4.81) and which is given by (8.4.82). For self-consistency the current perturbations must have the same rf dependence, $\exp j(\omega t - \beta z - m\theta)$, as the field that created it (see for instance (8.4.77) for the nature of rf field dependence). This enables one to take $\partial J_r/\partial\theta = -jm \, J_r$ so that one may write the integral of (8.4.82) as

$$\mathcal{I} = \int_0^a \frac{\partial}{\partial r}(rJ_\theta) J_m\{k_t r\} dr + \int_0^a jmJ_r \, J_m\{k_t r\} dr \ . \qquad (8.4.88)$$

Integrating the first term of the right hand side of (8.4.88) one gets

$$\mathcal{I} = [J_m\{k_t r\} r J_\theta]_0^a - \int_0^a rJ_\theta k_t J'_m\{k_t r\} dr + \int_0^a jmJ_r \, J_m\{k_t r\} dr \ . \qquad (8.4.89)$$

Since $J_\theta = 0$ both at r = a and b (the thin hollow beam being so located as discussed in the beginning of this section), the first term of the right hand side of (8.4.89) becomes zero when the upper and lower limits are substituted. Using the recurrence relations (see equations (a8.4.2) and (a8.4.6)),

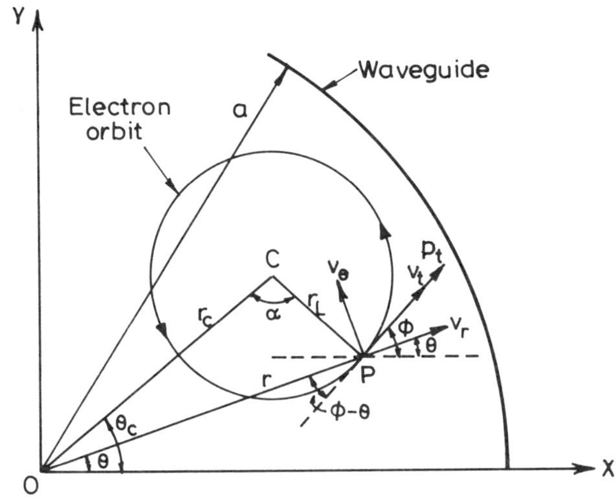

Fig. 8.4.6 Projection of the electron orbit on the cross-sectional plane of the waveguide showing the centre of the waveguide (0), the guiding centre of the gyrating electron (C), the Larmor radius (r_L), the real-space coordinates (r_c, θ_c) of the guiding centre, and the real-space (r, θ) and momentum-space (p_t, ϕ) coordinates of the instantaneous position (P) of the gyrating electron (the radial coordinates being measured from the centre of the waveguide 0 and the angular coordinates from the reference X-axis drawn through 0).

$$J_m\{x\} = \frac{x}{m}\frac{J_{m-1}\{x\} + J_{m+1}\{x\}}{2} \qquad \text{(a)}$$

and

$$J'_m\{x\} = \frac{J_{m-1}\{x\} - J_{m+1}\{x\}}{2}, \qquad \text{(b)}$$

(8.4.90)in which x is to be interpreted as $x = k_t r$ in the third and second terms of the right hand side of (8.4.89), respectively, and remembering that the first term has already become zero, one then obtains

$$\mathcal{I} = \frac{-k_t}{2}\int_0^a [(J_\theta - j J_r) J_{m-1}\{k_t r\} - (J_\theta + j J_r) J_{m+1}\{k_t r\}]\, r dr\,. \qquad (8.4.91)$$

Now, first putting (8.4.86) and (8.4.87) in (8.4.85) and then using the latter as well as the relation $\exp \pm j(\phi-\theta) = \cos(\phi-\theta) \pm j\sin(\phi-\theta)$, one gets

$$J_\theta - jJ_r = j|e|\int\left(\frac{p_t}{\gamma m_{eo}}\right)(\exp j(\phi-\theta))f_1\, p_t\, dp_t\, dp_z\, d\phi$$

and

$$J_\theta + jJ_r = -j|e|\int\left(\frac{p_t}{\gamma m_{eo}}\right)(\exp - j(\phi-\theta))\, f_1\, p_t\, dp_t\, dp_z\, d\phi$$

which may then be substituted in (8.4.91) to obtain

$$\mathcal{I} = \frac{-jk_t|e|}{2}\int_{r=0}^{a}\int_{p_t=0}^{\infty}\int_{p_z=-\infty}^{\infty}\int_{\phi=0}^{2\pi}[J_{m-1}\{k_t r\}\exp j(\phi-\theta) + J_{m+1}\{k_t r\}\exp - j(\phi-\theta)]\times$$

$$\times\, r\frac{p_t^2}{\gamma m_{eo}}f_1\, dr\, dp_t\, dp_z\, d\phi\,. \qquad (8.4.92)$$

The quantity within the bracket in the integrand of (8.4.92), which as such is a function of r, may be expressed as independent of r with the help of Graf's addition theorem of Bessel functions [74] as (Fig. 8.4.6) (see appendix *a8.4.2*):

$$J_{m-1}\{k_t r\}\exp j(\phi-\theta) + J_{m+1}\{k_t r\}\exp -j(\phi-\theta)$$

$$= 2j\exp jm(\theta_c-\theta)\sum_{h=-\infty}^{\infty}J'_h\{k_t r_L\}J_{m+h}\{k_t r_c\}\exp - jh\left(\frac{\pi}{2}-\theta_c+\phi\right), \qquad (8.4.93)$$

where h is an integer. Therefore, plugging (8.4.93) in (8.4.92),

$$\mathcal{I} = k_t|e|\int_{r=0}^{a}\int_{p_t=0}^{\infty}\int_{p_z=-\infty}^{\infty}\int_{\phi=0}^{2\pi}\exp jm(\theta_c-\theta)\sum_{h=-\infty}^{\infty}J'_g\{k_t r_L\}J_{m+h}\{k_t r_c\}\times$$

$$\times(\exp - jh(\frac{\pi}{2}-\theta_C+\phi))\frac{rp_t^2}{\gamma m_e}f_1\, dr\, dp_t\, dp_z\, d\phi\,. \qquad (8.4.94)$$

The perturbation part f_1 of the electron distribution function in (8.4.94) can be obtained by solving the following equation that follows from (8.4.83):

$$\left(\frac{\partial}{\partial t}+\frac{\mathbf{p}}{\gamma m_{eo}}.\nabla-|e|(\mathbf{E}+\mathbf{v}\times\mathbf{B}_0).\nabla_p\right)f_1=|e|(\mathbf{E}_1+\mathbf{v}\times\mathbf{B}_1).\nabla_p f_0\,, \qquad (8.4.95)$$

where, on the right hand side, the second-order term $(|e|(\mathbf{E}_1+\mathbf{v}\times\mathbf{B}_1).\nabla_p)\,f_1$ has been ignored. The solution of (8.4.95) is (see appendix *a8.4.3*):

$$f_1=\frac{|e|B_0}{k_t}\exp j(\omega t-\beta z)\exp(-jm\theta_c)\left[(\omega-\beta v_z)\frac{\partial f_0}{\partial p_t}+v_t\beta\frac{\partial f_0}{\partial p_z}\right]\times$$

$$\times\sum_{p=-\infty}^{\infty}\frac{J_{m+p}\{k_t r_c\}J'_p\{k_t r_L\}}{\omega-kv_z+p\omega_c/\gamma}\exp jp\left(\frac{\pi}{2}-\theta_c+\phi\right). \qquad (8.4.96)$$

Substituting (8.4.96) in (8.4.94),

$$\mathcal{J}=k_t|e|\exp jm(\theta_c-\theta)\int_{r=0}^{a}\int_{p_t=0}^{\infty}\int_{p_z=-\infty}^{\infty}\int_{\phi=0}^{2\pi}\sum_{h=-\infty}^{\infty}\sum_{p=-\infty}^{\infty}J'_h\{k_t r_L\}J_{m+h}\{k_t r_c\}\left(\exp-jh(\frac{\pi}{2}-\theta_c+\phi)\right)\times$$

$$\times\frac{r p_t^2|e|B_{z0}}{\gamma m_{e0}k_t}\exp j(\omega t-\beta z)\exp(-jm\theta_c)\left((\omega-\beta v_z)\frac{\partial f_0}{\partial p_t}+v_t\beta\frac{\partial f_0}{\partial p_z}\right)\times$$

$$\times\left(\frac{J_{m+p}\{k_t r_c\}J'_p\{k_t r_L\}}{\omega-\beta v_z+p\omega_c/\gamma}\right)\exp jp(\frac{\pi}{2}-\theta_c+\phi)\,dr\,dp_t\,dp_z\,d\phi\,. \qquad (8.4.97)$$

Now, if (8.4.97) is substituted in (8.4.81), one obtains

$$k_0^2-\beta^2-k_t^2=\frac{-2\mu_0|e|^2}{a^2K_{mn}}\int_{r=0}^{a}\int_{p_t=0}^{\infty}\int_{p_z=-\infty}^{\infty}\int_{\phi=0}^{2\pi}\sum_{h=-\infty}^{\infty}\sum_{p=-\infty}^{\infty}\exp j(p-h)\phi\;J'_h\{k_t r_L\}J_{m+h}\{k_t r_c\}\times$$

$$\times\exp j(p-h)(\frac{\pi}{2}-\theta_c)\frac{J_{m+p}\{k_t r_c\}J'_p\{k_t r_L\}}{\omega-\beta v_z+p\omega_c/\gamma}\left((\omega-\beta v_z)\frac{\partial f_0}{\partial p_t}+v_t\beta\frac{\partial f_0}{\partial p_z}\right)\frac{r p_t^2}{\gamma m_{eo}}dr\,dp_t dp_z d\phi\,. \qquad (8.4.98)$$

In (8.4.98), the integration with respect to ϕ involves terms in the summation of the following form, which may be evaluated as

$$\int_{\phi=0}^{2\pi} \exp j(p-h)\phi \, d\phi = \frac{\cos 2\pi(p-h) + j \sin 2\pi(p-h) - 1}{j(p-h)} \quad . \tag{8.4.99}$$

When p ≠ h, the right hand side of (8.4.99) becomes zero, whereas if p = h, it can be evaluated by L' Hospital's rule giving

$$\underset{(p-h)\to 0}{\mathrm{Lt}} \frac{\cos 2\pi(p-h) + j \sin 2\pi(p-h) - 1}{j(p-h)} = \underset{(p-h)\to 0}{\mathrm{Lt}} \frac{-2\pi \sin 2\pi(p-h) + j2\pi \cos 2\pi(p-h)}{j} = 2\pi ,$$

so that one may express (8.4.99) as

$$\int_{\phi=0}^{2\pi} \exp j(p-h)\phi \, d\phi = 2\pi \, \delta_{ph} , \tag{8.4.100}$$

where δ_{ph} is the Kronecker delta which becomes 0, if p ≠ h, and 1, if p = h.

In view of (8.4.100), therefore, all the terms corresponding to p ≠ h in the double summation of (8.4.98) will become zero thereby converting the double summation into a single summation corresponding to p = h. Thus, one may express (8.4.98), using (8.4.99) and (8.4.100), as:

$$k_0^2 - \beta^2 - k_t^2 = \frac{-4\pi\mu_0 |e|^2}{a^2 K_{mn}} \int_{r=0}^{a} \int_{p_t=0}^{\infty} \int_{p_z=-\infty}^{\infty} \sum_{p=-\infty}^{\infty} \frac{J_p'^2\{k_t r_L\} J_{m+p}^2\{k_t r_c\}}{\omega - \beta v_z + p\omega_c / \gamma} \times$$
$$\times \left((\omega - \beta v_z) \frac{\partial f_0}{\partial p_t} + v_t \beta \frac{\partial f_0}{\partial p_z} \right) \frac{r\, p_t^2}{\gamma m_{eo}} dr \, dp_t \, dp_z \quad . \tag{8.4.101}$$

The integration with respect to r in (8.4.101) may be evaluated by choosing the unperturbed or equilibrium electron distribution function considering a beam of zero guiding centre spread as [72], [74] (see Example 8.4.3, interpreting rdr within the integral as $r_c dr_c$ while by examining the trigonometrical relations between the sides and angles of the triangle OCP of Fig. 8.4.6 [74]):

$$f_0 = \frac{N_0}{2\pi r_c} \delta\{r_c - r_{c0}\} \, g\{p_t, p_z\}, \tag{8.4.102}$$

where N_0 is the number of electrons per unit axial length of the interaction structure (waveguide), where r_{c0} is the average beam radius, and the Dirac-delta function [4] $\delta(r_c - r_{c0})$ representation means that all the electrons of a beamlet have the same guiding centre without a spread located on a cylindrical surface defined by $r = r_{c0}$. $g\{p_t, p_z\}$ is the momentum distribution function which, for an idealised cold beam for which the energy spread is considered to be small, may be expressed as

$$g\{p_t, p_z\} = \frac{1}{2\pi p_t}\delta\{p_t - p_{t0}\}\,\delta\{p_z - p_{z0}\}\,, \tag{8.4.103}$$

a function that is normalised according to (see Example 8.4.3):

$$\int g\, d^3p = 1 \quad . \tag{8.4.104}$$

<u>Example 8.4.3</u> (a) Verify by integrating the momentum distribution function (8.4.103) that the function is normalised according to (8.4.104) and hence also verify by integrating the electron distribution function (8.4.102) that N_0 represents the number of electrons per unit axial interaction length.

Substituting $d^3p = dp_t\, p_t\, d\phi\, dp_z$ (8.4.86) into (8.4.104),

$$\int g\, d^3p = \int_{\phi=0}^{2\pi}\int_{p_t=0}^{\infty}\int_{p_z=-\infty}^{\infty} p\, g\{p_t, p_z\} d\phi\, dp_t\, dp_z$$

which on integration over ϕ becomes

$$\int g\, d^3p = \int_{p_t=0}^{\infty}\int_{p_z=-\infty}^{\infty} 2\pi p_t\, g\{p_t, p_z\}\, dp_t\, dp_z \ . \tag{8.4.105}$$

Substituting (8.4.103) in (8.4.105),

$$\int g\, d^3p = \int_0^{\infty}\delta\{p_t - p_{t0}\}dp_t \int_{-\infty}^{\infty}\delta\{p_z - p_{z0}\}dp_z \ . \tag{8.4.106}$$

Each of the integrals of the right hand side of (8.4.106) becomes unity in view of the unit-impulse property of the Dirac-delta function [4], thus giving

$$\int g\, d^3p = (1)(1) = 1$$

which agrees with (8.4.104).

Now, if one integrates the equilibrium electron distribution function over the momentum space one gets the dc number density n_0 as

$$n_0 = \int f_0\, d^3p\,. \tag{8.4.107}$$

Substituting (8.4.102) in (8.4.107),

$$n_0 = \int \frac{N_0}{2\pi r_c}\delta\{r_c - r_{c0}\}\, g\{p_t, p_z\} d^3p$$

which in view of (8.4.104), which we have already verified above, becomes

$$n_0 = \frac{N_0\, \delta\{r_c - r_{c0}\}}{2\pi r_c}\,. \tag{8.4.108}$$

With the help of (8.4.108), the number of electrons for the volume extending over a unit axial interaction length becomes

$$\int_0^{r_{c,max}=a-r_L} n_0\, 2\pi r_c\, dr_c = \int_0^{a-r_L} \frac{N_0\, \delta\{r_c - r_{c0}\}}{2\pi r_c} 2\pi r_c\, dr_c = N_0 \int_0^{a} \delta\{r_c - r_{c0}\} dr_c = (N_0)(1) = N_0\,,$$

in view of the unit-impulse property of the Dirac function [4], it being assumed that the outermost guiding-centre position could be one corresponding to the situation in which the electrons gyrating with radius r_L barely do not intercept the wall [74].

Thus the relation between ω and k_0 given by (8.4.101) now reduces to two integrations as follows, instead of three, as we can see by substituting f_0 from (8.4.102) in (8.4.101) and carrying out integration with respect to r while invoking the unit-impulse property of Dirac-delta function [4]:

$$k_0^2 - \beta^2 - k_t^2 = \frac{-2\mu_0 |e|^2 N_0}{a^2 K_{mn}} \int_{p_t=0}^{\infty} \int_{p_z=-\infty}^{\infty} \frac{p_t^2}{\gamma m_{e0}} \left((\omega - \beta v_z) \frac{\partial g\{p_t, p_z\}}{\partial p_t} \right.$$

$$\left. + v_t \beta \frac{\partial g\{p_t, p_z\}}{\partial p_z} \right) \sum_{p=-\infty}^{\infty} \frac{J_p'^2\{k_t r_L\} J_{m+p}^2\{k_t r_c\}}{\omega - \beta v_z + p\omega_c/\gamma} dp_t\, dp_z \tag{8.4.109}$$

By simple integration by parts which, however, involves lengthy mathematical steps, one may express (8.4.109) as (see appendix *a8.4.4*)

$$k_0^2-\beta^2-k_t^2=\frac{-2\mu_0|e|^2N_0}{K_{mn}a^2}\int_{p_t=0}^{\infty}\int_{p_z=-\infty}^{\infty}\sum_{p=-\infty}^{\infty}\left(\frac{\eta_t^2(\omega^2-\beta^2c^2)\,H_{m,p}\{k_t r_c,k_t r_L\}}{(\omega-\beta v_z+p\omega_c/\gamma)^2}\right.$$

$$\left.-\frac{(\omega-\beta v_z)Q_{m,p}\{k_t r_c,k_t r_L\}}{\omega-\beta v_z+p\omega_c/\gamma}\right)\left(\frac{p_t}{\gamma m_{eo}}\right)g\{p_t,p_z\}dp_t\,dp_z \qquad (8.4.110)$$

where

$$\eta_t=\frac{v_t}{c}=\frac{p_t}{\gamma m_{eo}c}, \qquad (a)$$

$$H_{m,p}=J^2_{m+p}\{k_t r_c\}\,J'^2_p\{k_t r_L\} \qquad (b)$$

and

$$Q_{m,p}=2H_{m,p}+2(k_t r_L)\,J^2_{m+p}\{k_t r_c\}\,J'_p\{k_t r_L\}\,J''_p\{k_t r_L\} \qquad (c) \quad (8.4.111)$$

Now, under cyclotron harmonic resonance condition (4.8.41), since one expects $\beta \approx \beta_{mn}$, one has

$$(\omega-\beta v_z-s\,\omega_c/\gamma)\to 0\,. \qquad (8.4.112)$$

Therefore, in view of (8.4.112), the denominators of the terms associated with $H_{m,p}$ and $Q_{m,p}$ in (8.4.110) each will tend to zero, if $p = -s$, thereby making the terms in the summation corresponding to $p = -s$ significantly large compared to the remaining terms (corresponding to $p \neq -s$) [72]. Therefore, retaining the terms corresponding to only $p = -s$ in (8.4.110) in which (8.4.103) is substituted for $g\{p_t, p_z\}$, one may write

$$k_0^2-\beta^2-k_t^2=\frac{-2\mu_0|e|^2N_0}{K_{mn}a^2}\int_{p_t=0}^{\infty}\int_{p_z=-\infty}^{\infty}\left(\frac{\eta_t^2(\omega^2-\beta^2c^2)H_{m,-s}\{k_t r_c,k_t r_L\}}{(\omega-\beta v_z-s\omega_c/\gamma)^2}\right.$$

$$\left.-\frac{(\omega-\beta v_z)Q_{m,-s}\{k_t r_c,k_t r_L\}}{\omega-\beta v_z-s\omega_c/\gamma}\right)\frac{p_t\,\delta\{p_t-p_{t0}\}\,\delta\{p_z-p_{z0}\}}{\gamma m_{e0}\,2\pi p_t}dp_t\,dp_z\,. \qquad (8.4.113)$$

Using the unit-impulse property of the Dirac-delta function, the integration in (8.4.113) can be carried out easily to finally yield the following dispersion relation [72]:

$$k_0^2 - \beta^2 - k_t^2 = \frac{-\mu_0 |e|^2 N_0}{\gamma\, m_{e0} \pi K_{mn} a^2} \left(\frac{\eta_t^2 (\omega^2 - \beta^2 c^2) H_{m,-s}\{k_t r_c, k_t r_L\}}{(\omega - \beta v_z - s\omega_c/\gamma)^2} - \frac{(\omega - \beta v_z) Q_{m,-s}\{k_t r_c, k_t r_L\}}{\omega - \beta v_z - s\omega_c/\gamma} \right). \tag{8.4.114}$$

Putting p = – s in (8.4.111(b)), one may write

$$H_{m,-s} = J^2_{-s+m}\{k_t r_c\}\, J'^2_{-s}\{k_t r_L\} = J^2_{-(s-m)}\{k_t r_c\}\, J'^2_{-s}\{k_t r_L\}$$

which, in view of the relations $J_{-n}\{x\} = (-1)^n J_n\{x\}$ and $J'_{-n}\{x\} = (-1)^{n+1} J'_n\{x\}$ which give $J_{-n}^2\{x\} = J_n^2\{x\}$ and $J'^2_{-n}\{x\} = J_n^2\{x\}$, respectively, becomes

$$H_{m,-s} = J^2_{s-m}\{k_t r_c\}\, J'^2_s\{k_t r_L\} \; . \tag{8.4.115}$$

Similarly, one may choose to express the function $Q_{m,-s}$ with the help of (8.4.111(c)) as:

$$Q_{m,-s} = 2\, J^2_{s-m}\{k_t r_c\}\, J'^2_s\{k_t r_L\} + 2\,(k_t r_L)\, J_{s-m}\{k_t r_c\}\, J'_s\{k_t r_L\} J''_s\{k_t r_L\}. \tag{8.4.116}$$

The first term of the right hand side of (8.4.114) dominates over its second term [76], [78] for a relativistic beam of significant transverse velocity (v_t) for which the value of η_t becomes significant. This is so also because the second term is proportional to, while the first term is inversely proportional to the square of, $\omega - \beta v_z - s\omega_c/\gamma \sim 0$ (see (8.4.41)). Thus, ignoring the second term in most of the practical situations, one may write the dispersion relation of a gyro-TWT, (8.4.114), in the following form:

$$k_0^2 - \beta^2 - k_t^2 = \frac{-\mu_0 |e|^2 N_0\, \eta_t^2\, H_{m,-s} (\omega^2 - \beta^2 c^2)}{\gamma\, m_{eo}\, \pi K_{mn}\, a^2 (\omega - \beta v_z - s\omega_c/\gamma)^2}$$

which may be re-arranged as

$$(k_0^2 - \beta^2 - k_t^2)(\omega - \beta v_z - s\omega_c/\gamma)^2 = \frac{-\mu_0 |e|^2 N_0\, \eta_t^2\, H_{m,-s}\, (\omega^2 - \beta^2 c^2)}{\gamma m_{eo}\, \pi K_{mn}\, a^2} \, . \tag{8.4.117}$$

For a weak coupling, setting the right hand side of (8.4.117) to zero, one gets the waveguide and beam modes decoupled, respectively, as:

$$k_0^2 - \beta^2 - k_t^2 = 0$$

and $$\omega - \beta v_z - s\omega_c / \gamma = 0 ,$$

the former, in view of (8.4.10), giving

$$\beta = \pm \beta_{mn} ,$$

as predicted by (8.4.68), while the latter giving

$$\beta = \beta_{e,gyro} = \frac{\omega - s\omega_c / \gamma}{v_z} ,$$

as predicted by (8.4.69) and (8.4.70).

Appendix a8.4.1 Simplification of the terms of the wave equation

Making use of the field expression (8.4.4) one gets

$$\frac{1}{r}\frac{\partial H_z}{\partial r} = \frac{k_t}{r} H_{zo} J'_m\{k_t r\} \exp j(\omega t - \beta z - m\theta) . \tag{a8.4.1}$$

Now using the recurrence relation of Bessel function

$$J'_m\{x\} = -J_{m+1}\{x\} + \frac{m}{x} J_m\{x\} \tag{a8.4.2}$$

one may write

$$J'_m\{k_t r\} = -J_{m+1}\{k_t r\} + \frac{m}{k_t r} J_m\{k_t r\} . \tag{a8.4.3}$$

Substituting (a8.4.3) in (a8.4.1),

$$\frac{1}{r}\frac{\partial H_z}{\partial r} = \frac{H_{zo}}{r}\left(-k_t J_{m+1}\{k_t r\} + \frac{m}{r} J_m\{k_t r\}\right) \exp j(\omega t - \beta z - m\theta) . \tag{a8.4.4}$$

Similarly, again using (8.4.4) one gets

$$\frac{\partial^2 H_z}{\partial r^2} = H_{zo}\left(-k_t^2 J'_{m+1}\{k_t r\} + \frac{m}{r} k_t J'_m\{k_t r\} - \frac{m}{r^2} J_m\{k_t r\}\right) \exp j(\omega t - \beta z - m\theta). \tag{a8.4.5}$$

Further, using the recurrence relation of Bessel functions

$$J'_m\{x\} = J_{m-1}\{x\} - \frac{m}{x} J_m\{x\} \tag{a8.4.6}$$

one may write (replacing m by m+1)

$$J'_{m+1}\{k_t r\} = J_m\{k_t r\} - \frac{m+1}{k_t r} J_{m+1}\{k_t r\} \,. \tag{a8.4.7}$$

Substitution of (a8.4.7) and (a8.4.3) in (a8.4.5) gives

$$\frac{\partial^2 H_z}{\partial r^2} = H_{zo}\left(-k_t^2 J_m\{k_t r\} + \frac{k_t}{r} k_t J_{m+1}\{k_t r\} + \frac{m^2}{r^2} J_m\{k_t r\} - \frac{m}{r^2} J_m\{k_t r\}\right) \exp j(\omega t - \beta z - m\theta) \tag{a8.4.8}$$

Adding (a8.4.8) and (a8.4.4), one then gets

$$\frac{\partial^2 H_z}{\partial r^2} + \frac{1}{r}\frac{\partial H_z}{\partial r} = -H_{zo}\left(-k_t^2 + \frac{m^2}{r^2}\right) J_m\{k_t r\} \exp j(\omega t - \beta z - m\theta)$$

which with the help of (8.4.4) may be read as

$$\frac{\partial^2 H_z}{\partial r^2} + \frac{1}{r}\frac{\partial H_z}{\partial r} = \left(-k_t^2 + \frac{m^2}{r^2}\right) H_z \,. \tag{a8.4.9}$$

Further using (8.4.4) it can be easily seen that

$$\frac{1}{r^2}\frac{\partial^2 H_z}{\partial \theta^2} + \frac{\partial^2 H_z}{\partial z^2} - \mu_0 \varepsilon_0 \frac{\partial^2 H_z}{\partial t^2} = \left(-\frac{m^2}{r^2} - \beta^2 + \omega^2 \mu_0 \varepsilon_0\right) H_z$$

which with the help of (8.4.3) may be read as

$$\frac{1}{r^2}\frac{\partial^2 H_z}{\partial \theta^2} + \frac{\partial^2 H_z}{\partial z^2} - \mu_0 \varepsilon_0 \frac{\partial^2 H_z}{\partial t^2} = \left(-\frac{m^2}{r^2} - \beta^2 + k_0^2\right) H_z \,. \tag{a8.4.10}$$

We have used (a8.4.9) and (a8.4.10) which are identical with (8.4.74) and (a8.4.75), respectively, to simplify the wave equation as discussed following (8.4.75).

Appendix a8.4.2 *Application of Graf's theorem of Bessel functions in the dispersion relation*

According to Graf's theorem of Bessel functions [74]:

$$(\exp \pm jm\theta_1) J_m\{x_1\} = \sum_{p=-\infty}^{\infty} (\exp \pm jp\theta_2) J_{m+p}\{x_2\} J_q\{x_3\} \,, \tag{a8.4.11}$$

where m, and p are integers, x_1, x_2 and x_3 represent sides of a triangle, and θ_1 and θ_2 represent the angles between x_1 and x_2 and between x_2 and x_3, respectively (Fig. a8.4.1).

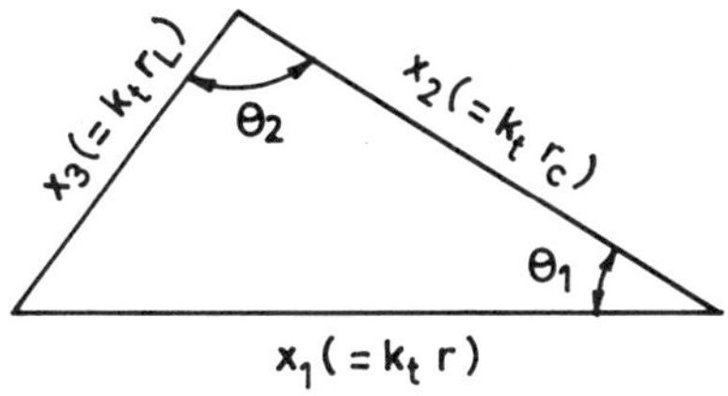

Fig. a8.4.1 Geometrical representation of the variables used in Graf's addition theorem of Bessel function [74].

In the present context, let x_1, x_2 and x_3 be $k_t r$, $k_t r_c$ and $k_t r_L$, respectively (Fig. a8.4.1). Then from Fig. 8.4.6. it can be easily seen that θ_1 (here, the angle between r and r_c)= $\theta_c - \theta$; also applying the basic geometrical considerations (such as the sum of the angles of a triangle is equal to π and the radial line through a point on the circumference of a circle makes an angle of $\pi/2$ with the tangent drawn on it) one can see θ_2 (here, the angle between r_c and r_L) = $\pi/2 - \theta_c + \phi$. Substituting the above values of x_1, x_2, x_3,, θ_1 and θ_2, one may now use (a8.4.11), taking its upper sign, to write

$$J_m\{k_t r\} = \exp - jm(\theta_c - \theta) \sum_{p=-\infty}^{\infty} J_{m+p}\{k_t r_c\} J_p\{k_t r_L\} \exp jp\left(\frac{\pi}{2} - \theta_c + \phi\right). \tag{a8.4.12}$$

Using (a8.4.12), let us write

$$J_{m-1}\{k_t r\} \exp j(\phi - \theta) + J_{m+1}\{k_t r\} \exp - j(\phi - \theta) \tag{a8.4.13}$$

$$= \exp j(\phi - \theta) \exp j(m-1)(\theta_c - \theta) \sum_{q=-\infty}^{\infty} J_{m-1+q}\{k_t r_c\} J_q\{k_t r_L\} \exp - jq\left(\frac{\pi}{2} - \theta_c + \phi\right)$$

$$+ \exp - j(\phi - \theta) \exp j(m+1)(\theta_c - \theta) \sum_{s=-\infty}^{\infty} J_{m+1+s}\{k_t r_c\} J_s\{k_t r_L\} \exp - js\left(\frac{\pi}{2} - \theta_c + \phi\right),$$

where q and s are integers. Choosing to put, in the first summation, $q-1 = h$ and also noting that $h \to \pm \infty$ when $q \to \pm \infty$, as well as, in the second summation, $s+1 = h$ and also noting that $h \to \pm \infty$ when $s \to \pm \infty$, one may express (a8.4.13) as:

$$J_{m-1}\{k_t r\} \exp j(\phi - \theta) + J_{m+1}\{k_t r\} \exp - j(\phi - \theta)$$

$$= \exp j(\phi - \theta) \exp j(m-1)(\theta_c - \theta) \sum_{h=-\infty}^{\infty} J_{m+h}\{k_t r_c\} J_{h+1}\{k_t r_L\} \exp - j(h+1)\left(\frac{\pi}{2} - \theta_c + \phi\right)$$

$$+ \exp - j(\phi - \theta) \exp j(m+1)(\theta_c - \theta) \sum_{h=-\infty}^{\infty} J_{m+h}\{k_t r_c\} J_{h-1}\{k_t r_L\} \exp - j(h-1)\left(\frac{\pi}{2} - \theta_c + \phi\right)$$

$$= \exp jm(\theta_c - \theta) \sum_{h=-\infty}^{\infty} \exp - j(\frac{\pi}{2}) J_{m+h}\{k_t r_c\} J_{h+1}\{k_t r_L\} \exp - jh\left(\frac{\pi}{2} - \theta_c + \phi\right)$$

$$+ \exp jm(\theta_c - \theta) \sum_{h=-\infty}^{\infty} \exp j(\frac{\pi}{2}) J_{m+h}\{k_t r_c\} J_{h-1}\{k_t r_L\} \exp - jh\left(\frac{\pi}{2} - \theta_c + \phi\right)$$

$$= \exp jm(\theta_c - \theta) \sum_{h=-\infty}^{\infty} J_{m+h}\{k_t r_c\}(-jJ_{h+1}\{k_t r_L\} + jJ_{h-1}\{k_t r_L\}) \exp - jh\left(\frac{\pi}{2} - \theta_c + \phi\right)$$

$$= \exp jm(\theta_c - \theta) \sum_{h=-\infty}^{\infty} j\, J_{m+h}\{k_t r_c\}(J_{h-1}\{k_t r_L\} - J_{h+1}\{k_t r_L\}) \exp - jh\left(\frac{\pi}{2} - \theta_c + \phi\right)$$

which with the help of the recurrence relation of Bessel functions, (8.4.90), becomes

$$J_{m-1}\{k_t r\} \exp j(\phi - \theta) + J_{m+1}\{k_t r\} \exp - j(\phi - \theta)$$

$$= 2j \exp jm(\theta_c - \theta) \sum_{h=-\infty}^{\infty} J'_h\{k_t r_L\} J_{m+h}\{k_t r_c\} \exp - jh\left(\frac{\pi}{2} - \theta_c + \phi\right) .$$

Appendix a8.4.3 *Solution of Vlasov's equation*

Substituting (8.4.84) in Vlasov equation (8.4.83) one may write the following equation for the perturbation part f_1 of the electron distribution function[66], [72]:

$$\frac{\partial f_1}{\partial t} + \frac{\boldsymbol{p}.\nabla f_1}{\gamma m_{eo}} - |e|\boldsymbol{E}_1.\nabla_p f_0 - |e|\boldsymbol{E}_1.\nabla_p f_1 - |e|(\boldsymbol{v} \times \boldsymbol{B}_0).\nabla_p f_1$$

$$-|e|(\boldsymbol{v} \times \boldsymbol{B}_1).\nabla_p f_0 - |e|(\boldsymbol{v} \times \boldsymbol{B}_1).\nabla_p f_1 = 0 ,$$

which ignoring higher-order terms may be written as

(a8.4.14) $$\frac{\partial f_1}{\partial t} + \frac{\boldsymbol{p}.\nabla f_1}{\gamma m_{eo}} - |e|(\boldsymbol{v} \times \boldsymbol{B}_0).\nabla_p f_1 = |e|(\boldsymbol{E}_1 + \boldsymbol{v} \times \boldsymbol{B}_1).\nabla_p f_0 ,$$

where $\boldsymbol{B}_0$ (= B_0 $\boldsymbol{a}_z$) is the external dc axial magnetic field and $\boldsymbol{E}_1$ and $\boldsymbol{B}_1$ represent the rf electric field intensity and magnetic flux density, respectively. One may now integrate (a8.4.14) along the unperturbed electron trajectories[66]. An unperturbed trajectory may be defined as[72]:

(a8.4.15) (a) $$\boldsymbol{p} = m_e \frac{d\boldsymbol{r}}{dt} = \gamma m_{eo} \frac{d\boldsymbol{r}}{dt}$$

(b) $$\frac{d\boldsymbol{p}}{dt} = -|e|\boldsymbol{v} \times \boldsymbol{B}_0 ,$$

the latter being the only force on the unperturbed trajectory caused by the external dc magnetic field.

With the help of (a8.4.15), one may then express (a8.4.14) as

(a8.4.16) $$\frac{\partial f_1}{\partial t}+\frac{d\boldsymbol{r}}{dt}.\nabla f_1+\frac{d\boldsymbol{p}}{dt}.\nabla_p f_1 = |e|(\boldsymbol{E}_1+\boldsymbol{v}\times\boldsymbol{B}_1).\nabla_p f_0 .$$

The left hand side of (a8.4.16) may now readily be identified as df_1/dt *so that one may write (a8.4.16) as*

(a8.4.17) $$\frac{df_1}{dt} = |e|(\boldsymbol{E}_1+\boldsymbol{v}\times\boldsymbol{B}_1).\nabla_p f_0 ,$$

which gives the rate of change of the distribution function as observed on the electron trajectory[74]. Interestingly, unlike (a8.4.14), (a8.4.17) now involves only one variable t, and which may be integrated along the unputurbed trajectory to give

(a8.4.18) $$f_1 = |e| \int_{t'=t_1=-\infty}^{t} (\boldsymbol{E}_1\{t'\}+\boldsymbol{v}\{t'\}\times\boldsymbol{B}_1\{t'\}).\nabla'_p f_0\, dt' ,$$

where t_1 *is the time before the electrons entered the electromagnetic field [67] which may be set [72], [74] as* $-\infty$*, that is, large in magnitude compared to any time at frequencies of interest. Along the unperturbed trajectory [74],*

(a8.4.19) (a) $$z\{t'\} = z + v_z\{t\}(t'-t) \quad \text{and}$$

(b) $$\phi\{t'\} = \phi + \omega'_c\{t\}(t'-t) ,$$

where ω'_c $(=\omega_c/\gamma)$ *is the relativistic cyclotron frequency. Differentiating (a8.4.19) with respect to* t' *while treating* t, z *and* ϕ *as constants one gets*

(a8.4.20) (a) $$v_z\{t'\} = v_z\{t\} \quad \text{and}$$

(b) $$\omega'_c\{t'\} = \omega'_c\{t\} .$$

Further, one may see, with the help of (8.4.28), (8.4.24) and (a8.4.20(b)),

(a8.4.21) $$v_t\{t'\} = v_t\{t\} ;$$

with the help of (a8.4.20), (a8.4.21) and (8.4.33a),

(a8.4.22) $$\gamma\{t'\} = \gamma\{t\} ;$$

and with the help of (8.4.31), (8.4.25), (8.4.20a), (a8.4.21) and (a8.4.22),

(a8.4.23) (a) $$p_z\{t'\} = p_z\{t\} \quad \text{and}$$

(b) $$p_t\{t'\} = p_t\{t\} .$$

Using (a8.4.23) then one may write

(a8.4.24) $$\nabla'_p f_0 = \frac{\partial f_0}{\partial p'_t}\boldsymbol{a}_t + \frac{\partial f_0}{\partial p'_z}\boldsymbol{a}_z = \frac{\partial f_0}{\partial p_t}\boldsymbol{a}_t + \frac{\partial f_0}{\partial p_z}\boldsymbol{a}_z = \nabla_p f_0 ,$$

where

(a8.4.25) $$\boldsymbol{a}_t = \cos\phi\ \boldsymbol{a}_x + \sin\phi\ \boldsymbol{a}_y$$

is the transverse unit vector (see Fig. 8.4.6).

Substituting (a8.4.24) in (a8.4.18), one may then write

(a8.4.26) $$f_1 = |e| \int_{-\infty}^{t} (\boldsymbol{E}_1\{t'\} + \boldsymbol{v}\times\boldsymbol{B}_1\{t'\}).\nabla_p f_0\ dt' .$$

For the TE-mode of the guide (for which the axial rf electric field intensities is zero) let us express

(a8.4.27) (a) $$\boldsymbol{E}_1 = E_{r1}\ \boldsymbol{a}_r + E_{\theta 1}\ \boldsymbol{a}_\theta \quad \text{and}$$

(b) $$\boldsymbol{B}_1 = B_{r1}\ \boldsymbol{a}_r + B_{\theta 1}\ \boldsymbol{a}_\theta + B_{z1}\ \boldsymbol{a}_z ,$$

where the subscript 1 in the right hand side refers to the rf components in the cylindrical system of coordinates. Similarly, let

(a8.4.28) $$\boldsymbol{v} = v_r\ \boldsymbol{a}_r + v_\theta\ \boldsymbol{a}_\theta + v_z\ \boldsymbol{a}_z .$$

Substituting (a8.4.27) and (a8.4.28) in (a8.4.26),

(a8.4.29) $$f_1 = |e| \int_{-\infty}^{t} [(E_{r1} + v_\theta B_{z1} - v_z B_{\theta 1})\boldsymbol{a}_r + (E_{\theta 1} + v_z B_{r1} - v_r B_{z1})\boldsymbol{a}_\theta$$
$$+ (v_r B_{\theta 1} - v_\theta B_{r1})\boldsymbol{a}_z].\nabla_p f_0\ dt' ,$$

Now, from the geometry of the problem (Fig. 8.4.6),

(a8.4.30) (a) $$v_r = v_t \cos(\phi - \theta) \quad \text{and}$$

(b) $$v_\theta = v_t \sin(\phi - \theta) ;$$

and also for TE-mode one may write (see problem 5.14)

(a8.4.31) (a) $$B_{r1} = -(\beta/\omega)\, E_{\theta 1} \quad \text{and}$$

(b) $$B_{\theta 1} = (\beta/\omega)\ E_{r1} .$$

Also, $\nabla_p f_0$ may be expressed with the help of (a8.4.24) and (a8.4.25) as

(a8.4.32) $$\nabla_p f_0 = \frac{\partial f_0}{\partial p_t}\cos\phi\ \boldsymbol{a}_x + \frac{\partial f_0}{\partial p_t}\sin\phi\ \boldsymbol{a}_y + \frac{\partial f_0}{\partial p_z}\boldsymbol{a}_z .$$

Therefore, substituting (a8.4.30) - (a8.4.32) into (a8.4.29), and noting that

$$\boldsymbol{a}_r \,.\, \boldsymbol{a}_x = \boldsymbol{a}_\theta .\, \boldsymbol{a}_y = \cos\theta$$

and

$$\boldsymbol{a}_r \,.\, \boldsymbol{a}_y = -\,\boldsymbol{a}_\theta .\, \boldsymbol{a}_x = \sin\theta,$$

one gets [72]:

$$f_1 = |e| \int_{-\infty}^{t} \left(\left(1 - \frac{\beta v_z}{\omega} \right) \frac{\partial f_0}{\partial p_t} + \frac{v_t \beta}{\omega} \frac{\partial f_0}{\partial p_z} \right) (E_{r1} \cos(\phi - \theta) + E_{\theta 1} \sin(\phi - \theta) dt'. \quad \text{(a8.4.33)}$$

Now, the quantity $E_{r1}\ \cos(\phi-\theta) + E_{\theta 1} \sin(\phi-\theta)$ *occurring in (a8.4.33) may be expressed with the help of (8.4.5) and (8.4.6) as:*

$$E_{r1} \cos(\phi - \theta) + E_{\theta 1} \sin(\phi - \theta) = \frac{-\omega B_{z0} m}{k_t^2 r} J_m\{k_t r\} \exp j(\omega t' - \beta z - m\theta) \cos(\phi - \theta)$$

$$+ \frac{j\omega B_{z0}}{k_t} J'_m\{k_t r\} \exp j(\omega t' - \beta z - m\theta) \sin(\phi - \theta)$$

which using the recurrence relation of Bessel function (8.4.90) becomes

$$E_{r1} \cos(\phi - \theta) + E_{\theta 1} \sin(\phi - \theta) = \frac{\omega B_{z0}}{2k_t} (-J_{m-1}\{k_t r\} - J_{m+1}\{k_t r\}) \cos(\phi - \theta) \exp - jm\theta \quad \text{(a8.4.34)}$$

$$\exp j(\omega t' - \beta z) + \frac{j\omega B_{z0}}{2k_t} (J_{m-1}\{k_t r\} - J_{m+1}\{k_t r\}) \sin(\phi - \theta) \exp - jm\theta \exp j(\omega t' - \beta z)$$

$$= \frac{\omega B_{zo}}{2k_t} \exp j(\omega t' - \beta z) \exp - jm\theta \, [-J_{m-1}\{k_t r\}(\cos(\phi - \theta) - j \sin(\phi - \theta))$$

$$-J_{m+1}\{k_t r\}(\cos(\phi - \theta) + j \sin(\phi - \theta))]$$

$$= \frac{-\omega B_{zo}}{2k_t} \exp j(\omega t' - \beta z) \exp - jm\theta \, [J_{m-1}\{k_t r\} \exp - j(\phi - \theta) + J_{m+1}\{k_t r\} \exp j(\phi - \theta)] \, .$$

As in appendix a8.4.2 let us apply Bessel's addition theorem (a8.4.11) to express (a8.4.34) as:

$$E_{r1} \cos(\phi - \theta) + E_{\theta 1} \sin(\phi - \theta) = \frac{-\omega B_{zo}}{2k_t} \exp j(\omega t' - \beta z) \exp - jm\theta \, [\exp - j(\phi - \theta) \times \quad \text{(a8.4.35)}$$

$$\times (\exp - j(m-1)(\theta_c - \theta) \sum_{q=-\infty}^{\infty} J_{m-1+q}\{k_t r_c\} J_q\{k_t r_L\} \exp jq(\frac{\pi}{2} - \theta_c + \phi)) + \exp j(\phi - \theta) \times$$

$$\times (\exp - j(m+1)(\theta_c - \theta) \exp - j(m+1)(\theta_c - \theta) \sum_{s=-\infty}^{\infty} J_{m+1+s}\{k_t r_c\} J_s\{k_t r_L\} \exp js(\frac{\pi}{2} - \theta_c + \phi))].$$

Choosing to put $q-1 = p$ and $s+1 = p$ and noting that when q and s each tend to $\pm\infty$, then p also tends to $\pm\infty$, one may write (a8.4.35) as

$$E_{r1}\cos(\phi-\theta)+E_{\theta 1}\sin(\phi-\theta) = \frac{-\omega B_{zo}}{2k_t}\exp j(\omega t'-\beta z)\exp- jm\theta\,[\exp- j(\phi-\theta)\times$$

$$\times\exp- j(m-1)(\theta_c-\theta)\sum_{p=-\infty}^{\infty}J_{m+p}\{k_t r_c\}\,J_{p+1}\{k_t r_L\}\exp j(p+1)(\frac{\pi}{2}-\theta_c+\phi)$$

$$+\exp j(\phi-\theta)\exp- j(m+1)(\theta_c-\theta)\sum_{p=-\infty}^{\infty}J_{m+p}\{k_t r_c\}\,J_{p-1}\{k_t r_L\}\exp j(p-1)(\frac{\pi}{2}-\theta_c+\phi)]$$

$$=\frac{-\omega B_{zo}}{2k_t}\exp j(\omega t'-\beta z)\exp- jm\theta\,[\exp- j(\phi-\theta)\exp- j(m-1)(\theta_c-\theta)\exp j\frac{\pi}{2}\exp j(-\theta_c+\phi)$$

$$\sum_{p=-\infty}^{\infty}J_{m+p}\{k_t r_c\}\,J_{p+1}\{k_t r_L\}\exp jp(\frac{\pi}{2}-\theta_c+\phi)\;+\exp(\phi-\theta)(\exp- j(m+1)(\theta_c-\theta)\times$$

$$\times\exp- j\frac{\pi}{2}\exp- j(-\theta_c+\phi)\sum_{p=-\infty}^{\infty}J_{m+p}\{k_t r_c\}\,J_{p-1}\{k_t r_L\}\exp jp(\frac{\pi}{2}-\theta_c+\phi))]\,,$$

which simplifies to

$$E_{r1}\cos(\phi-\theta)+E_{\theta 1}\sin(\phi-\theta)=\frac{-\omega\,B_{zo}}{2k_t}\exp j(\omega t'-\beta z)\exp(-jm\theta_c)$$

$$\times\sum_{p=-\infty}^{\infty}jJ_{m+p}\{k_t r_c\}\,(J_{p+1}\{k_t r_L\}-J_{p-1}\{k_t r_L\})\exp jp\,(\frac{\pi}{2}-\theta_c+\phi)$$

which further, using the recurrence relation of Bessel function (8.4.90(b)), becomes

(a8.4.36)
$$E_{r1}\cos(\phi-\theta)+E_{\theta 1}\sin(\phi-\theta)=\frac{j\,\omega\,B_{zo}}{k_t}\exp j(\omega t'-\beta z)\exp(-jm\theta_c)$$

$$\times\sum_{p=-\infty}^{\infty}J_{m+p}\{k_t r_c\}\,J'_p\{k_t r_L\}\exp jp(\frac{\pi}{2}-\theta_c+\phi)\;.$$

Substitution of (a8.4.36) in (a8.4.33) gives

(a8.4.37)
$$f_1 = |e|\int_{-\infty}^{t}\left(\left(1-\frac{\beta\,v_z}{\omega}\right)\frac{\partial f_0}{\partial p_t}+\frac{v_t\beta}{\omega}\frac{\partial f_0}{\partial p_z}\right)\left(\frac{j\,\omega\,B_{zo}}{k_t}\exp\, j(\omega t'-\beta z)\right)$$

$$\times\exp(-jm\theta_0)\sum_{p=-\infty}^{\infty}J_{m+p}\{k_t r_c\}\,J'_p\{k_t r_L\}\exp jp(\frac{\pi}{2}-\theta_c+\phi)\,dt'.$$

One may now easily identify the term dependent on t′ in the integrand of (a8.4.37) and carry out its integration as follows:

$$\int_{-\infty}^{t} exp\ j(\omega t' - \beta\ z)\ exp\ jp\left(\frac{\pi}{2} - \theta_c + \phi\right) dt' = \int_{-\infty}^{t} exp\ j[\omega t' - \beta\ z + p(\frac{\pi}{2} - \theta_c + \phi)]\ dt'$$

$$= \left[\frac{exp\ j(\omega t' - \beta\ z + p(\frac{\pi}{2} - \theta_c + \phi))}{j(\omega - \beta\ \frac{d\,z}{d\,t'} + p\frac{d\,\phi}{d\,t'})}\right]_{-\infty}^{t}$$

which, on substituting the upper and lower limits and putting $dz/dt' = v_z\{t'\} = v_z\{t\} = v_z$ *and* $d\phi/dt' = \omega'_c\{t'\} = \omega'_c\{t\} = \omega'_c = \omega_c/\gamma$ *(see (a8.4.20) and (8.4.24)), becomes*

$$\int_{-\infty}^{t} exp\ j(\omega t' - \beta\ z)\ exp\ jp\left(\frac{\pi}{2} - \theta_c + \phi\right) dt' = \frac{exp\ j(\omega t - \beta\ z)\ exp\ jp\left(\frac{\pi}{2} - \theta_c + \phi\right)}{j(\omega - \beta_z + p\omega_c/\gamma)}. \tag{a8.4.38}$$

In view of (a8.4.38), then (a8.4.37) is evaluated as:

$$f_1 = \frac{|e|\,B_{zo}}{k_t}\ exp\ j(\omega t - \beta\ z)\,exp - jm\,\theta_c\left((\omega - \beta\ v_z)\frac{\partial f_0}{\partial p_t}\right.$$

$$\left. + v_t\,\beta\frac{\partial f_0}{\partial p_z}\right)\sum_{p=-\infty}^{\infty}\frac{J_{m+p}\{k_t\,r_c\}\,J'_p\,\{k_t\,r_L\}}{\omega - \beta\ v_z + p\,\omega_c/\gamma}\ exp\ jp\left(\frac{\pi}{2} - \theta_c + \phi\right).$$

Appendix a8.4.4 Evaluation of the integral in the dispersion relation by the method of integration by parts

Let us divide the integral of (8.4.109) into two parts as follows

$$k_0^2 - \beta^2 - k_t^2 = \frac{-2\,\mu_0|e|^2\,N_0}{K_{mn}\,a^2}\,(\vartheta_1 + \vartheta_2), \tag{a8.4.39}$$

where ϑ_1 *and* ϑ_2 *, using the relation* $p_{z,t} = m_e\,v_{z,t} = \gamma m_{eo}\ v_{z,t}$*, may be expressed as:*

$$\vartheta_1 = \int_{p_t=0}^{\infty}\int_{p_z=-\infty}^{\infty}\left(\frac{p_t^2}{m_{eo}}\right)\left(\omega - \beta\frac{p_z}{\gamma\,m_{eo}}\right)\frac{\partial g}{\partial p_t}\sum_{p=-\infty}^{\infty}\frac{J_{m+p}^2\{x\}\,J'^2_p\{y\}}{\gamma\,\omega - \frac{\beta\,p_z}{m_{eo}} + p\omega_c}\,dp_t dp_z \tag{a8.4.40}$$

and

(a8.4.41) $$\mathcal{J}_2 = \int_{p_t=0}^{\infty}\int_{p_z=-\infty}^{\infty}\left(\frac{p_t^2}{m_{eo}}\right)\left(\frac{\beta\, p_t}{\gamma\, m_{eo}}\right)\frac{\partial g}{\partial\, p_z}\sum_{p=-\infty}^{\infty}\frac{J_{m+p}^2\{x\}\,J_p'^2\{y\}}{\gamma\,\omega-\dfrac{\beta\, p_z}{m_{eo}}+p\omega_c}\,dp_t dp_z,$$

where

(a8.4.42) (a) $$x = k_t\, r_c \quad and$$

(b) $$y = k_t\, r_L \ .$$

Now, let us take $\mathcal{J}_1$ *and* $\mathcal{J}_2$ *one by one. Integrating (a8.4.40) by parts with respect to* p_t, $\mathcal{J}_1$ *may be written as:*

(a8.4.43) $$\mathcal{J}_1 = \int_{p_z=-\infty}^{\infty}\left[\frac{p_t^2}{m_{eo}}(\omega-\beta\frac{p_z}{\gamma\, m_{eo}})\sum_{p=-\infty}^{\infty}\frac{J_{m+p}^2\{x\}\,J_p'^2\{y\}}{\gamma\,\omega-\dfrac{\beta\, p_z}{m_{eo}}+p\omega_c}g\right]_{p_t=-\infty}^{\infty} dp_z$$

$$-\int_{p_z=-\infty}^{\infty}\int_{p_t=0}^{\infty}\left[\frac{\partial}{\partial\, p_t}\left(\frac{p_t^2}{m_{eo}}(\omega-\beta\frac{p_z}{\gamma\, m_{eo}})\sum_{p=-\infty}^{\infty}\frac{J_{m+p}^2\{x\}\,J_p'^2\{y\}}{\gamma\,\omega-\dfrac{\beta\, p_z}{m_{eo}}+p\omega_c}\right)\right]g\ .$$

The first term of (a8.4.43) is zero, since, both at the upper and the lower limits $p_t = \pm\infty$, *the function* $g(p_t, p_z)$ *is zero. While differentiating the term within the integrand of its second term, it may be noted that* γ *depends on* p_t *and that* y *is dependent on* p_t *though* x *is independent of it, and that the following relations hold good:*

(a8.4.44) (a) $$\frac{\partial\gamma}{\partial\, p_t} = \frac{\eta_t}{m_{eo}C}$$ *(with the help of (8.4.33) and (8.4.111a)), and*

(b) $$\frac{\partial\, y}{\partial\, p_t} = k_t\frac{\partial\, r_L}{\partial\, p_t} = \frac{k}{m_{eo}\omega_c}$$ *(with the help of (a8.4.42), (8.4.28) and (8.4.31))*.

Thus putting the first term of (a8.4.43) equal to zero and evaluating the partial differentiation with respect to p_t *of its second term while remembering (a8.4.44),* $\mathcal{J}_1$ *becomes after a little simplification:*

(a8.4.45) $$\mathcal{J}_1 = \mathcal{J}_{1,1} + \mathcal{J}_{1,2} + \mathcal{J}_{1,3} + \mathcal{J}_{1,4}\ ,$$

where

(a8.4.46) $$\mathcal{J}_{1,1} = = -2\int_{p_z=-\infty}^{\infty}\int_{p_t=0}^{\infty}\sum_{p=-\infty}^{\infty}\left(\frac{p_t}{\gamma m_{eo}}\right)\frac{(\omega-\beta\, v_z)J_{m+p}^2\{x\}J_p'^2\{y\}}{\omega-\beta v_z+p\omega_c/\gamma}g\, dp_z\, dp_t$$

(a8.4.47) $$\mathcal{J}_{1,2} = -\int_{p_z=-\infty}^{\infty}\int_{p_t=0}^{\infty}\sum_{p=-\infty}^{\infty}\left(\frac{p_t}{\gamma m_{eo}}\right)\frac{\eta_t^2\,\beta\, v_z J_{m+p}^2\{x\}J_p'^2\{y\}}{\omega-\beta v_z+p\omega_c/\gamma}g\, dp_z\, dp_t$$

(a8.4.48)
$$\mathcal{I}_{1,3} = -\int_{p_z=-\infty}^{\infty}\int_{p_t=0}^{\infty}\sum_{p=-\infty}^{\infty} -2y\left(\frac{p_t}{\gamma m_{eo}}\right)\frac{(\omega-\beta v_z)J_{m+p}^2\{x\}J'_p\{y\}J''_p\{y\}}{\omega-\beta v_z+p\omega_c/\gamma} g\, dp_z\, dp_t$$

and

(8.4.49)
$$\mathcal{I}_{1,4} = \int_{p_z=-\infty}^{\infty}\int_{p_t=0}^{\infty}\left(\frac{p_t}{\gamma m_{eo}}\right)\frac{\eta_t^2\,\omega(\omega-\beta v_z)J_{m+p}^2\{x\}J_p'^2\{y\}}{(\omega-\beta v_z+p\omega_c/\gamma)^2} g\, dp_z\, dp_t \; .$$

Similarly, integrating (a8.4.41) by parts with respect to p_z $\mathcal{I}_2$ may be written as:

(a8.4.50)
$$\mathcal{I}_2 = \int_0^{\infty}\left(\frac{p_t^2}{m_{eo}}\frac{\beta p_t}{m_{eo}}\sum_{p=-\infty}^{\infty}\frac{J_{m+p}^2\{x\}\,J_p'^2\{y\}}{\gamma\left(\gamma\,\omega-\dfrac{\beta p_z}{m_{eo}}+p\omega_c\right)} g\right)_{p_z=-\infty}^{\infty}$$
$$-\int_{p_t=0}^{\infty}\int_{p_z=-\infty}^{\infty}\left(\frac{\partial}{\partial p_z}\left(\frac{p_t^2}{m_{eo}}\right)\left(\frac{\beta p_t}{m_{eo}}\right)\sum_{p=-\infty}^{\infty}\frac{J_{m+p}^2\{x\}\,J_p'^2\{y\}}{\gamma\left(\gamma\,\omega-\dfrac{\beta p_z}{m_{eo}}+p\omega_c\right)}\right) g \; .$$

The first terms of (a8.4.50), like that of (a8.4.43), is zero, since both at the upper and the lower limits $p_z = \pm\infty$, the function $g(p_t, p_z)$ is zero. While differentiating the term within the integrand of the second term it may be noted that γ depends on p_z, and that x and y are both independent of p_z, and that the following relations hold:

(a8.4.51)
$$\frac{\partial\gamma}{\partial p_z} = \frac{v_z}{m_0 c^2} \quad \text{(with the help of (8.4.33)).}$$

Thus putting the first term of (a8.4.50) equal to zero, and evaluating the partial differentiation with respect to p_z of its second term remembering (a8.4.51), $\mathcal{I}_2$ becomes after a little simplification:

(a8.4.52)
$$\mathcal{I}_2 = \mathcal{I}_{2,1} + \mathcal{I}_{2,2} \,,$$

where

(a8.4.53)
$$\mathcal{I}_{2,1} = \int_{p_t=0}^{\infty}\int_{p_z=-\infty}^{\infty}\sum_{p=-\infty}^{\infty}\frac{\eta_t^2(\omega\beta v_z-\beta^2C^2)\,J_{m+p}^2\{x\}\,J_p'^2\{y\}}{(\omega-\beta v_z+p\omega_c/\gamma)^2} g\, dp_t\, dp_z$$

and

(a8.4.54)
$$\mathcal{I}_{2,2} = \int_{p_t=0}^{\infty}\int_{p_z=-\infty}^{\infty}\sum_{p=-\infty}^{\infty}\left(\frac{p_t}{\gamma m_{e0}}\right)\frac{\eta_t^2\,\beta\, v_z J_{m+p}^2\{x\}\,J_p'^2\{y\}}{\omega-\beta v_z+p\omega_c/\gamma} g\, dp_t\, dp_z \, .$$

With the help of (a8.4.45) and (a8.4.52), one may then write

$$\mathcal{S}_1 + \mathcal{S}_2 = \mathcal{S}_{1,1} + \mathcal{S}_{1,2} + \mathcal{S}_{1,3} + \mathcal{S}_{1,4} + \mathcal{S}_{2,1} + \mathcal{S}_{2,2} \; . \tag{a8.4.55}$$

It may be noted from (a8.4.46)-(a8.4.49), (a8.4.53) and (a8.4.54) that the right hand side of (a8.4.55) may be divided into two groups — one containing $(\omega - \beta v_z + p\,\omega_c/\gamma)^2$ *and the other* $(\omega - \beta v_z + p\omega_c/\gamma)$ *in the denominator, thus enabling one to write (a8.4.55), shown within two parentheses, as follows:*

$$\mathcal{S}_1 + \mathcal{S}_2 = (\mathcal{S}_{1,4} + \mathcal{S}_{2,1}) + (\mathcal{S}_{1,1} + \mathcal{S}_{1,2} + \mathcal{S}_{1,3} + \mathcal{S}_{2,2}) \; . \tag{a8.4.56}$$

It is of interest to note, by examining (a8.4.47) and (a8.4.54), that

$$\mathcal{S}_{1,2} + \mathcal{S}_{2,2} = 0 \; . \tag{a8.4.57}$$

Therefore, substituting (a8.4.57) into (a8.4.56),

$$\mathcal{S}_1 + \mathcal{S}_2 = (\mathcal{S}_{1,4} + \mathcal{S}_{2,1}) + (\mathcal{S}_{1,1} + \mathcal{S}_{1,3}) \; . \tag{a8.4.58}$$

Now, substituting $\mathcal{S}_{1,4}$, $\mathcal{S}_{2,1}$, $\mathcal{S}_{1,1}$ *and* $\mathcal{S}_{1,3}$ *from (a8.4.49), (a8.4.53), (a8.4.46) and (a8.4.48), respectively, into (a8.4.58) one obtains after a little simplification:*

$$\begin{aligned} \mathcal{S}_1 + \mathcal{S}_2 = & \int_{p_t=0}^{\infty} \int_{p_z=-\infty}^{\infty} \Big(\sum_{p=-\infty}^{\infty} \eta_t^2 (\omega^2 - \beta^2 C^2) J_{m+p}^2\{x\} J_p'^2\{y\} \left(\frac{p_t}{\gamma m_{eo}} \right) \times \\ & \times g / (\omega - \beta v_z + p\omega_c/\gamma)^2 \Big) \, dp_t \, dp_z \\ & - \int_{p_t=0}^{\infty} \int_{p_z=-\infty}^{\infty} \Big(\sum_{p=-\infty}^{\infty} (\omega - \beta v_z)(2J_{m+p}^2\{x\} J_p'^2\{y\} + 2y J_{m+p}^2\{x\} J_p'\{y\} J_p''\{y\}) \left(\frac{p_t}{\gamma m_{eo}} \right) \times \\ & \times g / (\omega - \beta v_z + p\omega_c/\gamma) \Big) \, dp_t \, dp_z \; . \end{aligned} \tag{a8.4.59}$$

Now, if we substitute (a8.4.59) into (a8.4.39) and remember that x and y are given by (a8.4.42), we immediately get equation (8.4.110).

8.5 Gyro-TWT Gain Equation in the Conventional TWT Formalism

It will be quite useful for a practising microwave tube engineer to get a 'Pierce description' of a gyro-TWT [76] in terms of the familiar parameters of a conventional TWT (section *8.2.3*), such as the gain parameter (C), the velocity parameter (b), the interaction impedance (K), the length (N) of the device in terms of relevant wavelengths, etc., however, with interpretations appropriate to a gyro-TWT. To be precise, let us formulate a Pierce-type gain equation similar to one for a conventional TWT of the familiar form: G = A + BCN (see (8.2.116)).

For this purpose let us first proceed to solve the dispersion relation (8.4.114). Substituting k_t^2 from (8.4.10) in (8.4.114), dividing the latter by v_z^2 and choosing to put $\Gamma = j\beta$, corresponding to rf quantities varying with the axial distance as $\exp(-\Gamma z) = \exp(-j\beta z)$ (see section *8.2*), one may write the following alternative form of (8.4.114):

$$(\beta_{mn}^2 + \Gamma^2)(\beta_{e,gyro} + j\Gamma)^2 = \frac{-\mu_0 |e|^2 N_0 \eta_t^2 H_{m,-s} C^2 (k_0^2 + \Gamma^2)}{\gamma m_{e0} \pi K_{mn} a^2 v_z^2}, \tag{8.5.1}$$

where $\beta_{e,gyro}$ is given by (8.4.70). It may be mentioned that the dispersion relation (8.5.1) in which $H_{m,-s}$ is to be read from (8.4.115), agrees with that given elsewhere [76]. Let us now take [76]:

$$-\Gamma = -j\beta_{mn} + \beta_{mn,} C_{gyro}\delta , \tag{8.5.2}$$

with $C_{gyro}\delta << 1$ (which is similar to the relation $-\Gamma = -j\beta_e + \beta_e C\delta$ given by (8.2.56) for a conventional TWT, where $\beta_e = \omega/v_0$ is the beam propagation constant, v_0 being the dc beam velocity); and also let

$$\beta_{e,gyro} = \beta_{mn}(1 + b_{gyro} C_{gyro}) \tag{8.5.3}$$

with $b_{gyro}C << 1$ (which is similar to the relation $\beta_0 = \beta_e(1 + bC)$ given by (8.2.65) for a conventional TWT, β_0 being the cold-circuit axial propagation constant). Using (8.5.2), the term $(\beta_{mn}^2 + \Gamma^2)$ occurring in the left hand side of (8.5.1) becomes

$$\beta_{mn}^2 + \Gamma^2 = (\beta_{mn} + j\Gamma)(\beta_m - j\Gamma) = -j\,2\beta_{mn}^2 C_{gyro}\delta\left(1 + \frac{j\,C_{gyro}\delta}{2}\right)$$

$$\cong -j\,2\beta_{mn}^2 C_{gyro}\delta . \tag{8.5.4}$$

Similarly, using (8.5.2) and (8.5.3) the term $(\beta_{e,gyro} + j\Gamma)^2$ occurring in the left hand side of (8.5.1) becomes

$$(\beta_{e,gyro} + \Gamma)^2 = [\beta_{mn}(1 + b_{gyro} C_{gyro}) - \beta_{mn} + j\beta_{mn} C_{gyro} \delta_{gyro}]^2$$

$$= [\beta_{mn} C_{gyro} (b_{gyro} - j\delta)]^2 . \qquad (8.5.5)$$

Also, the term $(k_0^2 + \Gamma^2)$ occurring in the right hand side of (8.5.1) with the help of (8.5.2) becomes

$$k_0^2 + \Gamma^2 \cong k_0^2 - \beta_{mn}^2$$

which using (8.4.15) may be written as

$$k_0^2 + \Gamma^2 = k_{c,mn}^2 . \qquad (8.5.6)$$

Further, v_z^2 occurring in the right hand side of (8.5.1) may be expressed in terms of the beam voltage V_0 from the consideration of conservation of energy discussed following (8.4.33) that gives $|e| V_0 = (\gamma-1) m_{e0} v^2/(v^2/c^2)$ which, with the help of (8.4.23) and (8.4.32(a)), becomes

$$(1/2)\gamma m_{eo}(v_z^2 + v_t^2) = |e| V_0 (\gamma+1)/(2\gamma)$$

whence, taking $(\gamma+1)/(2\gamma) \cong 1$ for small values of γ, one gets

$$v_z^2 \cong \frac{2|e| V_0}{\gamma m_{e0} (1+\alpha_0^2)} , \qquad (8.5.7)$$

where

$$\alpha_0 = v_t / v_z \qquad (8.5.8)$$

is called the velocity pitch factor of the beam. Making use of (8.5.4)-(8.5.6), (8.5.7) and the relations $c = (\mu_0\varepsilon_0)^{-1/2}$, $\eta_z = v_z/c$, and

$$I_0 = N_0 |e| v_z , \qquad (8.5.9)$$

where I_0 is the dc beam current, one may express (8.5.1) as:

$$-2j\, \beta_{mn}^4\, C_{gyro}^3\, \delta\, (b_{gyro} - j\,\delta)^2 = \frac{-(\mu_0/\varepsilon_0)^{1/2} I_0 \eta_t^2 k_{c,mn}^2 (1+\alpha_0^2) H_{m,-s}}{2\pi K_{mn} a^2 V_0 \eta_z}$$

which may be arranged as

$$-j\, C_{gyro}^3\, \delta\, (b_{gyro} - j\delta)^2 = \frac{-K_{gyro} I_0}{4 V_0} , \qquad (8.5.10)$$

where K_{gyro} has the dimension of an impedance (that also ensures that the right hand side of (8.5.10) becomes dimensionless as its left hand side is), given by [76]:

$$K_{gyro} = \frac{(\mu_0/\varepsilon_0)^{1/2}\, \eta_t^2\, k_{c,mn}^2 (1+\alpha_0^2)\, H_{m,-s}}{\pi\, K_{mn}\, a^2\, \eta_z\, \beta_{mn}^4} . \tag{8.5.11}$$

Now one may choose to define C_{gyro} and δ vis-à-vis (8.5.2) such that C_{gyro} satisfies the relation

$$\frac{K_{gyro}\, I_0}{4\, V_0} = C_{gyro}^3 , \tag{8.5.12}$$

a relation that is similar to the relation $KI_0/4V_0 = C^3$ (see equation (8.2.55)) for a conventional TWT, where K is the interaction impedance of the slow-wave circuit. By analogy with the case of a conventional TWT, therefore, K_{gyro} may be called the interaction impedance of a gyro-TWT. Putting (8.5.12) into the right hand side of (8.5.10), one then gets

$$-j\, C^3\, \delta_{gyro} (b_{gyro} - j\,\delta)^2 = -\, C_{gyro}{}^3$$

which simplifies to:

$$\delta\, (\delta + j\, b_{gyro})^2 = j , \tag{8.5.13}$$

a result that corresponds to what was obtained for a conventional TWT as (8.2.73) which has to be read as $\delta^2(\delta + jb) = -j$ for $d = QC = 0$ (when the effects of circuit losses and space charges are neglected). For $b_{gyro} = 0$, one gets with the help of (8.4.70) and (8.5.3):

$$\omega - \beta_{mn}\, v_z - s\omega_c / \gamma = 0$$

that corresponds to the beam-mode resonance condition given by (8.4.71). Putting $b_{gyro} = 0$ in (8.5.13) one obtains

$$\delta^3 = j . \tag{8.5.14}$$

The solutions of (8.5.14) can be obtained by a method similar to what is followed in obtaining the solutions of (8.2.57) as (8.2.58) for a conventional TWT. The solutions of (8.5.14) thus obtained (beam-mode resonance) are:

$$\delta_1 = \frac{\sqrt{3}}{2} + j\frac{1}{2} , \quad \delta_2 = -\frac{\sqrt{3}}{2} + j\frac{1}{2} \text{ and } \delta_3 = -j . \tag{8.5.15}$$

The positive real part of δ_1 (growing-wave solution), namely, $\sqrt{3}/2$ given by (8.5.15) is found to be the same as that of δ_1 for a conventional TWT given by (8.2.58). For velocity parameters $b \neq 0$(conventional TWT) and $b_{gyro} \neq 0$ (gyro-TWT), x_1, the real part solution of δ_1, versus the velocity parameters, b and b_{gyro} for the conventional and gyro-TWTs, respectively, are plotted in Fig. 8.5.1. For $b = b_{gyro} = 0$, the same value of x_1, namely, $\sqrt{3}/2$ obtained for both the devices, as mentioned above, is indicated on the plot.

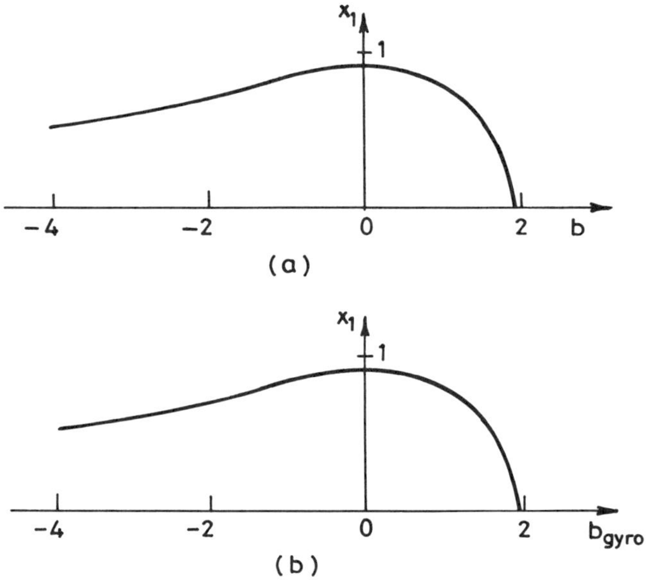

Fig. 8.5.1 The variation of x_1, the positive real part of the growing-wave solution δ_1 with the velocity parameter b, for a conventional TWT **(a)** and with the velocity parameter b_{gyro}, for a gyro-TWT **(b)**.

Putting $\delta_1 = x_1 + j\,y_1$, one may write (8.5.2) for the growing-wave component as:

$$-\Gamma_1 = -j\,\beta_{mn} + \beta_{mn}\,C_{gyro}\,\delta_1$$

$$= -j\,\beta_{mn} + \beta_{mn}\,C_{gyro}\,(x_1 + j\,y_1) = \beta_{mn}\,C\,x_1 - j\,\beta_{mn}\,(1 - C\,y_1)\,, \quad (8.5.16)$$

where x_1 is positive real (for instance $x_1 = \sqrt{3}/2$ for $b_{gyro} = 0$), an expression that corresponds to a similar expression (8.2.75) for a conventional TWT.

The gain in decibels of the gyro-TWT may then be written interpreting x_1 following the same method as used in section *8.2.4* as:

$$G = A + B_{guide}\,C_{gyro}\,N_{gyro,\,guide}\,, \quad (8.5.17)$$

where A in terms of δ_1, δ_2 and δ_3 continues to be given by (8.2.113) (here, QC = 0), where $N_{gyro,\ guide}$ is the interaction length expressed in number of guide wavelengths given by (see a similar expression (8.2.115) for a conventional TWT):

$$\beta_{mn}\ell = 2\pi N_{gyro,\ guide} \ . \tag{8.5.18}$$

Also, in (8.5.17) B_{guide} continues to be given by (8.2.117) as

$$B_{guide} = 40\pi(\log_{10} e)x_1 \cong 54.6\ x_1 \ . \tag{8.5.19}$$

One may also choose to express the gain formula (8.5.18) in terms of the interaction length in number of growing-wave wavelengths [79], $N_{gyro,\ growing}$, which is given by an expression similar to (8.5.18) as:

$$\beta_{growing}\ell = 2\pi N_{gyro,\ growing} \ , \tag{8.5.20}$$

where $\beta_{growing}$ may be written from (8.5.16) as

$$\beta_{growing} = \beta_{mn}(1 - C y_1) \ . \tag{8.5.21}$$

With the help of (8.5.18), (8.5.20) and (8.5.21) one gets the relation:

$$N_{gyro,guide} = \frac{N_{gyro,growing}}{1 - C y_1} \ . \tag{8.5.22}$$

Substituting (8.5.19) and (8.5.22) in (8.5.17), one gets an expression alternative to (8.5.17) as:

$$G = A + \frac{54.6\,x_1\,C_{gyro}\,N_{gyro,growing}}{1 - C y_1}$$

which may be put as [79]:

$$G = A + B_{growing}\,C_{gyro}\,N_{gyro,\ growing} \ , \tag{8.5.23}$$

where

$$B_{growing} = \frac{54.6\,x_1}{1 - C y_1} \ . \tag{8.5.24}$$

It may be noted that the gain parameter C_{gyro} occurring in the gyro-TWT gain expression is related to the interaction impedance K_{gyro} as (8.5.12), where K_{gyro} is given by (8.5.11). However, it will be of interest to compare K_{gyro} given by (8.5.11) with an impedance parameter K'_{gyro}, say, that can be defined, in terms of the rf electric field supported by and the power P propagating through the interaction structure, by an expression analogous to (5.5.4), the latter referring to a conventional TWT. Considering an azimuthally symmetric TE_{mn}-mode (m = 0) of the cylindrical-waveguide interaction structure, that is, a TE_{on}-mode, one may thus define K'_{gyro} as [76]:

$$K'_{gyro} = \frac{\left|E_{\theta,0n}\right|^2}{2\,\beta^2\,P}, \tag{8.5.25}$$

since, unlike in a conventional TWT, in such a mode (TE_{0n}) supported by the structure of a gyro-TWT, it is the rf azimuthal electric field, $E_{\theta,\,0n}$, rather than the rf axial electric field, E_z of the slow-wave structure, that interacts with the electron beam of the device. One may reasonably well take K'_{gyro}, obtainable from (8.5.25), instead of the more rigorous K_{gyro} given by (8.5.11), in the expression (8.5.12) for the gain parameter C_{gyro}, since it is found, typically for a TE_{01}-mode, that $K'_{gyro} \cong K_{gyro}$ over an approximately 10% bandwidth whose lower end is not close to the cut-off frequency of the waveguide [76].

The interaction impedance of the gyro-TWT is as important as it is in the case of a conventional TWT. It is found that the interaction impedance of a gyro-TWT decreases with frequency as it does in the case of a conventional TWT [76]. The choice of the cyclotron harmonic (beam-mode) number, s, and the waveguide-mode number, n, becomes important from the standpoint of attaining a reasonably high value of interaction impedance. A typical study of a TE_{0n}-mode gyro-TWT brings out many important features of interaction [76]. With the increase of s, the required external dc magnetic field is reduced (see (8.4.21)) but at the cost of the value of the interaction impedance. Also, with the increase of s, the value of the hollow beam radius r_c increases that could make difficult the confinement of the beam within the waveguide dimensions. Also, the range of r_c/a, for which the solution of the dispersion relation exists corresponding to a gain of the device, decreases with s, and, in the limit, giving no such solution for very high values of s. The waveguide size can be increased by choosing a higher-order waveguide mode number n (see Example 8.4.1) making it easier to contain an electron beam within the waveguide dimensions. With the increase of n, for a given value s, the value of the relative hollow beam radius r_c/a, however, could decrease to a value that would require a larger external dc magnetic field for confinement. The reduced hollow beam radius will also mean a reduced beam power. At a compromised value of $r_c/a \sim 0.4$-0.5, the interaction is found to optimise corresponding to a maximum value of the interaction impedance, if the beam mode order is matched to the waveguide-mode order (n = s). The major disadvantage of employing higher order waveguide modes to facilitate cyclotron-harmonic operation at n = s is that it introduces an increasingly difficult mode-suppression problem [76]. At this juncture, it is worth recalling the discussions on cyclotron-harmonic operation in section *8.4.2A*, with reference to second-harmonic operation, in which an electron was considered to orbit in an rf electric field that spatially reverses at the centre of the orbit. The waveguide mode n = 2 permits such field reversal near the mid-point of the centre and the wall of the waveguide cross section [67]. Therefore, the operation of the device using a TE_{02}-mode waveguide and a second harmonic beam resonance (n = s = 2) does ensure the placement of the beam at a suitable position within the waveguide dimensions and at the same time provide optimum interaction.

The grazing point operation of a gyro-TWT is usually selected at a frequency within 3% of the cut off frequency of the waveguide, ω_{cut} (see section *8.4.3*) [76]. The operation of the gyro-TWT at the intersection between the waveguide and beam-mode dispersion plots, therefore, requires that both ω_{cut} and the beam mode intercept frequency, $s\omega_c/\gamma$, should be increased simultaneously (Fig. 8.4.2), if the operating frequency has to be increased. ω_{cut} can be increased by reducing the waveguide size or by using higher-order waveguide modes, and the beam mode intercept frequency can be increased by increasing the external dc magnetic field or using a higher-order cyclotron harmonic mode. For wide-band operation of the device, it is necessary to maintain a wide-band coalescence between the waveguide and beam modes. This requires to optimise the controlling parameters such as the dc magnetic flux density (B_0), beam voltage (V_0), velocity pitch factor (α_0), waveguide mode number (n), axial beam velocity spread, etc [79]. Besides, for wide-band performance one may go for innovative schemes like simultaneous tapering of the waveguide cross section and the external magnetic field [80]. In such a scheme the gain and stability can be further enhanced by an inherent mechanism of two-stage amplification and a sever in the device [81]. Also, for wide bandwidths, one may consider the shaping the dispersion characteristics of the interaction structure. Thus one may load the cylindrical waveguide, say, by placing a circular-cross section dielectric rod at the axis of the guide [82], by dielectric-lining the guide [83], by placing a helix close to the guide wall [84], by placing along the structure annular discs at regular axial intervals [72], etc. It may be mentioned here that just as an attenuator, for instance, in the form of a lossy coating on helix-support dielectric rods is provided in a conventional TWT, (section *8.2.4A*), one may use a lossy-wall waveguide in a gyro-TWT to stabilise it against oscillations caused by reflections at the input and output ends [66]. The efficiency of a gyro-TWT can be increased by the usual multistage depressed collection technique (appendix *a8.2.3*) as is done in the case of a conventional TWT. However, in a gyro-TWT one must employ a method of converting the rotational energy of the beam electrons into their axial energy, that is, of reducing the value of α_0 before the electrons of the spent beam proceed for depressed collection. This can be done by employing a magnetic cusp at the entry point of the collector system which unwinds the encircling beam electrons [85].

In this section, the gyro-TWT gain expression is deduced giving a Pierce-type description of the device using the usual terminology of a conventional TWT. While drawing the analogy between the analytical results of these two devices, however, care must be taken to note the manner in which the basic parameters have been defined (appendix *a8.5.1*). The analysis has been developed here with particular reference to a conventional small-orbit gyro-TWT in which the axes passing through the guiding centers of the gyrating electrons are situated off from the waveguide axis. It may be mentioned that a modified version, namely, the large-orbit gyro-TWT has also come into existence in which the waveguide axis-circling electrons interact with a higher-order cylindrical waveguide mode at high-harmonic beam resonance which reduces the required dc magnetic field from an intense to a reasonably small value [86].

Appendix a8.5.1 Comparison between the conventional TWT and gyro-TWT interactions [76], [86]

	Conventional TWT	*Gyro-TWT*
Qualitative:		
wave	*slow*	*fast*
phase velocity	$< c$	$> c$
phase synchronism	$\omega/\beta \le v_0$	$\omega - \beta_{mn} v_z - s\omega_c/\gamma \ge 0$
bunching		
alignment	*axial*	*transverse*
force	*electric*	*electric*
mechanism	*non-relativistic*	*relativistic*
field responsible for energy extraction	*axial, electric*	*transverse, electric*
free energy	*axial electron velocity*	*transverse electron velocity*
Analytical:		
$-\Gamma\,(=-j\beta)$	$-\Gamma = -j\beta_e + \beta_e C\delta$	$-\Gamma = -j\beta_{mn} + \beta_{mn} C\delta$
β_e	$\beta_e = \omega/v_0$	$\beta_{e,\,gyro} = (\omega - s\omega_c/\gamma)/v_z$
b	$b = (\beta_0 - \beta_e)/\beta_e C$	$b_{gyro} = (\beta_{e,\,gyro} - \beta_{mn})/\beta_{mn} C_{gyro}$
C	$C = (KI_0/4V_0)^{1/3}$	$C_{gyro} = (K_{gyro} I_0/4V_0)^{1/3}$
cubic equation in δ *(no structure loss, no space-charge effects)*	$\delta^2(\delta + jb) = -j$	$\delta(\delta + jb_{gyro})^2 = j$
solutions of cubic equation	$\delta_1, \delta_2, \delta_3$	$\delta_1, \delta_2, \delta_3$
positive real part solution of cubic equation in δ	$\delta_1 = x_1 + j y_1$	$\delta_1 = x_1 + j y_1$
N	$N = \beta_e l/2\pi$	$N = N_{gyro,\,growing}$
A *(no space charge effect)*	$A = 20 \log_{10} \lvert 1/(1-\delta_2/\delta_1)(1-\delta_3/\delta_1) \rvert$	$A = 20 \log_{10} \lvert 1/(1-\delta_2/\delta_1)(1-\delta_3/\delta_1) \rvert$
B	$B = 54.6\, x_1$	$B = B_{growing} = 54.6\, x_1/(1 - Cy_1)$
G	$G = A + BCN$	$G = A + B_{growing} C_{gyro} N_{gyro,\,growing}$

8.6 Summary

By applying the fundamental principles of time-varying fields developed in chapter *4* the mechanism of a practical device in which an electron beam is used as one of its constituents can easily be understood. Four electron beam devices, namely, the TWT, the double-stream amplifier, the beam-plasma amplifier and the gyro-TWT have been studied here. The first of these devices uses an interaction structure which is essentially a slow-wave structure that supports an electromagnetic wave whose phase velocity is made less than the speed of light in free space, c, by the axial periodicity of the structure. In the second of these devices, another electron beam is used as a medium that plays the role that the slow-wave structure does in the TWT. In the third, the medium that plays this role is a plasma. In the fourth device considered, an annular relativistic electron beam of electrons in helical trajectories interacts in gyro-resonance with an electromagnetic wave supported by a conventional hollow-metal-pipe waveguide whose phase velocity is greater than c. Thus the first three of the four growing-wave devices considered here belong to the family of *slow-wave* devices while the last one is essentially a *fast-wave* device.

The electron beam as an isolated system supports two space-charge waves which are neither of growing-wave nor of attenuating-wave type. One of these is the fast space-charge wave while the other is the slow space-charge wave, their phase velocities being respectively greater and less than the dc beam velocity. The amount by which the phase velocities of these waves differ from the dc beam velocity depends upon the plasma frequency of beam electrons which is the natural frequency of oscillation of electrons when they are perturbed from their equilibrium position in the neutralising background of positive ions. For a beam of finite transverse cross-sectional area which is bounded in a metallic enclosure, the value of the plasma frequency is reduced by a factor, called the plasma-frequency reduction factor, from its value for a beam of infinite cross-sectional area. Excited by a narrow-gap velocity modulator at the input plane, these space-charge waves with their dissimilar velocities will set up standing waves in relevant beam quantities. The nodes of the beam current density standing wave coincide with the antinodes of the beam velocity standing wave, and vice-versa, the input plane being one of the locations of the beam current density node and beam velocity anti-node. Further, the beam current density and velocity standing waves are axially separated from each other by quarter of a plasma wavelength, and the distance between the consecutive nodes or antinodes for either of the current density and velocity standing waves is equal to half the plasma wavelength.

Next a very important result is found from Chu's power conservation theorem which forms the basis of energy exchange between an electron beam and a nearby circuit. It is found that the slow space charge wave on the beam carries a negative kinetic power density the excitation of which is accompanied by the delivery of power from the beam to the circuit. In a growing-wave device like the TWT, therefore, the beam velocity is made close to but greater than the circuit phase velocity to ensure that the slow space charge wave is coupled to the circuit wave. In such a device, therefore, the spatial growth of the amplitude of the slow space-charge wave becomes responsible for the corresponding growth of the circuit wave.

Pierce's method of finding the dispersion relation of a TWT is to find the ratio of the circuit voltage to rf beam current in the form of the circuit equation and also to find the same ratio in the form of the electronic equation. These two forms of the ratio are then equated to obtain the dispersion relation of the TWT. In one approach, the circuit equation is found with the help of the voltage and current telegraphist's equations, taking care to incorporate into the current telegraphist's equation the modification due to the presence of the modulated electron beam the effect of which manifests itself in the form of a displacement current that is considered to be totally linked up with the circuit from the beam, making a simplifying assumption that the beam is thin and located close to the circuit. The circuit equation so formed involves the characteristic impedance of the transmission-line equivalent of the slow-wave structure. In a more general and realistic approach the effect of an element of the modulated beam is simulated by an infinitesimal current generator sending two circuit waves in opposite directions. The electric field intensity at a circuit point is found by integrating the contributions from all such infinitesimal current generators to the left and to the right of the point and then adding them to the contribution from the injected input signal. The approach yields the circuit equation which is found as identical with that found by the first approach except that now the interaction impedance replaces the characteristic impedance.

The electronic equation is found basically from the force equation of an electron subjected to the circuit-electric and space-charge fields. The dispersion relation of the TWT, which is obtained by equating the expressions for the ratio of the circuit voltage-to-rf beam current, obtained respectively in the forms of the circuit and electronic equations, is essentially a fourth-degree equation. The four solutions of this dispersion relation becomes quite easy to obtain if it is assumed that they are close to the beam propagation constant and that there would be three forward waves and one backward wave present in the beam-wave coupled system, in view of the nature of the waves which are independently supported by the circuit and the beam when they are considered as isolated systems — the electron beam supporting two forward space-charge waves and the slow-wave circuit supporting one forward wave and also one backward wave if the circuit is not properly matched. The solutions are expressed in terms of the Pierce's gain parameter C, which depends upon the interaction impedance of the slow-wave structure, beam voltage and beam current.

It is easy to get the analytical solutions of the dispersion relation if it is assumed that the dc electron beam velocity and the circuit phase velocity are synchronous, the structure losses are nil, and the space-charge effects are absent, that is, the electrons experience only the circuit electric field. On examining the solutions in this simplified situation, it is found that one of the forward waves exponentially grows with the axial distance and has a phase velocity less than the dc beam velocity. The second of the forward waves is exponentially attenuated with the axial distance and the rate of its attenuation is same as the rate of the growth of the first wave; also, the phase velocity of the second wave is found to be the same as that of the first wave. The third forward wave neither grows nor attenuates with the axial distance and its phase velocity is greater than the dc beam velocity. The fourth solution of the dispersion relation that corresponds to a backward wave also neither grows nor attenuates with the axial distance in the backward direction, and its phase velocity in this direction is very slightly greater than the dc beam velocity.

One may numerically obtain the solution of the dispersion relation to include the effects of lack of synchronism, structure losses and space charge. These effects are estimated in terms of Pierce's velocity, loss and space-charge parameters b, d and QC, respectively. Some of the typical findings are as follows. There exists an optimum value of b, which is positive corresponding to a beam having a dc velocity greater than the circuit phase velocity, which corresponds to the maximum growth rate of the forward growing-wave component. With the increase of d, the growth rate of the growing-wave component decreases. The growth rate also decreases with QC. Also with the increase of QC, the optimum value of b for the maximum growth rate increases, and the allowable range of b corresponding to a growing wave becomes narrower.

The TWT gain equation can be found by considering the contribution from only three forward-wave components, it being assumed that the circuit is so matched that the fourth wave, which is a backward wave, is not generated. The contribution to the output circuit voltage predominantly comes from the exponentially-growing forward-wave circuit voltage component. The input circuit voltage is constituted by all the forward-wave circuit voltage components. Also we invoke the condition that at the entry of the interaction region the beam is unmodulated making the sums of the three forward-wave velocity and current density components are each equal to zero. Also each of the velocity components can be expressed in terms of the corresponding circuit voltage component using the force equation of an electron and similarly each of the current density components can be expressed in terms of the corresponding circuit voltage component using the electronic equation. This enables one to express the growing-wave component of circuit voltage at the input in terms of the total input circuit voltage, the latter comprised of all the forward-wave circuit voltage components. The procedure finally leads to the expression of the output circuit voltage in terms of the input circuit voltage whence the gain of the device can be found. In general, the expression for the gain in decibels is found in terms of the space-charge parameter QC and the three forward-wave solutions of the dispersion relation of the TWT. The first part of this expression can have a negative value and thus represent the launching loss of the device. The second part of the gain expression contains the real part solution of the growing-wave component, the interaction length expressed in number of beam wavelengths and the Pierce's gain parameter which, in turn, depend upon the interaction impedance of the slow-wave structure and the dc beam voltage and current. Since the solutions of the dispersion relation depend upon Pierce's parameters b, d and QC, the gain of the TWT also does depends upon these parameters.

The in-built attenuator in a TWT prevents the device from going into oscillations caused by mismatches at the input and output couplers or at circuit severs if the circuit is split into a number of sections. The analysis of the attenuator is easily carried out by assuming that the circuit voltage is attenuated to an insignificant value by an attenuator providing a high cold attenuation and that the rf beam velocity and current density modulations remain unaffected in the passage through the attenuator. The analysis shows that, although the cold attenuation of the attenuator is high, yet its hot attenuation value is rather low, or in other words, the presence of the attenuator reduces the overall gain of the TWT only to a small extent.

An overview of various subassembly design considerations brings out many important points relevant to the performance improvement of the TWT (appendix *a8.2.3*). The particular performance that needs attention depends upon the system in which the device is used. For instance, in electronic warfare system applications the bandwidth of the TWT is an important consideration; for television broadcasting satellite communication/space applications the efficiency and the life of the TWT become important.

The helix TWTs have a wider bandwidth potential. By anisotropically or inhomogeneously loading the helix, its dispersion can be controlled to achieve wide-band performance. Typically, an envelope provided with radially inward metal vanes can be used with its dimensions optimised to give the required anisotropic loading; the tapered-geometry support rods can provide the required inhomogeneous loading. Two-section helix — one section with conventional loading and the other with anisotropic loading — has also been designed for ultra-band performance. The negative dispersion characteristics which correspond to phase velocity increasing with frequency yields a wider-band performance than the dispersion-free characteristics of the helix. The negative dispersion that makes Pierce's velocity parameter (b) constant with frequency also reduces the harmonic content, AM/PM distortion etc. of the device.

The efficiency of the TWT can be increased by velocity profiling the interaction structure, say, by reducing the pitch in the case of a helix TWT or by providing a velocity-jump cavity in the case of a coupled-cavity, both near the output end. The multi-stage depressed collector system is also another method of efficiency improvement. The method is to sort the electrons according to their energies in the spent beam and collect them at appropriate potentials in number of stages such that they 'soft-land' on the collector. The average power of the TWT can be increased by improving the thermal management of the interaction structure. A coupled-cavity structure is better than a helix from thermal dissipation standpoint. In a helix TWT the thermally-conducting support materials such as boron nitride, anisotropic pyrolytic boron nitride (APBN), beryllia, diamond etc. should be used. The brazed-helix technology that reduces the thermal contact resistances at the interfaces between the helix and the supports and between the supports and the envelope has considerably improved the thermal management of the helix. For a pulsed TWT, the use of an NIG gun also improves the average power capability of the device. The peak power capability of the TWT can be increased by increasing the dc beam voltage. As per the required dc beam voltage, therefore, one has to choose the interaction structure. Thus the centipede structure has a higher peak power capability than the coupled-cavity structure while the latter has a higher peak power capability than the helical structure. The peak power capability of the helix can, however, be increased by increasing its diameter and then suppressing the backward-wave oscillation in such a large diameter helix with the help of a resonant attenuator or a tapered-pitch fast space-charge wave coupling type attenuator or by a conical-helix attenuator. The band-edge oscillation caused by asymmetry in helix supports with respect to azimuthal positioning or support cross sections can be suppressed by providing a distributed loss along the interaction structure. For high-power and high-efficiency capabilities, the structure is also severed into a number of sections provided with matched loads at the severe points.

The operating frequency of the TWT can be increased to the millimeter-wave range by decreasing the transverse dimension of the interaction structure. This demands high-precision and high-tolerance tube parts. Also a large dc focusing magnetic field is required to confine the thin beam of a millimeter-wave TWT, such a beam also containing thermal electrons formed by a low-perveance gun used for the purpose. Increased cathode emission density to compensate for the small cathode diameter poses a reliability problem. The miniaturised parts also make the thermal management more difficult. In the case of a helix-TWT, the brazed-helix technology is difficult to implement. The helix supported by plasma-sprayed beryllia on the helix surface with pressure fitting has been used for the purpose. Also ridged helical waveguides have been tried out.

The coupling of rf power in and out of the interaction structure is also an important design consideration that has to be commensurate with the design of the interaction structure itself. The helical tapered-waveguide, elliptical-cavity and impedance-matched coaxial-cable couplers are used for this purpose. The reduction of mismatches at the input and output couplers as well as at the ends of the sections into which the interaction structure could be split is necessary to reduce the gain and phase ripples. The mismatches can be reduced by providing, a tapered loss in the structure, for instance, in a coupled-cavity TWT, by providing small cavities with lossy ceramic buttons the protrusions of which are suitably varied along the interaction length. The gain of the device or that of each section of the device, if it is severed into a number of sections, should also be reduced for the reduction of gain and phase ripples. It is also found that the input power level with respect to saturation has much relevance to AM/PM conversion of the device, the latter also being frequency-dependent, becoming higher at the high frequency end of the tube's pass band. The phase compensation in a helix TWT can be accomplished by using a helix section of increased pitch section. For the improvement of gain and phase characteristics the dynamic velocity taper has also been implemented.

The key to getting a long-life space-TWT is to increase the life of the cathode and use an electronic power conditioner provided with protection measures against alarming extremes.

Thus the performance of the TWT can be increased by optimising the design of subassemblies. Also, with modification in the design, the tube can be used for non-conventional uses, for instance, as a multi-mode tube using dual-perveance or tetrode guns. Also, the performance of the TWT can be controlled by external means. For example, a number of pulsed TWT's can be combined to obtain a higher pulse power over a wide bandwidth as well as for a higher intrinsic reliability. Similarly, the electronic efficiency of a broadband TWT can be increased by the injection of a harmonically-related signal with a proper amplitude and phase relative to the signal to be amplified.

The device that has been taken up next for study is the double-stream amplifier (DSA) — a growing wave device in which two electron beams of comparable volume charge densities but slightly different dc velocities interact with each other. In this device, the slow-wave structure is dispensed with and its role is actually played by one of the beams. The absence of an interaction circuit relaxes the requirement of confining the electron beam close to the circuit for a better interaction. Also, the risk of the interaction structure being destroyed by the beam intercept current, the problem of maintaining wide-band synchronisation between the circuit phase velocity and the beam velocity, etc. do not arise. The amplification range and the optimum condition for maximum gain of the device can be stated in terms of an inhomogeneity factor which is defined as the product of two ratios, namely, the ratio of the signal frequency-to-the geometric mean of the plasma frequencies of the electrons of the two beams and the ratio of the dc velocity difference between the two beams-to-the mean of the dc beam velocities.

The analysis is carried out to derive the dispersion relation of the DSA considering that an electron of each beam experiences the space-charge field of both the beams and that there is no variation of rf magnetic field over the beam cross section. This makes the sum of the convection and displacement current densities due to the time-varying space-charge electric field for each beam equal to zero. The examination of the dispersion relation shows that the growing-wave solution exists over a range of inhomogeneity factor and that the growth rate becomes maximum at an optimum value of the inhomogeneity factor. It is found that the operating frequency is more sensitive to the relative dc beam velocity difference than to the relative values of beam plasma frequency (or charge density) of the two beams. Although the device has the advantage of being of high gain and wide bandwidth, yet it has some limitations. The cathode of a DSA is more complicated than that of a TWT. The potential depression in a beam of high current density as required for high operating frequencies stops the collector current. The reduction of dc velocity difference between the two beams for increasing the operating frequency becomes limited by the velocity spread of electrons. Also the maximum available output in this device is restricted by the energy difference between the two beams and this limits the device efficiency. The output stage of the DSA, say, a helix ultimately prevents the device from superseding the TWT in performance since the device may be looked upon as a TWT with enhanced bunching due to the two-stream interaction.

The beam-plasma amplifier (BPA) is another device similar to a DSA in which the role of a slow-wave structure is played by a medium — a plasma. The oscillation of beam electrons in this device is aided by the total space-charge in the presence of the plasma. The BPA may be treated as a DSA in which one of the beam velocities is taken as zero. The dispersion relation of the BPA thus obtained predicts amplification range extending upto the plasma frequency of the plasma electrons at which the predicted growth rate becomes infinity. However, if a finite plasma temperature is considered, the dispersion relation gets modified, which predicts a finite maximum growth rate at a value somewhat less than the plasma frequency of plasma electrons, and also then the amplification range extends beyond this plasma frequency value.

In a BPA the millimeter-wave amplification is realised by creating a high-density plasma. The device provides wide-band tunability through the adjustment of plasma electrodynamic characteristics. The absence of almost no rf heating is an advantage of the device. The device, unlike a TWT, requires no synchronism giving continuous bunching without efficiency degradation. The device gain being relatively independent of beam density, a large diameter beam can be used in a BPA. This relaxes the requirement of high-density cathodes and a large gun convergence. The power of the device can be increased by using a high density beam penetrating through a plasma that can be isolated from a metallic enclosure. With proper care some of the problem areas can be handled. For instance, using a negative-glow mode discharge the noise level of the BPA can be reduced; the damage of the cathode by the bombardment of positive ions may be prevented by the employment of ion discharge holes and hollow cathodes; oscillations and fluctuations in the device can be suppressed by the choice of geometries and other working parameters like pressure and magnetic field.

Finally, the analysis is developed for a growing-wave device, namely, the gyro-TWT which is essentially a fast-wave device, unlike the TWT, DSA and BPA which belong to the family of slow-wave devices. In a gyro-TWT, an annular relativistic beam of electrons in helical trajectories is formed by a magnetron injection gun and they are made to interact with a TE-mode supported by a hollow-pipe waveguide. For high operating frequencies in the millimeter-wave band, the size of the waveguide, say, a cylindrical waveguide becomes much larger than that of the slow-wave structure, say, a helix of a conventional TWT. Also, it is possible to place the electron beam in a gyro-TWT far from the wall of the guide. The interaction structure becomes much simpler and the thermal management much easier in a gyro-TWT than in a TWT. In a gyro-TWT the transverse kinetic energy of electrons is converted into electromagnetic energy unlike in a TWT in which, the axial kinetic energy is converted; also the bunching in a gyro-TWT is transverse and is based on relativistic effects unlike in a TWT in which the bunching is axial and non-relativistic.

The gyro-TWT operates at grazing intersection between the beam-mode cyclotron and waveguide-mode resonances near the cut-off frequency of the waveguide. The required dc magnetic field becomes reduced if the device is operated at a cyclotron-harmonic resonance. For second cyclotron-harmonic resonance, a cylindrical guide may be used in the TE_{02}-mode in which the centre of gyration of electrons is located at the waveguide cross section, approximately midway between the centre and the wall of the guide, where the azimuthal rf electric field spatially reverses. The phase bunching of electrons takes place because those electrons which gain energy from the rf electric field become heavier, their angular velocities decrease and they accumulate a phase lag, while those which lose energy become lighter, their angular velocities increase and they accumulate a phase lead. The resonance condition is so tuned that the bunched electrons find themselves continuously in the decelerating rf electric field.

The dispersion relation has been derived for a TE-mode gyro-TWT. For this purpose one may write the wave equation for the axial component of rf magnetic field in the presence of the electron beam. Under the tenuous-beam approximation it is assumed that the spatial structure of the waveguide-mode is unaffected by the presence of the beam. This enables one to substitute into the wave equation the cold-beam field expressions in which, however, the axial propagation constant is taken as one corresponding to the beam-wave coupled system. Multiplying the wave equation by an appropriate function of radial coordinate and integrating it over the waveguide cross section between the centre of the guide and its wall, the dispersion relation is obtained. The relation involves the axial and the radial components of rf beam convection current density each of which is obtained as the integration over the momentum space of the product of the perturbation part of the electron distribution function in phase space and the relevant rf beam velocity component. The perturbation part of the distribution function in turn is obtained by solving Vlasov's equation which is valid for the assumed tenuous beam. The latter involves the unperturbed part of the distribution function expressed in terms of Dirac-delta functions considering a cold beam of small energy and small guiding-centre spread. The integral form of dispersion relation involves a summation of which, however, only one term that corresponds to a beam-mode resonance becomes dominant over the remaining terms. The integration in the dispersion relation involved is finally evaluated invoking the unit-impulse property of the Dirac-delta function. Thus the dispersion of the gyro-TWT is obtained in terms of the number of electrons per unit axial length, waveguide and magnetic field parameters and functions of guiding-centre radius of gyrating electrons and Larmor radius. For weak coupling, the dispersion relation of the beam-wave coupled system of the gyro-TWT is decoupled into the beam- and waveguide-mode dispersion relations.

For microwave-tube engineers the statement of the results of the gyro-TWT in terms of the familiar Pierce's conventional TWT parameters such as the velocity or synchronisation parameter, space-charge parameter, interaction impedance, gain parameter etc. proves to be quite useful. Thus the solutions of the dispersion relation of the gyro-TWT are found assuming them to be close to the cold waveguide propagation constant as they are assumed for the conventional TWT to be close to the beam propagation constant. This enables one to express the gyro-TWT gain in a form similar to the well known Pierce's G = A + BCN formula for a conventional TWT.

A higher cyclotron harmonic device calls for a reduction in the required dc magnetic field. However, the use of a higher-order cyclotron harmonic number causes a reduction in the value of the interaction impedance. Also, it could increase the value of the hollow-beam radius making the confinement of the beam within the waveguide dimensions difficult; it also reduces the range of solution expressed in terms of the hollow-beam radius relative to the waveguide radius, in the limit giving no solutions for high harmonic numbers. The use of higher waveguide numbers permits larger waveguide sizes but reduces the hollow beam radius calling for a larger dc magnetic field and reduced beam power. At a compromised location of the beam, midway between the axis and the wall of the waveguide, the maximum interaction impedance is realised if the beam-mode order is matched to the waveguide-mode order. The choice of higher-order waveguide modes, however, makes the suppression of higher harmonics more difficult. The use of $n = s = 2$ permits the electron beam to be located approximately midway between the axis and the wall of the waveguide where the rf electric field spatially reverses at the centre of orbit of the gyrating electrons. The operating frequency of the gyro-TWT may be increased by either decreasing the size of the waveguide or increasing higher waveguide-mode numbers and simultaneously increasing the dc magnetic field or increasing the beam-mode number. For a wide bandwidth, the controlling parameters such as the dc magnetic field, beam voltage, velocity pitch factor, beam velocity spread, etc. should be optimised. Also, the methods like the simultaneous tapering of waveguide size and dc magnetic field or the dielectric or helix-loading for the control of dispersion characteristics etc. may be employed. The efficiency may be increased by first subjecting the spent beam to a magnetic cusp to convert the rotational energy of beam electrons into their axial energy and then to a multi-stage depressed collector system.

Problems

8.1 A 3 kV electron beam of diameter 1.5 mm and current 70 mA supports space-charge waves at 6 GHz. Calculate (a) the plasma frequency of beam electrons, making an infinite beam-cross section approximation (b) the beam velocity (c) the fast space-charge wave phase velocity (d) the slow space-charge wave phase velocity and (e) the distance between the consecutive nodes of the standing-wave pattern formed by the slow and fast space-charge waves.

Answer: (a) 0.783 GHz (b) 3.25×10^7 m/s (c) 3.73×10^7 m/s (d) 2.87×10^7 m/s and (e) 0.19 m.

8.2 To what extent will the phase propagation constant of the slow space-charge wave of problem 8.1 be modified for a bounded beam for which the plasma frequency reduction factor is 0.4.

Answer: by a factor of 0.93.

8.3 Calculate (a) Pierce's gain parameter C and (b) all the four propagation constants Γ_1, Γ_2, Γ_3, and Γ_4 in a TWT, assuming the rf quantities to vary as exp $(j\omega t - \Gamma z)$, taking b = QC = d = 0, beam voltage = 2.5 kV, beam current = 40 mA and the operating frequency = 8 GHz.

Answer: (a) C = 0.043 and (b) $\Gamma_1 = -63.4 + j\ 1735.7$, $\Gamma_2 = 63.4 + j\ 1735.7$, $\Gamma_3 = j\ 1625.8$, $\Gamma_4 = -j\ 1697$, all in radian/m.

8.4 i) Following the method developed in section *8.2.4A* to estimate the effects of the circuit attenuator of a TWT, deduce the following general expressions in terms of δ_1, δ_2 and δ_3 suitable for the case of $b\neq 0$, $QC \neq 0$, $d \neq 0$:

(a)
$$V_1^b = \left(1+\frac{4QC}{\delta_1^2}\right)\left(\frac{1}{(1-\delta_2/\delta_1)(1-\delta_3/\delta_1)}\right)V_{in}\exp-(j\,2\pi N_1)\left(\frac{(1/\delta_1)(1/\delta_3)+(1/\delta_2)-(1/\delta_1)}{(1/\delta_1-1/\delta_2)\,(1/\delta_3-1/\delta_1)}\times\right.$$

$$\left.\times\exp(2\pi CN_1\delta_1) - \frac{1}{(1-\delta_2/\delta_1)(1-\delta_3/\delta_1)}\exp(2\pi\,CN_1\delta_2) - \frac{1}{(1-\delta_2/\delta_1)(1-\delta_3/\delta_1)}\exp(2\pi\,CN_1\delta_3)\right)$$

(b)
$$V_1^a = \left(1+\frac{4QC}{\delta_1^2}\right)\left(\frac{1}{(1-\delta_2/\delta_1)(1-\delta_3/\delta_1)}\right)V_{in}\exp-(j\,2\pi N_1)\ \exp(2\pi CN_1\delta_1)$$

(c)
$$\text{hot attenuation} = 20\left|\frac{(1/\delta_1-1/\delta_2)(1/\delta_3-1/\delta_1)}{(1/\delta_1)(1/\delta_3)+(1/\delta_2)-(1/\delta_1)}\right|\text{dB},$$

where δ_1, δ_2 and δ_3 are the solutions of the cubic equation (8.2.73).

ii) Show from (i)-(c) that the hot attenuation of the circuit attenuator for the special case of b = QC = d = 0 is 3.52 dB as obtained following (8.2.145).

8.5 For a 45 dB gain-TWT operating in the frequency band 5.9 - 6.4 GHz with beam voltage 7.5 kV and beam current 200 mA, calculate the approximate interaction length, taking the interaction impedance as 36.7 dB, hot attenuation as 6 dB and b = QC = d = 0 [18].

Answer: 17.2 cm.

8.6 A TWT operates at 6 GHz with beam parameters as follows: beam voltage = 2.95 kV, beam current = 66 mA, beam radius = 0.74 mm, plasma-frequency reduction factor = 0.5. Calculate Pierce's parameters b, QC and d taking the SWS parameters as: phase velocity = 3.1×10^7 m/s, interaction impedance = 65 ohms, attenuation constant = 1 naper/m.

Answer: b = 0.6, QC = 0.20, d = 0.01.

8.7 Show that a gain ~ 40 dB is obtainable from the interaction length 14 cm of the tube of parameters given in problem 8.6, taking into account the negative contribution of the attenuator from problem 8.4.

8.8 A gyro-TWT uses a cylindrical waveguide in the TE_{01}-mode at the fundamental beam resonance and operates at a frequency 3% above the waveguide cut-off. Choosing the device parameters as: operating frequency = 35 GHz, beam voltage = 70 kV, velocity pitch factor = 1.5, dc magnetic flux density = 12.87 kG and beam voltage = 70 kV, calculate (a) waveguide cut-off frequency (b) waveguide radius (c) relativistic mass factor γ (d) non-relativistic cyclotron frequency (e) relativistic cyclotron frequency and (f) Larmor radius.

Answer: (a) 33.95 GHz (b) 5.39 mm (c) $\gamma = 1.137$ (d) 36.01 GHz (e) 31.68 GHz and (f) 0.62 mm.

8.9 Find the grazing-intersection frequency and dc magnetic flux density for a gyro-TWT operating with a beam voltage of 70.82 kV (a) in the TE_{01}-waveguide mode at the fundamental cyclotron resonance and (b) in the TE_{02}-waveguide mode at the second cyclotron-harmonic resonance. Choose the cut-off frequency value as 33.95 GHz from the solution of problem 8.8.

Answer: (a) 35.31 GHz, 13.27 kG (b) 35.31 GHz, 6.64 kG.

8.10 A gyro-TWT operates at a frequency 36 GHz in the cylindrical TE_{01}-mode with the following parameters: beam voltage (V_0) = 70kV, beam current (I_0) = 9.76 A, velocity pitch factor(α_0) = 1.5, waveguide radius (a) = 5.37 mm, hollow-beam radius (r_c) = 0.48a and Larmor radius (r_L) = 0.12a. Taking the applied dc magnetic flux density as 98% of the value corresponding to grazing intersection, calculate (a) the Pierce's velocity parameter (b_{gyro}) (b) the ratio of the axial dc beam velocity-to-the velocity of light in free space (v_z/c) (c) the ratio of the transverse dc beam velocity-to-the velocity of light in free space (v_t/c) (d) the applied dc magnetic flux density (B_0) (e) the interaction impedance (K_{gyro}) (f) Pierce's gain parameter (C_{gyro}) and (g) gain per cm with launching loss ignored.

Answer: (a) $b_{gyro} = 0.899$ (b) $v_z/c = 0.264$ (c) $v_t/c = 0.396$ (d) $B_0 = 13.08$ kG (e) $K_{gyro} = 523.32$ ohm (f) $C_{gyro} = 0.263$ and (g) gain per cm = 4.15 dB/cm.

References

[1] M. Chodorow and C. Susskind, *Fundamentals of Microwave Electronics* (Mc Graw-Hill, New York, 1964).

[2] R. F. Soohoo, *Microwave Electronics* (Addison-Wesley, Reading 1971).

[3] S. Ramo, "Space-charge and field waves in an electron beam", *Phys. Rev.* **56** (1939) 276-283.

[4] E. C. Jordan and K. G. Balmain, *Electromagnetic Waves and Radiating Systems* (Prentice Hall, Englewood Cliffs, 1968).

[5] A. S. Gilmour, Jr., *Principles of Traveling Wave Tubes* (Artech House, Boston, 1994).

[6] R. G. Carter, *Electromagnetic Waves/Microwave Components and Devices* (Chapman and Hall, London, 1990).

[7] R. W. Gould and A. W. Trivelpiece, "A new mode of wave propagation on electron beams", in *Proc. Symp. Electronic Waveguides* (Polytechnic Press, Brooklyn, 1958).

[8] H. A. Haus, "Electron beam waves in microwave tubes", *Ibid.*

[9] H. A. Haus and D. Bobroff, "Small-signal power theorem for electron beams", *J. Appl. Phys.* **28** (1957), 694-703.

[10] L. J. Chu, "A kinetic power theorem", paper presented at the IRE-PGED Electron Tube Research Conference, Durham, New Hampshire, June, 1951.

[11] J. W. Klüver, "Small-signal power conservation theorem for irrotational electron beams", *J. Appl. Phys.* **29** (1958), 618-622.

[12] R. E. Collin, *Foundations for Microwave Engineering* (Mc Graw-Hill Kogakusha, Tokyo, 1966).

[13] L. Tonks and I. Langmuir, "Oscillations in ionised gases", *Phys. Rev.* **33** (1929) 195-210.

[14] G. M. Branch and T. G. Mihran, "Plasma-frequency reduction factors in electron beams", *IRE Trans on Electron Dev.* **ED-2** (1955) 3-11.

[15] V. N. Shevchik, G. N. Shvedov and A. V. Soboleva, *Wave and Oscillatory Phenomena in Electron Beams at Microwave Frequencies* (Pergamon Press, Oxford, 1966).

[16] J. R. Pierce, *Traveling-Wave Tubes* (D. Van Nostrand, New York, 1950).

[17] J. F. Gittins, *Power Travelling-Wave Tubes* (American Elsevier, New York, 1965).

[18] Om P. Gandhi, *Microwave Engineering and Applications* (Pergamon Press, New York, 1981).

[19] R. G. E. Hutter, *Beam and Wave Electronics in Microwave Tubes* (D. Van Nostrand, Princeton, 1960).

[20] *Hughes TWT and TWTA Handbook* (A bulletin for business circulation) Electron Dynamics Div. Hughes Aircraft Co. California.

[21] R. Strauss, J. Bretting, and M. Metivier, "Travelling-wave tubes for communication satellites", *Proc. IEEE* **65** (1977) 387-400.

[22] J. L. Putz and M. J. Cascone, "Effective use of dispersion shaping in broadband helix TWT circuits", *I.E.D.M.Tech. Digest*, 1979, p.p. 422-424.

[23] J. P. Rymer and M. J. Cascone, "Three octaves with one TWT, *I.E.D.M.Tech. Digest*, 1982, p.p. 30-31.

[24] A. Scott and M. J. Cascone, "What is new in helix TWTs", *I.E.D.M.Tech. Digest,* 1978, p.p. 526-529.

[25] L. Wislow, "Phase velocity dispersion shaping as a design parameter in traveling-wave tubes", *I.E.D.M.Tech. Digest,* 1977, p.p. 350A-350C.

[26] L Dombro, J. Long, and L. Roberts, "The velocity-tapered helix in traveling-wave tubes for satelite communications", *Radio Electron and Comm. Syst.* **27** (1984) 75-79.

[27] H. G. Kosmahl, "Modern multi-stage depressed collectors — a review", *Proc. IEEE* **70** (1982) 1325-1334.

[28] J. W. Hussen and C. Susskind, "Improvement of beam-tube performance by collector potential depression and a novel design", *IRE Trans. Electron Dev.* **ED-7** (1960) 282-288.

[29] G. Fleury, J. C. Kuntzmann and P. Lafuma, "High power brazed-helix telecommunication TWTs", *I.E.D.M.Tech. Digest*, 1977 p.p. 116-119.

[30] P. Gosset and E. D. Maloney, "Brazed-helix TWT's for TV-broadcasting and high-power telecommunication satellites", Thomson-CSF (TP-095), Electron Tube Div., Paris, paper presented at the national telecomm. conf., Washington DC, 1979.

[31] A. R. Jung, "10 kw and up from a helix TWT", *I.E.D.M.Tech. Digest,* 1978, p.p. 530-533.

[32] C. E. Hobrecht, "Study of resonant loss for helix TWTs", Technical Report RADC-TR-79-35, March 1979, Rome Air Development Centre, New York.

[33] B. Epsztein, "Slow-Wave structures in microwave tubes", *I.E.D.M.Tech. Digest*, 1984, p.p. 486-489.

[34] S. S. S. Agarawala and C. Dattatreyan, "Development of helical couplers and coupled helix attenuators for an S-band TWA tube", *J. Inst. Electron Telecomm. Engrs.* **3** (1965) 502-509.

[35] J. S. Cook, R. Kompfner and H. Suhl, "Non-reciprocal loss in traveling-wave tubes using ferrite attenuators", *Proc. IRE* **42** (1954) 1188-1189.

[36] F. K. Muller, "Traveling-wave tubes with ferrite attenuator", *IRE Trans. Electron Devices* **ED-7** (1961) 284-289.

[37] E. L. Lien, "Stop-bands produced by asymmetrical support rod systems in helix structures", *I.E.D.M.Tech. Digest*, 1979, p.p. 412-415.

[38] L. R. Falce, "Improved technique for producing pyrolytic graphite attenuator film for microwave tubes", *IEEE conference Record* (8th Conference on Tube Techniques), Sept., 1966, p.p. 157-162.

[39] R. Campbell, "Recent developments in low noise mm-wave TWTA's", *Microwave J.* Dec (1975) 45-46, 59.

[40] O. Sauseng, A. E. Manoly, and A. Hall, "Thermal properties and power capability of helix structures for millimeter waves", *I.E.D.M.Tech. Digest*, 1978, p.p. 534-537.

[41] C. Liss, R. Harper, and M. P. Puri, "Helical waveguide millimeter wave TWT", *I.E.D.M.Tech. Digest*, 1988, p.p. 374-377.

[42] J. S. Cook, R. Rompfner, and C. F. Quate, "Coupled helices", *Bell Syst. Tech. J.* **35** (1956) 127-178.

[43] T. S. Chen, "Design and performance of coupled-helix transducer for traveling-wave tubes", *J. Electron and control* **6** (1959) 289-306.

[44] V. N. Singh, B. N. Basu, B. B. Pal, and N. C. Vaidya, "Equivalent circuit analysis of a system of coupled helical transmission lines in a complex environment", *J. Appl. Phys.* **54** (1983) 4141-4146.

[45] G. T. Konrad and S. K Cho, "Elliptical-cavity coupler for travelling-wave tubes, *IEEE Trans. Electron Dev.* **10** (1963) 85-89.

[46] E. F. Belohoubek, "Helix support structure for ultra-wide-band traveling-wave tubes", *RCA Rev.* **26** (1965) 106-117.

[47] H. C. Limburg, J. A. Davis, I. Tammaru, J. P. Vaszari, and J. Wilson, "Reducing the gain and phase variation in high power MMW TWT's", *I.E.D.M.Tech. Digest,* 1988, p.p. 381-384.

[48] H. G. Kosmahl and J. C. Paterson, "TWT amplifier exhibits linear power transfer characteristics and improved efficiency", *Microwave Syst. News* **14**, June (1984) 83-93.

[49] E. Ezura and T. Kano, "Measured and theoretical nonlinear phase distortion in traveling-wave tubes", *IEEE Trans. Electron Dev.* **ED-22** (1975) 890-897.

[50] C. Bremenson, M. Palazo, and R. Neger, "Linearising TWTAs in satellite transponders", *Conf. Proc. AIAA 8th Comm. Satellite Conf.*, Orlando, 1980.

[51] L. I. Yarrington and E. D. Maynard, "The status of dual-mode tubes", *Microwave J.*, May (1973), 15-20.

[52] J. R. Hechtel and B. J. Hamak, "A dual-mode electron gun having non-intercepting grids", *I.E.D.M.Tech. Digest*, 1973, p.p. 171-174.

[53] N. J. Dionne, "Harmonic generation in octave bandwidth traveling-wave tubes, *IEEE Trans. Electron Dev.* **ED-17** (1970) 365-372.

[54] O. Sauseng and W. E. Garrigus, "Reduction of intermodulation distortion with harmonic injection for wideband traveling wave tubes", *I.E.D.M.Tech. Digest*, 1975, p.p. 411-414.

[55] J. R. Pierce and W. B. Hebenstreit, "A new type of high frequency amplifier, *Bell Syst. Tech J.* **28** (1949) 33-51.

[56] A. V. Hollenberg, "Experimental observation of amplification by interaction between two electron streams", *Bell Syst. Tech J.* **28** (1949) 52-58.

[57] A. V. Haeff, "The electron-wave-tube — a novel method of generation and amplification of microwave energy", *Proc. IRE* **37** (1949) 4-10.

[58] L. S. Nergaard, "Analysis of a simple model of a two-beam growing-wave tube", *RCA Rev.* **9** (1948) 585-601.

[59] G. A. Bernashevsky, Z. S. Voronov, T. I. Iziumova, and A. S. Tchernov, "Investigation of double-stream electron wave systems", in *Proc. Symp. Electronic Waveguides* (Polytechnic Press, Brooklyn, 1958).

[60] P. M. Phillips, E. G. Zaidman, H. P. Freund, A. K. Ganguly, and N. R. Vanderplaats, "Review of two-stream amplifier performance", *IEEE Trans. Electron Dev.* **37** (1990) 870-877.

[61] G. D. Boyd, L. M. Field and R. W. Gould, "Interactions between an electron stream and an arc discharge plasma", in *Proc. Symp. Electronic Waveguids* (Polytechnic Press, Brooklyn 1958).

[62] G. D. Boyd, R. W. Gould, and L. M. Field, "Interaction of a modulated electron beam with a plasma", *Proc. IRE* **49** (1961) 1906-1916.

[63] E. V. Bogdanov, V. J. Kislov and Z. S. Chernov, "Interaction between an electron beam and a plasma", in *Proc. Symp Millimeter Wave Generation* (Polytechnic Press, Brooklyn 1959).

[64] N. B. Chakrabarti and B. N. Basu, "Difference frequency generation using non-linear interaction between a modulated electron beam and a collisionless plasma", *Int. J. Electron* **34** (1973) 65-74.

[65] P. Chorney, "Recent advances in beam-plasma amplifiers", *IEEE Convention Record,* **part 5** p.p. 5-16, New York, 1966.

[66] R. S. Symon and H. R. Jory, "Cyclotron Resonance Devices" in *Advances in Electronics and Electron Physics,* **55** (Academic, New York 1981).

[67] A. S. Gilmour, Jr., *Microwave Tubes* (Artech House, Boston, 1986).

[68] T. K. Ishii, *Practical Electron Devices* (Academic, San Diego, 1990).

[69] J. M. Baird and W. Lawson, "Magnetron injection gun (MIG) design for gyrotron applications", *Int. J. Electron* **61** (1986) 953-967.

[70] P. Sprangle and A. T. Drobot, "The linear and self consistent non-linear theory of the electron cyclotron maser instability", *IEEE Trans. Microwave Theory Tech.* **MTT 25** (1977) 528-544.

[71] K. R. Chu and A. T. Drobot, "Theory and single wave simulation of the gyrotron traveling-wave amplifier operating at cyclotron harmonics", NRL Memorandum Report 3788, August 1978, Naval Research Lab, Washington DC.

[72] A. Rekiouak and B. R. Cheo, "Wide band gyrotron traveling wave amplifier analysis", Report Poly WRI -1531-88, Weber Research Institute, New York.

[73] G. Dohler and D. Gallagher, "The multi-signal theory of the cyclotron maser and other gyrotron-type devices, *IEEE Trans. Electorn Dev.* **35** (1988) 1730-1745.

[74] K. R. Chu and A. T. Lin, "Gain and bandwidth of the gyro-TWT and CARM amplifiers", *IEEE Trans. Plasma Sc.* **PS-16** (1988) 90-104.

[75] K. R. Chu, A. T. Drobot, H. H. Szu, and P. Sprangle, "Theory and simulation of gyrotron traveling-wave amplifier operating at cyclotron harmonics", *IEEE Trans. Microwave Theory Tech.* **MTT-28** (1980) 313-317.

[76] A. J. Sangster, "Small-signal analysis of the travelling-wave gyrotron using Pierce parameters", *IEEE Trans. Microwave Theory Tech.* **MTT-28** (1980) 313-317.

[77] A. A. Vlasov, "The vibrational properties of an electron gas", *Zh. eksp. teor. fiz.* **8** (1938) 291-318.

[78] C. S. Kou, Q. S. Wang, D. B. Mc Dermott, K. R. Chu and N. C. Luhmann, "High power harmonic gyro-TWT's — part-1: linear theory and oscillation study", *IEEE Plasma Sc.* **20** (1992) 155-161.

[79] A. J. Sangster, "Small-signal bandwidth characteristics of a travelling-wave gyrotron amplifier", *Int. J. Electron* **51** (1981) 583-594.

[80] L. R. Bernett, Y. Y. Lau, K. R. Chu, and V. L. Granatstein, "An experimental wideband gyrotron traveling-wave amplifier", *IEEE Trans. Electron Dev.* **ED-28** (1981) 872-878.

[81] A. K. Ganguli and S. Ahn, "Large-signal theory of a two-stage wideband gyro-TWT", *IEEE Trans. Electron Dev.* **ED-31** (1984) 474-488.

[82] H. S. Uhm, J. Y. Choe, and S. Ahn, "Theory of a gyrotron amplifier in a waveguide with inner dielectric material, *Inst. J. Electron* **51** (1981) 521-532.

[83] K. R. Chu, A. K. Ganguli, V. L. Granatstein, J. L. Hirshfield, S. Y. Park and J. M. Baird, "Theory of a slow-wave cyclotron amplifier", *Inst. J. Electron* **51** (1981) 493-502.

[84] H. S. Uhm, "Kinetic description of free streaming mode in an electron beam propagating through a tape helix waveguide", *Phys. Fluids* **26** (1983) 848-855.

[85] A. Singh, "Millimeter wave gyrotron operation at high efficiency and power levels", *Ind. J. Radio & Space Phys.* **19** (1990) 297-305.

[86] D. S. Furuno, D. B. Mc Dermott, C. S. Kou, N. C. Luhmann and P. Vitello, "Operation of a large-orbit high-harmonic gyro-traveling-wave tube amplifier", *IEEE Trans. Plasma Sc.* **PS-18** (1990) 313-319.

INDEX